78 181 64

Radars Volume 1

Monopulse Radar

David K. Barton

THE ARTECH RADAR LIBRARY
David K. Barton, Editor

Electronic Scanning Radar Systems (ESRS) Design Handbook
Peter J. Kahrilas
Consulting Engineer, Raytheon Company

Phased Array Antennas
Arthur A. Oliner
Head of the Department of Electronics and Electrophysics,
Polytechnic Institute of New York
George H. Knittel
Staff Member, MIT Lincoln Laboratory

Radar Detection and Tracking Systems
Shahen A. Hovanessian
Senior Scientist, Hughes Aircraft Company

Radar Signal Simulation
Richard L. Mitchell
Vice-President, MARK Resources

Radar System Analysis
David K. Barton
Consulting Scientist, Raytheon Company

Radar Technology
Eli Brookner
Consulting Scientist, Raytheon Company

RADARS – in five volumes
- I. **Monopulse Radar**
- II. **The Radar Equation**
- III. **Pulse Compression**
- IV. **Radar Resolution and Multipath Effects**
- V. **Radar Clutter**

David K. Barton
Consulting Scientist, Raytheon Company

Significant Phased-Array Papers
Robert C. Hansen
President, R.C. Hansen Inc.

Synthetic Aperture Radar
John J. Kovaly
Consulting Engineer, Raytheon Company

Affiliated Title:
RF Radiometer Handbook
C.W. McLeish and G. Evans
Staff Members, National Research Council of Canada

Radars

Volume 1

Monopulse Radar

David K. Barton
Consulting Scientist
Raytheon Company
Bedford, Mass.

Second Printing

ARTECH HOUSE, INC.
610 Washington Street
Dedham, Massachusetts 02026

Printed and bound in the United States of America.

Library of Congress Catalog Card Number: 74-82597
International Standard Book Number: 0-89006-030-4

International Standard Book Number (Volume Two): 0-89006-031-2
International Standard Book Number (Volume Three): 0-89006-032-0
International Standard Book Number (Volume Four): 0-89006-033-9
International Standard Book Number (Volume Five): 0-89006-034-7
International Standard Book Number (Volumes One through Five): 0-89006-035-5

TABLE OF CONTENTS

INTRODUCTION vii

ANNOTATED BIBLIOGRAPHY viii

1. "Accurate Angle Tracking by Radar" 1
   R.M. Page, *Naval Research Laboratory Report* (December 1944).
2. "Final Engineering Report on Angular Accuracy Improvement" 17
   G.M. Kirkpatrick, *General Electric Company Report* (August 1952).
3. "Guided Missile Instrumentation Radar" 105
   I. Stokes and D.K. Barton, *RCA Corporation*, internal publication (December 1956).
4. "Accuracy of a Monopulse Radar" 111
   D.K. Barton, *IRE Conf. Proc.*, 3rd National Conv. on Military Electronics, pp. 179-186 (July 1959).
5. "Multiple Target Resolution of Monopulse Versus Scanning Radars" 119
   S.F. George and A.S. Zamanakos, *Proc. NEC 15*, pp. 814-823 (October 1959).
6. "Amplitude - and Phase-Sensing Monopulse System Parameters" 129
   W. Cohen and C.M. Steinmetz, *Microwave Journal*, pp. 27-33, (October 1959)
7. "Amplitude - and Phase-Sensing Monopulse System Parameters" 137
   W. Cohen and C.M. Steinmetz, *Microwave Journal*, pp. 33-38 (November 1959)
8. "A Monopulse Cassegrainian Antenna" 143
   R.W. Martin and L. Schwartzman, *IRE Conv. Record, Pt. 1*, pp. 96-102 (1960).
9. "Precision Tracking with Monopulse Radar" 151
   J.H. Dunn and D.D. Howard, *Electronics 35*, No. 17, pp. 51-56 (April 1960).
10. "Distribution Functions for Monopulse Antenna Difference Patterns" . . . 157
    O.R. Price and R.F. Hyneman, *IRE Trans.* AP-8, No. 6, pp. 567-576 (November 1960).
11. "Design of a Twelve-Horn Monopulse Feed" 167
    L.J. Ricardi and L. Niro, *IRE Conv. Record*, Pt. 1, pp. 49-56 (1961).
12. "A Monopulse Antenna Having Independent Optimization of the Sum and Difference Modes" 175
    P.W. Hannan and P.A. Loth, *IRE Conv. Record*, Pt. 1, pp.57-60(1961).
13. "Maximum Gain in Monopulse Difference Mode" 179
    P.W. Hannan, *IRE Trans.* AP-9, No. 3, pp. 314-315 (May 1961).
14. "A Note on the Spatial Information Available from Monopulse Radar" . . 181
    D.B. Anderson and D.R. Wells, *Proc. 5th Natl. Conv. on Military Electronics, IRE-PGMIL*, pp. 268-278 (June 1961).
15. "Thermal-Noise Errors in Simultaneous-Lobing and Conical-Scan Angle-Tracking Systems" 219
    J.A. Develet, Jr., *IRE Trans.*, SET-7, No. 2, pp. 42-51 (June 1961).

16. **"Optimum Feeds for All Three Modes of a Monopulse Antenna"** **229**
P.W. Hannan, *IRE Trans.* AP-9, No. 5, pp. 444-461 (September 1961).
17. **"'The Future of Pulse Radar for Missile and Space Range Instrumentation"** **247**
D.K. Barton, *IRE Trans.* MIL-5, No. 4, pp. 330-351 (October 1961).
18. **"Phase-Amplitude Monopulse System"** **269**
W. Hausz and R.A. Zachary, *IRE Trans.* MIL-6, No. 2, pp. 140-146 (April 1962).
19. **"Monopulse Difference Slope and Gain Standards"** **277**
R.R. Kinsey, *IRE Trans.* AP-10, No. 3, pp. 343-344 (May 1962).
20. **"A Relationship Between Slope Functions for Array and Aperture Monopulse Antennas"** **278**
G.M. Kirkpatrick, *IRE Trans.* AP-10, No. 3, p. 350 (May 1962).
21. **"New Performance Records for Instrumentation Radar"** **279**
J.T. Nessmith, *Space/Aeronautics,* pp. 86-93 (December 1962).
22. **"SCAMP - A New Ratio Computing Technique with Application to Monopulse"** **285**
W.L. Rubin and S.K. Kamen, *Microwave Journal,* pp. 83-90 (December 1964).
23. **"The Use of 'Complex Indicated Angles' in Monopulse Radar to Locate Unresolved Targets"** **291**
S.M. Sherman, *Proc. NEC 22,* pp. 243-248 (1966).
24. **"Monopulse Operation with Continuously Variable Beamwidth by Antenna Defocusing"** **297**
H.W. Redlien, *IEEE Trans.* AP-16, No. 4, pp. 415-423 (July 1968).
25. **"Monopulse Networks for Series Feeding an Array Antenna"** **307**
A.R. Lopez, *IEEE Trans.* AP-16, No. 4, pp. 436-440 (July 1968).
26. **"Contour Pattern Analysis of a Monopulse Radar Cassegrainian Antenna"**
D.D. Howard, *Microwave Journal,* pp. 61-63 (December 1968). **313**
27. **"Complex Indicated Angles Applied to Unresolved Radar Targets and Multipath"** **317**
S.M. Sherman, *IEEE Trans.* AES-7, No. 1, pp. 160-170 (January 1971).

MONOPULSE RADAR

Selected Reprints from the Literature
Compiled and Edited by David K. Barton

INTRODUCTION

Monopulse tracking radar had its origin during World War II in the laboratories of the U.S. Navy, the General Electric Company, and the Bell Telephone Laboratories. The technique of simultaneous lobing had previously been used in direction finding and search radar, but its application to tracking started at the Naval Research Laboratory in 1940 and reached a practical state of development in 1943-44. Monopulse radars appeared in the field, for test programs, in 1948, both in ground missile guidance and in airborne fire control applications. Since that time, monopulse tracking has been applied in dozens of different systems where high precision is an important objective. The principal contribution of the monopulse techniques was in elimination of target scintillation error, in which amplitude modulation of the signal was converted to a spurious reading of pointing angle, generating noise in the tracking loop. Higher efficiency and freedom from mechanical scan limitations were also advantageous.

The term "monopulse" was first proposed by H.T. Budenbom of Bell Telephone Laboratories in 1946, and has been accepted in U.S. usage in preference to "simultaneous lobing" and the British "static split" terminology. The word is descriptive of the basic concept, in which complete angular pointing data can be derived from a single received pulse in radar, and the same term is applied to passive tracking antennas in which the signal may be a continuous carrier or even random noise. The common principle is the simultaneous generation of a pair of antenna patterns in the angular coordinate to be measured, such that the complex ratio of the two received signals is a measure of the target angle relative to the antenna axis, independent of signal amplitude. In "amplitude comparison" systems, the relative phase between the signals remains constant, except for a reversal of sign at the axis, and the target direction is measured by the ratio of amplitudes. In "phase comparison," the amplitudes are essentially constant and the phase angle of the ratio is measured. Combination systems have been built with amplitude comparison in one coordinate and phase comparison in the other. Most two-coordinate systems, however, use three patterns, a common sum (or reference) pattern on the axis, and two difference (or error) patterns whose null planes cross at a right angle on the axis.

The literature on monopulse radar, prior to 1953, was restricted to classified reports. Two of the most significant of those reports, recently declassified but not previously available to most radar engineers, are included in this volume. The report by R.M. Page is the first description of the sum-and-difference system, which overcame many of the problems of gain and phase balance which had impeded prior attempts at monopulse instrumentation. The longer report by G.M. Kirkpatrick presents the results of a two-year theoretical and experimental study conducted at General Electric Company under a sponsorship of the U.S. Army Signal Corps. In this report, for the first time, a theoretical basis was laid for determining the limits to performance of a monopulse antenna. The optimum sum and difference illuminations, patterns, gains, and slopes derived by Kirkpatrick have since served as the reference against which practical designs could be compared, much as the "matched filter" provides the reference for signal-to-noise calculations of different signal processors. These concepts appeared in a brief paper in the IRE Transactions in 1953, but the bulk of the work has remained in locked files since 1952.

The Kirkpatrick paper is significant both because of its thoroughness and because it anticipates several key ideas which took many years to appear in practical equipment: use of the electrical error signal to correct the antenna shaft output data in an open-loop process having wide data bandwidth; use of narrow Doppler filtering in sum and difference channels to extend tracking range in noise and to avoid clutter; generation of sum and difference illumination by excitation of high-order modes within the feed horn; normalization of the error signal by limiting and phase detection of sum-plus-j-difference signals; use of RF or IF commutation to avoid bias from circuit instabilities; and use of the off-axis measurement to extend low-angle tracking capability while holding the beam above the horizon. Over the past twenty years, most of the material has appeared in one or another form in books and professional papers, often with attribution to the 1953 Kirkpatrick paper, but the original report will remain valuable both for its organized derivations and presentation and for historical reasons.

Subsequent papers in this volume are arranged chronologically, and present descriptions of specific radars, applications to precise instrumentation, design of feed horns and networks, analysis of resolution and accuracy, and procedures for monopulse signal processing. The selection has been based in some cases on the frequency with which the papers are cited and used in other literature, and in others on providing practical information regarding feed components or complete systems which may not have received wide enough circulation. A more complete bibliography on the subject of monopulse radar is also included. The bibliography is divided into subject headings, within which entries are given chronologically. Many references to government and contractor reports, not listed in the bibliography because of their questionable availability, will be found as footnote references in the Reprinted Papers.

ANNOTATED BIBLIOGRAPHY

A. Books and general papers covering monopulse radar.

Page, R.M., "Monopulse Radar, *IRE Conv Rec 55,* Pt 8, p 132 (abstract only, based on NRL Report, Reprint Paper 1).

Rhodes, D.R., *Introduction to Monopulse,* New York, McGraw-Hill Book Co (1959). First basic text on monopulse principles.

Cohen, W. and Steinmetz, C.M., "Amplitude- and Phase-sensing Monopulse System Parameters," *Microwave Journal,* Oct 59, pp 27-33 and Nov 59, pp 33-38. Analysis of monopulse system performance, including effects of antenna and receiver imperfections. Reprint Papers 6, 7.

Dunn, J.H. and Howard, D.D., "Precision tracking with Monopulse Radar," *Electronics 35,* No 17, Apr 22, 1960, pp 51-56. General description of monopulse radars and antennas including evergy distribution and horn apertures in focal plane. Reprint Paper 9.

Skolnik, M.I., *Introduction to Radar Systems,* New York, McGraw-Hill Book Co (1962). Basic radar text, including monopulse principles and description of typical tracking radars.

Hausz, W. and Zachary, R.A., "Phase-Amplitude Monopulse System," *IRE Trans MIL-6,* No 2, Apr. 62, pp 140-146. First description of combination phase-amplitude monopulse, applied to AN/APG-25 airborn fire-control radar. Reprint paper 18.

Barton, D.K., *Radar System Analysis,* Englewood Cliffs, N.J., Prentice-Hall, Inc. (1964) Basic radar text, including detailed analysis of performance of monopulse trackers from a systems viewpoint.

Adomian, G., "Monopulse Radar Tracking Techniques," *IEEE Trans MIL-9,* Nos 3-4, Jul-Oct 65, pp 293-294. Brief history, without description or data, of work in COSRO and randomly-lobed sequential scan.

Carpentier, M.H., *Radars: New Concepts,* New York, Gordon and Breach (1968), translated from the French second edition.

Barton, D.K. and Ward, H.R., *Handbook of Radar Measurement,* Englewood Cliffs, N.J. Prentice-Hall, Inc. (1969).

Skolnik, M.I., *Radar Handbook,* New York, McGraw-Hill Book Co (1970).

B. Monopulse Antenna and Feed Designs

Taylor, D. and Westcott, C.H., "Divided Broadside Aerials with Applications to 200 mc/s Ground Radiolocation Systems," *IEE Jour 93,* Pt IIIA, 1946, pp 588-597.

Watts, C.B., Jr., "Simultaneous Radiation of Odd and Even Patterns by a Large Array," *Proc IRE 40,* No 10, Oct 52, pp 1236-1239. Waveguide arrays fed from both ends to generate even and odd patterns for direction finding; bandwidth limitation noted.

Kirkpatrick, G.M., "Aperture Illuminations for Radar Angle-of-Arrival Measurements," *IRE Trans PGAE-9,* Sep 53, pp 20-27. Brief summary of concepts and results of GE/Signal Corps study. Reprint Paper 2.

Shelton, J.P., "Improved Feed Design for Amplitude Monopulse Radar Antenna," *IRE Natl Conv Rec 59,* Pt 1, pp 93-102. Design of four- and nine-element feeds with coupling to excite even and odd modes, optimizing sum and difference illuminations.

Martin, R.W. and Schwartzman, L., "A Monopulse Cassegrain Antenna," *IRE Conv Rec 60,* Pt 1, pp 96-102. Design principles of Cassegrainian antenna with monopulse feed. Reprint Paper 8.

Price, O.R. and Hyneman, R.F., "Distribution Functions for Monopulse Antenna Difference Patterns," *IRE Trans AP-8,* No 60, pp 567-576. Synthesis of optimum difference illuminations for narrow difference lobes and low sidelobes. Reprint Paper 10.

Ricardi, L.J., and Niro, L., "Design of a Twelve-Horn Monopulse Feed." *IRE Conv Rec 61,* Pt 1, pp 49-56. Description and test data on 12-horn feed with high slope and efficiency and low sidelobes. Reprint Paper 11.

Hannan, P.W. and Loth, P.A., "A Monopulse Antenna Having Independent Optimization of the Sum and Difference Modes," *IRE Conv Rec 61,* Pt 1, pp 57-60. Description and test data of multimode, multi-level feed, overcoming previous compromises between sum and defference illuminations. Reprint Paper 12.

Hannan, P.W., "Maximum Gain in Monopulse Difference Mode," *IRE Trans AP-9,* No 3, May 61, pp 314-315. Derivation of aperture illumination to provide maximum gain at difference-pattern peaks. Reprint Paper 13.

Hannan, P.W., "Optimum Feeds for All Three Modes of a Monopulse Antenna," *IRE Trans AP-9,* No 5, Sep 61, pp 444-461. Basic theory and practice of modern horn feeds for lens or reflector antennas. Definitions of terms and design curves included. Reprint Paper 16.

Kelly, D.C. and Goebels, F.J., Jr., "Annular Slot Monopulse Antennas," *IRE Conv Rec 62,* Pt 1, pp 71-80. Design and Test Data on slotted radial waveguide antenna.

Kinsey, R.R., "Monopulse Difference Slope and Gain Standards," *IRE Trans AP-10,* No 3, May 62, pp 343-344. Brief discussion of relative merits of difference peak gain and difference slope as criteria for optimization. Reprint Paper 19.

Kirkpatrick, G.M., "A Relationship Between Slope Functions for Array and Aperture Monopulse Antennas," *IRE Trans AP-10,* No 3, May 62, p 350. Brief discussion of angle sensitivity as a function of squint angle in arrays and aperture antennas. Reprint Paper 20.

Wheeler, H.A., "Antenna Beam Patterns which Retain Shape with Defocusing," *IRE Trans AP-10,* No 5, Sep 62, pp 573-580. Choice of even and odd iiluminations based on probability function to permit change in beamwidth without destruction of monopulse characteristics or change in pattern shape.

DeVito, G., "A New Type of Monopulse Vernier Obtained by the Excitation of Both H11 and E01 Modes in a Circular Waveguide," *IRE Trans AP-10,* No 6, Nov 62, p 781. Brief description and test data on circularly polarized single-horn feed generating difference pattern with odd mode.

Nester, W.H., "The Optimum Excitation for a Phase Comparison Arrays," *Microwave Journal,* Apr 63, pp 67-73. Derivation of single aperture illumination which is optimum in the sense that it provides maximum accuracy in thermal noise when two halves are driven in-phase for sum and out-of-phase for difference.

LaPage, B.F., Blaisdell, E.W. and Ricardi, L.J., "West Ford Antenna and Feed System," *Proc IEEE 52,* No 5, May 64, pp 589-598. Description and test-data on large tracking antenna for space research and astronomy.

Howard, D.D., "Single Aperture Monopulse Radar Multimode Antenna Feed and Homing Device," *Proc 8th Natl Conv on Military Electr, IEEE,* Sep 14-16, 64, pp 259-263. Description and test data on single horn for complete two-coordinate monopulse operation.

Ross, G. and Schwartzman, L., "Continuous Beam Steering and Null Tracking with a Fixed Multiple-Beam Antenna Array System," *IEEE Trans AP-12,* No 5, Sep 64, pp 541-551. Generation of sum and difference beams by combining outputs of a matrix-beam array antenna system.

Keeping, K.J., "Design and Construction of a Multimode Circularly Polarized Monopulse Tracking Feed for High-Power Application in a Cassegrain Reflector System," *IEEE Conv Rec, 65,* Pt 5, pp 101-109. Design and Test data on feed for Haystack radar.

Cooper, D.N., "A New Circularly Polarized Monopulse Feed System," *Proc IEEE 53,* No 9, Sep 65, pp 1252-1254. Brief description of single-horn feed using circular polarization and multiple modes to generate sum and difference patterns.

Franklin, S.B., Hilbers, C.L. and Kosydar, W.E., "A Wideband Two-Channel Monopulse Technique," *Proc 9th Natl Conv on Military Electr, IEEE,* Sep 22-24, 65, pp 174-176. Difference channels combined in phase quadrature for amplification in one AGC-controlled receiver.

Chadwick, G.G., and Shelton, J.P., "Two Channel Monopulse Techniques - Theory and Practice," *Proc 9th Natl Conv on Military Electr, IEEE,* Sep 22-24, 65, pp 177-181. Principles of two-channel monopulse, using phase-quadrature combination of difference channels. Examples given using helix array, waveguide feed, and dualmode spiral antenna.

Drabowitch, S.W., "Multimode Antennas," *Microwave Journal,* Jan 66, pp 41-51. Design data on multimode monopulse feed horns.

Mavroides, W.G., Dennett, L.G. and Door, L.S., "3-D Radar Based in Phase-in-Space Principle," *IEEE Trans AES-2,* No 3, May 66, pp 323-331. Phase-comparison height finder using traveling-wave array feed.

Sinnott, J., "Patterns of Out-of-Phase Taylor Semicircular Apertures," *IEEE Trans AP-14,* No 3, May 66, pp 390-391. Equations and tab tables describing pattern and slope of difference channel generated by out-of-phase excitation of two halves of Taylor-weighted array.

Webster, R.E., "The Ultimate S-Band Monopulse Focal-Point Feed," *IEEE Trans AES-2,* No 6, (suppl), Nov 66, pp 61-67, (Aerospace Systems Conv Record). Design and test data on five-element feed using central horn and four surrounding dipoles.

Kahrilas, P.J. and Jahn, D.M., "Hardpoint Demonstration Array Radar," *IEEE Trans AES-2,* No 6, (suppl), Nov 66, pp 286-299, (Aerospace Systems Conv Record). Description and test data on optically-fed array system.

Profera, C.E., Jr. and Yorinks, L.H., "A High Efficiency Dual Frequency Multimode Monopulse Antenna Feed System," *IEEE Trans AES-2,* No 6, (supl), Nov 66, pp 314-322 (Aerospace Systems Conv Record). Design and test of five-horn S- and C-band monopulse feed.

Takeshima, T., "A Slot Array Antenna for Monopulse Tracking Radar," *Microwave Journal 9,* No 12, Dec 66, pp 63-65. Synthesis of center-fed slotted-line array for sum and difference pattern generation.

Pelchat, G.M., "Relationships Between Squinted, Sum and Difference Radiation Patterns of Amplitude Monopulse Antennas with Mutual Coupling Between Feeds," *IEEE Trans AP-15,* No 4, Jul 67, pp 519-526. Theoretical and practical considerations in selecting squint angle of beams to form monopulse pattersn.

Taylor, R.E., "Satellite Tracking Simultaneous-Lobing Monopulse Receiving System with Polarization Diversity Capability," *IEEE Trans AES-3,* No 4, Jul 67, pp 664-680. Description of four-element helical array for passive tracking on satellites.

Powers, E.J., "Utilization of the Lambda Function in the Analysis and Synthesis of Monopulse Antenna Difference Pattersn," *IEEE Trans AP-15,* No 6, Nov 67, pp 771-777. Mathematical models of monopulse patterns based on the lambda function.

Walker, S.H. and Osborn, J.D., "Investigation of a Nonambiguous Monopulse Antenna," *Microwave Journal 10,* No 13, Dec 67, pp 39-44. Synthesis of monopulse patterns having no ambiguous difference-channel nulls outside the main lobe.

Bayliss, E.T., "Design of Monopulse Antenna Difference Patterns with Low Sidelobes," *BSTJ 47,* No 5, May-Jun 68, pp 623-650. Synthesis of difference patterns having maximum slope subject to sidelobe constraint simular to Taylor's analysis for sum patterns.

Redlien, H.W., "Monopulse Operation with Continuously Variable Beamwidth by Antenna Defocusing," *IEEE Trans AP-16,* No 4, Jul 68, pp 415-423. Computations of gain, beamwidth and slope of monopulse antennas defocussed to broaden beam. Reprint Paper 24.

Lopez, A.R., "Monopulse Networks for Series Feeding an Array Antenna," *IEEE Trans AP-16,* No 4, Jul 68, pp 436-440. Description of dual ladder feeds for separate sum and difference excitation of constrained array antennas. Reprint Paper 25.

Howard, D.D., "Contour Pattern Analysis of a Monopulse Radar Cassegrain Antenna," *Microwave Journal,* Dec 68, pp 61-63. Presentation of AN/FPQ-6 antenna patterns as contour plots and 3-D drawings. Reprint Paper 26.

Kelly, A.J., "A Note of the Mathematical Equivalence of Analytic Signals and Monopulse Antennas," *Proc IEEE 57,* No 6, Jun 69, pp 1208-1209. Calls attention to analogies between aperture/pattern and signal spectrum/waveform, and to existence of tables applicable to difference patterns.

Taylor, M., Keough, R.A. and Moeller, A.W., "Beam Broadening of a Monopulse Tracking Antenna by Feed Defocusing," *IEEE Trans AP-18,* No 5, Sep 70, pp 622-627. Possibility of beam broadening up to five times the focussed width, by simple feed displacement.

Bullock, L.G., Oeh, G.R. and Sparagna, J.J., "An Analysis of Wide-Band Microwave Monopulse Direction-Finding Techniques," *IEEE Trans AES-7,* No 1, Jan 71, pp 188-203. Use of wideband spiral feeds in phase and amplitude comparison systems.

Melancon, L., "A Limited Scan Antenna," *IEEE Eascon Rec 71,* pp 190-192. Brief description of limited-scan monopulse array for GCA application.

Sanzgiri, S.M., Butler, J.K. and Voges, R.C., "Optimum Aperture Monopulse Excitations," *IEEE Trans AP-20,* No 3, May 72, pp 275-280. Synthesis procedure for optimization of illumination, given sidelobe envelope constraint.

Bridge, W.M., "Cross Coupling in a Five Horn Monopulse Tracking System," *IEEE Trans AP-20,* No 3, Jul 72, pp 436-442. Effect of depolarized signal components on operation of circularly polarized monopulse antenna.

Berger, H., "On the Optimum Squint-Angles of Amplitude Monopulse Radar and Beacon Tracking System," *IEEE Trans AES-8,* No 4, Jul 72, pp 545-547. Angle sensitivity as a function of squint angle between overlapping beams, considering radar and beacon targets and mutual coupling between feeds.

Appelbaum, A.J. et. al., "A Flat-Feed Technique for Phased Arrays," *IEEE Trans AP-20,* No 5, Sep 72, pp 582-588. Theory and test data on radical flat-feed power divider using multimode launcher to excite sum and difference illuminations.

Brown, G.S., "A Simplification in the Analysis of Four- and Five-Horn Fed Cassegrainian Reflectors when the Horns Have Nearly Symmetric Patterns," *IEEE Trans AP-21,* No 3, May 73, pp 382-384. Closed-form results given for feeds using circularly symmetric patterns.

C. Monopulse Signal and Data Processing

Cheetham, R.P. and Mulle, W.A., "Enhanced Real-Time Data Accuracy for Instrumentation Radars by Use of Digital-Hydraulic Servos," *IRE Wescon Record, 58,* Pt 4, pp 239-253. Description of off-axis error correction and tracking at wide bandwidth, using hydraulic servo to drive analog data devices from AN/FPS-16.

Storke, F.P., "A Monopulse Instrumentation System," *Proc IRE 49,* No 8, Aug 61, pp1328-1329. Description of Conical-Scan-On-Receive-Only (COSRO) system which derives scan by adding a modulated difference signal to the sum signal prior to the receiver.

Rubin, W.L. and Kamen, S.K., "SCAMP - A Single-Channel Monopulse Radar Signal Processing Technique," *IRE Trans MIL-6,* No 2, Apr 62, pp 146-152. Description and analysis of instantaneous normalization processor for multiple-target monopulse measurement. See Reprint Paper 22.

Korff, M., Brindley, C.M. and Lowe, M.H., "Multiple-Target Data Handling with a Monopulse Radar," *IRE Trans MIL -6,* No 4, Oct 62, pp 359-366. Off-axis measurement of multiple targets within monopulse beam, using AN/FPS-16 example.

Rubin, W.L. and Kamen, S.K., "SCAMP - A New Ratio Computing Technique with Application to Monopulse," *Microwave Journal,* Dec 64, pp 83-90. Description and analysis of instantaneous normalization processor using limiter in a frequency-multiplexed common IF amplifier to handle multiple targets in a monopulse beam. Reprint Paper 22.

Abel, J.E., George, S.F., and Sledge, O.D., "The Possibility of Cross Modulation in the SCAMP Signal Processor," *Proc IEEE 53,* No 3, Mar 65, pp 317-318. Possible source of errors and menas to avoid them in common-limiter normalization processor (Rubin and Kamen, 1964).

Adler, S.B., "Performance of a Traveling Wave Maser for Monopulse Radars," *IEEE Trans AES-2,* No 6, (suppl), Nov 66, pp 408-413 (Aerospace Systems Conv Record). Applicaiton of masers to monopulse receivers, maintaining phase and amplitude balance.

Schwartzkopf, D.B., "A High Precision Monopulse Receiver," *Microwave Journal 10,* No 9, Aug 67, pp 92-101. Design of solid-state receiver using limiter normalization and supplementary sidelobe indicator channel.

Peebles, P.Z., "Signal Processors and Accuracy of Three-beam Monopulse Tracking Radar," *IEEE Trans AES-5,* No 1, Jan 69, pp 52-57. Analysis of performance for three-beam amplitude or phase comparison systems using fixed beam matrix.

Taylor, H.P., Haroules G.G., and Brown, W.E., "Analysis of a Multiplicative Feed System for Monopulse Tracking Applications," *IEEE Trans AES-5,* No 4, Jul 69, pp 581-588. Discusses multiplicative processing of signals from paired feeds in four-horn configuration, and concludes that sidelobe levels are lower than with conventional sum-and-difference monopulse.

D. Analysis of Tracking Performance

Budenbom, H.T., "Monopulse Automatic Tracking and the Thermal Bound," *IRE-PGMIL Conv Record,* Jun 19, 57, pp 387-392. Qualitative analysis of thermal noise error determined to be proportional to beamwidth and to $snr^{-1/2}$, experimental data conforming slope of error vs. snr.

Mechler, E.A., Porter, J.W. and Yavne, R.O., "A Guidance Radar for Satellites, Space Vehicles and Ballistic Missiles," *Proc 2nd Natl Conv on Military Electr, IRE-PGMIL,* Jun 16-18, 58, pp 213-218. Angular rate accuracy of AN/FPS-16.

Dunn, J.H. and Howard, D.D., "The Effects of Automatic Gain Control Performance on the Tracking Accuracy of Monopulse Radar Systems," *Proc IRE 47,* No 3, Mar 59, pp 430-435. Analysis of interaction between glint error and dynamic lag error, as controlled by speed of normalization process in receiver.

Barton, D.K., "Sputnik II as Observed by C-Band Radar," *IRE Conv Rec 59,* Pt 5, pp 67-73. Early results of AN/FPS-16 track of Sputnik, showing presence of corner reflector and other features of target signature analysis.

Dunn, J.H., Howard, D.D. and King, A.M., "Phenomena of Scintillation Noise in Radar Tracking Systems," *Proc IRE 47,* No 5, May 59 pp 855-863. Analysis and experimental data on target fluctuation and resulting errors in tracking radar. Describes glint in terms of phase-front tilt.

Barton, D.K., "Accuracy of a Monopulse Radar," *Proc 3rd Natl Conv on Mil Electr, IRE-PGMIL,* Jun 29-Jul 1, 59, pp 179-186. Complete error analysis of tracking radar, applied to AN/FPS-16, with experimental results. Reprint Paper 4.

Manasse, R., "Maximum Angular Accuracy of Tracking a Radio Star by Lobe Comparison," *IRE Trans AP-3,* No 1, Jan 60, pp 50-56. Basic Theory of ideal measurement of angle using linear-odd illumination of aperture to minimize thermal noise. Includes "detector loss" term for low-snr case.

Skolnik, M.I., "Theoretical Accuracy of Radar Measurements," *IRE Trans ANE-7,* No 4, Dec 60, pp 123-129. Basic Theory of radar accuracy in thermal noise. Results for Angle measurement apply to scanning beams, and are not generally correct for monopulse with tapered illumination: see Handbook of Radar Measurement, pp 23-34.

Querido, H.B., "Amplitude Comparison Error of a Signal Received by Two Circularly-Polarized Antennas Due to Off-Axis Ellipticity," *IRE Trans AP-9,* No 2, Mar 61, pp 222. Error Component caused by different ellipticity of squinted circularly polarized patterns for signals off axis.

Brennan, L.E., "Angular Accuracy of a Phased Array Radar," *IRE Trans AP-9,* No 3, May 61, pp 268-276. Analysis of angle errors from thermal noise and phase-shifter error in phased array antenna.

Develet, J.A., Jr., "Thermal-Noise Errors in Simultaneous-Lobing and Conical-Scan Angle-Tracking Systems," *IRE Trans SET-7,* No 2, Jun 61, pp 42-51. Compares monopulse and conical-scan accuracy in thermal noise, at both high and low snr, using squinted Gaussian-beam models. Reprint Paper 15.

Nester, W.H., "A Study of Tracking Accuracy in Monopulse Phased Arrays," *IRE Trans AP- 10,* No 3, May 62, pp 237-246. Effect of amplitude and phase errors in array on tracking accuracy, for both on-axis and off-axis targets.

Sharenson, S., "Angle Estimation Accuracy in Monopulse Phased Arrays," *IRE Trans ANE-9,* No 3, Sep 62, pp 175-179. Analysis of thermal noise error for off-axis targets, disclosing existence of additional noise component caused by sum-channel noise in normalization of difference/sum. See further correspondence (1971).

Stegen, R.J., "The Null Depth of a Monopulse Tracking Antenna," *IEEE Trans AP-12,* No 5, Sep 64, pp 645-646. Use of phase-sensitive monopulse error detector to eliminate error caused by non-zero null.

Kosowsky, L.H. and Chanzit, L., "A Comparison of the Angular Accuracy of an Amplitude Monopulse Radar and a Phase Interferometer Radar in the Presence of Noise," *Proc 8th MiL-E-Con,* Sep, 14-16, 64, pp 297-301. Interferometer found more efficient in thermal noise than off-axis monopulse measurement.

Rainal, A.J., "Theoretical Accuracy of Monopulse Radar Receivers," *IEEE Trans AES-2,* No 2, Mar 66, pp 234-235. Analysis of three monopulse antenna-processor configurations with regard to sensitivity on random signals in noise. See full paper (May 1966) and correspondence (Bishop and Sheehan, 1967.)

Rainal, A.J., "Monopulse Radars Excited by Gaussian Signals," *IEEE Trans AES-2,* No 3, May 66, pp 337-345. Distributions of thermal noise error for phase and amplitude comparison monopulse with random signal inputs.

Resnick, S., "A Note on the Angular Accuracy of a Monopulse Radar Using Logarithmic Normalization," *Proc IEEE 54,* No 8, Aug 1966, pp 1111-1112. Extends Sharenson's results (1962) to monopulse estimators using logarithmic normalization.

McGinn, J.W., "Thermal Noise in Amplitude Comparison Monopulse Systems," *IEEE Trans AES-2,* No 5, Sep 66, pp 550-556. Maximum likelihood estimate of angle of arrival compared to monopulse process.

Riggs, R.F., "The Angular Accuracy of Monopulse Radar in the presence of clutter," *Proc NEC 22,* 66, pp 237-242. Extension of angle error equations to include signal-to-clutter ratio for low-elevation targets.

Bishop, D.R., and Sheehan, J.A., Comments on "Monopulse Radars Excited by Gaussian Signals," *IEEE Trans AES-3,* No 3, May 67, p 574. Explains apparent asymmetries in plotted results of Rainal (1966).

Mosca, E., "Systematic Tracking Inaccuracy in Monopulse Phased Arrays Produced by Analog Phase Shifters with Nonlinear Transfer Characteristics," *Proc IEEE 55,* No 6, Jun 67, pp 1091-1092. Effect of sinusoidal phase

Kehr, D.E., "Variance Analysis of the Angle Output Voltage for an Amplitude Monopulse Receiver," *IEEE Trans AES-4,* Jul 68, pp 535-540. Thermal noise error for system using separate squinted antennas, each with its own receiver and envelope detector.

Buck, G.J., "Phase Monopulse Error Due to Nonuniform Noise Background," *IEEE Trans AES-5,* No 4, Jul 69, pp 606-610. Analysis bias error caused by gradient in background noise level of space.

Mosca, E., "Angle Estimation in Amplitude Comparison Monopulse Systems," *IEEE Trans AES-5,* No 2, Mar 69, pp 205-212. Analysis bias and noise errors in maximum likelihood estimates of angle.

Ewell, G.W., "Monopulse Antenna Performance When Receiving Nonuniform Waves," *IEEE Trans AES-7,* No 3, May 71, pp 561-564. Modification of plane phase-front theory when radar is in near-field of complex target.

Blachman, N.M., "The Effect of Noise on Bearings Obtained by Amplitude Comparison," *IEEE Trans AES-7,* No 5, Sep 71, pp 1007-1009. Effect of thermal noise on ratio measurements as used in amplitude-comparison monopulse.

Lipman, M.A., Comment on "Variance Analysis of the Angle Output Voltage for an Amplitude Monopulse Receiver," *IEEE Trans AES-7,* No 5, Sep 71, pp 1009-1011. Correction to Kehr (1968).

Gustafsson, S.G., "Difference Pattern Null Depth and Angular Accuracy," *IEEE Trans AP-20,* No 1, Jan 72, pp 99-100. Restatement of ability of phase-sensitive detector to eliminate error caused by quadrature difference signal.

Berger, H., "Angular Accuracy of Amplitude Monopulse Off-Boresight Radar," *Proc IEEE 59,* No 3, Mar 71, pp 411-412. Re-Analysis of Sharenson's off-axis error estimate (1962) using squinted beams, concluding that correlated noise cannot lead to zero error and that difference slopes are lower than given by Sharenson.

Sherman, S.M., Comments on "Angular Accuracy of Amplitude Monopulse Off-Boresight Radar," *Proc IEEE 59,* No 11, Nov 71, pp 1637-1638. Comments on Berger (1971), reaffirming Sharenson's analysis with minor correction.

Berger, H., Reply on "Angular Accuracy of Amplitude Monopulse Off-Boresight Radar," *Proc IEEE 59,* No 11, Nov 71, p 1638. Reply to Sherman, emphasizing importance of correlated noise and more rigirous derivations which led to lower difference slopes in original correspondence.

Sharenson, S. and Barton, D.K., Further comment on "Angular Accuracy of Amplitude Monopulse Off-Boresight Radar," *Proc IEEE 59,* No 11, Nov 71, pp 1638-1639. Explains discrepancy in difference slopes between Berger and Sharenson (1962) is caused by use of different beam-width measures rather than mathematical rigor and that noise correlation, while leading to zero error in certain cases, is of little practical importance.

Lind, G., Comments on "Difference Pattern Null Depth and Angular Accuracy," *IEEE Trans AP-21,* No 1, Jan 73, p 132. Comments on Gustafsson (1962), calling attention to need for reasonable balance between phase errors before and after monopulse comparator. Reply emphasizes need to consider relative difficulty of controlling two error sources.

Carver, K.R., Cooper, W.K. and Stutzman, W.L., Beam-Pointing Errors of Planar-Phased Arrays," *IEEE Trans AP-21,* No 2, Mar 73, pp 199-202. Pointing error vs. element phase error and number of elements for various illumination tapers.

Sahmel, R.H. and Manasse, R., "Spatial Statistics of Instrument-Limited Angular Measurement Errors in Phased Array Radar," *IEEE Trans AP-21,* No 4, Jul 73, pp 524-532. Analysis of spatial correlation of quantization error and other beam steering errors in phased arrays.

E. Multiple Target Resolution.

George, S.F. and Zamanakos, A.S., "Multiple-Target Resolution of Monopulse Versus Scanning Radars," *Proc NEC 15,* pp 814-823. Analysis of minimum separation between targets necessary to permit measurement of each. Reprint Paper 5.

Anderson, D.B. and Wells, D.R., "A Note of the Spatial Information Available from Monopulse Radar, *Proc 5th Natl Conv on Military Electr, IRE-PGMIL,* Jun 27-29, 61, pp 268-278. Extension of monopulse theory to measurement of target shape and rotation. Reprint Paper 14.

Hausz, W., "Angular Location, Monopulse and Resolution," *Microwave Journal,* Feb 64, pp 60-65. Possible use of monopulse beam sharpening to resolve closely spaced targets.

McGillem, C.D., "Monopulse Radar Air-to-Ground Ranging," *IEEE Trans AES-2,* No 3 May 66, pp 303-315. Application of mono-

pulse to identify range to surface at specified elevation angle within beam.
Sherman, S.M., "The Use of 'Complex Indicated Angles' in Monopulse Radar to Locate Unresolved Targets, *Proc NEC 22,* 66, pp 243-248. Description and analysis of procedure to measure angular separation of two unresolved targets by observing quadrature component of difference/sum ratio. Reprint Paper 23.
Pollon, G.E., "On the Angular Resolution of Multiple Targets," *IEEE Trans AES-3,* No 1, Jan 67, pp 145-148. Cramer-Rao bound for simultaneous estimation of positions of two unresolved targets.
Pollon, G.E. and Lank, G.W., "Angular Tracking of Two closely Spaced Radar Targets," *IEEE Trans AES-4,* No 4, Jul 68, pp 541-550. Implementation of maximum likelihood estimator for two targets using coupled monopulse tracking loops.
Peebles, P.Z. and Berkowitz, R.S., "Multiple-Target Monopulse Radar Processing Techniques," *IEEE Trans AES-4,* No 6, Nov 68, pp 845-854. Multiple-target location based on solution of simultaneous equations using Taylor expansion of antenna patterns.
Barton, D.K., Comments on "Angular Tracking of Two Closely Spaced Radar Targets," *IEEE Trans AES-5,* No 2, Mar 69, pp 350-351. Points out difference between resolution of two targets and mutual nulling approach of Pollon and Lank (1968).
Sherman, S.M., "Complex Indicated Angles Applied to Unresolved Radar Targets and Multipath," *IEEE Trans AES-7,* No 1, Jan 71, pp 160-170. Review of complex monopulse measurement with application to two-target and multipath problems. Reprint Paper 27.
Peebles, P.Z., Jr. and Goldman, L., Jr., "Radar Performance with Multipath Using the Complex Angle," *IEEE Trans AES-7,* No 1, Jan 71, pp 171-178. Analysis of accuracy when using complex difference/sum ratio to locate target over reflecting surface.
Berkowitz, R.S. and Sherman, S.M., "Information Derivable from Monopulse Radar Measurements of Two Unresolved Targets," *IEEE Trans AES-7,* No 5, Sep 71, pp 1011-1012. Possibility of locating midpoint of two targets and slope of line joining them by measurement of difference/sum ratios in two usual tracking coordinates.
Howard, D.D., Nessmith, J. and Sherman, S.M., "Monopulse Tracking Errors Due to Multipath: Causes and Remedies," *IEEE Eascon Record 71,* pp 175-182. Survey of multipath error causes and remedies, calling attention to amplitude gradient across antenna as a further source of error.

F. Descriptions of Complete Monopulse Radars and Applications.

Stokes, I. and Barton, D.K., "Guided Missile Instrumentation Radar," *RCA Internal Publication,* Dec 1956, pp 26-31. Reprint Paper 3.
Barton, D.K., "Application of Precision Tracking Radar to Location, Control and Data Transmission from an Unmanned Observation Platform," *Proc 1958 Natl Conf on Aeronautical Electronics, IRE-PGANE,* pp 301-308. Description of AN/FPS-16 with possible reconnaissance or satellite applications.
Barton, D.K. and Sherman, S.M., "Pulse Radar for Trajectory Instrumentation," *Proc 6th Natl Flight Test Instr Symp, Instr Soc of America,* 1960. Survey of test range requirements and comparison of diverse techniques for trajectory measurement, leading to choice of AN/FPS-16 and related monopulse radars.
Bonney, L.D., "Radar Ballistic Instrumentation at White Sands Missile Range," *IRE Trans MIL-4,* No 4, Oct 60, pp 583-586. Brief history and description of application of AN/FPS-16 to instrumentation of missile range.
Barton, D.K., "The Future of Pulse Radar for Missile and Space Range Instrumentation," *IRE Trans MIL-5,* No 4, Oct 61, pp 330-351. History, theoretical and practical limitations, future possibilities for monopulse instrumentation radars. Reprint Paper 17.
Lerch, C.S., Jr., "Phased Array Radars for Satellite Tracking," *IRE Conv Record,* 62, Pt 5, pp 50-57. General description of array techniques illustrated by ESAR, an experimental radar including multiple-beam clusters for tracking.
Nessmith, J.T., "New Performance Records for Instrumentation Radar," *Space/Aeronautics,* Dec 62, pp 86-93. Description of large monopulse instrumentation radar used in missile re-entry measurements on Kwajalein Atoll. Reprint Paper 21.
Anders, J.V. et. al., "The Precision Tracker," *BSTJ 42,* No 4, Jul 63, pp 1309-1356. Tracker used in Telstar experiment.
Barton, D.K., "Recent Developments in Radar Instrumentation," *Astronautics and Aero-*

space Eng, Jul 63, pp 54-59. Survey of existing radar instrumentation equipment and networks, with possible improvements through coherent Doppler and trilateration procedures.

Jacobson, R.I., "The ARIS Instrumentation Radar," *Proc 7th Natl Conv on Military Electr, IEEE-PGMIL,* Sep 9-11, 63, pp 60-63. General description of instrumentation tracker using two-channel monopulse.

Curry, G.R., "Introduction to the TRADEX Radar System," *Lincoln Lab Tech Rep No 357,* 15 Jul 64. Description of large monopulse instrumentation radar used in missile re-entry measurements on Kwajalein Atoll. See also Reprint Paper 21.

Mertens, L.E. and Tabeling, R.H., "Tracking Instrumentation and Accuracy on the Eastern Test Range," *IEEE Trans SET-11,* No 1, Mar 65, pp 14-23. Estimated accuracy of range instrumentation, including AN/FPS-16 and AN/FPQ-6 radars.

Kirkpatrick, G.M., "Use of Airborne Monopulse Radar as a Low Approach Aid," *IEEE Trans AES-2,* No 3, May 66, pp 353-359. Application of beam sharpening to display of discrete reflectors, with test data from AN/APS-61 radar.

Kahrilas, P.J., "HAPDAR - An Operational Phased Array Radar," *Proc IEEE 56,* No 11, Nov 68, pp 1967-1975. Description and patterns of experimental optically-fed array.

Johnson, C.M., "Ballistic-Missile Defense Radars," *IEEE Spectrum,* Mar 70, pp 32-41. General description of missile site radar and perimeter acquisition radar of Safeguard system, both multifunction phased arrays.

NAVAL RESEARCH LABORATORY
Washington 20, D. C.

Radio Division - Fire Control Section

Report RA 3A 222A

28 December 1944

Retyped with references deleted, May 1973

ACCURATE ANGLE TRACKING BY RADAR

By

R. M. Page
Head, Fire Control Section

Abstract

The problem of accurate angle tracking by radar is briefly reviewed. Circuit stability and modulation of signals are explained as two chief sources of error. Various proposals for reduction of these errors are enumerated and the fatal weakness of each is pointed out. A new system for simultaneous comparison of lobe energy is described which promises attractive advantages in simplicity, stability, accuracy, speed, and immunity to jamming.

Introduction

1. For the past two years this Laboratory has been making special studies of automatic tracking by radar. During this period the Radars Mark 4 Mod 2, Mark 5, Mark 6, Mark 7 Mod 0, and Mark 12 Mod 2 have been studied and tested by the Automatic Tracking Group. This Group participated in the final stages of development on some of these automatic radar sets. In addition, automatic corrector systems have been developed and tested for the Radars Mark 4 and Mark 12. Out of this experience has grown a considerable fund of information on the problems of automatic tracking and on accurate angle tracking in general. This report reviews the factors responsible for most of the tracking errors in these systems, describes proposals for neutralization of various ones of these factors, and presents a solution which promises a favorable compromise among these factors.

2. In order to track in angle it is necessary to know at all times (1) whether or not an angular error is present, and (2) the direction of this error when present. This is accomplished by resolving the error signals from the target into two paired channels, one channel of each pair containing all of the plus errors, and the other all of the minus errors. These pairs of channels are then compared, tracking being accomplished by so orienting the antenna as to equalize the signals in the paired channels. Since the equality of these signals indicates coincidence of the radar line of sight with the target, the amplification of the error signals by the apparatus must be identical in the two channels paired together. A simple means of accomplishing this is to operate on the two channels alternately with the same electronic equipment. Ultimately, however, it is necessary somewhere in the circuit to separate plus and minus errors and present these two in a balanced network of some kind. The conventional lobe-switch radar as illustrated in Plate 1 and the conical scan radar as illustrated in Plate 2 are typical examples.

Sources of Error

3. In systems of this type there have been found two sources of error so pronounced as to mask those from all other sources. These are the shift of balance in electronic circuits and the modulation of the signal independently of the scanning process. It is at once obvious that an unbalanced circuit condition at or following the point of signal separation will give an indication of coincidence when an angular error actually exists. This phenomenon has been a serious weakness in all but the most recently developed fire control radar sets, and even yet exists to some extent as a defect.

4. The effect of modulation of the signal independently of the scanning process is less obvious, but simple to explain. The primary signals used alternately in the two channels of a pair are derived by some form of scanning of the target. This process produces a modulation of the signal, which is modified by the angular position of the target relating to the radar line of sight. This modulation is utilized to determine the angular position of the target relative to the line of sight of the radar antenna. If the signal becomes modulated from some other cause, the additional modulation may give rise to confusion or error in the determination of the angular position of the target.

5. There are several mechanisms by which targets modulate radar signals. Most targets appear to be at least partially polarized in that they reflect signals in one plane of polarization more effectively than in other planes of polarization. Moving targets appear to exhibit a most favorable plane of polarization which rotates in random fashion. This gives rise to apparent amplitude modulation of the echo signal. Aircraft targets with propellers give rise to amplitude modulation of the reflected signal due to rotation of the propellers. This well-known phenomenon has been the subject of several special studies and constitutes the major source of radar echo modulation from these targets. Enemy targets may introduce intentionally a modulation of the signal for the purpose of confusing or falsifying the angle information in the tracking radar. On weak echoes the signals are modulated by noise. Errors in angle determination arise when the

unwanted modulation of the signal contains frequencies, or within the system produces frequencies, at or near the frequencies which are used in determining the position of the target. Propeller modulation gives errors when the propeller tip frequency, or occasionally the propeller shaft frequency, or any of the harmonics thereof, or a difference frequency between any of these frequencies and the radar pulse recurrence rate or its harmonics, approaches the lobing or scanning frequency or any of its harmonics. Random target polarization and noise may contain and produce all frequency components that are used in determining target position.

6. Angle errors due to the modulation of the signal by the target are a function not only of the frequency and percentage of modulation, but also of the beam width of the radar itself. The steeper the slope at the crossover point in a lobed or scanned radar, and the narrower the primary beam width of the radar, the less will be the error due to a certain percentage of modulation. However, with a beam width of $1/2^{\circ}$ errors from this cause may still be as high as $1/4^{\circ}$ and in some cases will oscillate with peak to peak amplitude of as much as $1/2^{\circ}$ at frequencies which are fatal to the fire control solution.

Methods of Minimizing Errors

7. A number of palliatives have been found for reducing the effects of errors from these causes. Errors due to shift in balance of circuits have been reduced to the vanishing point by separation of plus and minus error signals at the highest possible energy level with consequent use of a minimum of amplification following the separation point, and by suppression of the effect on circuit performance of vacuum tube characteristics. In one laboratory system, the necessity of matching circuit components has been entirely eliminated and dependence of performance on tube characteristics reduced to the vanishing point. Interaction between train and elevation which may arise from target polarization may be eliminated from a rotating dipole system by using a nutating scan in which the plane of polarization does not rotate, or by simply using a fixed lobe switching system. Suppression of one axis by excessively strong signals on the other through the action of

automatic gain control (AGC) may be eliminated by the use of a fast AGC and derivation of error signals from the AGC voltage, a device utilized in radar MK-35. Effects of propeller modulation may be reduced by varying the lobing rate or the scan rate of the radar so as to increase the beat frequency between components of the propeller modulation and of the lobing frequency. They may be further reduced by the use of very high-speed lobing synchronized with the repetition rate together with a small variation in lobing rate. The effects of this and other periodic modulation may be further reduced by aperiodic lobing, synchronism being maintained between lobing and transmitter pulsing. Since both the high-speed lobing and aperiodic lobing are outside the limits of any known mechanical lobing system, an ionic system of lobing making use of controlled gas switches is being developed for these devices.

5

8. It will be noted that in any scanned or lobed system, variation of scanning or lobing rate is necessary ultimately to escape the effects of signal modulation at lobing frequency, whether that modulation be intentional or otherwise. Under such conditions it is desirable, if not necessary, to maintain synchronism between lobing rate and pulse recurrence rate. To reduce the hazard of jamming it is imperative that the transmitted signal not be lobed. All these requirements add complication to the system. In the ultimate of lobed systems, where lobing and pulsing are synchronized at one pulse per lobe, and the inter-lobe spacing varied in random fashion with purely aperiodic devices, circuit complexity mounts to discouraging proportions. Furthermore, the broad banding of error circuits necessitated by variation of lobing rate opens the way to interference on harmonics of lobing frequencies. In fact, the more refinement that is made in lobing systems the more extensively is the weakness of such systems revealed.

Simultaneous Lobe Comparison - Multiple Receiver Systems

9. The effect of modulation of the signal by the target may be eliminated by the simultaneous comparison of energy or phase in opposite lobes. This is not a new approach: Consideration of the idea in its more rudimentry form brings to mind such names as Bellini-Tosi, Watson-Watts, and others

of the instantaneous direction finding field. This laboratory had a dual receiver radar system operating on such principles in 1940. The General Electric Co. has recently done considerable work along similar lines, which has been reported. From a theoretical standpoint, this approach has some very attractive aspects. Errors due to signal modulation are eliminated. The problems of integrating energy over a large number of pulses are simplified since there is no switching, and long time-constant hold-over circuits are unnecessary. Attempts at this solution made heretofore have, however, some rather serious disadvantages. In the first place, it has been necessary to balance either gain or phase in two high-gain amplifiers. Here the problem of shift in balance of circuits is met at its worst. Devices have been developed for automatically adjusting differentially the gain between two channels, this adjustment being accomplished between pulses during the time when the radar receiver is not receiving echo information. This device, however, adds complication to the circuit in that an additional function must be performed with precision by electronic equipment. Such a system is illustrated in Plate 3.

6

R-F Phase Comparison System

10. Another system that has been developed depends on matching phase in two channels in such a way that angular misalignment of the antenna appears in the form of phase difference between the outputs of two amplifying channels. This system depends on an extremely high degree of stability of relative phase shift between the two high-gain channels and requires two separate antenna systems for each axis. These two weaknesses are fatal for Naval application.

Simultaneous Lobe Comparison - Signals Delayed From One Paired Lobe

11. Another proposal is that signals received simultaneously on the two lobes be separated in time by a delay circuit, the echo received from one lobe being delayed relative to the echo received from the other lobe by approximately one pulse width. This system presents the problem of balancing the attenuation of the delay circuit by the insertion in the other channel

of an attenuating circuit which has no delay. If this is done at the intermediate frequency, it is necessary that crystal mixers and the initial I-F stage be balanced. If it is done at R-F, the sensitivity of the system suffers from excessive attenuation at the input. Such systems are illustrated in Plates 4 and 5.

Null System

12. Still another solution is the well-known null method in which two channels are opposed at the input to the receiver, and the system balanced on a null. This system is used in some Japanese radars. It has two disadvantages:

(a) When there is no error there is also no signal, and

(b) In the presence of an error there is no indication of phase.

Simultaneous Lobe Comparison - NRL Modified "Null" System

13. It is proposed here that a modified "null" method with a separate sense channel be used for simultaneous lobe comparison in radar tracking. By such a method a solution is sought which will eliminate the effects of a signal modulation by the target and of certain other types of jamming, at the same time obtaining a serviceable degree of circuit stability. The system may be described briefly as follows:

14. A single antenna is to be used having five (5) lobes. One central lobe will be used for transmission and for range and reference signal. Two lobes symmetrically displaced horizontally about this central lobe will be used for train and another two similarly disposed about the central lobe vertically will be used for elevation. Since the train and elevation channels are identical, only the train channel will be described. The two train lobes will be opposed at the antenna and the energy difference will be fed into a receiver. This receiver will consist of a mixer with local oscillator and a high-gain I-F amplifier. The central lobe will be fed to an identical receiver with its mixer excited from the same local oscillator. The train receiver will contain no signal when the antenna is on target. If the antenna is off target in train, a signal will appear in the train receiver and the phase of that signal

relative to the phase of the reference signal in the range receiver will depend on the direction of the displacement of the antenna. The phase of one amplifier will be so shifted that for displacement in one direction the outputs of the two receivers will be in phase, and for displacement in the opposite direction, they will be in anti-phase. The outputs of the two receivers will be combined in two mixers - one adding and one subtracting the two outputs. The outputs of the two mixers will be opposed. The resultant output will be a video error signal which is zero when the antenna is on target, plus when it is off in one direction, and minus when it is off in the opposite direction. At the same time the output of the range or reference receiver will give an indication of the presence of the signal which will be a maximum when on target. The system is shown in Plate 6.

15. An analysis of this system shows that the balance point for train is independent of signal amplitude in either receiver and is also independent of the relative phase shift between the two receivers, so long as it does not closely approach 90°. The sensitivity of the system will be a function of the absolute signal level at the output of the receivers and of the relative phase shift between the two receivers. However, a relative phase shift of 25° will reduce the sensitivity less than one db, while the dependence of sensitivity on signal level can be removed by another device, the circuit for which is outside the scope of this report. It is anticipated, therefore, that this system will not require automatic gain control. Having no lobing or scanning, the system is immune to deceptive angle jamming from the target. Since opposite lobes are compared in a strictly linear device and error information is obtained from unintegrated video pulses, the system is immune to off-target jamming. The system may be easily modulated with "Jitterbug" modulation, and may be made immune to "Leopard" jamming. Since no lobing modulation will be present in the signal, utilization of Doppler modulation for tracking through window and ground clutter is vastly simplified. This leaves continuous noise modulation jamming as the only known type of jamming which can affect the system, and with the high power and beam concentration available in the X-band, jamming by this method will be exceedingly difficult, if not operationally impossible. It is conceivable that

this system may ultimately be considerably more simple than any of the present lobe switching or scanning systems, and presumably more accurate, particularly when applied to automatic control.

16. There are several uncertain factors in the development of such a system which are undergoing investigation. One is the practicability of utilizing five lobes simultaneously in the antenna structure in which each lobe retains full advantage of the entire antenna aperture for directivity and gain. A second is the practicability of design of a mixer and high gain IF amplifier for radar pulses that will be stable in phase against variations in supply voltage, signal amplitude, gain adjustment, and tube characteristics. A third factor is the practicability of combining three such receivers in a single reasonably compact chassis without crosstalk among the three receivers. It is felt that if these three problems can be solved, the rest of the system is straightforward. With regard to the first, some preliminary calculations and tests have been made which show high promise for the proposed utilization of five lobes simultaneously from one antenna without any loss of energy or directivity for each lobe. With regard to the phase stability in the I-F amplifiers, measurements made so far indicate that this problem need not be unduly difficult. The matter of the elimination of cross-talk between high-gain receivers mounted side by side in a chassis is largely one of engineering and appears to be no more difficult than some other problems that have already been worked out by this Group. Work is proceeding to check this factor.

Applications of NRL Simultaneous Lobe Comparison System

17. If this system proves to be an successful as it now promises to be, it will have a considerable significance in the field of fire control radar. It should make angle-tracking radar accurate enough for precision control of blind firing. It should make it fast enough for controlling guided missiles and also for controlling anti-guided missile fire. Having no lobing, it is more readily adaptable to modulation for other purposes such as "jitterbug", communication, and remote control, and for the same reasons is more suitable for tracking through "window", and ground clutter. It is immune

to all present known practical forms of angle jamming except multiple target confusion when two or more targets are at exactly equal ranges and within the beam, in which case both targets might be brought under fire. There also is the possibility that with this system automatic tracking may be made simple enough to be reliable in service.

10

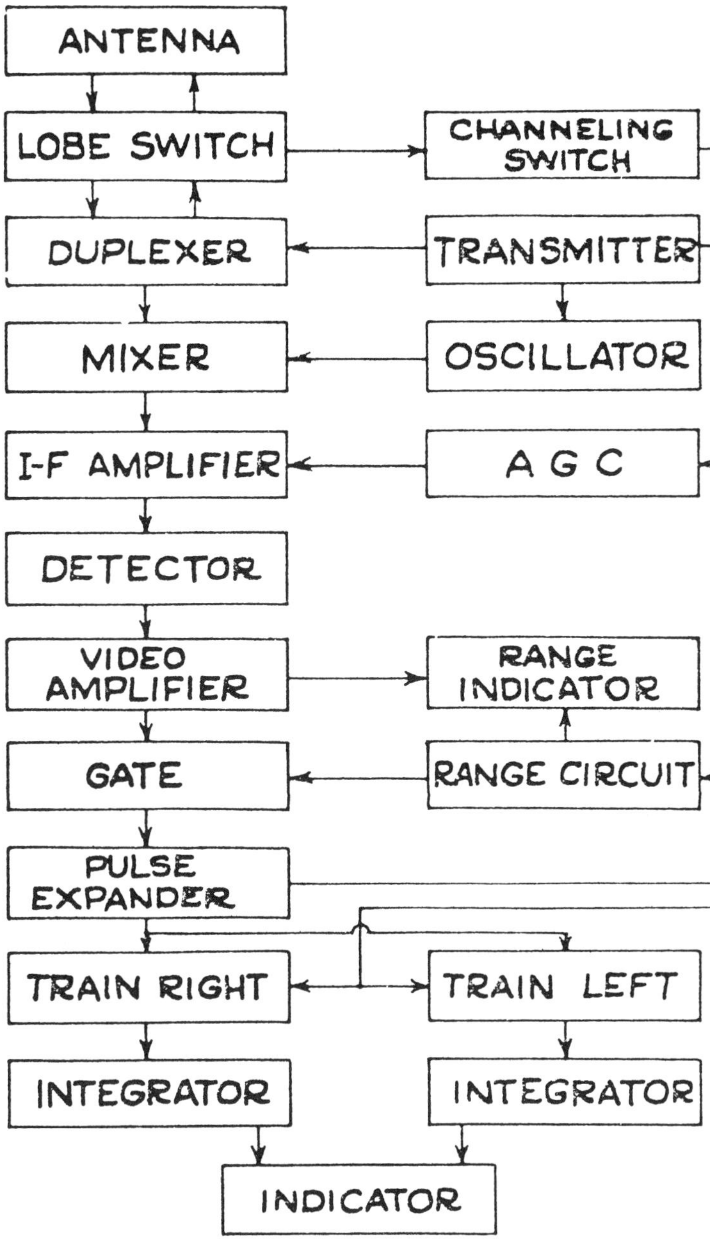

LOBE SWITCH SYSTEM

PLATE 1

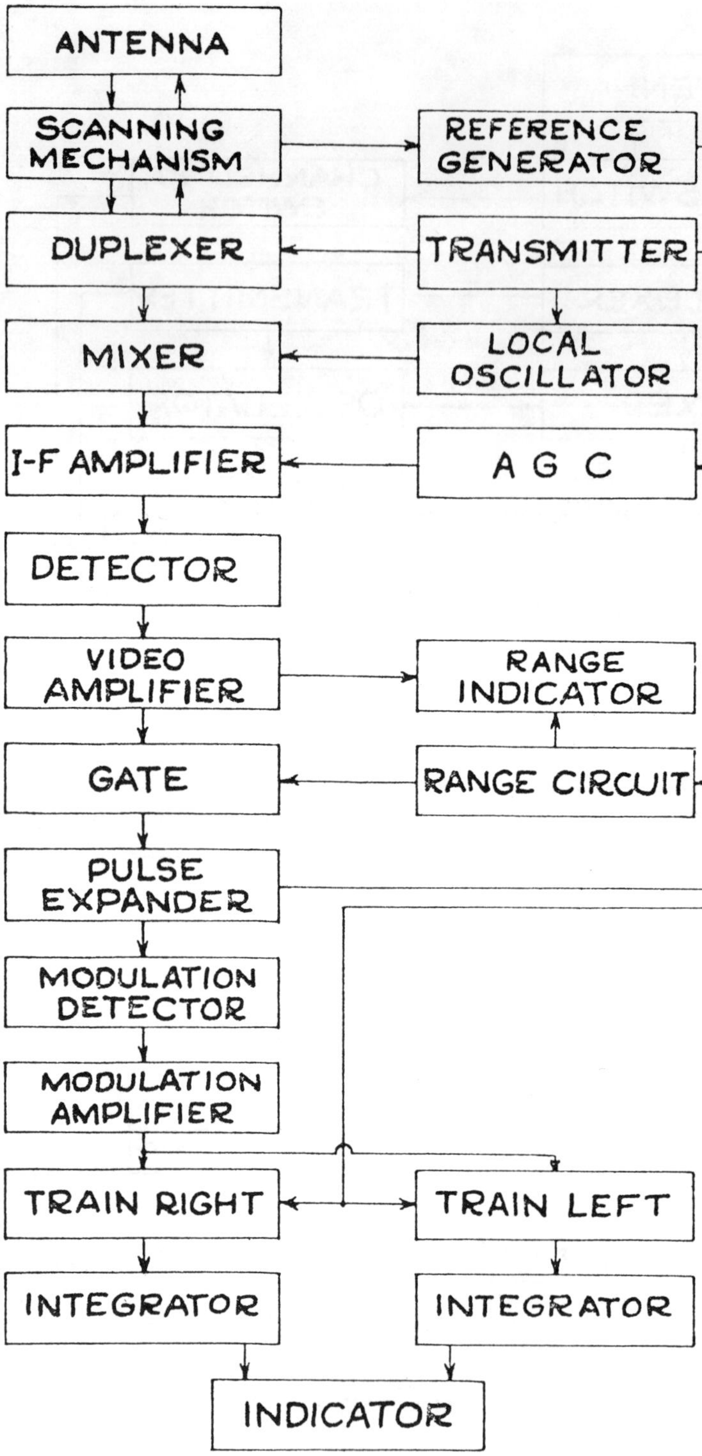
ANTENNA
SCANNING MECHANISM
REFERENCE GENERATOR
DUPLEXER
TRANSMITTER
MIXER
LOCAL OSCILLATOR
I-F AMPLIFIER
A G C
DETECTOR
VIDEO AMPLIFIER
RANGE INDICATOR
GATE
RANGE CIRCUIT
PULSE EXPANDER
MODULATION DETECTOR
MODULATION AMPLIFIER
TRAIN RIGHT
TRAIN LEFT
INTEGRATOR
INTEGRATOR
INDICATOR

CONICAL SCAN SYSTEM

PLATE 2

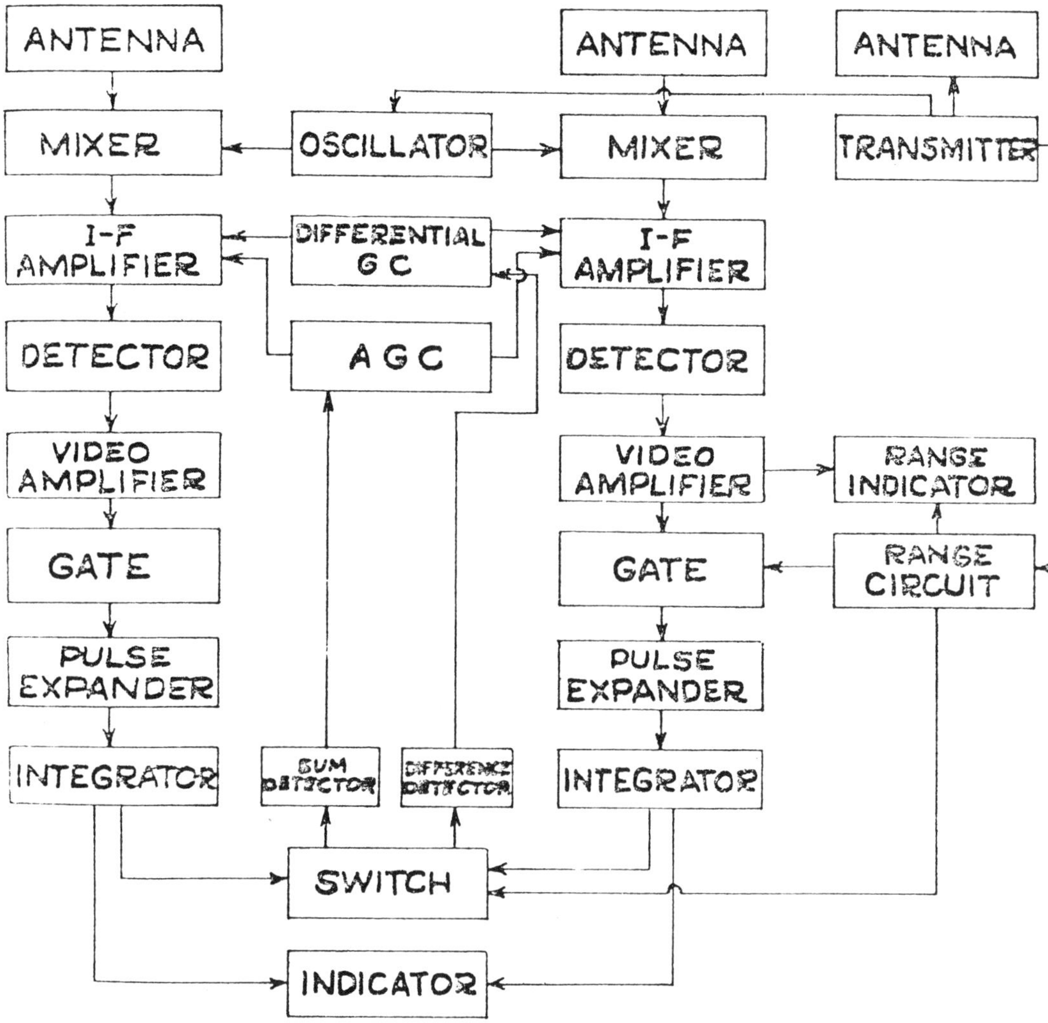

SIMULTANEOUS LOBE COMPARISON MULTIPLE RECEIVER SYSTEM

PLATE 3

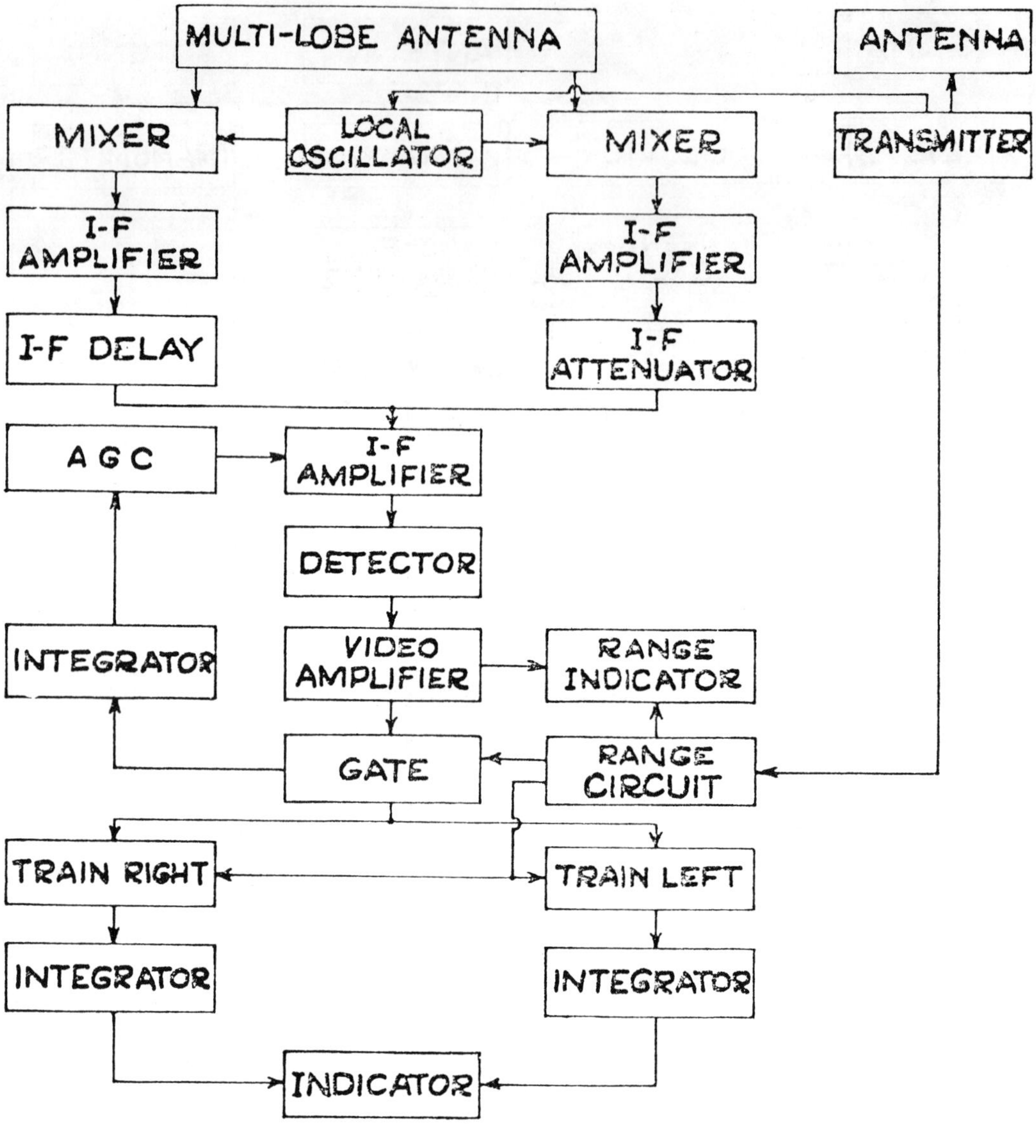

SIMULTANEOUS LOBE COMPARISON
I-F SIGNAL DELAYED FROM ONE PAIRED LOBE

PLATE 4

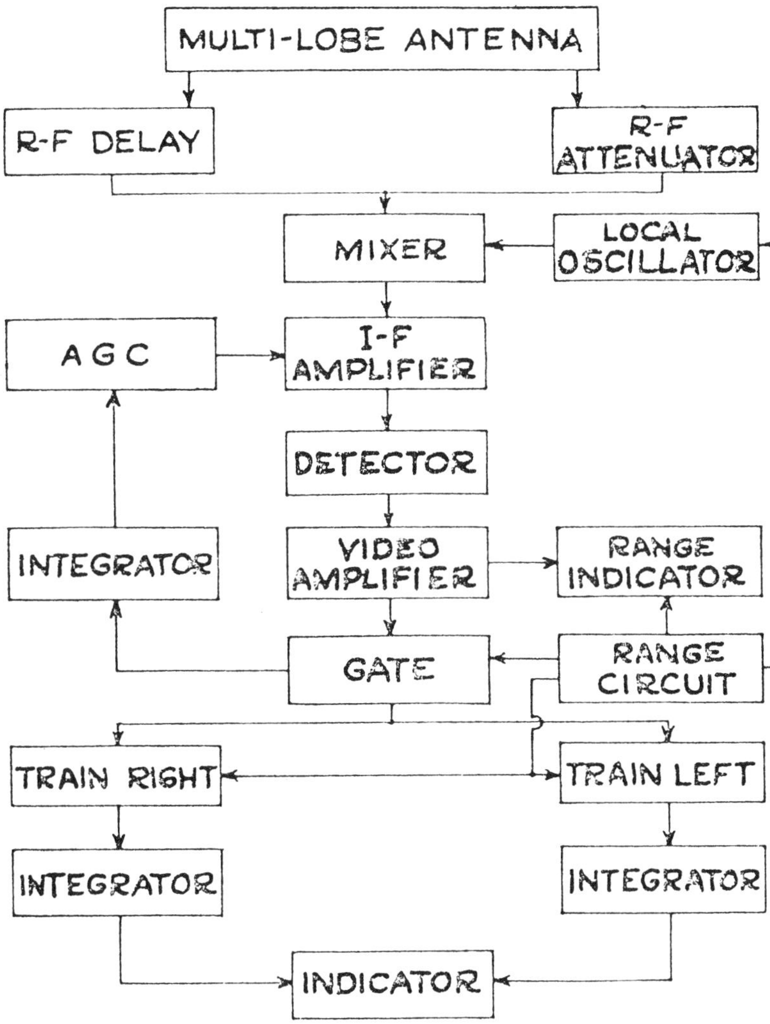

SIMULTANEOUS LOBE COMPARISON
R-F SIGNAL DELAYED FROM ONE PAIRED LOBE

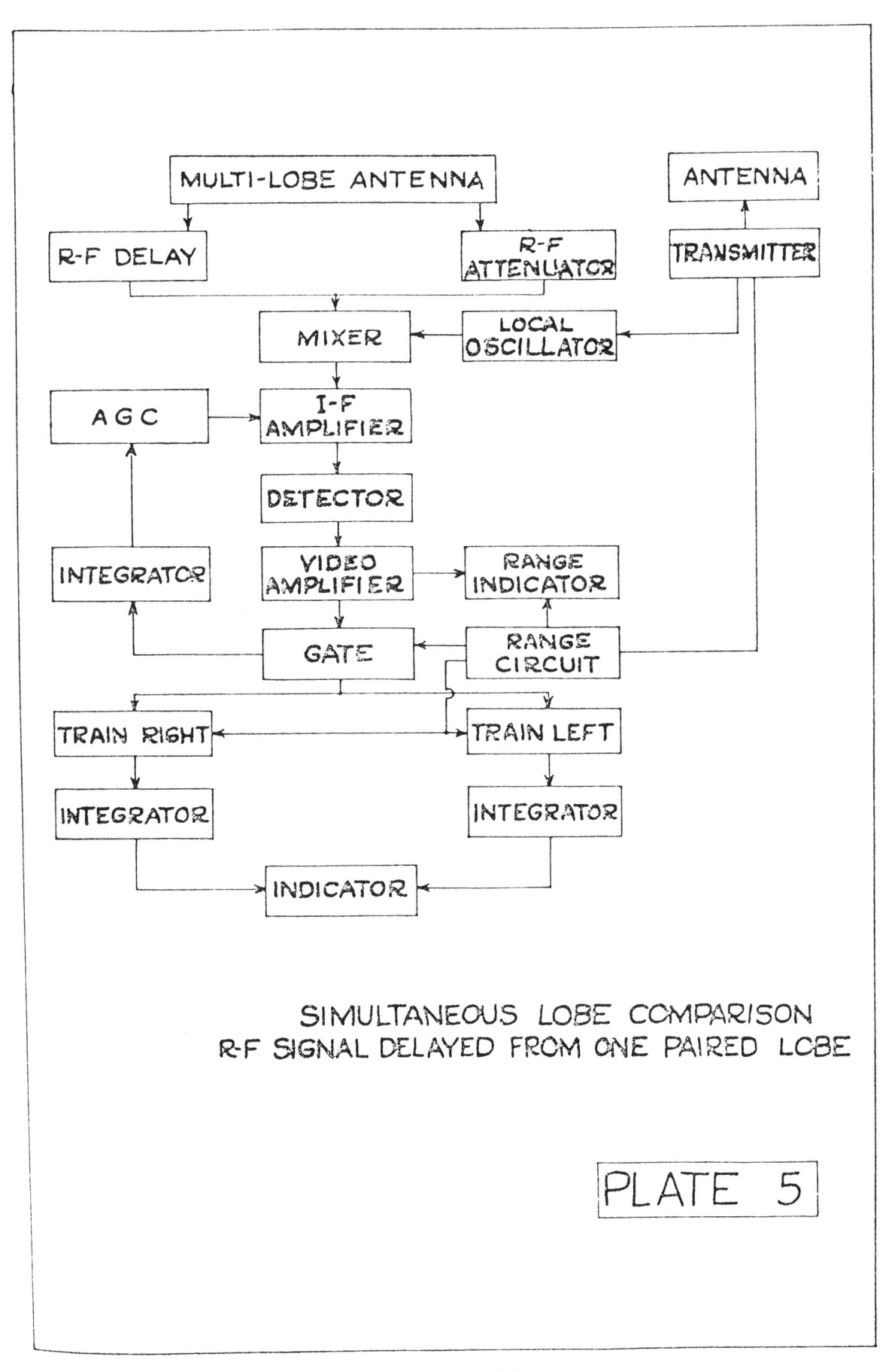

PLATE 5

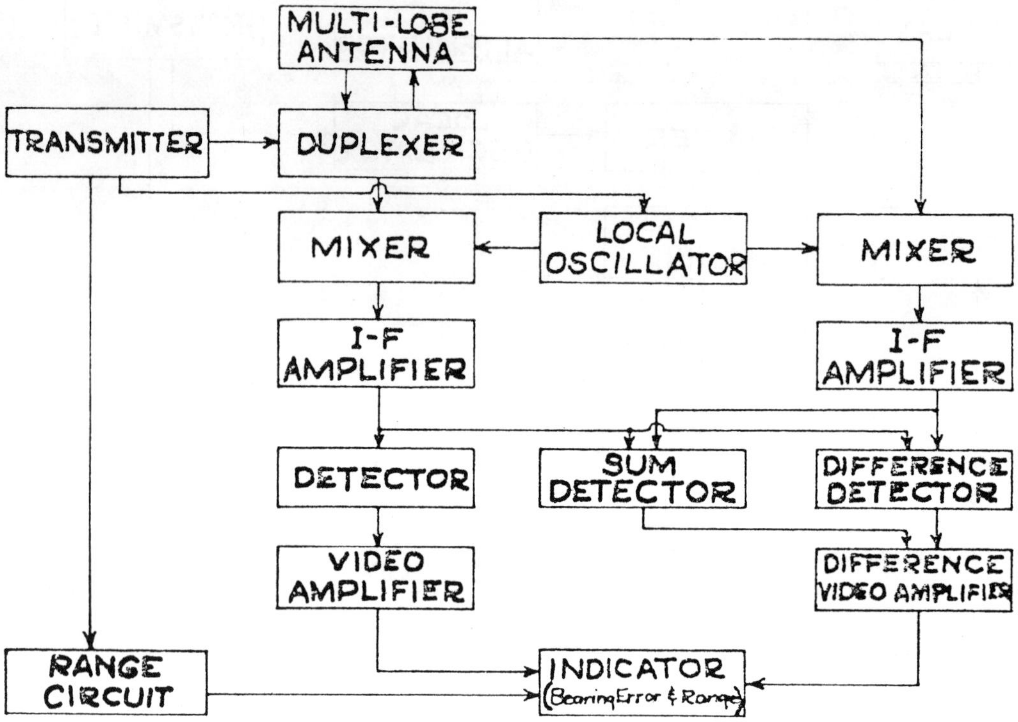

NRL
NEW PROPOSED SYSTEM

SIMULTANEOUS LOBE COMPARISON
MODIFIED "NULL" SYSTEM

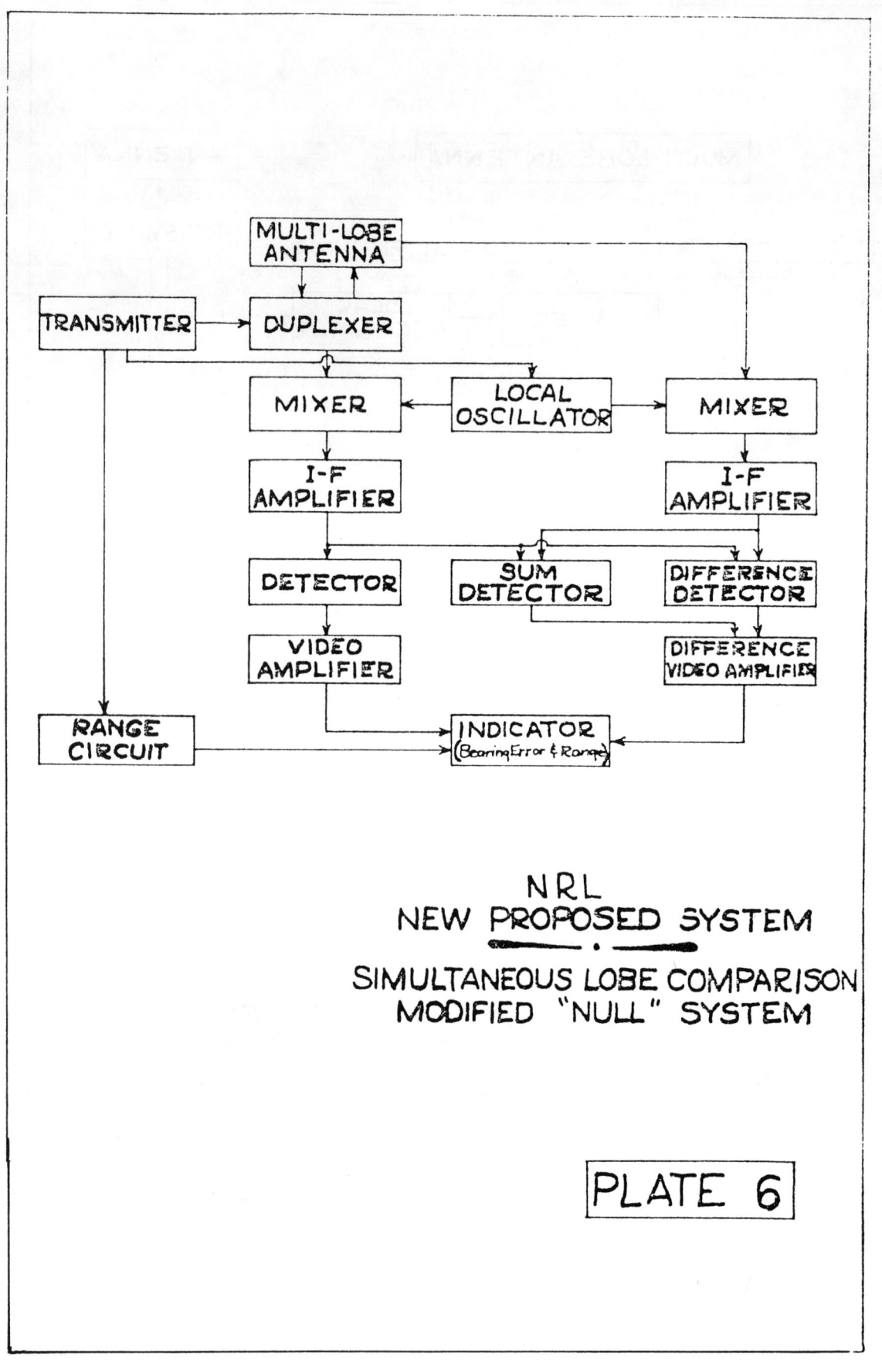

PLATE 6

GENERAL ELECTRIC

ELECTRONICS LABORATORY

SYRACUSE, NEW YORK

17

Contract No. D.A. 36-039-sc-194

Signal Corps Project 22-122-B-O

Department of Army Contract No. 3-14-03-051

S.C. Spec. EC11T36 SR

FINAL ENGINEERING REPORT ON ANGULAR ACCURACY IMPROVEMENT

Covering the Period
14 November 1950
30 June 1952

Report Dated
1 August 1952

Submitted:

G. M. Kirkpatrick

Nomenclature

θ = angle from z axis in far field
θ_{az} = angle from z axis in xz plane
θ_{el} = angle from z axis in yz plane
$(\theta_{az})_N$ = RMS noise angle equivalent of thermal noise voltage
α = normalized azimuth angle $= \frac{2\pi d \sin\theta_{az}}{\lambda}$
ε = normalized elevation angle $= \frac{2\pi h \sin\theta_{el}}{\lambda}$
α_b, ε_b = normalized squint angle measured from antenna axis
$\alpha_N = \frac{2\pi d (\theta_{az})_N}{\lambda}$ = normalized equivalent noise angle
λ = wavelength
d = aperture width
h = aperture height
A(x, y) = antenna aperture illumination
$A_o(y), A_e(y) = A_o(1, y), A_e(1, y)$ = odd and even aperture functions
$A_e'(x) = d/dx\, A_e(x)$
$f(\alpha, \varepsilon)$ = one way voltage far field antenna pattern = $\mathcal{F}\,[A(x, y)]$
$f_o(\varepsilon), f_e(\varepsilon) = f_o(0, \varepsilon), f_e(0, \varepsilon)$ = odd and even far field patterns (one way)
$f_o'(\varepsilon) = \frac{d}{d\varepsilon} f_o(0, \varepsilon)$ = slope of odd far field
$G(\alpha, \varepsilon)$ = far field power gain of one way pattern at angle (α, ε)
G = power gain at $\alpha, \varepsilon = 0$
$G_o = \frac{4\pi h d}{\lambda^2}$ = maximum value of G, A(x, y) = 1
G_r, G_t = receive and transmit gains in round trip radar operation
$K(\alpha, \varepsilon) = \frac{\partial G^{1/2}(\alpha, \varepsilon)}{\partial \varepsilon}$ = slope factor of one way far field voltage gain
K = slope factor at $\alpha, \varepsilon = 0$
$K_o = \left[\frac{G_o}{12}\right]^{1/2}$ = maximum value of K, A(x, y) = x or y
K_a = slope factor for amplitude plane of combination phase-amplitude radar antenna
K_p = slope factor for phase plane of combination phase-amplitude radar antenna
$KG^{1/2}$ = round trip angular sensitivity factor (proportional to voltage)
BW = beamwidth of one way even far field pattern in radians
η_i = spillover loss factor
η_o = odd loss factor = η_r unless otherwise specified
η_e = even loss factor = η_t unless otherwise specified
P = power

P_t = RF power transmitted
P_r = RF power received
Δ = round trip received voltage, an odd function
Σ = round trip received voltage, an even function
ECS = signal independent of target size and range—a ratio
L = product of RF plumbing and collimator loss factors
L_r, L_t = receive and transmit plumbing and collimator loss factors
L_o, L_e = odd and even plumbing and collimator loss factors
H = several parameters defined by $\left[\frac{A_{eff} P_t \lambda^2}{(4\pi)^3 R^4}\right]^{1/2}$
A_{eff} = effective target area
E_x, E_y = x and y components of the electric field
$\psi_{az} = \tan^{-1} \frac{f_o(\alpha)}{f_e(\alpha)}$
p = antenna parameter
$\mathcal{F}$ = Fourier transform defined by equations (B-3) and (B-4)
S = RMS signal amplitude
N = RMS noise voltage
N_Σ = RMS thermal noise added to Σ channel
N_Δ = RMS thermal noise added to Δ channel
N(t) = instantaneous noise voltage
v = voltage
v_R = reference voltage amplitude
v_o = voltage output of divider or phase detector
v_Σ = voltage output of Σ channel
Z = impedance
Z_Δ = terminating impedance of Δ RF output
Z_Σ = terminating impedance of Σ RF output
Z_m = mutual impedance to represent coupling
Z_l = load impedance
g = conductance of detector load
T = 293 degrees Kelvin
NF = overall noise figure
β = IF bandwidth
β_n = narrow filter bandwidth
τ = pulse length
f_r = pulse repetition frequency
μ_Δ = gain of Δ channel relative to Σ channel

I. ABSTRACT

This study was initiated to critically examine available monopulse radar techniques and to propose alternative circuits or antennas where deficiencies are clearly indicated. In particular, considerable progress has been made in developing techniques for obtaining a more useful error signal. When the radar error signal has certain desirable characteristics, it is termed an electrical correction signal (ECS).

The report is divided into four sections: (1) Analysis of factors in monopulse antenna design, (2) effect of thermal noise and ratio circuits on angular accuracy, (3) monopulse radar components, and (4) experimental results. The analysis of symmetrical monopulse antennas has been simplified. Tables of odd and even aperture functions enable a particular design to be compared to a hypothetical "ideal" monopulse antenna. The range at which a specified RMS error is exceeded can be predicted. The "equivalent angular error" is related to the thermal noise and radar parameters. Component parts are discussed from the viewpoint of providing alternative designs. Most of the experimental work was on the antenna. Many phase, amplitude and ECS curves are presented for a 4 horn feed—lens antenna. A cascaded 6BN6 IF limiter, and an extended range phase detector are reported.

B. Detailed Objectives

This project has as its goal a comprehensive theoretical study of possible means of improving the angular accuracy of a radar. Simultaneous lobing or monopulse type systems are to be considered. The analysis is to include a study of sources of error in system components—antennas circuits, mixers, IF amplifiers, ratio and phase detectors, log amplifiers, instantaneous AGC, data storage, etc. This research and study will be directed toward the selection of a system or systems giving improved accuracy of the target position data.

In general, the project is to be a three-phase program of research and development:

1. Study and analysis.
2. Development and test of breadboard components.
3. Incorporation of components into existing radar systems where said phases may overlap as desired to form a demonstration system.

IV. FACTUAL DATA

A. Introduction

In the development of radar techniques to find target positions, pulse to pulse variations of the target echo have imposed an inherent limitation. Simultaneous lobing techniques, also called monopulse, have been intensively developed in the past several years to overcome this limitation of sequential, low lobing rate radars. With the present demand for better angular data on fast moving, small targets, additional radar techniques are needed. This study was initiated to critically examine available techniques and to propose alternative circuits or antennas where deficiencies are clearly indicated. In particular, considerable progress has been made during this study in techniques for obtaining a more useful error signal. When the radar error signal has certain characteristics, it is termed an electrical correction signal, usually abbreviated to ECS. The desired qualities of ECS voltages are given in the following table.

1. The error voltage is independent of target size, A_{eff}, and range, R.	$ECS \neq f(R, A_{eff})$
2. The error voltage is linearly proportional to the error angle, θ.	$ECS = \rho\theta$

Electrical Correction Signals (ECS) are Radar Error Signals with above Qualifications

A brief historical survey is presented here for those readers with a knowledge of sequential lobing radar but unfamiliar with monopulse radar. The Radiation Laboratory Series is an excellent source of general background material on radar. In particular, one method of sequential lobing, conical scan, is discussed in Vol. 25, "Theory of Servo-mechanisms," pages 212 and 291.

Work on monopulse techniques in the General Electric Company was initiated prior to 1943. At that time, a phase comparison radar,[1] shown in the sketch below, was completed and tested. Analysis of photographic data indicated that the effects of pulse to pulse amplitude fluctuations were minimized by the use of three receivers with "fast" AGC. With three receivers an error output is obtained in both azimuth and elevation for each transmitted pulse, hence the name monopulse.

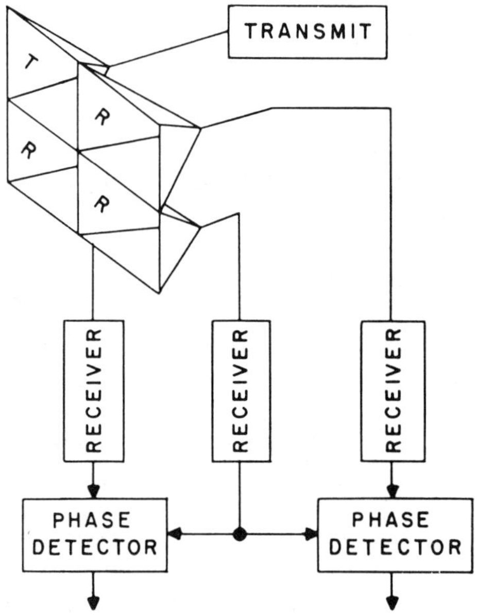

Phase Comparison Radar

With the development of microwave hybrid circuits it became possible to make RF signal combinations. This led to monopulse circuits[2,3] which utilize the antenna aperture more efficiently, and stabilize the boresight.[4] With balanced hybrid circuits the difference output (Δ) is zero when the target is on the antenna axis. Therefore, slight phase and gain variations in the receiver channels cannot change the boresight from the antenna axis.

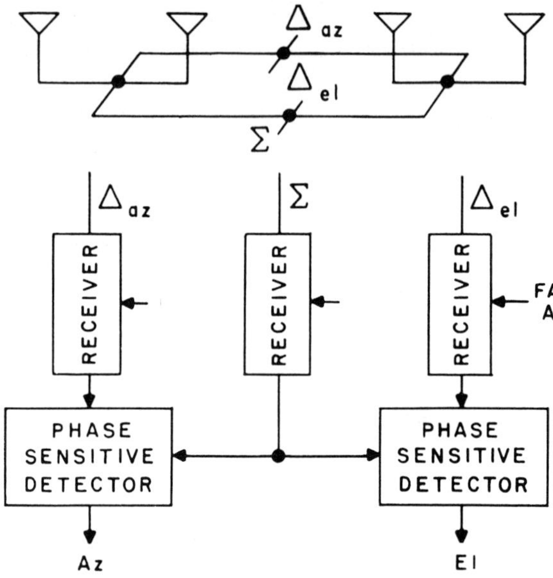

Amplitude Comparison Radar

The antenna aperture is fed with four feed horns which produce "squinted" beams in the far field. If the four feed horns and the microwave hybrid circuits are considered as a unit, the analysis of the antenna is more straightforward than if an attempt is made to consider the squinted beams produced by individual feeds. This will be taken up at length in a later section.

A tracking radar which uses phase comparison on one axis and amplitude comparison on the other, devised by

[1] J. P. Blewett, S. Hansen, R. Troell, G. M. Kirkpatrick, "The Multilobe Tracking System," General Electric Company Research Laboratory Report, Schenectady, New York, January 5, 1944.

[2] J. E. Trevor and A. E. Hastings, "Analysis and Specifications of Simultaneous Lobing System TAB," NRL Report 2554, 1 July 1945.

[3] H. T. Budenbom, "Preliminary Studies on Alternate Systems and Components for Monopulse Angle Tracking," Bell Telephone Laboratories Memorandum MM46-2730-38, July 8, 1946.

[4] The boresight is the indicated direction of the radar axis when the error signal Az and El is at a null.

General Electric Company engineers,[5] was reported in 1947. This monopulse radar retains the advantages of the amplitude comparison radar, and in addition simplifies the RF plumbing and reduces to two the number of receivers required.

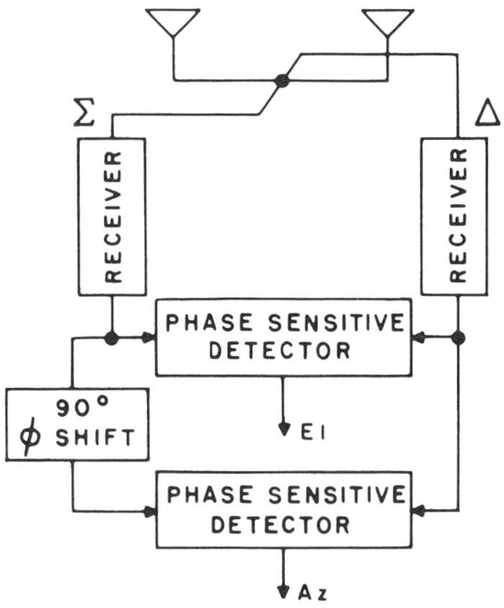

Combination Phase-Amplitude Radar

The combination phase-amplitude radar has been tested extensively, and test results show considerable improvement in tracking accuracy over a nearly equivalent conical scan radar.

[5]"Automatic Gun Laying Turret, Tail Aero X 5 A," Design Study Report Vol. II, General Eelectric Company Report GET-1763-2. An objective of the Aero X 5 A project was to reduce the size and complexity of a monopulse tracking radar for an airborne application. W. Hausz, of the General Electric Company Electronics Laboratory, proposed the final form of the combination phase-amplitude system.

The possibility of applying monopulse techniques[6] to a scanning radar to improve the visual presentation has also received considerable attention. The apparent width of the antenna pattern can be reduced by using an ECS voltage to deflect each indication of target return to the correct position on the indicator. This is illustrated with a sketch of a B scope presentation.

A phase comparison circuit for producing the deflection voltage is shown in block diagram form. This antenna is not suitable for most applications because of high side lobes.

This completes the brief historical survey of monopulse.

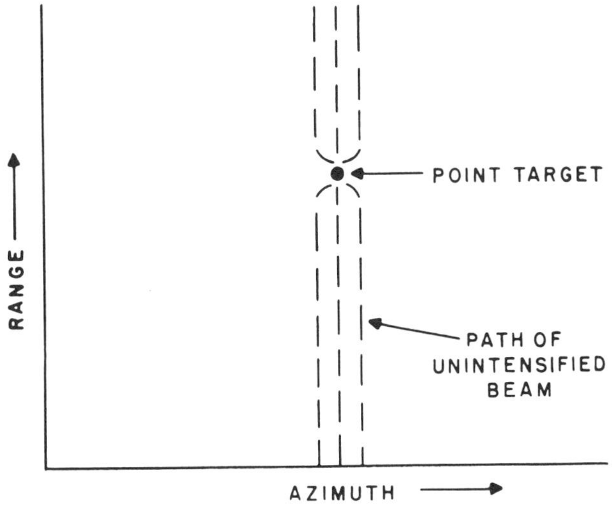

Beam Sharpening of B Scope Presentation

[6]"Interim Report No. 2 on ALSTAR System," Contract No. AF28(099)-42, General Electric Company, 10 August 1949 (and earlier reports of higher classification on Contract W-33-038-ac-20412).

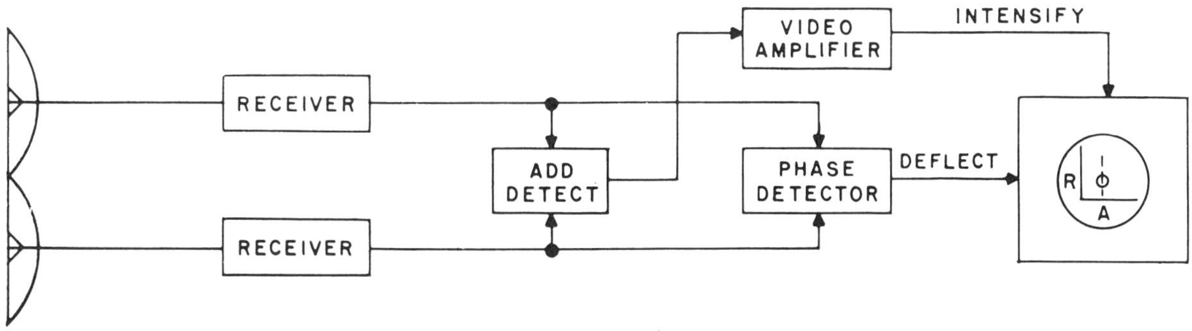

Circuit for Beam Sharpening

The trends in monopulse radar techniques indicated by the subsequent analysis and experimental results will be reviewed briefly.

1. The antenna is recognized as being a key part of a monopulse radar. Instead of adhering to the early division of monopulse radar antennas into phase and amplitude types a more general theory has evolved. The direction finding properties of a radar antenna result from the addition of an odd far field pattern to the usual even far field found in search radars. An investigation of the properties of various odd and even far field functions and combinations of these functions provides a basic theory of monopulse antennas.

2. By appropriate methods of analysis it is possible to calculate the maximum range with a given angular accuracy of a tracking radar, just as it is possible to calculate the range of detection of a search radar.

3. Using the odd and even aperture functions, it is possible to specify an "ideal" antenna, which can be used for comparing actual designs just as a noise figure is a measure of actual receiver performance compared to an ideal receiver.

4. Many antenna arrangements are available to obtain an ECS in one or two axes. Among the applications of ECS are (a) improved presentation, (b) improved data for track-while-scan, (c) overcome limitations of mechanical portions of a tracking system such as back-lash and servo jitter, (d) low angle tracking.

5. If the error signal is modulated at an audio frequency at an early stage in the circuits, the effects of gain variations, phase shifts, and DC drift in the amplifiers and detectors can be minimized. A method of modulating the RF error signal using a ferrite switch has been devised. This arrangement gives the advantages of an AC error such as is obtained inherently with conical scan radars without the disadvantages of susceptibility to jamming and amplitude fluctuations in the received signals.

These five trends involve concepts which are discussed at length in subsequent sections. To further clarify the brief listing of the trends in monopulse, some illustrations will be given as explanation. These are primarily for those who do not desire to follow the more detailed discussions in later sections.

To show that the terminology of amplitude comparison or phase comparison for a monopulse radar is not very rigorous, an example of the conversion of one type to the other will be used. The difference output (Δ) of an amplitude comparison radar is in-phase with the sum (Σ) signal, while the difference output of a phase comparison radar is in time quadrature with the sum signal. One type of difference output is easily converted to the other by a ninety degree phase shift. An amplitude comparison radar converted to a phase output is shown in the sketch above.

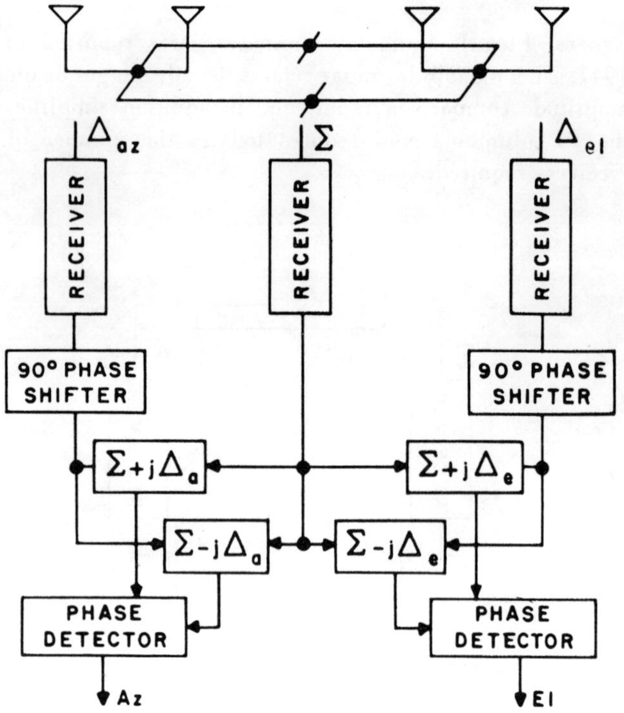

Inverse Tangent Radar

The addition of the sum and difference in quadrature could also be performed by appropriate RF additions and subtractions which would make this amplitude comparison radar even more like a phase comparison set. Since the phase detector outputs are proportional to the inverse tangent of the ratio Δ/Σ, this modification of an amplitude comparison set is called an inverse tangent radar. With certain antenna far field patterns, the error output voltage meets the requirements for an ECS.

The second illustration will deal with the addition of an ECS to an antenna position signal to overcome defects of the antenna drive mechanism. This is also shown in block diagram form on page 23.

As long as the antenna servo keeps the target in the beam, the ECS can correct the antenna position to give the true target position. The limitation of noise originating in the receivers or in the received target signal is the same with or without the ECS addition.

The amount in db by which a typical monopulse antenna is of lower performance than the ideal antenna, is illustrated by some figures determined for the experi-

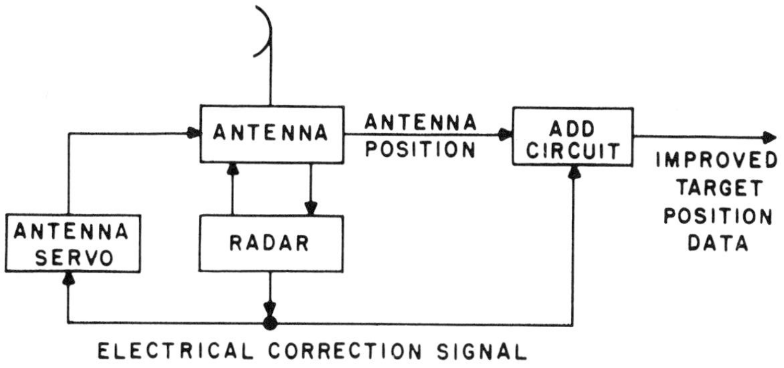

Target Position Data Improvement

mental set. The ideal transmit antenna which is used as a reference is assumed to have uniform aperture illumination and no spillover losses. The derivation of the ideal receive antenna is discussed in Section B 1 b. The noise figure for a good X band receiver is included to show that the receiver is a greater limitation on performance than the antenna relative to the ideal in each case.

Antenna (even transmit pattern)	— 1.7 db
Antenna (odd receive pattern)	— 3.9
RF Plumbing Loss	— 0.5
Antenna Total	— 6.1 db
Receiver Noise Figure	—12.0
Receiver plus Antenna	—18.1 db

The above total for the antenna of —6.1 db would appear to offer some margin for improvement. However. a critical examination of aperture feeds has not revealed any practical designs which are better.

The conversion of the difference signal to an audio frequency can be accomplished by the addition of a phase modulator and RF sum and difference device to the antenna. The sketch below is for one axis of a monopulse radar.

This sketch is perhaps oversimplified, in that the transmitter, mixers, and many other components have been omitted. However, it shows where the RF phase modulator is introduced, and a more detailed discussion is reserved to a later section.

An even greater simplification of diagrams along functional lines is possible. The antenna should provide reference (sum) and error (difference) signals, and the circuitry which follows the antenna should accurately obtain the ratio of the difference to the sum. The ratio circuit is necessary to eliminate range and target size from the radar error signal. The subsequent theory sections have been divided into these two main groupings.

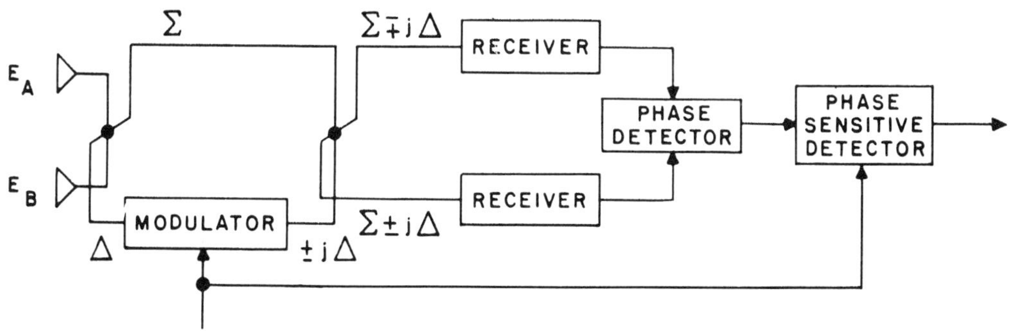

Delta Signal Modulation

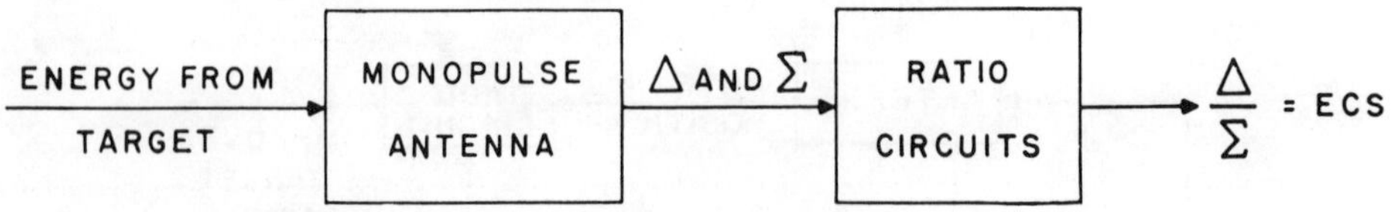

Monopulse Radar

B. Analysis of Factors in Monopulse Antenna Design

The analysis of a monopulse antenna can be made in two ways. These two methods have been used previously[7] to solve specific monopulse antenna problems. The material presented here differs in two respects from earlier work on antenna analysis. First, tables are introduced to enable a variety of problems to be solved rapidly, and second, a slope factor K is defined for odd far field functions.

The two methods of antenna analysis will be discussed briefly, before a more thorough description of the second method is given. No generality is lost by discussing only the second method thoroughly, as it includes the mathematical expressions necessary for the first method.

The first method is an extension of single feed antenna analysis to multiple feeds used in monopulse radar. The pattern produced by each feed is considered to be independent of the presence of other feeds. Thus we have the term "squint" introduced to indicate the amount by which a beam is deflected from the antenna axis. It was recognized at an early date that the two feeds have mutual coupling, and the gain must be determined from the total aperture field resulting from both feeds.

[7] J. F. P. Martin, "Radiation Characteristics of Certain Antenna Systems for Angle Tracking," Bell Telephone Laboratories Report MM47-2730-7, January 27, 1947. On pages IX-1 and IX-2 the two methods of analysis are described briefly.

A second method of antenna analysis which leads to several useful concepts deals directly in terms of aperture illuminations and ignores the problem of how the illuminations are produced. To obtain angular information it is desirable to have an odd far field as well as the even field encountered in single feed antennas. By the use of a microwave hybrid circuit, odd and even far field functions can be separated. The sketch on page 25 illustrates the use of a magic-T.

The two inputs, labeled even and odd, can be considered essentially independent since an input to one produces no output on the other (provided no reflecting surface is present in front of the antenna). This is not true if the two feeds are considered independently as there is a finite, though small, mutual coupling.

The problem of how aperture functions are produced is left to another section of this report, and antenna analysis by the use of various odd and even aperture functions is taken up. The first point of interest is the relationship between the aperture functions and the far field patterns. For convenience, only rectangular apertures will be considered in the analysis.

The integral relation between the aperture function and the far field can be found in a number of texts. A simplified form is

$$E = \frac{je^{-j\frac{2\pi R}{\lambda}}}{2\lambda R}(1+\cos\theta)\iint A(x,y)\, e^{j\left(\frac{\alpha x}{d}+\frac{\varepsilon y}{h}\right)}\, dx\, dy \tag{B-1}$$

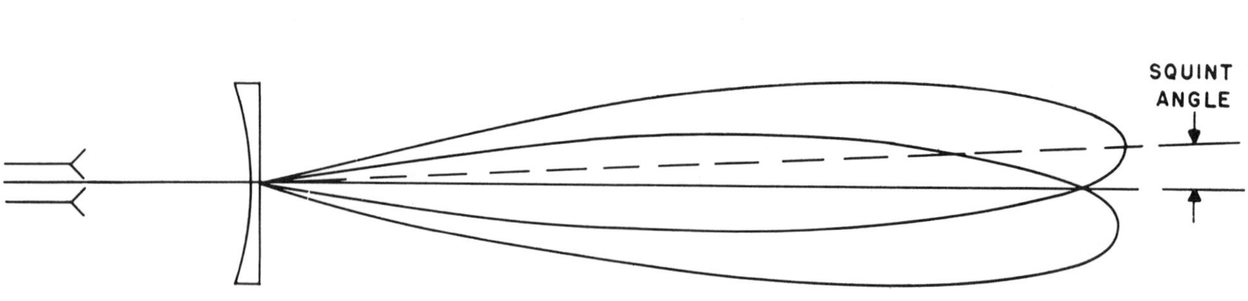

Squinted Beams

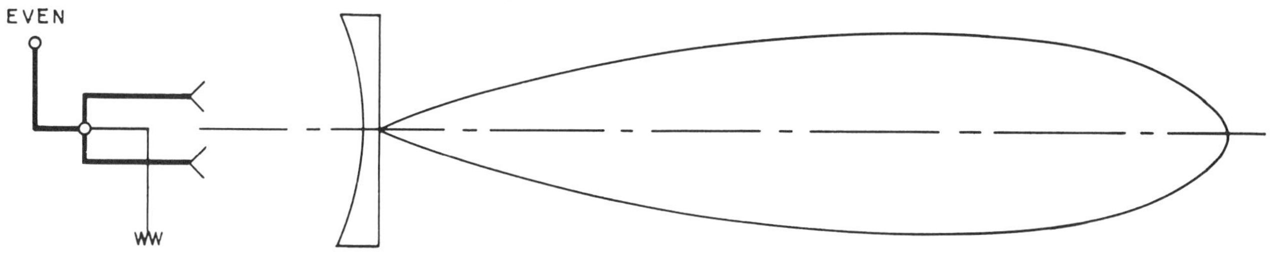

Even Far Field Pattern

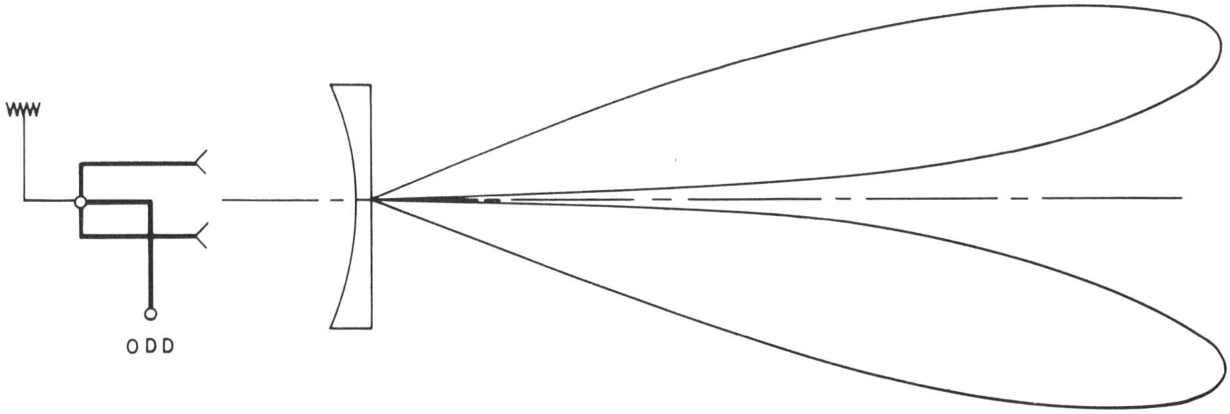

Odd Far Field Pattern

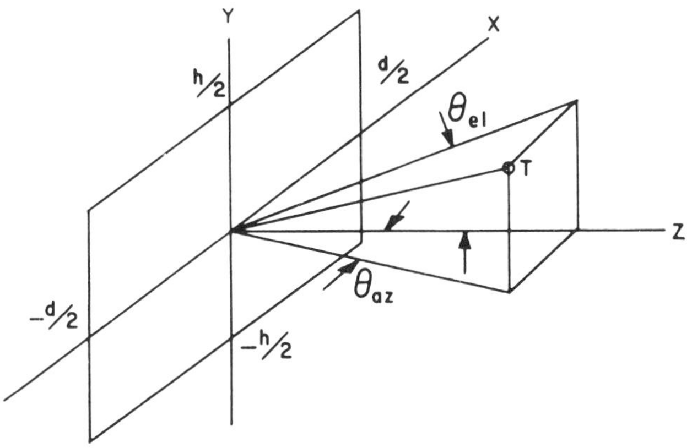

Antenna Aperture

25

where

$$\alpha = \frac{2\pi d}{\lambda} \cos\theta \tan\theta_{az} \doteq \frac{2\pi d\, \theta_{az}}{\lambda}$$

$$\varepsilon = \frac{2\pi h}{\lambda} \cos\theta \tan\theta_{el} \doteq \frac{2\pi h\, \theta_{el}}{\lambda}$$

For narrow beams the variation with θ can be neglected. The aperture field is assumed factorable, so:

$$E = \frac{j e^{-j\frac{2\pi R}{\lambda}}}{2\lambda R} (1 + \cos\theta)\, f(\varepsilon)\, f(\alpha) \qquad \text{(B-2)}$$

where

$$f(\varepsilon) = \int_{-h/2}^{+h/2} A(y)\, e^{j\frac{\varepsilon y}{h}}\, dy \qquad \text{(B-3)}$$

$$f(\alpha) = \int_{-d/2}^{d/2} A(x)\, e^{j\frac{\alpha x}{d}}\, dx \qquad \text{(B-4)}$$

The above equations are equivalent to Fourier transforms, as the aperture field beyond the edge of the aperture is zero. The integral relationship can therefore be indicated by

$$f(\varepsilon) = \mathfrak{F}\, [A(x)] \qquad \text{(B-5)}$$

$$f(\alpha) = \mathfrak{F}\, [A(y)] \qquad \text{(B-6)}$$

The A(x) and A(y) functions can be separated into odd and even terms,[8] for example, any aperture function can be expressed in a Fourier Series.

$$A(x) = \Sigma [A_e(x) + A_o(x)] \qquad \text{(B-7)}$$

$$A(y) = \Sigma [A_e(y) + A_o(y)] \qquad \text{(B-8)}$$

The Fourier transform of an even function is another even function, and the transform of an odd function is another odd function

$$f_e(\varepsilon) = \mathfrak{F}\, [A_e(y)] \qquad \text{(B-9)}$$

$$j f_o(\varepsilon) = \mathfrak{F}\, [A_o(y)] \qquad \text{(B-10)}$$

As shown above, the transform of a real even is real, and the transform of a real odd term is imaginary. Various combinations of complex time functions are possible.

There are a number of basic even and odd functions for which it is desirable to have plots available.

[8] W. L. Murdock and J. S. Kerr, "Relations Between the Far Field and the Illumination of Antenna Apertures," General Electric Company, Electronics Laboratory, TIS 51E234, November 1, 1951.

Even Functions

Name	A(x)	$f_e(\alpha)$
Uniform Figure 1	1	$\frac{d \sin \alpha/2}{\alpha/2}$ (B-11)
11.4 db Taper (r = 0.826) Figure 2	$\cos \frac{\pi r x}{d}$	$d \left[\frac{\left(\frac{\pi r}{2}\right) \sin\left(\frac{\pi}{2} r\right) \cos \alpha/2 - \alpha/2 \sin(\alpha/2) \cos \frac{\pi r}{2}}{\left(\frac{\pi r}{2}\right)^2 - (\alpha/2)^2} \right]$ (B-12)
Cosine (r = 1.0) Figure 3	$\cos \frac{\pi x}{d}$	$d \frac{\pi}{2} \frac{\cos \alpha/2}{(\pi/2)^2 - (\alpha/2)^2}$ (B-13)
Cosine Squared Figure 4	$\cos^2 \frac{\pi x}{d}$	$d/2 \frac{\sin \alpha/2}{\alpha/2} \cdot \frac{\pi^2}{\pi^2 - (\alpha/2)^2}$ (B-14)

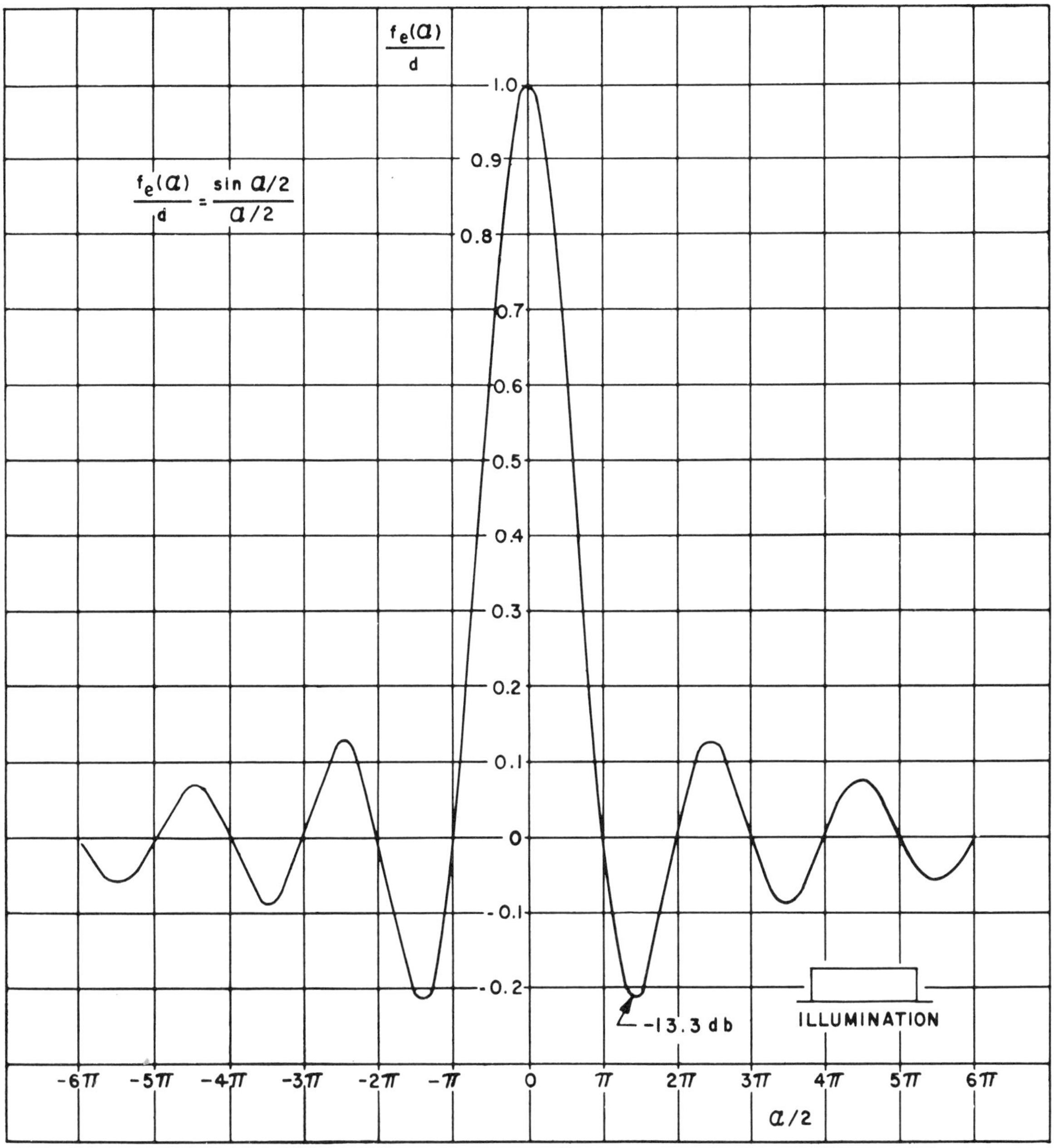

Figure 1. Far Field of Uniform Even Aperture Function.

$$\frac{f_e(\alpha)}{d} = \frac{\left(\frac{\pi r}{2}\right)\sin\left(\frac{\pi r}{2}\right)\cos\alpha/2 - \alpha/2 \sin\alpha/2 \cos\left(\frac{\pi r}{2}\right)}{\left(\frac{\pi r}{2}\right)^2 - (\alpha/2)^2}$$

WHERE $\left(\frac{\pi r}{2}\right) = 1.3$

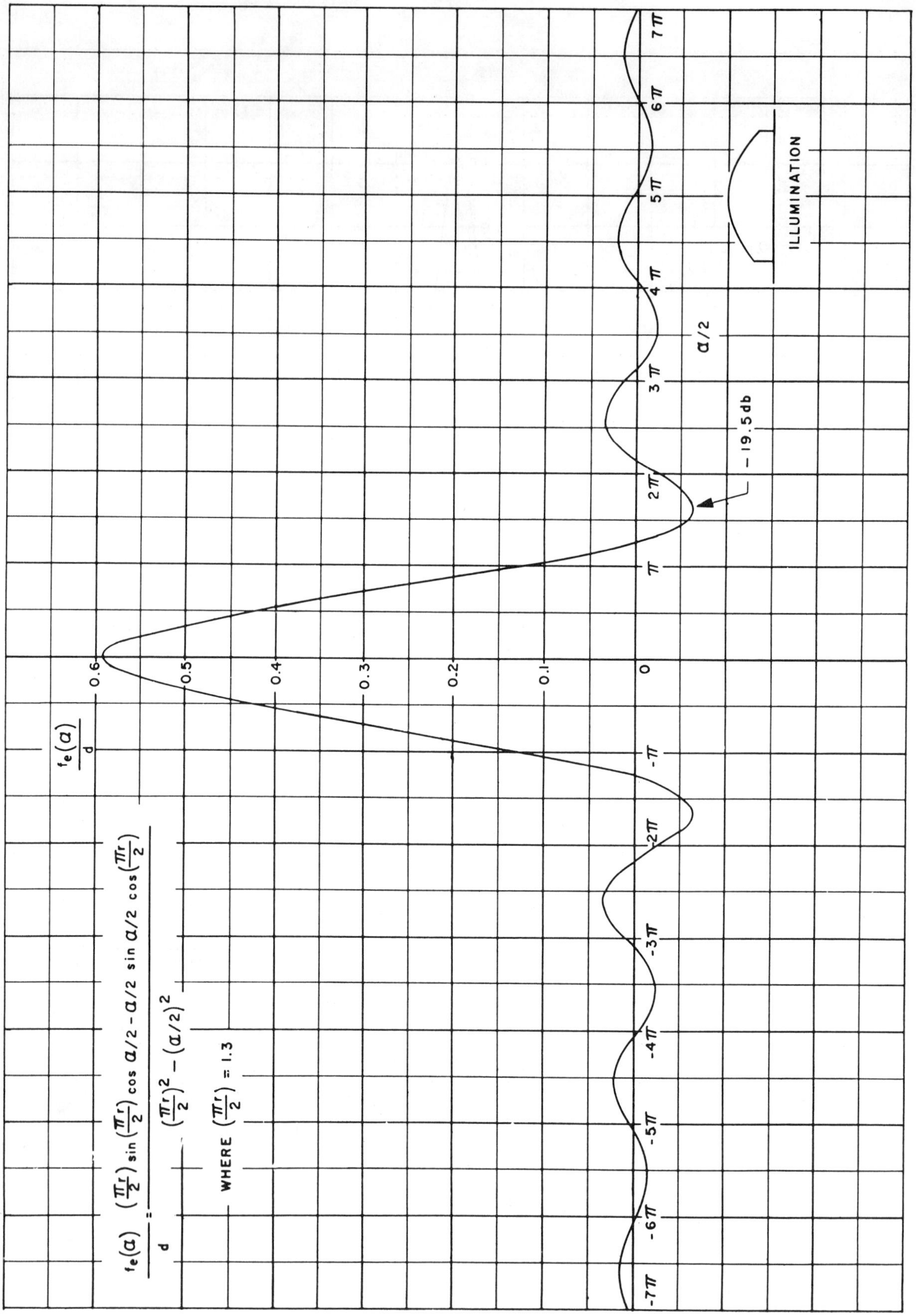

Figure 2. Far Field of 11.4 db Tapered Even Aperture Function.

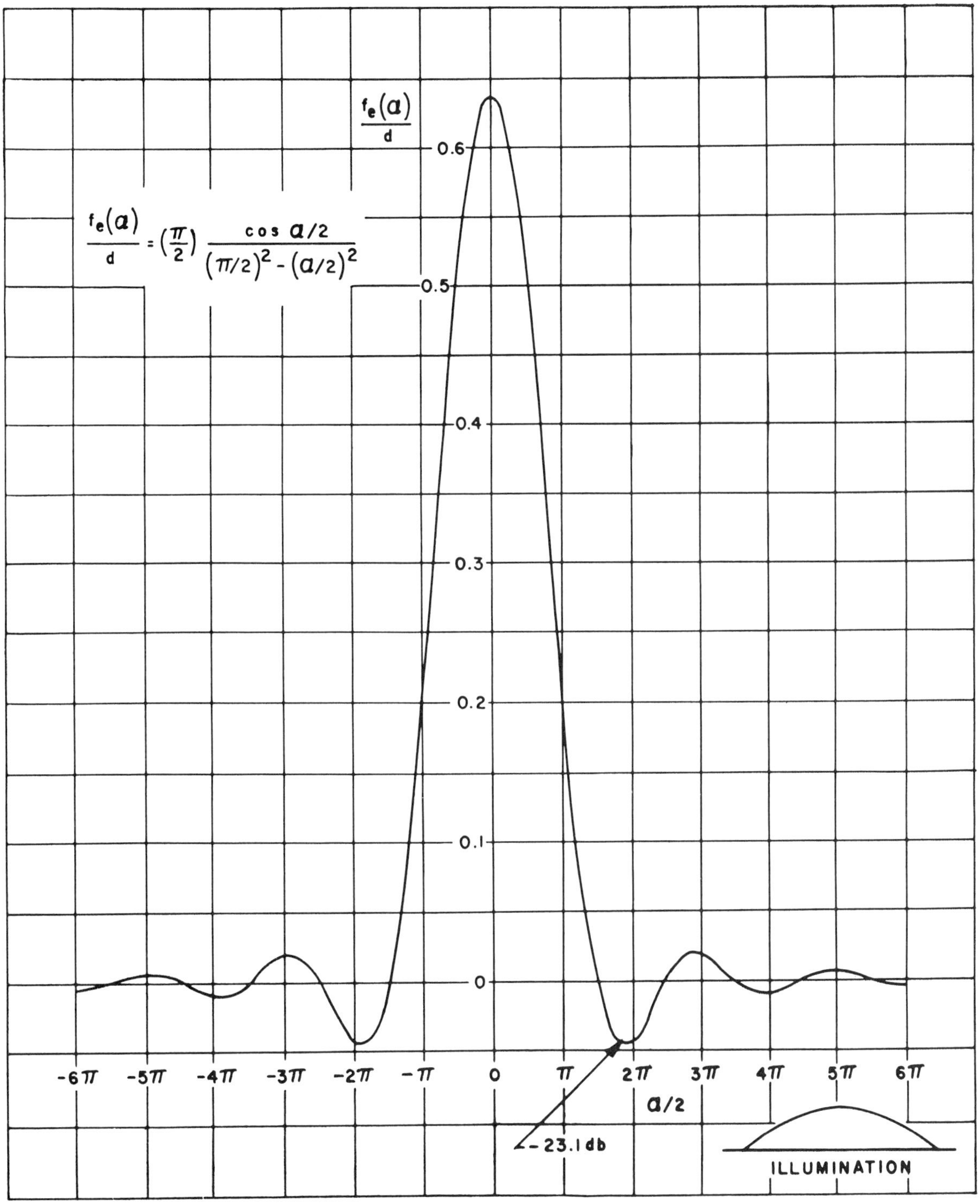

Figure 3. Far Field of Cosine Even Aperture Function.

$$\frac{f_e(a)}{d} = \frac{1}{2} \frac{\sin a/2}{a/2} \cdot \frac{\pi^2}{\pi^2 - (a/2)^2}$$

−31.7 db

ILLUMINATION

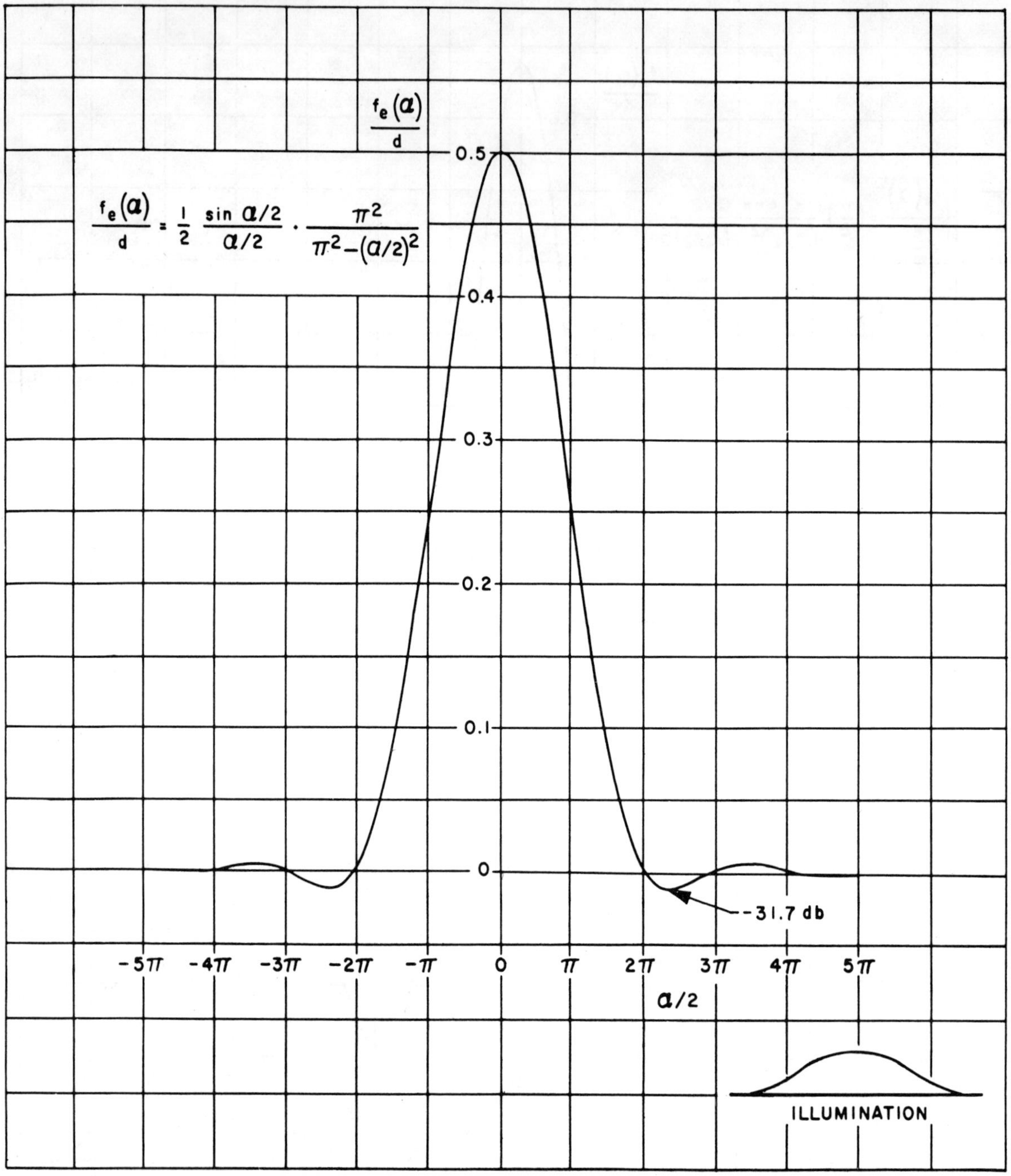

Figure 4. Far Field of Cosine Squared Even Aperture Function.

1. Antenna Gain

Since the even and odd functions can be independent, the gains of various even and odd functions will be considered separately. A table of the relative gains will be very useful to evaluate the relative (and absolute) direction finding properties of an antenna. Since the properties of even aperture and far field functions are better known than those of odd functions, the even functions will be taken up first.

a. Gain, G, of Even Aperture Functions.—It is desirable to make the gain of the even far field function as near a maximum as is consistent with other requirements. It is well known that the greatest gain for a constant phase even far field[9] is obtained with uniform illumination of the aperture. It is also known that in theory[10] an antenna may be made arbitrarily directive by suitable choice of the current distributions. In practice the ohmic losses resulting from the large currents required has prevented the development of "super-gain" antennas. Thus the gain of an aperture with uniform illumination will be considered the maximum practical value.

The gain function is defined[11] as the ratio of the power radiated in a given direction per unit solid angle to the average power radiated per unit solid angle:

$$G(\theta_{az}, \theta_{el}) = \frac{P(\theta_{az}, \theta_{el})}{\frac{P_t}{4\pi}} \qquad \text{(B-19)}$$

Thus G $(\theta_{az}, \theta_{el})$ expresses the increase in power radiated in a given direction by the antenna over that from an isotropic radiator emitting the same power. The maximum value $(\theta_{az} = \theta_{el} = 0)$ of the gain function (for constant phase) is

$$G = \frac{4\pi}{\lambda^2} \frac{|\int_s A(x,y)\, dx\, dy|^2}{\int_s |A(x,y)|^2\, dx\, dy} \qquad \text{(B-20)}$$

For the case of uniform illumination over the aperture, $A(x,y) = 1$, the gain, G_o, is by inspection

$$G_o = \frac{4\pi hd}{\lambda^2} \qquad \text{(B-21)}$$

It is conventional to express the decrease in gain resulting from the use of even aperture illuminations other than uniform as a decrease in the aperture efficiency. It is convenient to tabulate the relative efficiencies on the basis of uniform illumination on one axis and some other type of illumination on the other axis. This has been done in Table I.

$$\text{Relative Gain in decibels} = 10 \log_{10} \frac{G}{G_o} \qquad \text{(B-22)}$$

b. Slope, K, of Odd Aperture Functions.—The greatest gain for an even far field is obtained with uniform illumination, however, no criterion is available in the literature to evaluate odd patterns. It is proposed to use the

[9] S. Silver et al, "Microwave Antenna Theory and Design," Radiation Laboratory Series, Volume 12, McGraw-Hill Book Company, Inc. 1949, p. 177.

[10] H. J. Riblet, "Note on the Maximum Directivity of an Antenna," Proceedings of the IRE, May, 1948, pp. 620-623.

[11] See p. 2 of footnote 9.

ODD FUNCTIONS

Name	A(x)	$f_o(\alpha)$	
Linear Figure 5	x	$\frac{d}{2}\left[\frac{\sin \alpha/2}{(\alpha/2)^2} - \frac{\cos \alpha/2}{\alpha/2}\right]$	(B-15)
Uniform—odd Figure 6	$-1, \ -d/2 < x < 0$ $1, \ 0 < x < d/2$	$d\left[\frac{1 - \cos \alpha/2}{\alpha/2}\right]$	(B-16)
Double-Angle Sine Figure 7	$\sin \frac{2\pi x}{d}$	$\frac{d\,\pi \sin \alpha/2}{\pi^2 - (\alpha/2)^2}$	(B-17)
Cosine—odd Figure 8	$-\cos \frac{\pi x}{d}, \ -d/2 < x < 0$ $\cos \frac{\pi x}{d}, \ 0 < x < d/2$	$d\left[\frac{(\alpha/2) - \pi/2 \sin \alpha/2}{(\pi/2)^2 - (\alpha/2)^2}\right]$	(B-18)

ILLUMINATION $A_e(y)$ $\left(A_e(X)=1\right)$		RELATIVE GAIN AT $\epsilon = 0$ $db = 10 \log_{10} \frac{G}{G_0}$
1. UNIFORM	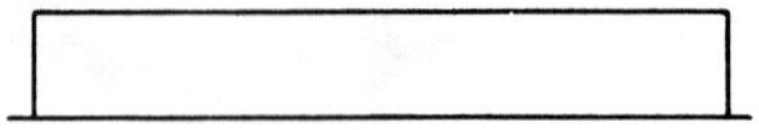	0 db $G_0 = \frac{4\pi hd}{\lambda^2}$
2. 11.4 db TAPER $\cos \frac{\pi r x}{d}$ $r = 0.826$	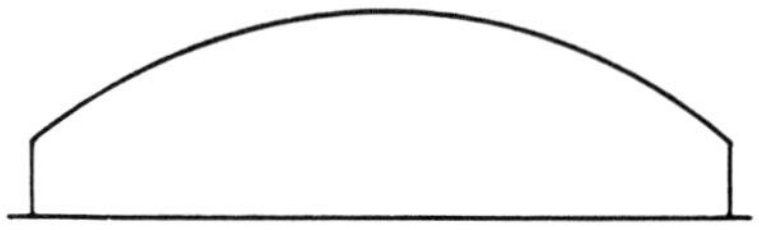	− 0.41 db
3. COSINUSOIDAL	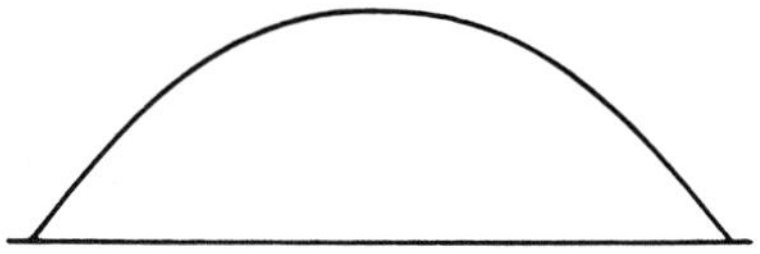	−0.92 db
4. SINE − SINE	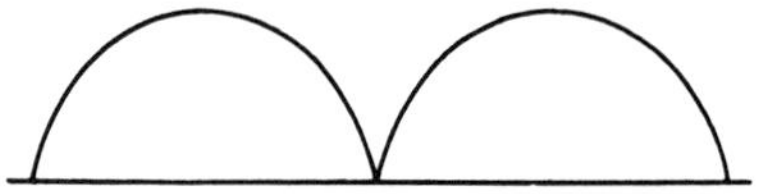	−0.92 db
5. COSINE SQUARED	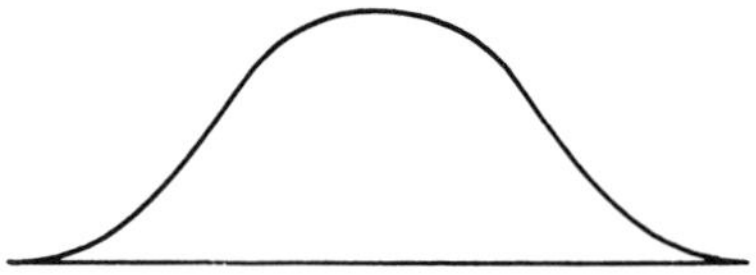	−1.76 db
6. TRIANGULAR	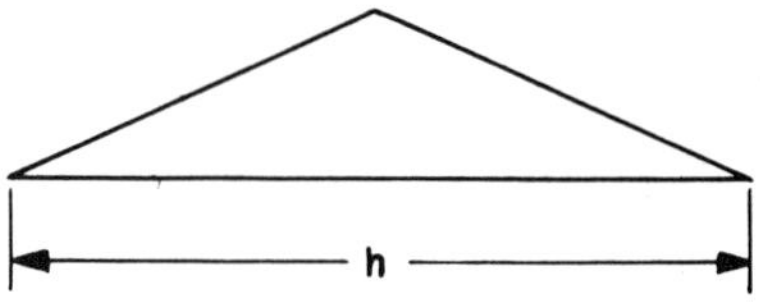	−1.26 db

Table I. Relative Gain of Even Aperture Functions.

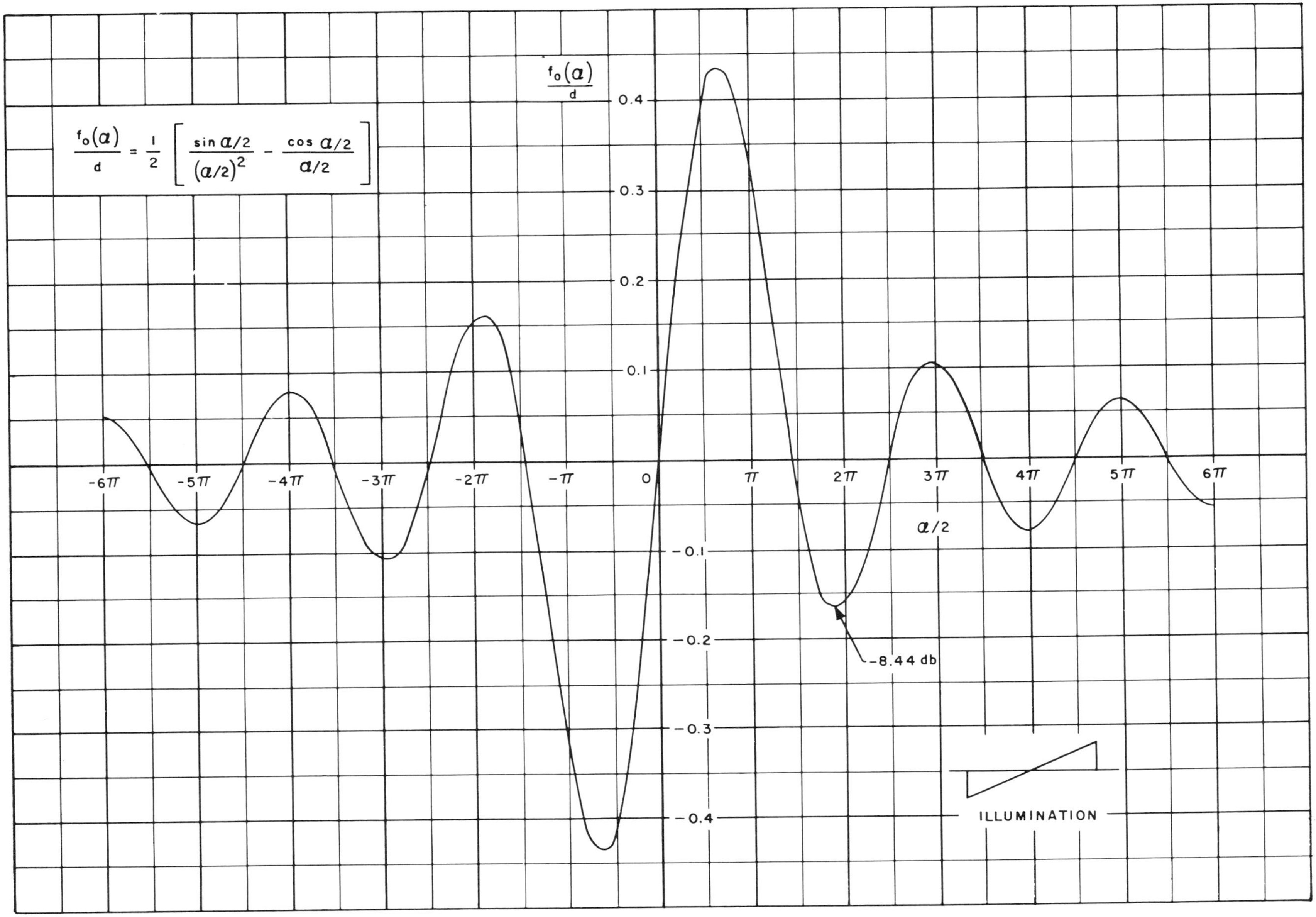

Figure 5. Far Field of Linear Odd Aperture Function.

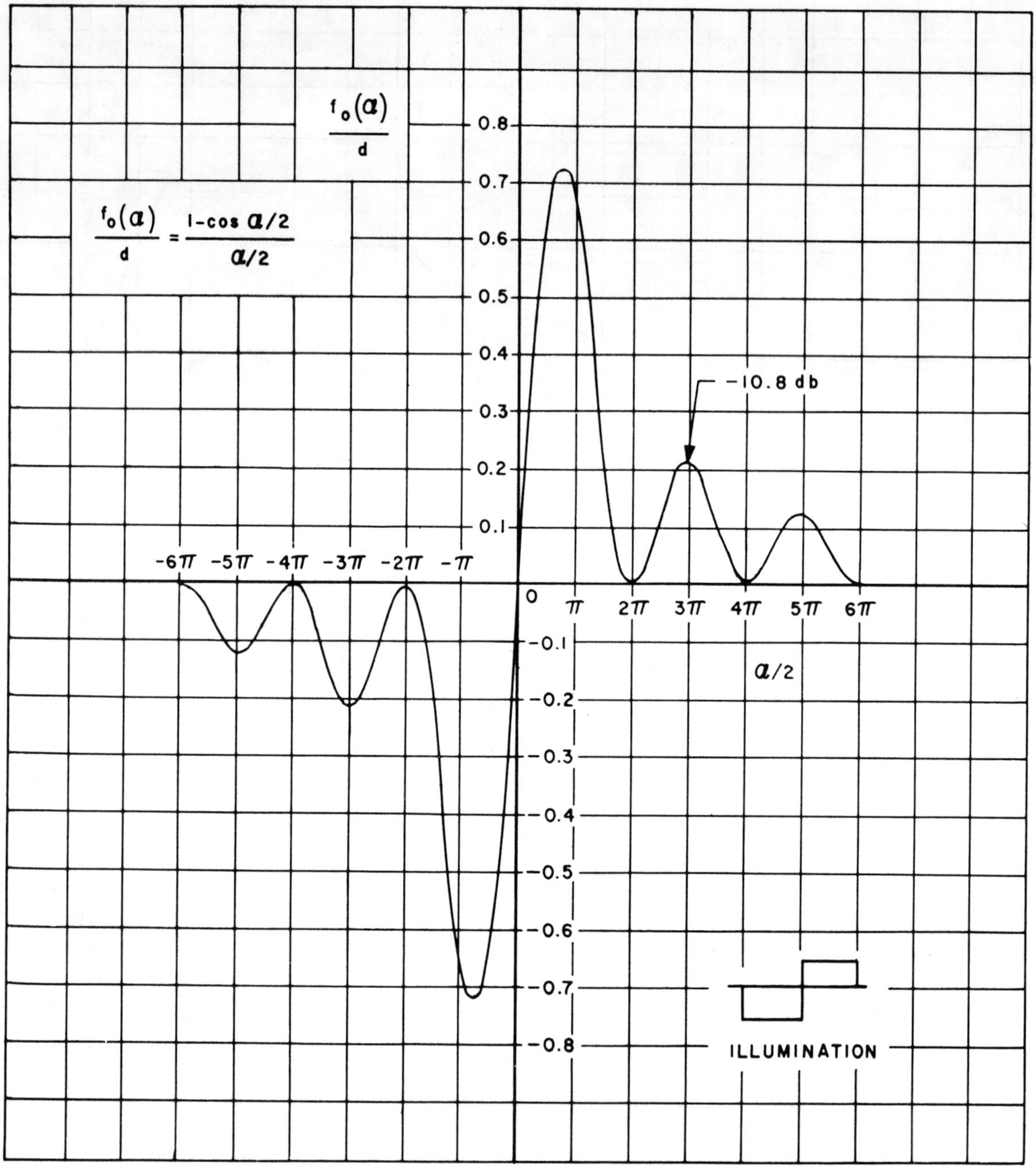

Figure 6. Far Field of Uniform Odd Aperture Function.

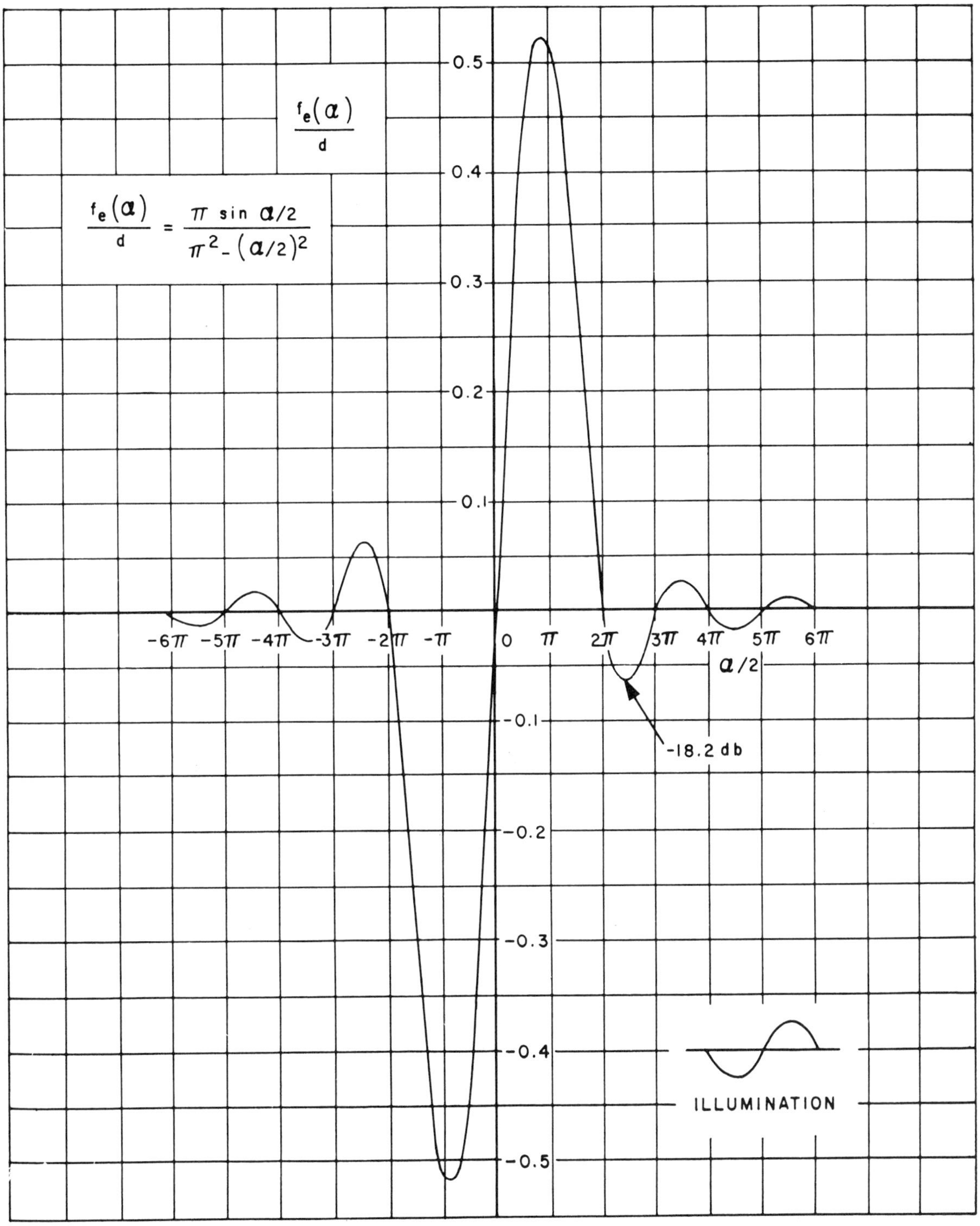

Figure 7. Far Field of Double-Angle Sine Odd Aperture Function.

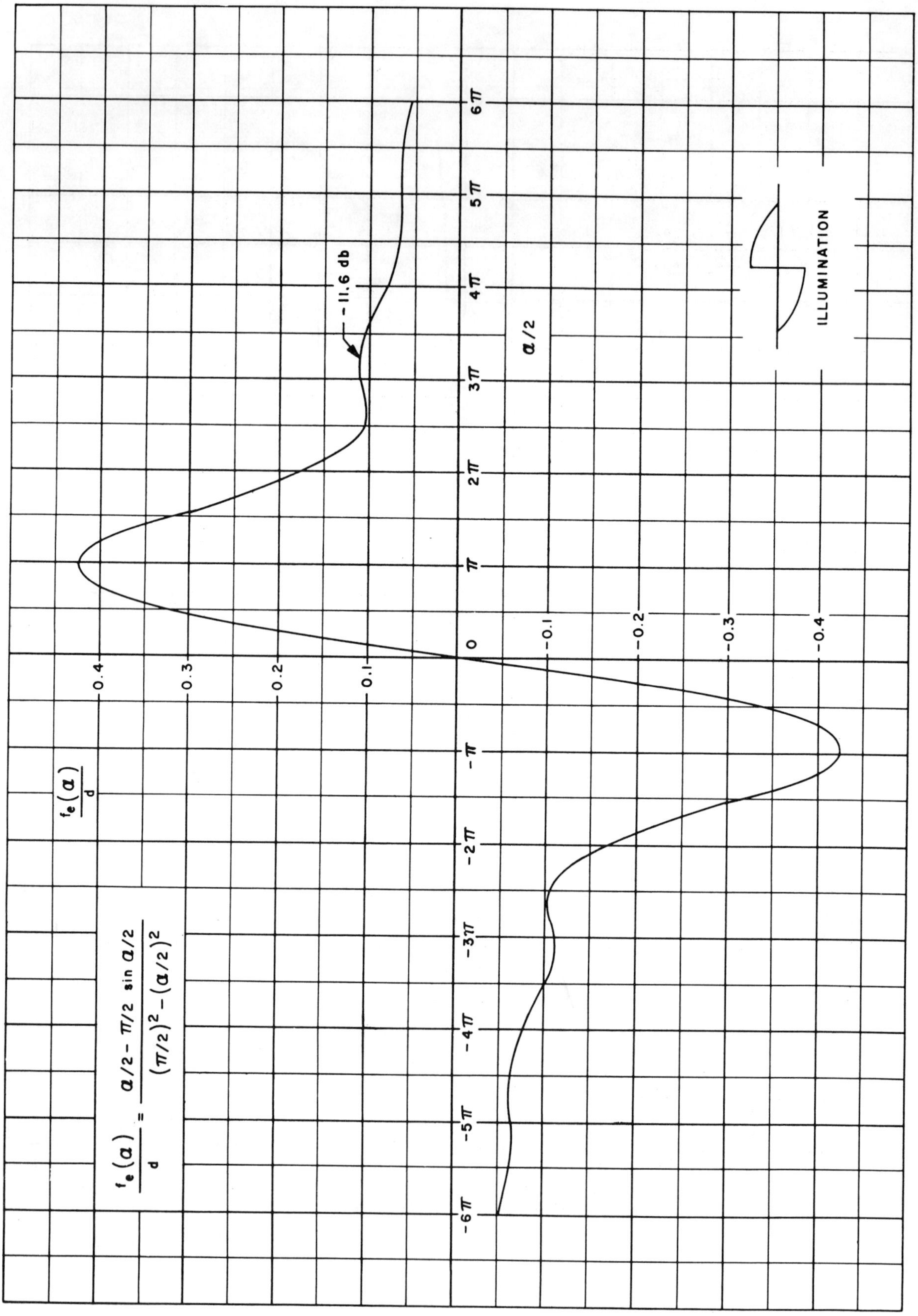

Figure 8. Far Field of Cosine-Odd Aperture Function.

slope of the odd far field as a figure of merit of the direction finding properties of an antenna. Using this criterion, the greatest slope in the far field is produced by a linear aperture function with odd symmetry about the center of the aperture.

The above result was obtained through the use of the Calculus of Variations. Since the derivation is not available in the literature, it will be repeated here. While the odd far field is used only during the receive period in a radar, it will clarify the problem to consider the use of the odd far field for transmission. This antenna is to radiate the available power in such a manner that the slope of the far field pattern on the axis of symmetry is a maximum. Then by the Reciprocity Theorem,[12] this should be the antenna which will give the maximum rate of change of current with angle in the odd function output if the transmitter and receiver are exchanged. The use of the odd input for transmission is illustrated in the sketch below:

$$f_o'(0) = \int_{-h/2}^{h/2} jA_o(y) \frac{y}{h} dy \qquad \text{(B-24)}$$

Using the method of the Calculus of Variations, let

$$A_o(y) = A_1(y) + \nu A_2(y) \qquad \text{(B-25)}$$

where ν is a parameter independent of (y), and $A_2(y)$ is an arbitrary function which vanishes at the limits $h/2$ and $-h/2$. If this expression is substituted into equation (B-24), the result is

$$f_o'(0) = \frac{j}{h} \int_{-h/2}^{h/2} [A_1(y) + \nu A_2(y)]\, y\, dy \qquad \text{(B-26)}$$

Since we desire $A_1(y)$ to give a maximum $f_o'(0)$, then equation (B-26) has a maximum at $\nu = 0$,

$$0 = \frac{\partial f_o'(0)}{\partial \nu} = \frac{j}{h} \int_{-h/2}^{h/2} A_2(y)\, y\, dy \qquad \text{(B-27)}$$

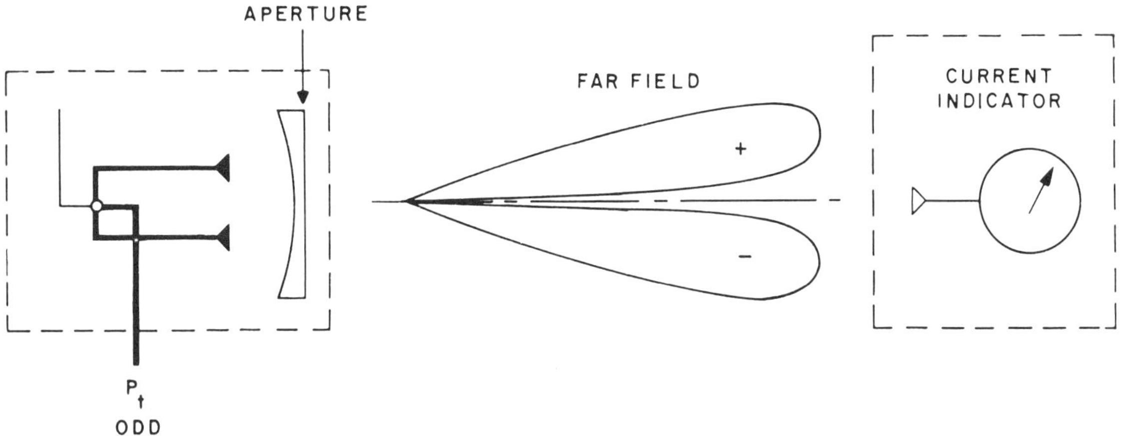

Device to Compare Slopes of Odd Far Fields

For this derivation it is assumed that the aperture is of unit length and has uniform illumination in the x direction, and is of width h in the y direction. From equation (B-4) the slope of the odd field is

$$f_o'(\varepsilon) = \frac{d\,[f_o(\varepsilon)]}{d\varepsilon} = \int_{-h/2}^{h/2} jA_o(y) \frac{y}{h} \cos \frac{\varepsilon y}{h} dy \qquad \text{(B-23)}$$

On the axis of symmetry $\varepsilon = 0$, equation (B-23) becomes

A constraint is a constant power input, P_t. The power output[13] for a unit length in the x direction is

$$P_t = a^2 \int_{-h/2}^{h/2} |A_o(y)|^2 dy \qquad \text{(B-28)}$$

or using equation (B-25)

$$P_t = a^2 \int_{-h/2}^{h/2} |A_1(y) + \nu A_2(y)|^2 dy \qquad \text{(B-29)}$$

[12]See p. 19 of footnote 2.

[13]See p. 177 of footnote 9.

The derivative of equation (B-29) with respect to ν will determine the relation between $A_1(y)$ and $A_2(y)$.

$$\frac{\partial P_t}{\partial \nu} = 0 = 2a^2 \int_{-h/2}^{h/2} [A_1(y) + \nu A_2(y)] A_2(y) \, dy \quad (B\text{-}30)$$

Since equations (B-27) and (B-30) are each equal to zero their sum must also be equal to zero, thus

$$\int_{-h/2}^{h/2} \left\{ 2a^2 A_2(y) [A_1(y) + \nu A_2(y)] + j\frac{y}{h} A_2(y) \right\} dy = 0 \quad (B\text{-}31)$$

and as ν approaches zero,

$$2a^2 A_2(y) A_1(y) + j\frac{y}{h} A_2(y) \to 0 \quad (B\text{-}32)$$

This equation, (B-32), can be satisfied either with $A_2(y) = 0$, or

$$A_1(y) = \frac{-j}{2a^2 h} y \quad (B\text{-}33)$$

Thus the maximum slope of the odd far field is obtained with a linear variation in the aperture field about the axis of symmetry.

This is a result of considerable significance for the evaluation of monopulse radar antennas. The actual odd illuminations should be compared to the "ideal" linear aperture field, and the even to the "ideal" uniform aperture field. It will be immediately recognized that other factors such as impedance matching, spillover losses, sidelobes, and linearity of the electrical correction signal must also be given consideration.

Since the slope of the antenna power gain function does not have a derivative for the axis of symmetry of an odd function, the square root of the power gain (proportional to voltage) will be used. The maximum slope is

$$K = \frac{\partial G^{1/2}}{\partial \varepsilon} = \left(\frac{4\pi}{\lambda^2}\right)^{1/2} \frac{\int_{-h/2}^{h/2} A(y) \frac{y}{h} dy \int_{-d/2}^{d/2} A(x) dx}{[\int_s |A(x) A(y)|^2 dx \, dy]^{1/2}}. \quad (B\text{-}34)$$

If $A(y) \cdot A(x) = A_o y$ then the slope factor is maximum and

$$K_o = \frac{\partial G^{1/2}}{\partial \varepsilon}\bigg|_{\varepsilon=0} = \left[\frac{h \, d\pi}{3\lambda^2}\right]^{1/2} \quad (B\text{-}35)$$

It is sometimes convenient to evaluate K_o in terms of G_o,

$$K_o = \left[\frac{G_o}{12}\right]^{1/2} \quad (B\text{-}36)$$

The slope factors for other odd aperture functions are tabulated in Table II. The decibel figures for relative slope are calculated by the equation,

$$\text{relative slope in decibels} = 20 \log_{10} \frac{K}{K_o} \quad (B\text{-}37)$$

(See Appendix VIII).

c. Spillover Losses.—Thus far the analysis has dealt only with the aperture illuminations, and the losses incident to producing these aperture functions have been neglected. Because of the large (approximately 3 db) spillover losses for some odd illuminations, it is important to consider spillover in comparing various antennas for radar direction finding.

Because the subject of producing even illuminations has been dealt with extensively in the literature[14] it will be taken up first. The losses which are experienced with a particular radar antenna will depend upon the physical configuration of:

(1) the primary source (horn, dipole feeds, etc.) and

(2) the secondary radiator (lens, reflector, etc.).

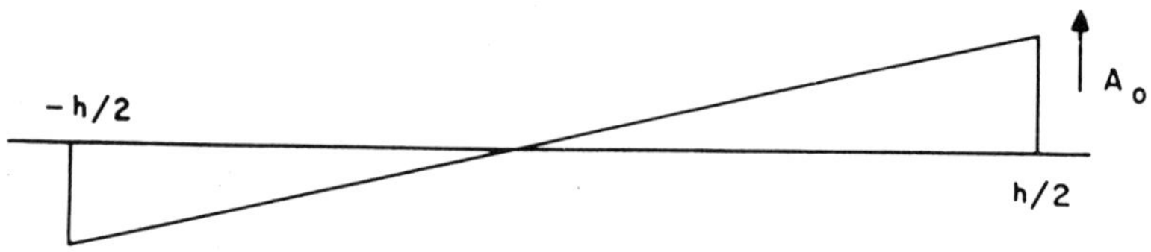

Aperture Illumination for Maximum Slope Odd Far Field

---
[14] See p. 426 of footnote 2.

ILLUMINATION $A_o(y)$ $\left(A_e(X) = 1\right)$		RELATIVE SLOPE AT $\epsilon = 0$ $db = 20 \log_{10} \frac{K}{Ko}$
1. LINEAR		0 db $K_o = \left[\frac{h d \pi}{3\lambda^2}\right]^{1/2}$
2. SINUSOIDAL	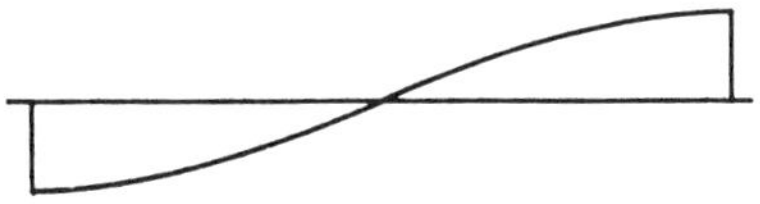	−0.06 db
3. CUBIC	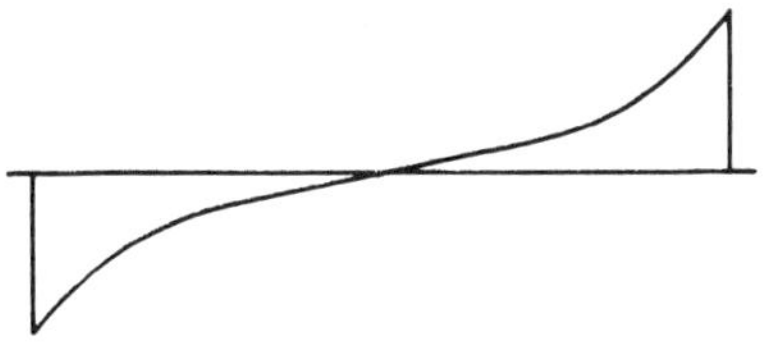	−0.76 db
4. UNIFORM	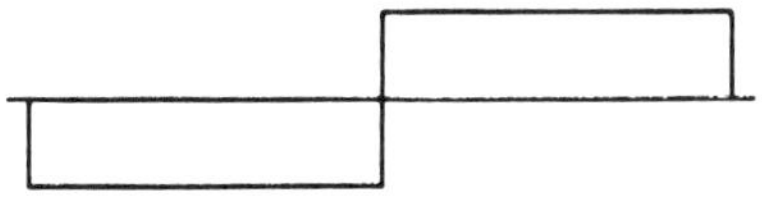	−1.25 db
5. DOUBLE ANGLE SINE	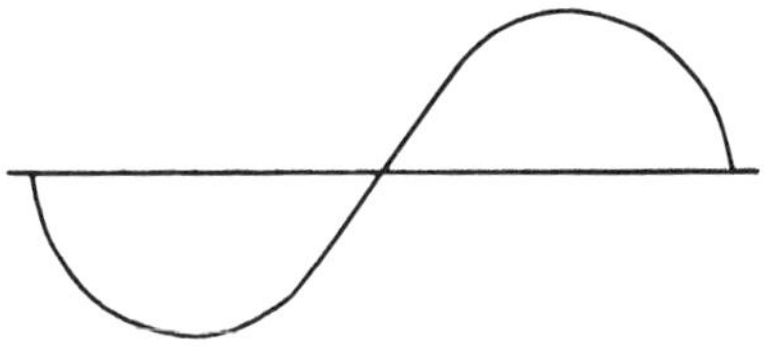	−2.16 db
6. COSINUSOIDAL	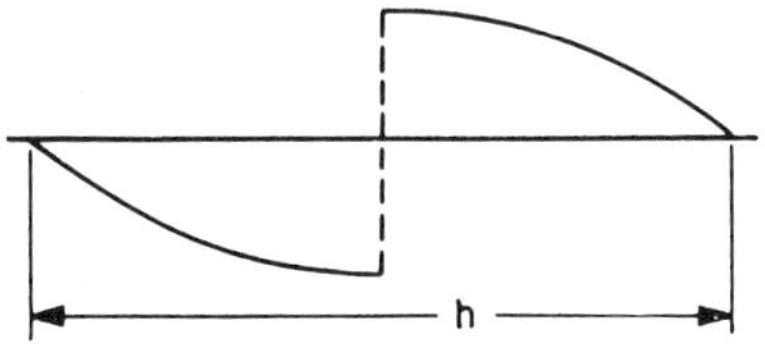	−4.96 db

Table II. Relative Slope of Odd Aperture Functions.

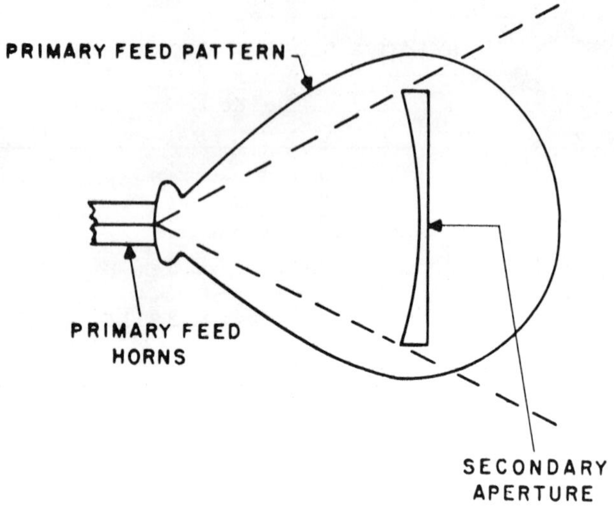

Illumination of Lens by Even Feed Horn Pattern

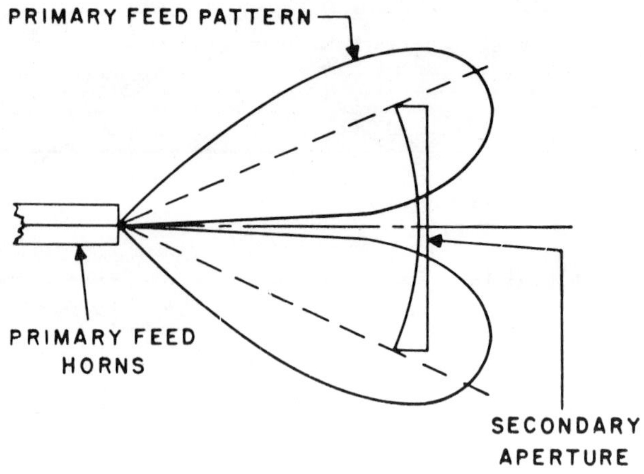

Illumination of Lens by Odd Feed Horn Pattern

If the primary source pattern is available, either from approximate analysis or measured patterns, the loss can be evaluated. However, the exact integration of the feed horn patterns is quite difficult as no approximations can be made for the trigonometric functions in equation (B-1). It is desirable to use a short focal length secondary aperture, thus a wide angle primary feed pattern is necessary. If the far field and aperture field of the feed are known then the spillover loss factor is

$$\eta_i = \frac{\int_{\substack{\text{Sec.}\\\text{Aperture}}} |E^2|\, R^2 \sin\theta \, d\theta d\phi}{\int_{\substack{\text{Pri.}\\\text{Aperture}}} |A\,(x,y)\,|^2 \,dx\, dy} \qquad \text{(B-38)}$$

An approximate db spillover loss for one axis of a typical four horn feed is given in Figure 9 as a function of the secondary radiator focal length (FL) to height or width (h, d) ratio.

Other investigations[15] have shown that it is difficult to minimize the odd feed spillover, and still obtain an efficient even illumination. The sketch of illumination of a lens by an odd feed horn pattern makes the difficulty apparent.

The approximate db spillover loss for an odd feed pattern from a four horn feed is also given on Figure 9.

[15] "Optimum Design Criterion for Simultaneous Lobing Antennas," NRL Report R-3451. Also see p. IX-11 of footnote 7.

The graphs of various odd and even functions, Figures 1 through 8, can also be used to estimate the spillover. The product solution is not sufficiently accurate to evaluate the numerator of equation (B-38), and the problem should be treated as three dimensional.

If sidelobes are not a consideration, the overall gain of an antenna for an even far field can approach the theoretical gain, G_o, to within a fraction of a decibel. However, an antenna for an odd far field is not likely to be closer than three decibels to the theoretical gain K_o. The subject of feeds is taken up more extensively in the section on Monopulse Feeds.

2. Electrical Correction Signals (ECS)

If an error signal has certain characteristics it can be used to measure the angular positions of point targets within a radar beam. These characteristics are:

$$\text{ECS} \neq F(R) \qquad \text{(B-39)}$$

$$\text{ECS} \neq F(A_{eff}) \qquad \text{(B-40)}$$

$$\text{ECS} = p\alpha \qquad \text{(B-41)}$$

The first two, equations (B-39) and (B-40), can be satisfied by the ratio of two far field patterns which are the same function of range, R, and effective target areas, A_{eff}. The last, equation (B-41), requires that the ratio be made up of one function having odd symmetry and the other even symmetry in the far field. To avoid any infinities in the useful range, the ratio should be of the form

$$\text{ECS} = F\left(\frac{f_o(\alpha)}{f_e(\alpha)}\right) = p\alpha. \qquad \text{(B-42)}$$

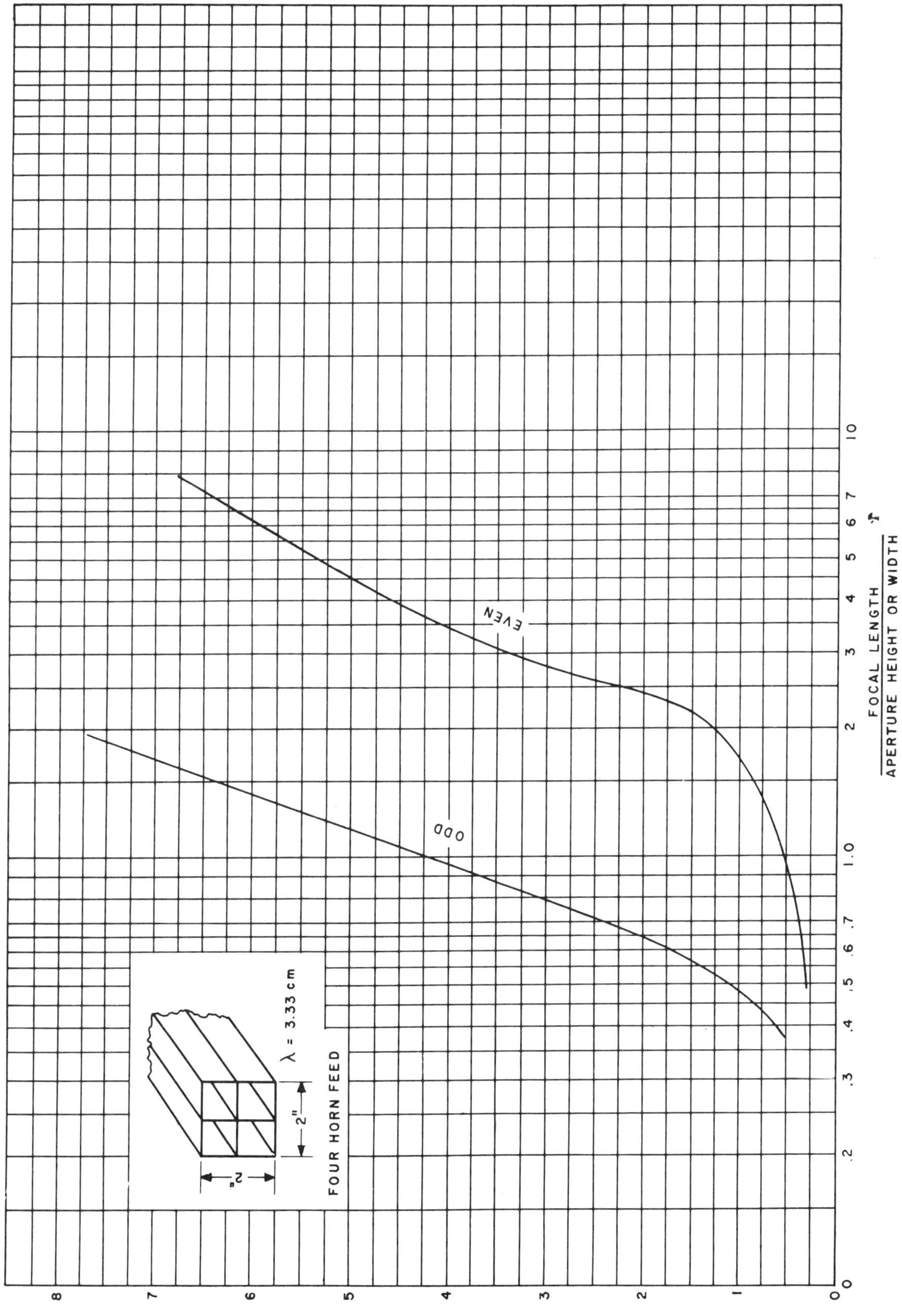

Figure 9. Approximate Spillover Loss.

The form of the aperture functions which will produce far field patterns, $f(\alpha)$, to satisfy the above relationship will be taken up next. There are at least three which lead to either linear or approximately linear ECS. These are

a. (1) Linear Phase
   (2) Approximately Linear Phase

b. (1) Linear Ratio
   (2) Approximately Linear Ratio

c. (1) Linear Logarithmic Ratio
   (2) Approximately Linear Logarithmic Ratio

a. (1) Linear Phase.—The functional relationship of equation (B-42) is

$$\text{ECS} = \tan^{-1}\left(\frac{f_o(\alpha)}{f_e(\alpha)}\right) = p\alpha \qquad \text{(B-43)}$$

This function can be obtained if the odd far field is in quadrature with the even.

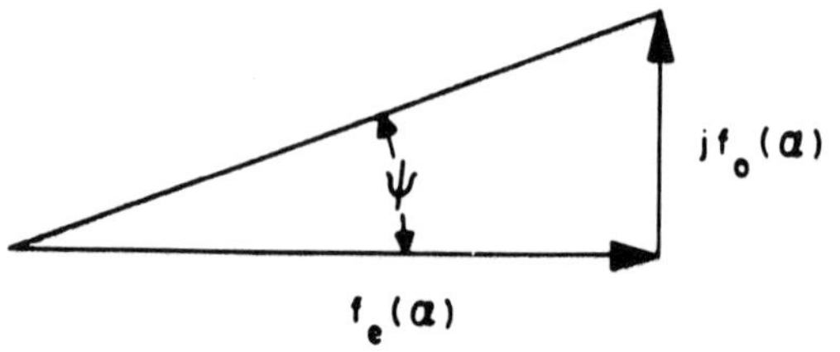

Odd and Even Functions Added in Quadrature

It is obvious from equation (B-43) that it must be possible to write $f_o(\alpha)$ and $f_e(\alpha)$ as

$$f_o(\alpha) = f(\alpha) \sin p\alpha \qquad \text{(B-44)}$$

$$f_e(\alpha) = f(\alpha) \cos p\alpha \qquad \text{(B-45)}$$

It can be shown that the only method by which the above odd and even functions can be obtained is to take the sum and difference of two far field patterns produced by two identical aperture functions with separated phase centers. The odd and even aperture functions which will satisfy equations (B-44) and (B-45) can be deduced with the use of the previous statement, namely $A(x) = A_o(x) + A_e(x)$. Certain suitable aperture and far field functions are given in Table III.

(2) Approximately Linear Phase.—Since the phase angle is ambiguous for $\psi > \pm\pi$ radians, seldom is there need for concern about the linearity of phase beyond $\pm\pi$ radians. In wide beam systems such as (a) in Table III some method must be included to eliminate the ambiguities.

With interest centered on obtaining linear phase angles of less than π radians over the portion of the beam between the half power points, it is not difficult to find odd and even far field functions which will give approximately linear phase. Plots for a number of inverse tangent ratios are shown in Figures 10 and 11.

Almost all odd far field patterns can be approximated over a small range near $\alpha = 0$ by

$$f_o(\alpha) \doteq f_1(\alpha) \sin p_1\alpha, \qquad \text{(B-46)}$$

and nearly all even functions by

$$f_e(\alpha) \doteq f_2(\alpha) \cos p_2\alpha. \qquad \text{(B-47)}$$

While there are many $f_o(\alpha)$ and $f_e(\alpha)$ functions which will not satisfy (B-43) exactly, a suitable adjustment of relative gain in the even and odd amplifier channels will allow an optimum linearity to be obtained. The ratio of $f_1(\alpha)$ to $f_2(\alpha)$ can be written in polynomial form as

$$\frac{f_1(\alpha)}{f_2(\alpha)} = a_o + a_1\alpha^2 + a_2\alpha^4 + \ldots \qquad \text{(B-48)}$$

For $p_1 = p_2$, a_o should be unity and all the other coefficients zero. The approximate effect of varying a_o if $p_1 = p_2$ is shown in the sketch.

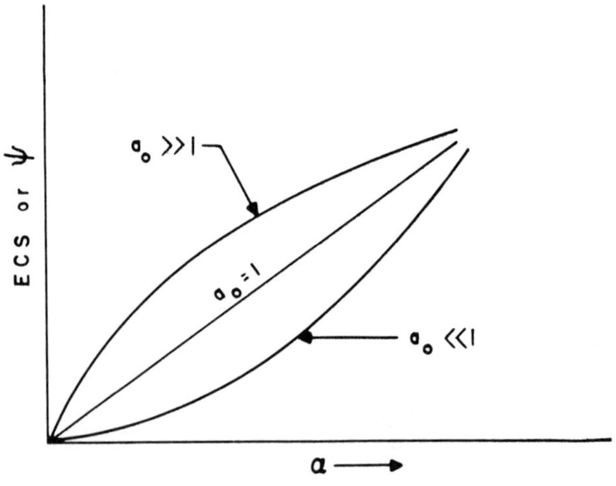

Effect on Linearity of Varying Relative Gain

The value of p_1 relative to p_2 can be controlled only by changing the aperture illumination. However, the effect will be similar to changing the relative gain a_o since

One of Pair of Symmetrical Aperture Functions with Separated Phase Centers $A_e(x) - A_o(x)$	$A_e(x)$ and $f_e(\alpha)$	$A_o(x)$ and $f_o(\alpha)$	Phase Angle $\psi = \tan^{-1}\left[\frac{f_o(\alpha)}{f_e(\alpha)}\right]$
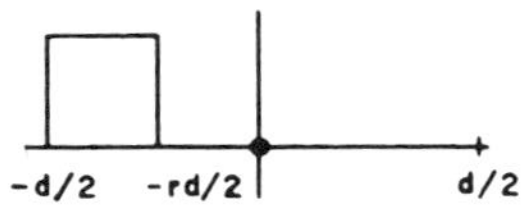 (a) UNIFORM DISPLACED $-1 < r < 1$	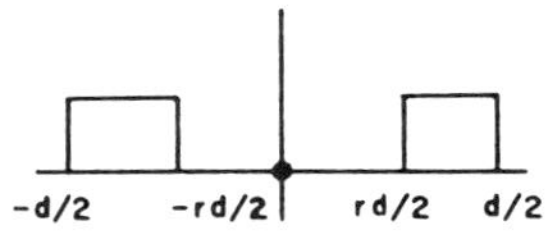 $f_e(\alpha) = \frac{d}{\frac{\alpha}{4}} \sin\left[\frac{\alpha}{4}(1-r)\right] \cos\left[\frac{\alpha}{4}(1+r)\right]$	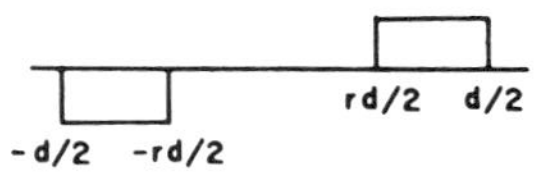 $f_o(\alpha) = j\frac{d}{\frac{\alpha}{4}} \sin\left[\frac{\alpha}{4}(1-r)\right] \sin\left[\frac{\alpha}{4}(1+r)\right]$	$\psi = \frac{\alpha}{4}(1+r)$
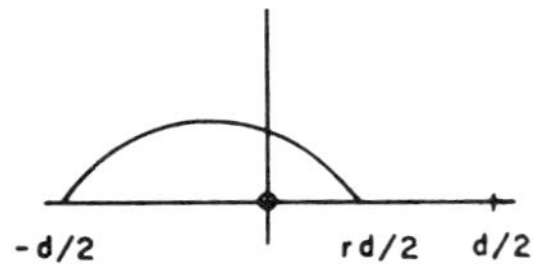 (b) COSINE OVERLAPPED $-1 < r < 1$	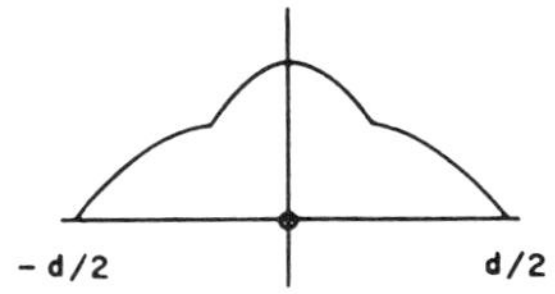 $f_e(\alpha) = \frac{\pi}{2} \frac{d(r+1)\cos\frac{\alpha}{4}(1+r)}{\left(\frac{\pi}{2}\right)^2 - \left[\frac{\alpha}{4}(1+r)\right]^2} \cos\frac{\alpha}{4}(1-r)$	 $f_o(\alpha) = j\frac{\pi}{2} \frac{d(r+1)\cos\frac{\alpha}{4}(1+r)}{\left(\frac{\pi}{2}\right)^2 - \left[\frac{\alpha}{4}(1+r)\right]^2} \sin\frac{\alpha}{4}(1-r)$	$\psi = \frac{\alpha}{4}(1-r)$
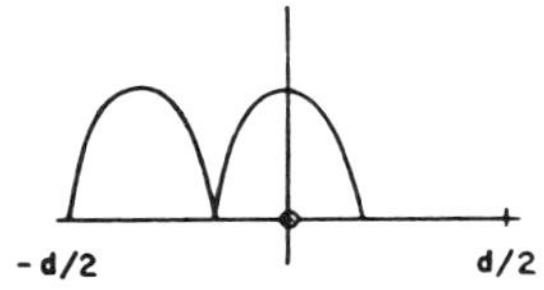 (c) DOUBLE COSINE OVERLAPPED	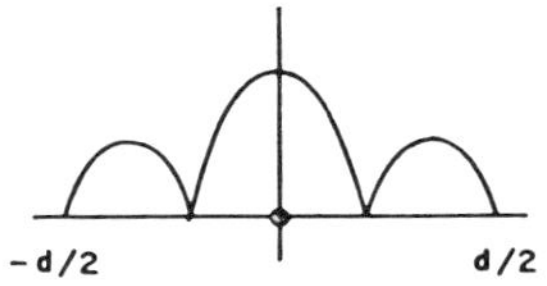 $f_e(\alpha) = \frac{\pi d}{3} \frac{\cos^2\frac{\alpha}{6}}{\left(\frac{\pi}{2}\right)^2 - \left(\frac{\alpha}{6}\right)^2} \cdot \cos\frac{\alpha}{6}$	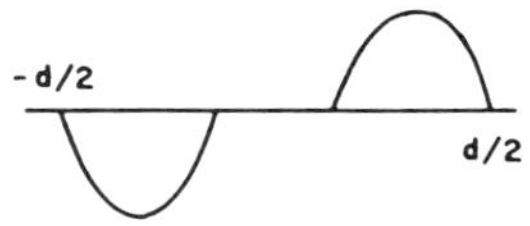 $f_o(\alpha) = j\frac{\pi d}{3} \frac{\cos^2\frac{\alpha}{6}}{\left(\frac{\pi}{2}\right)^2 - \left(\frac{\alpha}{6}\right)^2} \sin\frac{\alpha}{6}$	$\psi = \frac{\alpha}{6}$

Table III. Aperture Functions which Give Linear Phase.

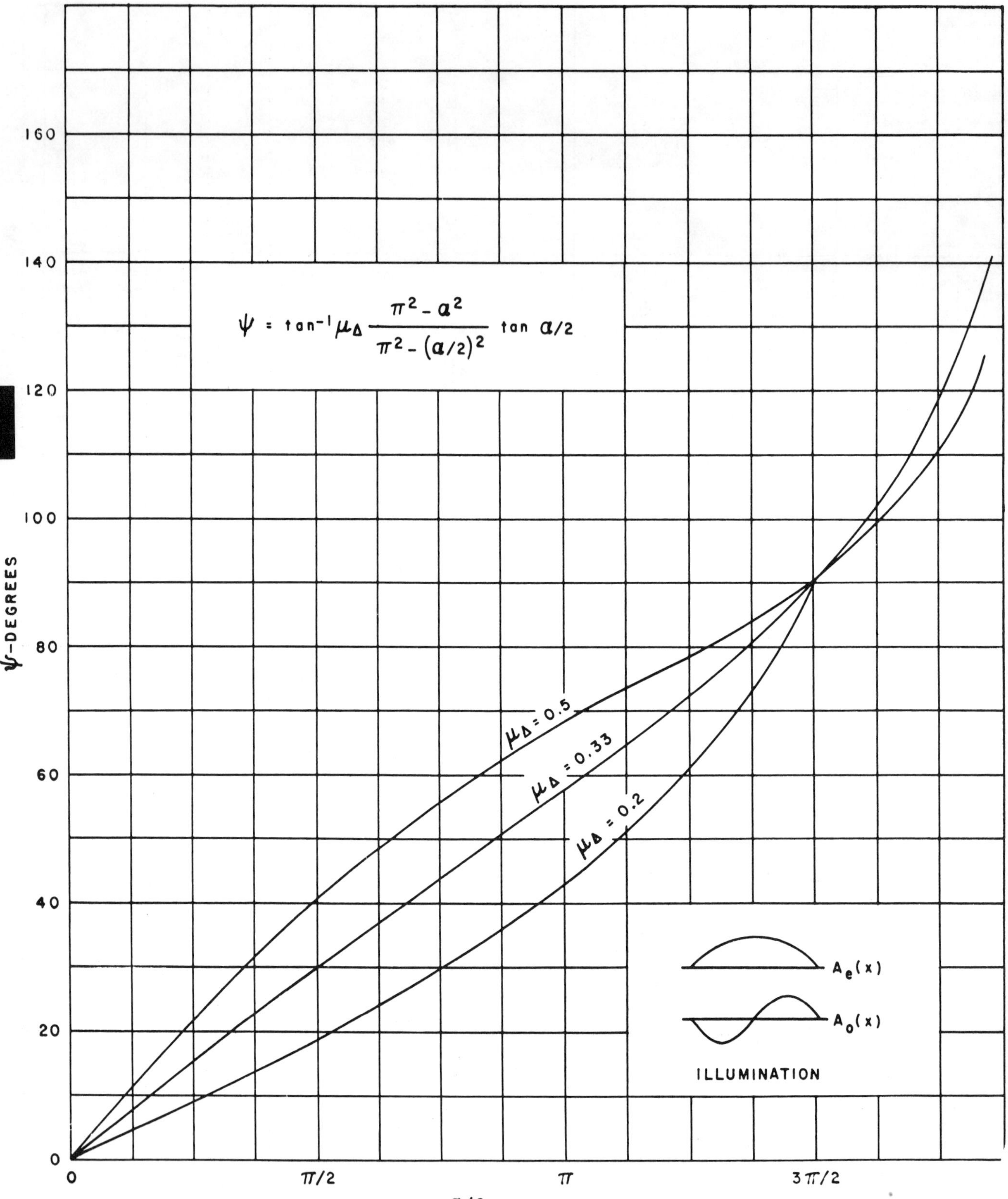

Figure 10. ECS Curves for Approximately Linear Phase Antenna (Variable Gain).

$$\psi = \tan^{-1} \mu_\Delta \left[\frac{a/2 \tan \frac{r\pi}{2} - \frac{r\pi}{2} \tan a/2}{\frac{r\pi}{2} \tan \frac{r\pi}{2} - a/2 \tan a/2} \right]$$

WHERE $\mu_\Delta = 1$

$\frac{r\pi}{2} = 85.9°$
$\frac{r\pi}{2} = 64.3°$
$\frac{r\pi}{2} = 42.8°$

$A_e(x)$
$A_o(x)$
ILLUMINATION

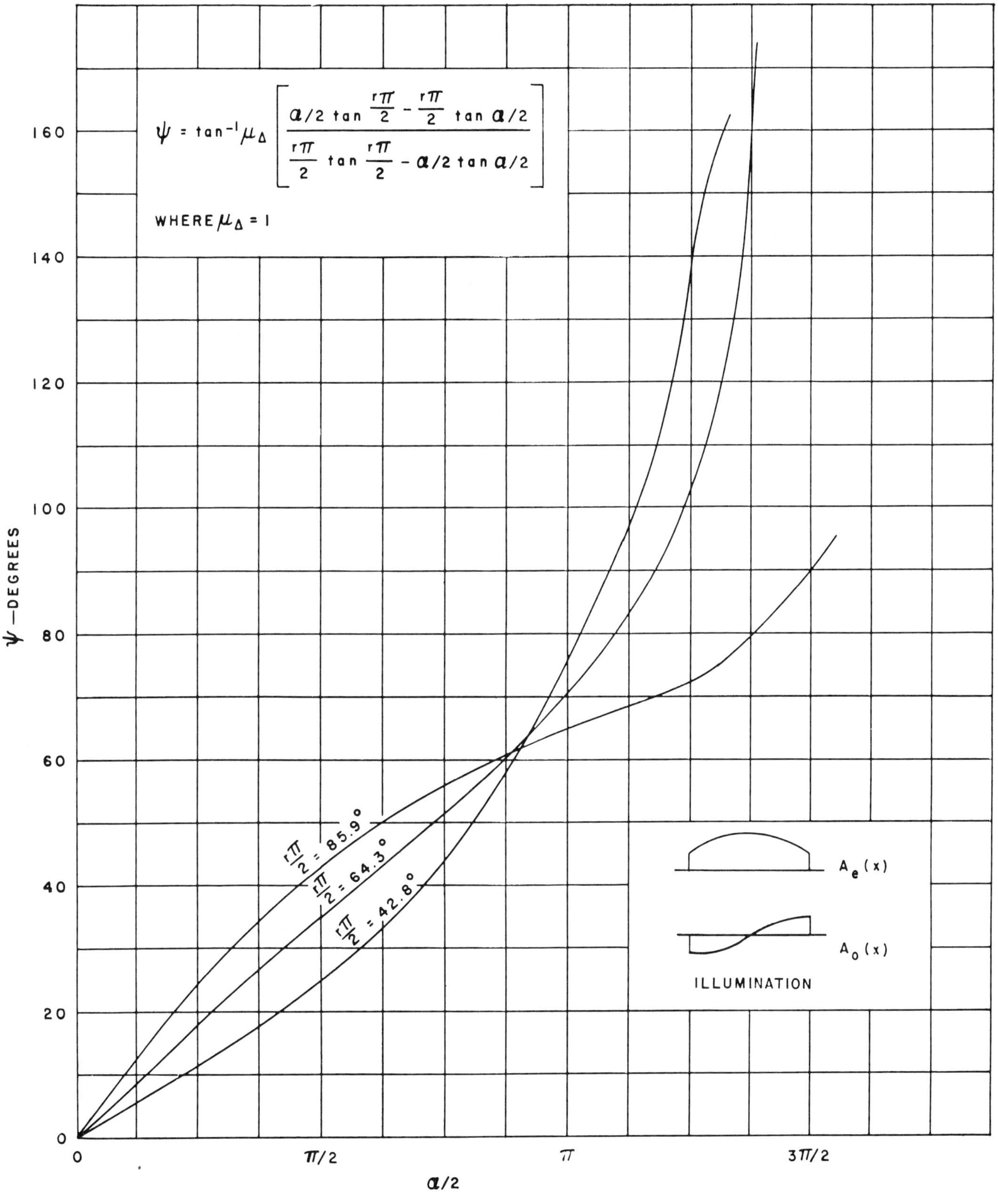

Figure 11. ECS Curves for Approximately Linear Phase Antenna (Variable Taper).

$$\sin p_1\alpha = \sin (p_2 - \delta)\,\alpha = \sin p_2\alpha \cos \delta\alpha + \cos p_2\alpha \sin \delta\alpha$$

$$= \sin p_2\alpha\,[\cos \delta\alpha + \cot p_2\alpha \sin \delta\alpha] \qquad \text{(B-49)}$$

$$= \sin p_2\alpha \left[1 + \frac{\delta}{p_2}\right]_{\alpha\to 0}$$

thus a slight difference in p_1 and p_2 can be adjusted by making a change in a_o by the reciprocal of the factor $\frac{p_2 + \delta}{p_2}$.

The effect of adding the odd function to the even function at angles other than 90 degrees has not been explored thoroughly. For example if the odd function is added at 45 degrees, as shown in the sketch, the odd function also appears in the denominator as in equation (B-50). The decrease in the angle ψ_1 would appear to approximately cancel the increase in the angle ψ_2.

46

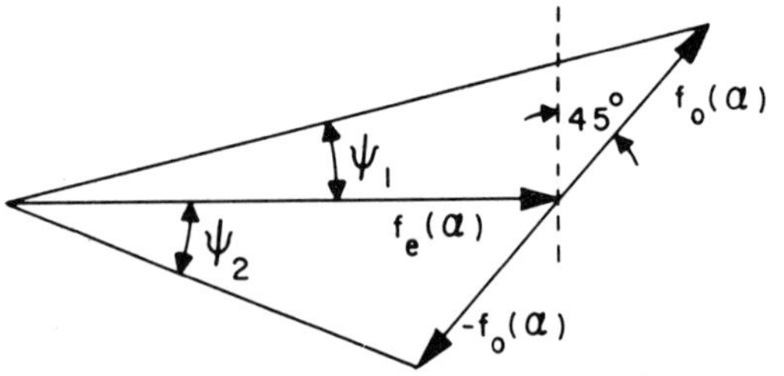

Odd and Even Functions Added at $\gamma = 45$ Degrees

$$\psi_1 = \tan^{-1} \frac{\dfrac{f_o(\alpha)}{\sqrt{2}}}{f_e(\alpha) + \dfrac{f_o(\alpha)}{\sqrt{2}}} \qquad \text{(B-50)}$$

b. (1) Linear Ratio.—A linear ratio is defined as

$$\text{ECS} = \frac{f_o(\alpha)}{f_e(\alpha)} = p\alpha \qquad \text{(B-51)}$$

The conditions on the odd and even aperture functions to obtain a linear ratio are:

$$A_o(x) = \frac{d}{dx} A_e(x) \qquad \text{(B-52)}$$

$$A\left(\pm \frac{d}{2}\right) = 0 \qquad \text{(B-53)}$$

A derivation of the above conditions is given in Appendix I. Pairs of even and odd aperture functions which meet the above requirements are shown in Table IV.

(2) Approximately Linear Ratios.—While functions of the type described above give linear ratios it is not practical with circuits to evaluate the ratios at the points where

$$\frac{f_o(\alpha)}{f_e(\alpha)} \to \frac{0}{0}, \qquad \text{(B-54)}$$

as the signal in both channels goes to zero. Equation (B-51) indicates that the poles and zeros must coincide to give the desired output. Thus in a practical case a linear ratio is obtained over a limited range of the even far field pattern. The effect of small errors in $f_o(\alpha)$ or $f_e(\alpha)$ has not been explored.

c. (1) Linear Logarithmic Ratio.—If the functional relationship in equation (B-42) is logarithmic, another method of obtaining a linear error signal can be defined.

$$\text{ECS} = \log \frac{f_1(\alpha)}{f_2(\alpha)} = \log f_1(\alpha) - \log f_2(\alpha) = p\alpha \qquad \text{(B-55)}$$

Obvious equations for $f_1(\alpha)$ and $f_2(\alpha)$ are

$$f_1(\alpha) = f(\alpha)\, e^{\frac{k\alpha}{2}} \qquad \text{(B-56)}$$

$$f_2(\alpha) = f(\alpha)\, e^{\frac{-k\alpha}{2}} \qquad \text{(B-57)}$$

Substitution of equations (B-56) and (B-57) into equation (B-55) will show that the desired result, $p\alpha$, is obtained. While mathematically, $f(\alpha)$, can have any form, for practical purposes it should have no zeros in the range of interest of the variable α. One type of antenna pattern which meets these requirements is the Gaussian. However, an infinite aperture is required to produce a truly Gaussian pattern so the region of useful ECS is restricted to the central portion of the antenna pattern.

If Gaussian patterns are assumed for $f_1(\alpha)$ and $f_2(\alpha)$, at squint angles of plus and minus α_b, the equations are:

$$f_1(\alpha) = e^{p_1(\alpha + \alpha_b)^2} = e^{p_1(\alpha^2 + \alpha_b^2)}\, e^{2p_1\alpha_b\alpha}, \text{ and} \qquad \text{(B-58)}$$

$$f_2(\alpha) = e^{-p_1(\alpha - \alpha_b)^2} = e^{-p_1(\alpha^2 + \alpha_b^2)}\, e^{-2p_1\alpha_b\alpha} \qquad \text{(B-59)}$$

so

$$f(\alpha) = e^{p_1(\alpha^2 + \alpha_b^2)} \text{ and} \qquad \text{(B-60)}$$

$$k = 4p_1\alpha_b. \qquad \text{(B-61)}$$

Even $A_e(x)$ $f_e(\alpha)$	Odd $A_o(x) = \frac{d}{dx} A_e(x)$ and $f_o(\alpha)$	Ratio $\frac{f_o(\alpha)}{f_e(\alpha)}$
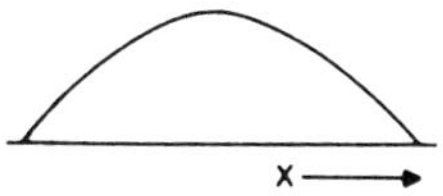 COSINE $f_e(\alpha) = \frac{d \cos \alpha/2}{(\pi/2)^2 - (\alpha/2)^2} \cdot \frac{\pi}{2}$	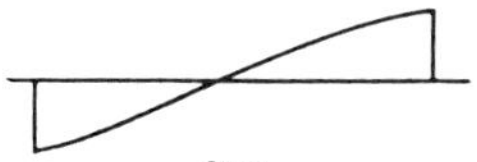 SINE $f_o(\alpha) = \frac{d \cos \alpha/2}{(\pi/2)^2 - (\alpha/2)^2} \cdot \frac{\alpha}{2}$	$\frac{\alpha}{\pi}$
 COSINE SQUARED $f_e(\alpha) = \frac{d\,\pi \sin \alpha/2}{\pi^2 - (\alpha/2)^2} \cdot \frac{\pi}{\alpha}$	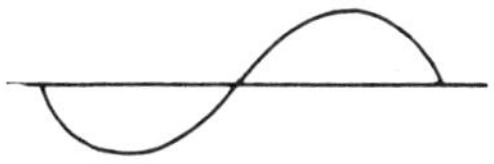 DOUBLE ANGLE SINE $f_o(\alpha) = \frac{d\,\pi \sin \alpha/2}{\pi^2 - (\alpha/2)^2}$	$\frac{\alpha}{\pi}$
 PARABOLIC $\left[1 - \left(\frac{2x}{h}\right)^2\right]$ $f_e(\alpha) = \frac{d}{\left(\frac{\alpha}{2}\right)}\left[\frac{\sin \alpha/2}{\alpha/2} - \cos \alpha/2\right] \cdot \frac{4}{\alpha}$	 LINEAR $f_o(\alpha) = \frac{d}{\frac{\alpha}{2}}\left[\frac{\sin \alpha/2}{\alpha/2} - \cos \alpha/2\right]$	$\frac{\alpha}{4}$
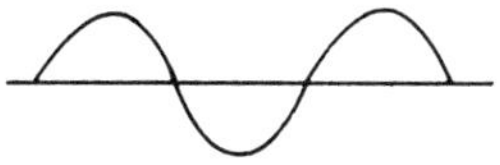 TRIPLE ANGLE COSINE $f_e(\alpha) = \frac{d \cos \alpha/2}{\left(\frac{3\pi}{2}\right)^2 - \left(\frac{\alpha}{2}\right)^2} \cdot \frac{3\pi}{2}$	 TRIPLE ANGLE SINE $f_o(\alpha) = \frac{d \cos \alpha/2}{\left(\frac{3\pi}{2}\right)^2 - \left(\frac{\alpha}{2}\right)^2} \cdot \frac{\alpha}{2}$	$\frac{\alpha}{3\pi}$

Table IV. Aperture Functions which Give Linear Ratios.

(2) *Approximately Linear Logarithmic Ratio.*—An investigation of the use of simple harmonic functions in the aperture[16] to produce far field functions which approximate Gaussian patterns was reported in the third quarterly engineering report. The aperture functions used were of the form

$$A(x) = a_0 + a_1 \cos x + a_2 \cos 2x \quad \text{(B-62)}$$

The use of two terms of equation (B-62), $a_2 = 0$, enabled a three-point match in the far field and three terms, a five-point match. The aperture functions are shown in Figure 12, and the points of match in the far field in Figures 13 and 14. With a selected squint angle the logarithm of the ratio is linear to more than two radians. The ECS characteristics are shown in Figure 15.

d. Combination Phase-Amplitude Antenna and ECS.—The sum and difference far fields of the combination phase-amplitude antenna are not of a separable form. That is, the far fields cannot be expressed as a product of a function of one coordinate, α, times a function of the other coordinate, ε. From Section B4b, equations (B-84) and (B-85), the sum and difference are

$$\text{sum} = \sqrt{2}\,[f_e(\alpha)\, f_e(\varepsilon) + j\, f_o(\alpha)\, f_o(\varepsilon)] \quad \text{and} \quad \text{(B-63)}$$

$$\text{difference} = \sqrt{2}\,[f_e(\alpha)\, f_o(\varepsilon) + j\, f_o(\alpha)\, f_e(\varepsilon)]. \quad \text{(B-64)}$$

The ratio of signals (B-64) and (B-63) will not satisfy equation (B-41) as the ratio will be a function of both α and ε. There are two methods of obtaining a ratio in which the functions of α and ε can be separated. The round trip signals are

$$\Sigma = \frac{k_1 A_{eff}^{1/2}}{R^2}\,[f_e(\alpha)\, f_e(\varepsilon) + j\, f_o(\alpha)\, f_o(\varepsilon)]^2 \quad \text{(B-65)}$$

$$\Delta = \frac{k_1 A_{eff}^{1/2}}{R^2}\,[f_e(\alpha)\, f_e(\varepsilon) + j\, f_o(\alpha)\, f_o(\varepsilon)]\,[f_e(\alpha)\, f_o(\varepsilon) + j\, f_o(\alpha)\, f_e(\varepsilon)] \quad \text{(B-66)}$$

k_1 = a constant for a particular radar
A_{eff} = effective target area
R = range to target
Σ = round trip sum signal
Δ = round trip difference signal

The first method requires that the Σ and Δ signals be added and subtracted. This gives signals proportional to the antenna signals, equations (B-82) and (B-83). If the absolute magnitude is taken, then the dependence upon phase is eliminated.

$$|\Sigma + \Delta| = \left| \frac{k_1 A_{eff}^{1/2}}{R^2}\,[f_e^2(\alpha)\, f_e^2(\varepsilon) + f_o^2(\alpha)\, f_o^2(\varepsilon)]^{1/2}\,[f_e^2(\alpha) + f_o^2(\alpha)]^{1/2}\,[f_e(\varepsilon) + f_o(\varepsilon)] \right| \quad \text{(B-67)}$$

$$|\Sigma - \Delta| = \left| \frac{k_1 A_{eff}^{1/2}}{R^2}\,[f_e^2(\alpha)\, f_e^2(\varepsilon) + f_o^2(\alpha)\, f_o^2(\varepsilon)]^{1/2}\,[f_e^2(\alpha) + f_o^2(\alpha)]^{1/2}\,[f_e(\varepsilon) - f_o(\varepsilon)] \right| \quad \text{(B-68)}$$

In a ratio, as can be produced by AGC of the absolute magnitudes, all the terms cancel except $f_o(\varepsilon)/f_e(\varepsilon)$.

$$\frac{|\Sigma + \Delta| - |\Sigma - \Delta|}{|\Sigma + \Delta| + |\Sigma - \Delta|} = \frac{f_o(\varepsilon)}{f_e(\varepsilon)} \quad \text{(B-69)}$$

The phase axis information is contained in the phase angle between the reconstituted antenna signals. $\Sigma + \Delta$ and $\Sigma - \Delta$. Therefore, these signals should be phase detected.

The second method of obtaining a ratio which is a function of but one angle is to alter the sum antenna pattern until the term $j\, f_o(\varepsilon)\, f_e(\varepsilon)$ in equation (B-65) is negligible. Then the ratio of Δ to Σ is

$$\frac{\Delta}{\Sigma} = \frac{f_o(\varepsilon)}{f_e(\varepsilon)} + j\,\frac{f_o(\alpha)}{f_e(\alpha)} \quad \text{(B-70)}$$

With a symmetrical sum pattern ($j\, f_o(\varepsilon)\, f_e(\varepsilon) = 0$) AGC action can produce the ratios indicated by equation (B-70). The two components of the error can be separated by using a large reference signal in the second detectors so that only the in-phase component will be detected.

3. Sidelobes

The limitations on maximizing the gain and at the same time minimizing the beamwidth and sidelobes of

[16] W. L. Murdock, "Relation Between Far Field Specifications and the Aperture Illumination of a Transducer," General Electric Company, Electronics Laboratory, TIS 51E221, 29 June 1951.

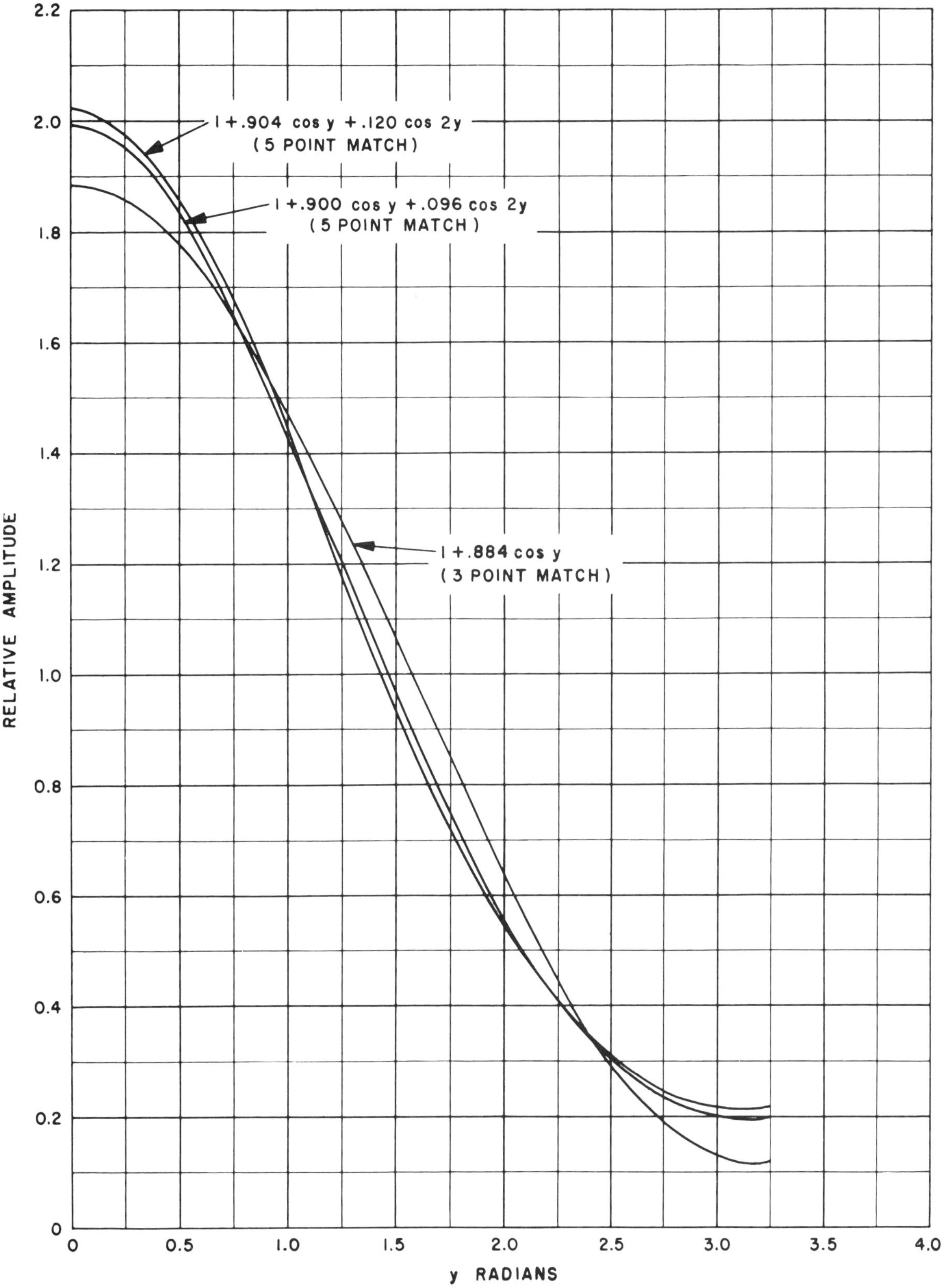

Figure 12. Required Antenna Aperture Illuminations to Obtain Synthesized Antenna Patterns.

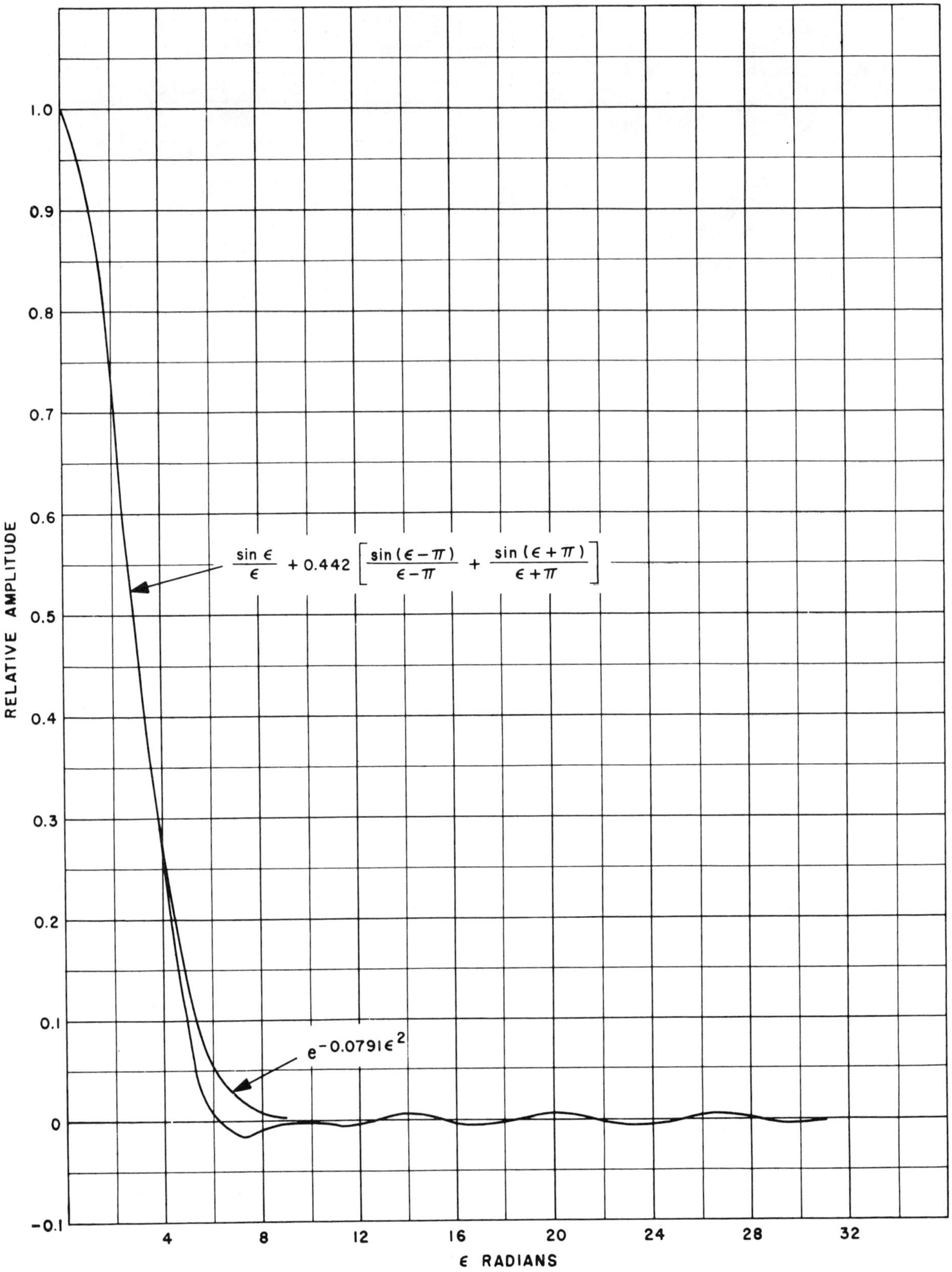

Figure 13. Antenna Pattern Approximation for Three Point Curve Matching Case.

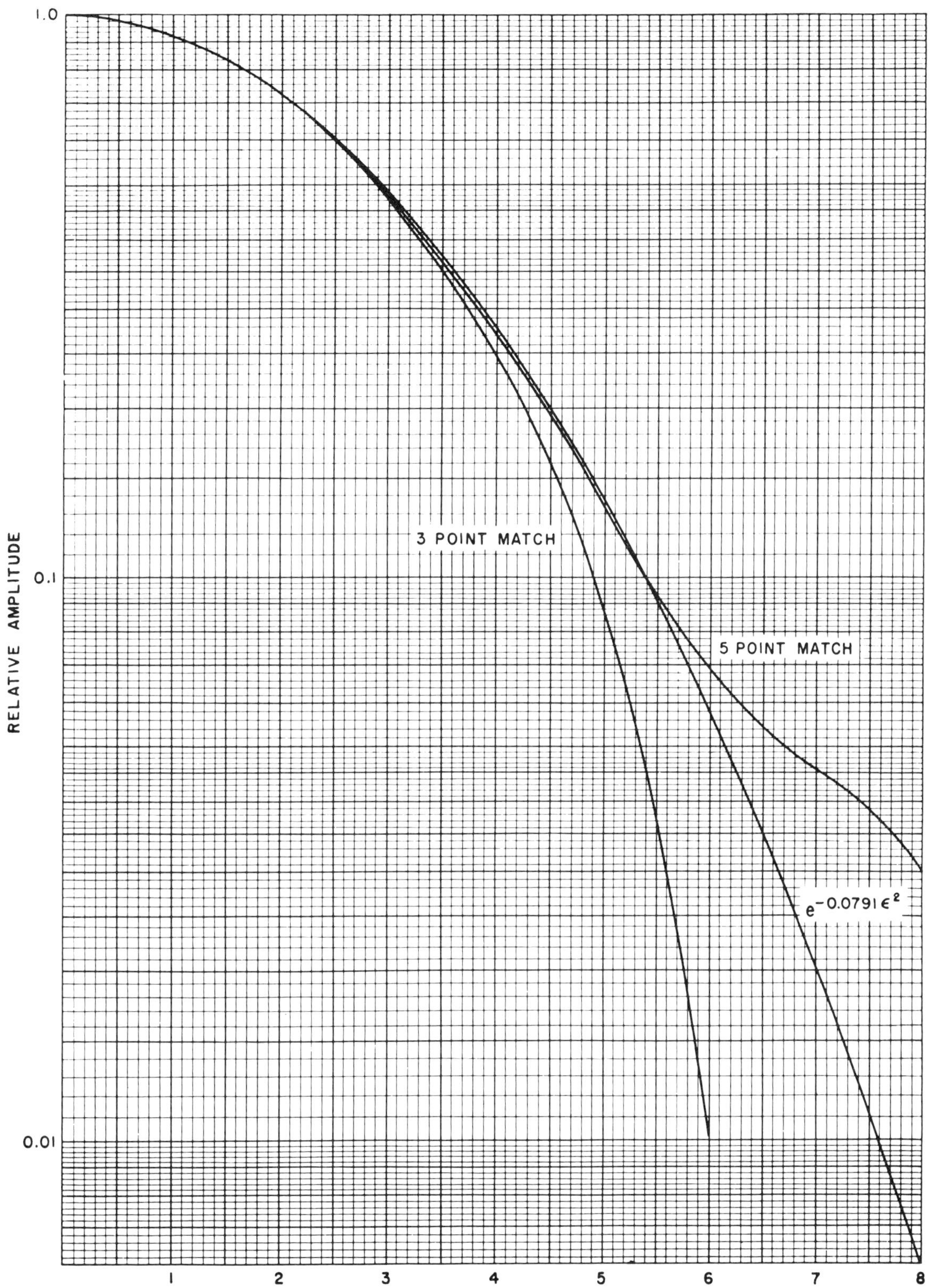

Figure 14. Comparison of Antenna Pattern Approximations to the Exponential.

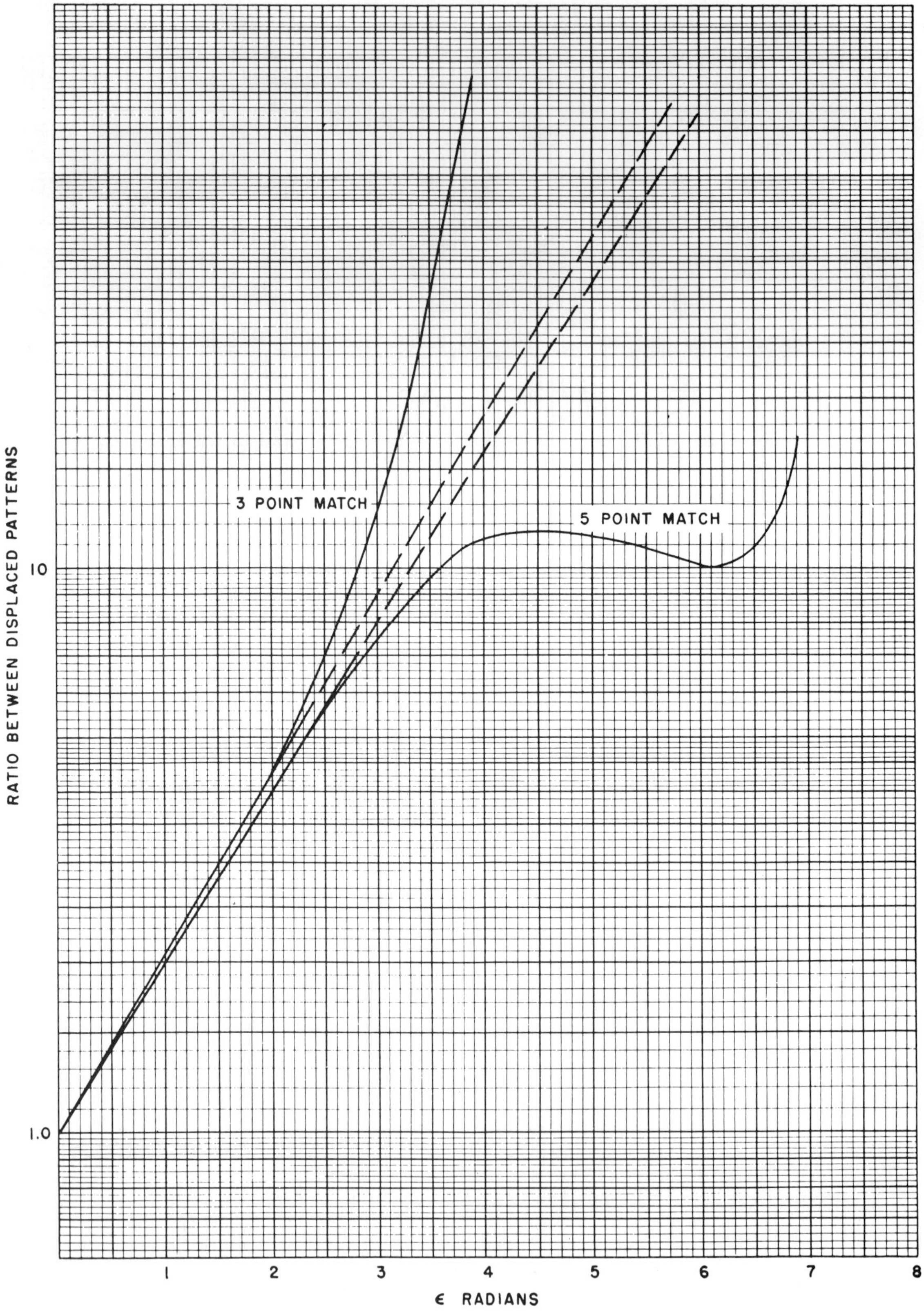

Figure 15. ECS Characteristics for Antenna Pattern Approximations.

even far field patterns is discussed at length in the literature. Tapered illuminations such as cosine, Figure 3, or cosine squared, Figure 4, minimize the sidelobes at the expense of antenna gain and beamwidth.

The effect of adding an odd far field pattern to an even pattern is not so well understood. However, there is one salient point. The addition of an odd far field to an even will increase the gain at angles off the antenna axis. The antenna axis is considered to be the axis of symmetry of the even or the odd far fields considered separately.

For example, if an odd field is added in quadrature with an even, the total field is

$$f(\alpha) = f_e(\alpha) + j\, f_o(\alpha), \quad \text{or} \qquad \text{(B-71)}$$

$$f(\alpha) = f_e(\alpha)\sqrt{1 + \tan^2 \psi}\ \underline{/\psi} \qquad \text{(B-72)}$$

$$\text{where } \tan \psi = \frac{f_o(\alpha)}{f_e(\alpha)} \qquad \text{(B-73)}$$

From equation (B-72) it is apparent that if $f_o(\alpha)$ has any value other than zero, the amplitude of the far field at points off the antenna axis is increased. It is also to be noted that the pattern is even whether the odd term is added at plus or minus ninety degrees.

If the odd field is added and subtracted from the even without phase shift the total field is

$$f(\alpha) = f_e(\alpha)\left[1 \pm \frac{f_o(\alpha)}{f_e(\alpha)}\right] \qquad \text{(B-74)}$$

When the odd far field is added to the even field the point of maximum gain is shifted to one side of the antenna axis, and when the odd field is subtracted the maximum gain is on the other side of the axis. Thus two squinted beams are produced.

To further illustrate the effect of adding an odd to an even far field function some examples will be given.

a. Linear or Approximately Linear Phase.—To show that the addition of an odd far field in quadrature increases the beamwidth and sidelobe level, three curves are plotted on Figure 16. The curve marked $f_e(\alpha)$ is the far field for the full aperture with a 7.3 db tapered illumination. Equation (B-12) with $r = 0.72$ is the formula for this even far field.

The same far field with an odd field added in quadrature gives the curve marked approximately linear phase, Figure 16. These odd and even functions were chosen since they give the most nearly linear phase curve, Figure 11. The even plus the quadrature odd field is 42% wider in beamwidth than the even field alone and the first sidelobe is 5.4 db higher.

The curve marked linear phase is the same as that for $f_e(\alpha)$, except that only half the aperture is used. Thus the beamwidth is twice (100% increase) and the sidelobe level the same as that of $f_e(\alpha)$.

These curves, Figure 16, show that the use of approximately linear phase offers a desirable radar antenna design. A signal proportional to $f_e(\alpha)$ is available to intensify a visual presentation or for range tracking and has a sidelobe level of —19 db, and a beamwidth only 11% greater than that obtained with uniform illumination of the aperture.

If still lower sidelobe levels are desired on the combined even and quadrature odd far fields, more taper must be used on the aperture illuminations. For a double angle sine odd illumination (Figure 7) and a cosine even (Figure 3) the far field pattern has been calculated and is plotted on Figure 17. The first sidelobe is —23.1 db and the beamwidth is 40% greater than that for a cosine illumination of the aperture. The odd function $f_o(\alpha)$, Figure 7, was multiplied by 1/3 before addition in quadrature to the even function, (Figure 3). This multiplier of 1/3 was used because it results in a phase curve which is linear over the greatest extent about the axis as is shown in Figure 10.

In the list of the sources of angular error, section C, clutter signal from undesired fixed ground targets is included. If the maximum allowable angular error is δ_a, then the clutter signal must be less than

$$S_c = \delta_a \left.\frac{\partial \Delta}{\partial \alpha}\right|_{a \to 0}, \qquad \text{(B-75)}$$

if the clutter signal adds to the desired signal in the worst possible phase relationship.

Since desired radar targets are usually moving relative to the ground clutter, it may be necessary to take advantage of the doppler frequency shift of the moving target to obtain adequate angular accuracy. If the clutter signal is less than the desired moving target signal, the cyclic relative phase will produce a cyclic angular error which usually is of too high a frequency to pass the narrow filter in the tracking circuits. However, if the clutter signal is greater than the desired target signal, then it is the target signal which will be eliminated by the tracking filter. If the desired target signal is so small that the clutter cannot be minimized sufficiently by the antenna pattern, then coherent doppler tracking techniques must be employed. Circuits for doppler tracking are described in the quarterly engineering reports on "Anti-Clutter Techniques," Signal Corps Contract DA 33-039-sc-5446.

b. Linear Ratios.—From section 2 on ECS, the conditions on the even and odd aperture illumination for a linear ratio of the far field patterns are

$$f_e(\alpha) = \frac{\dfrac{\dfrac{r\pi}{2}\sin\dfrac{r\pi}{2}\cos\alpha/2 - \alpha/2\sin\alpha/2\cos\dfrac{r\pi}{2}}{\left(\dfrac{r\pi}{2}\right)^2 - (\alpha/2)^2}}{\dfrac{\sin\dfrac{r\pi}{2}}{\dfrac{r\pi}{2}}}$$

$$\left|f_e(\alpha) + jf_o(\alpha)\right| = f_e(\alpha)\sqrt{1 + \tan^2\psi}$$

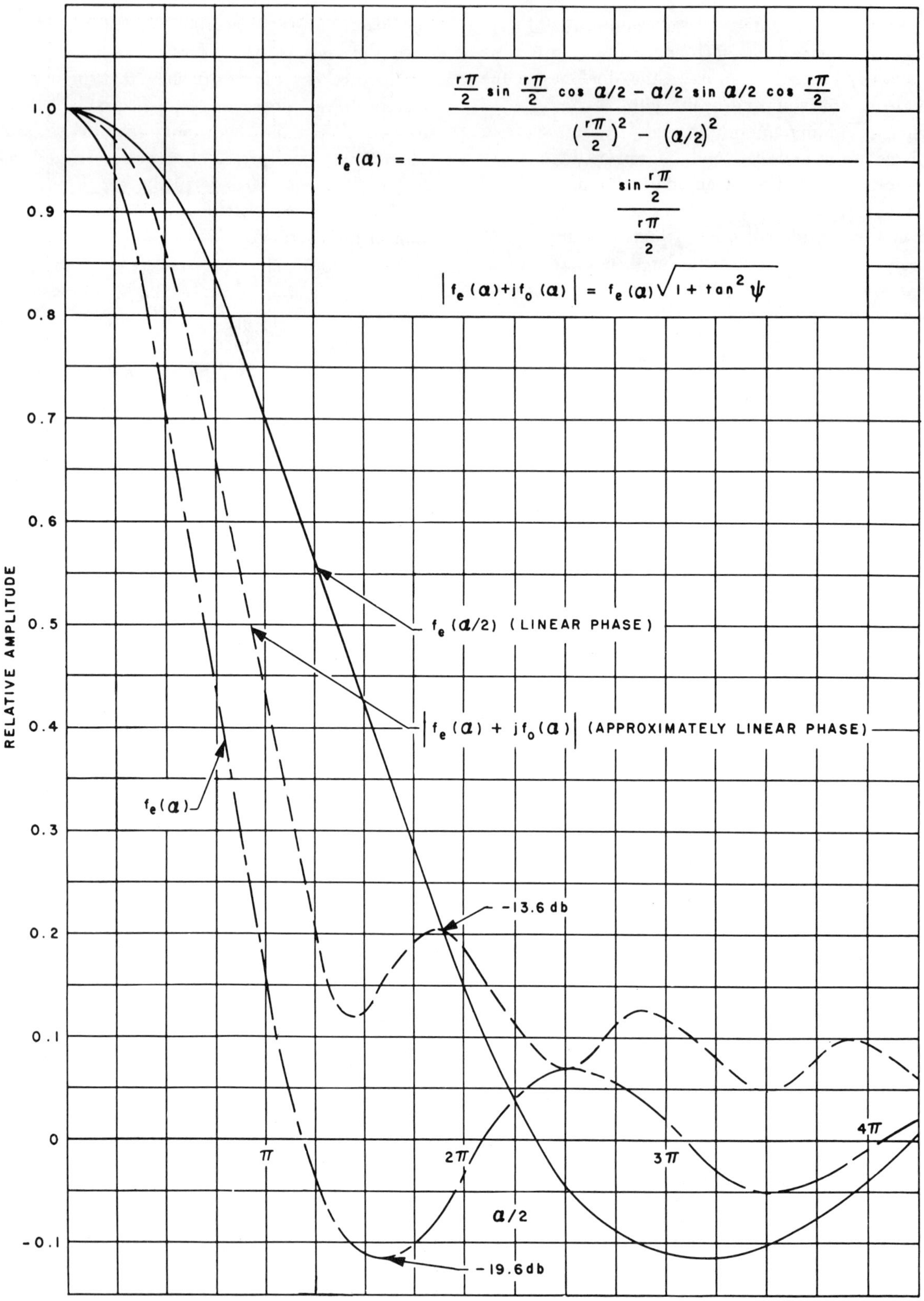

Figure 16. Combinations of Far Field Functions which Give Linear and Approximately Linear Phase.

$$\left| f_e(a) + 1/3\, j f_0(a) \right| = \frac{\sqrt{\left[\dfrac{\cos a/2}{\pi^2 - a^2}\right]^2 + \left[1/3\, \dfrac{\sin a/2}{\pi^2 - (a/2)^2}\right]^2}}{0.10132}$$

−23.1 db

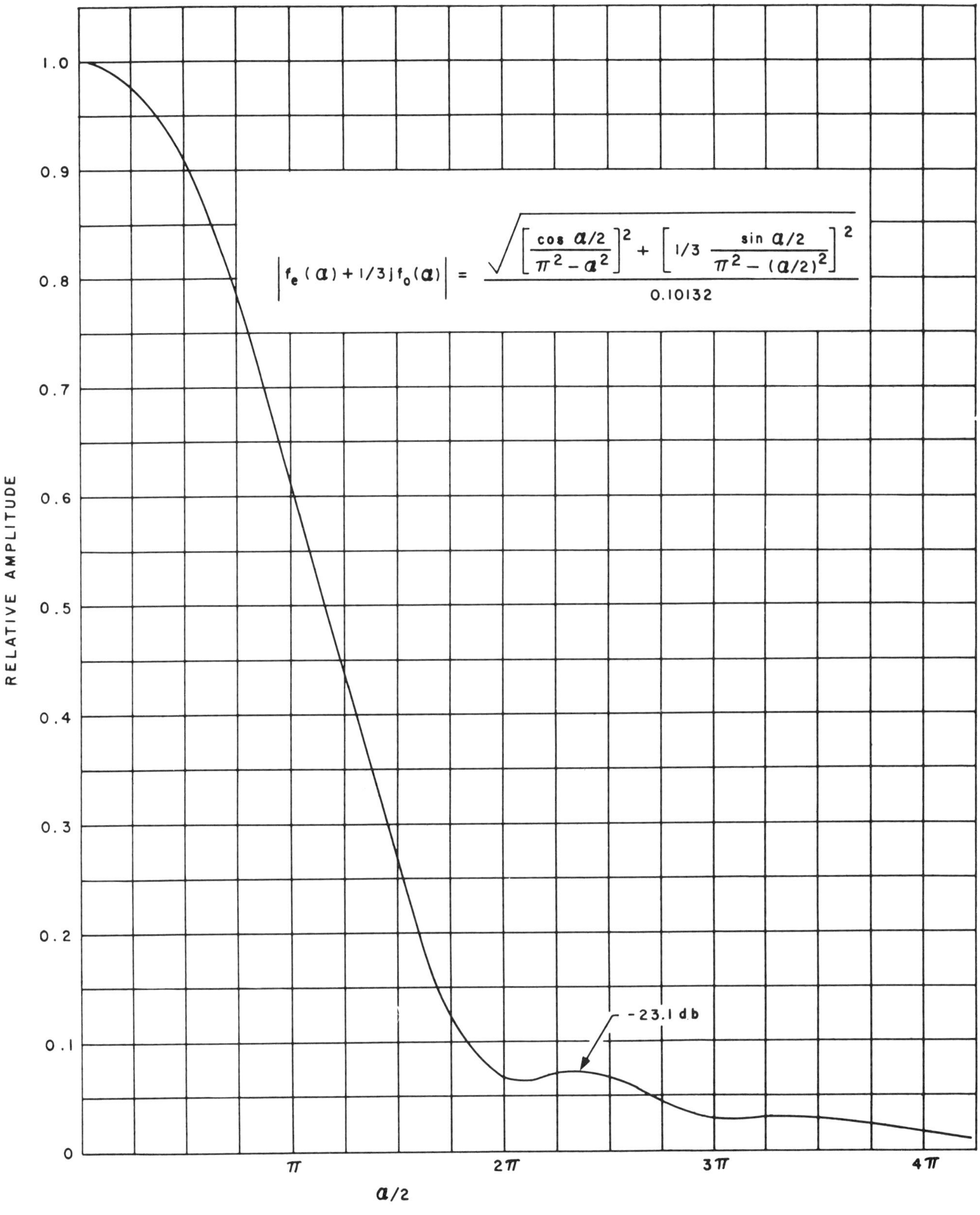

Figure 17. Combination of Far Field Functions which Result in Low Side Lobes.

$$A_e(\pm d/2) = 0 \quad \text{and}$$
$$A_o(x) = A_e'(x). \qquad \text{(B-76)}$$

The first requirement implies an infinite tapered even aperture function, and the second an abrupt discontinuity at the edges of the odd aperture function. Thus it is to be expected that the side lobe ratio of the combined even and odd far fields will be worse than that of the even field alone. For example, if the even aperture function is cosine and the odd is sine then the total far field is given by equation (B-74). From Table IV the ratio of $f_o(\alpha)/f_e(\alpha)$ is α/π so equation (B-74) can be written for this particular case as

$$f(\alpha) = f_e(\alpha)\ [1 + \alpha/\pi] \qquad \text{(B-77)}$$

By algebraic manipulation $f(\alpha)$ is

$$f(\alpha) = \frac{d \sin(\pi/2 - \alpha/2)}{(\pi/2 - \alpha/2)}, \qquad \text{(B-78)}$$

if $A_e(x) = \cos \frac{\pi x}{d}$.

As is shown by Figure 1, the first sidelobe of this far field equation (B-78) is 9.8 db higher than that of $f_e(\alpha)$, Figure 3. This confirms the previous statement that the addition of an odd field deteriorates an even field.

The even pattern $f_e(\alpha)$ for a linear ratio has low sidelobes, and it may be possible to take full advantage of the low sidelobes of $f_e(\alpha)$ when tracking moving targets in the presence of ground clutter. The mode of operation would be to use much less amplification in the difference channel than in the sum channel receivers prior to the second detector. The linearity of the ratio is not altered by the use of a reduced difference channel gain as is shown by Table IV. Then even though the clutter signal may be several times the magnitude of the target error signal in the difference channel, the cyclic phase shift of clutter relative to the target signal will enable a filter to eliminate the clutter. The success of this circuit adjustment depends upon the clutter being small relative to the target signal in the sum channel. Since sidelobe levels of actual antennas are rather unpredictable as to gain relative to maximum gain, it would be desirable to record the difference and sum channel clutter signals at typical sites.

4. Comparison of Antennas for Direction Finding

There are many factors which should be considered for full comparison of radar direction finding antennas. Listed in an approximate order of importance some of these are:

(1) Efficiency of far field patterns as measured by $KG^{1/2}$

(2) Spillover, lens, and other RF losses

(3) Sidelobes

(4) Beamwidth

(5) Complexity of antenna and circuits

(6) Bandwidth

It is not possible to keep these factors separate and changing one may affect several of the others. Since the efficient use of transmitter power is of major importance all the emphasis in this section will be upon factors one and two, efficiency of far field patterns and spillover losses.

a. Symmetrical Monopulse Antennas.—As in all previous analyses only a square aperture will be considered. The basis for comparison will be the "ideal" antenna with uniform even and linear odd aperture functions as described in Sections 1a and 1b. The antenna efficiency relative to the "ideal" in decibels is:

$$\text{Ant Efficiency} = \text{Transmit efficiency} + \text{Receive efficiency} \qquad \text{(B-79)}$$

$$\text{Transmit Efficiency} = 10 \log_{10} \left.\frac{G}{G_o}\right|_{x\ \text{axis}} + 10 \log_{10} \left.\frac{G}{G_o}\right|_{y\ \text{axis}} + 10 \log_{10} \eta_e|_{x\ \text{axis}} + 10 \log_{10} \eta_e|_{y\ \text{axis}} \qquad \text{(B-80)}$$

$$\text{Receive Efficiency} = 20 \log_{10} \left.\frac{K}{K_o}\right|_{x\ \text{axis}} + 10 \log_{10} \left.\frac{G}{G_o}\right|_{y\ \text{axis}} + 10 \log_{10} \eta_o|_{x\ \text{axis}} + 10 \log_{10} \eta_e|_{y\ \text{axis}} \qquad \text{(B-81)}$$

The relative one axis values for several aperture illuminations are tabulated in Tables I and II. The spillover losses have been estimated with the aid of Figure 9.

The results for symmetrical aperture functions are tabulated in Table V. In the preparation of this table an effort has been made to give a breakdown into factors which can be readily checked by experiment or analysis. Also all of the antennas for which spillover losses are given are believed to be realizable designs.

b. Combination (Phase-Amplitude) Monopulse Antenna.—The illuminations used with the combination antenna cannot be expressed in separable form. The aperture is considered to be square and divided down the middle (see sketch on page 58).

One half of the aperture is illuminated so as to produce a beam which is tilted upwards from the antenna axis. The other half produces a downward tilted beam. Since the aperture field of each half of the aperture is separable the one way far field of each half can be represented symbolically as

Previous Terminology (if any)		Uniform Phase	11.4 db Taper Phase	Cosine Phase		Uniform Amplitude	Cosine Amplitude
Suggested Terminology	"Ideal"	Linear Phase	Linear Phase	Linear Phase	Approximately Linear Phase	Linear Ratio	Linear Ratio
Even Aperture Function $A_e(x)$, $A_e(y)$ see Table I	Uniform	Uniform	Double 11.4 db Taper (not shown in Table I)	Double Cosine	11.4 db Taper	Cosine	Cosine Squared
Odd Aperture Function $A_o(x)$ see Table II	Linear Odd	Uniform Odd	Double 11.4 db Taper (not shown in Table II)	Sine Odd	Linear Odd	Sine Odd	Double Angle Sine
(1) Transmit Gain $10 \log_{10} \left[\frac{G}{G_o}\right]$ x axis	0.00	0.00	−0.41	−0.92	−0.41	−0.92	−1.76
y axis	0.00	0.00	−0.41	−0.92	−0.41	−0.92	−1.76
(2) Transmit Spillover x axis	0.00	—	−0.45	−0.30	−0.45	−0.30	−0.20
$10 \log_{10} \eta_{it}$ y axis	0.00	—	−0.45	−0.30	−0.45	−0.30	−0.20
(3) Receive $20 \log_{10} \frac{K}{K_o}$ x axis	0.00	−1.25	−1.66	−2.16	−0.00	−0.06	−2.16
$10 \log_{10} \frac{G}{G_o}$ y axis	0.00	0.00	−0.41	−0.92	−0.41	−0.92	−1.76
(4) Spillover x axis	0.00	—	−0.45	−0.30	−3.00	−3.00	−0.30
$10 \log_{10} \eta_{ir}$ y axis	0.00	—	−0.45	−0.30	−0.45	−0.30	−0.20
Relative Angular Sensitivity at $\alpha = \varepsilon = 0$ equals [(1) + (2) + (3) + (4)]	0.00	−1.25	−4.69	−6.12	−5.58	−6.72	−8.34
Relative Σ Signal at $\alpha = \varepsilon = 0$ equals 2[(1) + (2)]	0.00	0.00	−3.44	−4.88	−3.44	−4.88	−7.84

Table V. Comparison of Symmetrical Monopulse Antennas.

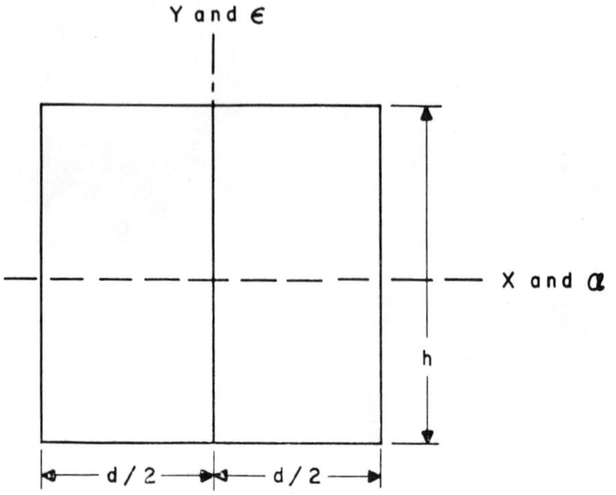

Combination Antenna Aperture

$$f(\alpha, \varepsilon) \text{ right} = [f_e(\alpha) + j f_o(\alpha)] [f_e(\varepsilon) + f_o(\varepsilon)] \quad \text{(B-82)}$$

$$f(\alpha, \varepsilon) \text{ left} = [f_e(\alpha) - j f_o(\alpha)] [f_e(\varepsilon) - f_o(\varepsilon)] \quad \text{(B-83)}$$

The RF sum of equations (B-82) and (B-83) is

$$f(\alpha, \varepsilon)_r + f(\alpha, \varepsilon)_l = \sqrt{2} [f_e(\alpha) f_e(\varepsilon) + j f_o(\alpha) f_o(\varepsilon)]. \quad \text{(B-84)}$$

The RF difference of equations (B-82) and (B-83) is

$$f(\alpha, \varepsilon)_r - f(\alpha, \varepsilon)_l = \sqrt{2} [f_e(\alpha) f_o(\varepsilon) + j f_o(\alpha) f_e(\varepsilon)] \quad \text{(B-85)}$$

The one-way sum pattern, equation (B-84), contains the undesired term $j f_o(\alpha) f_o(\varepsilon)$ which adds sidelobes to the symmetrical pattern $f_e(\alpha) f_e(\varepsilon)$, and decreases the gain. The one-way difference pattern, equation (B-85), contains odd functions on both axes simultaneously though in time quadrature.

The equations for calculating K and $G^{1/2}$ for the combination antenna will be found in Appendix II. The effect on angular sensitivity of various aperture illuminations is given in Table VI.

Comparison of Tables V and VI shows that a symmetrical antenna should give better sensitivity than a combination antenna. For example, in Table V a 11.4 db taper even and linear odd gives an angular sensitivity of —5.6 db relative to the "ideal" antenna. This is 6 db better than the amplitude comparison plane and 4.7 db better than the phase comparison plane of the combination antenna. The reason for the better sensitivity of the symmetrical antenna is brought out in equations (B-84) and (B-85). The undesired term $j f_o(\alpha) f_o(\varepsilon)$ requires power which could be concentrated in $f_e(\alpha) f_e(\varepsilon)$ in a symmetrical design. The presence of an odd pattern in each plane simultaneously means that the gain of each K factor, K_a and K_p, is less than if only an odd pattern for one plane were present in equation (B-85).

The gain of the combination antenna can be improved by overlapping the illumination along the x axis. This will increase the sensitivity of amplitude comparison with some decrease of phase comparison sensitivity. However, as is shown in Table VI, the sensitivity of the phase comparison plane is higher before overlap. Overlap is being utilized in the development of the antenna for the AN/TPQ-5 of Signal Corps Contract DA-36-039-sc-5408.

It should be possible to design a combination antenna which will completely eliminate the undesired term in equation (B-84). The overall angular sensitivity with $j f_o(\alpha) f_o(\varepsilon)$ eliminated is approximately 3 db instead of 6 db lower than a completely symmetrical (4 feed) design. The advantage of getting azimuth and elevation errors in one RF difference signal appears to require about 3 db more transmitter power than a two RF difference signal antenna. It is assumed that the even far field patterns would be the same in both cases.

c. Sequential Lobing Antenna.—The same equations, Appendix II, as were developed for the amplitude axis of the combination antenna can be used to predict sequential lobing antenna performance. The far field pattern produced by a squinted beam can be represented symbolically as

$$f(\alpha, \varepsilon) = f_e(\alpha) [f_e(\varepsilon) + f_o(\varepsilon)]. \quad \text{(B-86)}$$

The effect of the conical scan is introduced by modulating the even far field term in equation (B-86) sinusoidally with the odd term, and introducing a $f_o(\alpha)$ term in time quadrature to $f_o(\varepsilon)$.

$$f(\alpha, \varepsilon) = [f_e(\alpha) + f_o(\alpha) \sin \omega_m t] [f_e(\varepsilon) + f_o(\varepsilon) \cos \omega_m t] \quad \text{(B-87)}$$

The round trip pattern is the square of the one-way pattern, or

$$\begin{aligned} f^2(\alpha, \varepsilon) = {} & f_e^2(\alpha) f_e^2(\varepsilon) + 2 f_e^2(\varepsilon) f_e(\alpha) f_o(\alpha) \sin \omega_m t \\ & + 2 f_e^2(\alpha) f_e(\varepsilon) f_o(\varepsilon) \cos \omega_m t + f_e^2(\varepsilon) f_o^2(\alpha) \sin^2 \omega_m t \\ & + f_e^2(\alpha) f_o^2(\varepsilon) \cos^2 \omega_m t + f_o^2(\alpha) f_o^2(\varepsilon) \sin^2 \omega_m t \cos^2 \omega_m t \end{aligned} \quad \text{(B-88)}$$

The usual method of separating the even and odd terms is to detect the received signals and then use a bandpass filter centered at ω_m. Noise power of $kT \beta$ (NF) is added to the antenna signal in the receiver. Some of the terms in equation (B-88) do not contribute to the useful output. For an error on the α axis the desired terms are

Aperture Illumination (each half)	Conical Scan Antenna* and Amplitude Axis (ε) of Combination Antenna			Phase Axis (α) Combination Antenna			Σ Signal		
	$20 \log_{10} \frac{K_a G^{1/2}}{K_o G_o^{1/2}}$	Spillover $20 \log_{10} \eta$	Total	$20 \log_{10} \frac{K_p G^{1/2}}{K_o G_o^{1/2}}$	Spillover $20 \log_{10} \eta$	Total	$20 \log_{10} \frac{G}{G_o}$	Spillover $20 \log_{10} \eta$	Total
Uniform	— 6.6	—	— 6.6	— 6.7	—	— 6.7	—5.5	—	— 5.5
11.4 db Taper	— 9.8	—1.8	—11.6	— 8.5	—1.8	—10.3	—7.2	—1.8	— 9.0
Cosine	—12.9	—1.2	—14.1	—10.1	—1.2	—11.3	—8.9	—1.2	—10.1

* Note:
The above table can be used to compare radar tracking accuracies when using the combination antenna and the symmetrical antennas, Table V, but a more complete analysis is necessary before a conical scan radar can be compared with monopulse radar (See Section C4).

Table VI. Comparison of Combination and Conical Scan Antennas to "Ideal" Antenna.

$$f^2(\alpha, \varepsilon) = f_e^2(\alpha)\, f_e^2(\varepsilon) + 2\, f_e^2(\varepsilon)\, f_e(\alpha)\, f_o(\alpha) \sin \omega_{mt} \tag{B-89}$$

Equation (B-89) shows the manner in which the received signal (round trip) will vary with α, ε, and time. In Appendix V the effect of G and K on the received signal is shown to be

$$S \approx G + 2\, KG^{1/2}\, d\alpha \sin \omega_{mt} \tag{B-90}$$

S = RMS signal amplitude

$KG^{1/2}$ = angular sensitivity factor of antenna

The value of $KG^{1/2}$ for the sequential lobing antenna is the same as $K_aG^{1/2}$ for the combination phase-amplitude antenna, Appendix II. However, the comparison of the tracking accuracy cannot be based upon the antenna alone, and the comparison is completed in Section C when the linear detector-boxcar-filter action is included. The factor of two in equation (B-90) is present because both the transmit and receive patterns have odd far field components.

C. Angular Accuracy

A number of authors[17,18,19] have dealt with the factors which limit the accuracy of an automatic tracking radar. Briefly these factors are:

(1) Thermal noise (receiver noise)

(2) External noise (glint and amplitude fluctuations from complex targets)

(3) Ground clutter

(4) Systematic errors

The effects of the first two factors are illustrated by a curve taken from page 71 of footnote 17. From the sketch below it is obvious that the thermal errors limit the maximum range of a radar. This section deals with methods of evaluating the thermal errors on a quantitative basis, and also methods of determining the maximum range for a particular antenna size, wavelength, aperture illumination, etc.

The starting point for a quantitative analysis of radar performance is the radar range equation,

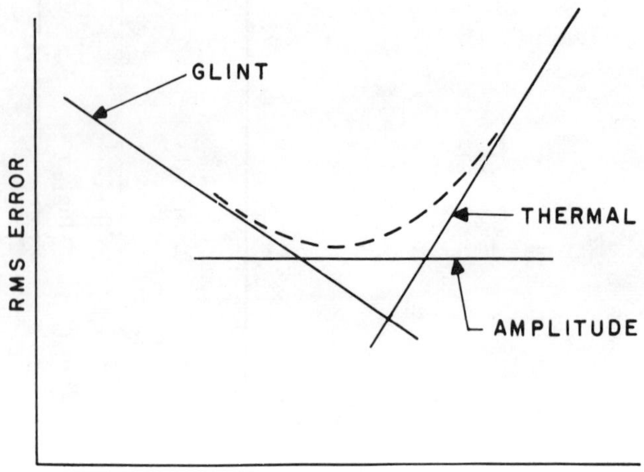

Combined Effects of Noise on Radar Angular Tracking Accuracy

$$P_r = \left(\frac{P_t\, G_t\, \eta_t\, L_t}{4\pi\, R^2}\right)\left(\frac{A_{eff}}{4\pi\, R^2}\right)\left(\frac{\lambda^2\, G_r\, \eta_r\, L_r}{4\pi}\right) \tag{C-1}$$

P_t = transmitter power output in watts

G_r, G_t = aperture gain as defined by equation (B-19)

η_r, η_t = spillover loss factor defined by equation (B-38)

L_r, L_t = product of RF plumbing and collimator loss factors

R = range to target (ft)

A_{eff} = effective target area $(ft)^2$

λ = wavelength (ft)

The subscripts r and t are used to denote the receive and transmit functions respectively. G, η, and L will in general have different values for receive and transmit.

The effective target area, A_{eff}, is defined more fully as 4π times the ratio of the power per unit solid angle scattered back toward the transmitter, to the power density in the wave incident on the target. For the limiting case ($\lambda << a$) the cross section approaches the geometrical section πa^2, where a is the radius of a conducting sphere of equivalent effective area A_{eff}.

Equation (C-1) is satisfactory to determine the radar performance when the transmitting and receiving gains result from even aperture functions. However, in a monopulse radar the direction-finding aperture function is odd so P_r is zero on the antenna axis. It is necessary to utilize the slope of the received voltage function to evaluate the direction finding properties of the antenna. First, the round trip function is

---

[17] O. H. Winn, et al, "Effects of Noise on Automatic Tracking Radars," Contract W33-038-ac-18098, General Electric Company, July 27, 1949.

[18] H. T. Budenbom, "Monopulse Radar Noise," Bell Telephone Laboratories, Inc. Reports MM-49-2750-3, also 17, Project Nike, 1949.

[19] C. E. Brockner, "Angular Jitter in Conventional Conical Scanning, Automatic Tracking Radar Systems," Proc. of IRE, p. 51, January, 1951.

$$\Delta = (P_r Z_\Delta)^{1/2} = \left(\frac{P_t Z_\Delta G_t \eta_{it} L_t}{4\pi R^2}\right)^{1/2} \left(\frac{A_{eff}}{4\pi R^2}\right)^{1/2} \left(\frac{\lambda^2 G_r \eta_{ir} L_r}{4\pi}\right)^{1/2} \quad \text{(C-2)}$$

Δ = round trip odd function, RMS voltage

Z_Δ = terminating impedance of Δ RF output

Near the antenna axis the slope of the Δ function is nearly constant so Δ can be approximated by

$$\Delta = \left(\frac{\partial \Delta}{\partial \alpha}\right) \alpha \qquad (\alpha \to 0) \qquad \text{(C-3)}$$

Many of the parameters in equation (C-2) are not a function of α. It is convenient to replace them with a constant, H. With the indicated operations on Δ performed, equation (C-3) becomes

$$\Delta = (Z_\Delta \eta_{ir} \eta_{it} L_r L_t)^{1/2} H \left[\frac{\partial G_r^{1/2}}{\partial \alpha} G_t^{1/2} + \frac{\partial G_t^{1/2}}{\partial \alpha} G_r^{1/2}\right] \alpha \qquad (\alpha \to 0) \qquad \text{(C-4)}$$

$$H = \left[\frac{P_t A_{eff} \lambda^2}{(4\pi)^3 R^4}\right]^{1/2} \qquad \text{(C-5)}$$

The transmit pattern of a monopulse antenna is an even function, consequently the derivative of G_t is zero on the antenna axis. A similar equation to (C-4) can be written for the ε axis. The derivative of the square root of the receive voltage gain is identified as K in equation (B-34). Near the antenna axis

$$\Delta(\alpha) = Z_\Delta^{1/2} H (\eta_{ir} \eta_{it} L_r L_t)^{1/2} K G^{1/2} \alpha \quad \text{or} \qquad \text{(C-6)}$$

$$\Delta(\varepsilon) = Z_\Delta^{1/2} H (\eta_{ir} \eta_{it} L_r L_t)^{1/2} K G^{1/2} \varepsilon. \qquad \text{(C-7)}$$

K = slope of odd one way voltage gain, $(\alpha, \varepsilon) = 0$

G = power gain of even one way pattern, $(\alpha, \varepsilon) = 0$

The sum signal received simultaneously, with the two difference signals, is

$$\Sigma = Z_\Sigma^{1/2} H \eta_{it} L_r^{1/2} L_t^{1/2} G \qquad \text{(C-8)}$$

Σ = round trip even function, RMS voltage

Z_Σ = terminating impedance, usually equal to Z_Δ

The principal source of thermal noise is assumed to be the frequency conversion from RF to IF. It is assumed that a noise N_Δ is added to the Δ channel and a noise N_Σ is added to the Σ channel. It is also assumed that the noise voltages N_Σ and N_Δ are uncorrelated. If balanced mixers are used this should be a satisfactory assumption. The mean square noise voltages are

$$\overline{N_\Delta^2} = \overline{N_\Sigma^2} = P_N Z\,(NF) = k\,T\,\beta\,Z\,(NF) \qquad \text{(C-9)}$$

k = Boltzman's constant, 1.38×10^{-23} joules/degree Kelvin

T = 293 degrees Kelvin

β = IF bandwidth in cps

NF = receiver noise figure

1. Derivation of Equivalent Noise Angle, α_N

The RMS values of the signal and noise voltages can be determined from equations (C-6) through (C-9). The effect of the second detector on the signal to noise ratio will be analyzed. Three cases will be considered: (1) target on antenna axis, (2) target off the antenna axis with linear ratio circuit, and (3) target off the antenna axis with limiter type of ratio circuit. The received signal is assumed to be CW, and the effect of using pulsed transmissions will be considered in sub-section 2.

a. Case I—Target on Antenna Axis, $\frac{\Sigma}{N_\Sigma} >> 1$.—In general, a ratio device will be employed ahead of the detectors in the IF channels. However, as will be shown in Cases II and III, the ratio circuit has no effect on the error signal to noise ratio when the target is on the antenna axis. The equivalent circuit for the linear detection of CW signals is shown in the sketch.

The linear detection of a CW signal plus noise is analyzed in Appendix IIIa. It is essential to recognize that the noise voltages at the two detectors are uncorrelated, or

$$\overline{(N_\Delta + N_\Sigma)(N_\Sigma - N_\Delta)} = 0 \qquad \text{(C-10)}$$

$$\overline{(N_\Sigma + N_\Delta)^2} = \overline{(N_\Sigma - N_\Delta)^2} = 2\,\overline{N_\Delta^2} \qquad \text{(C-11)}$$

The output of the subtraction circuit (see sketch) is a DC error voltage plus an RMS noise voltage.

$$V = \frac{2}{\pi}\left[\sqrt{2}\;\Delta + N_\Delta\right] \text{for} \left(\frac{\Sigma}{N_\Sigma} >> 1\right) \qquad \text{(C-12)}$$

$$V = V_{DC} + V_N \qquad \text{(C-13)}$$

where

V = RMS total error signal

V_{DC} = DC component

V_N = noise component

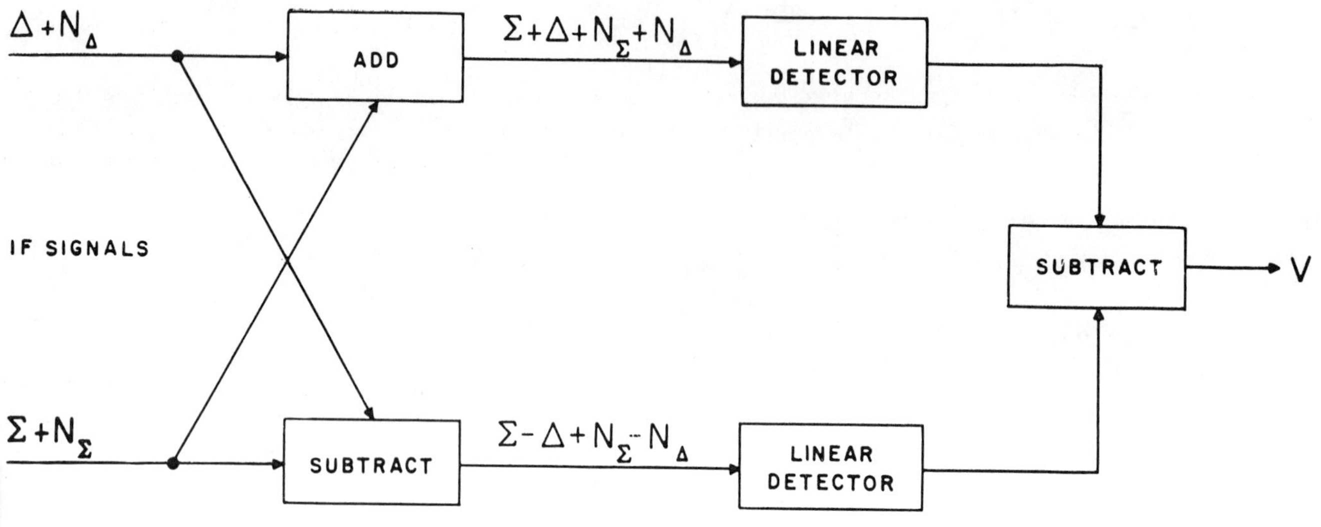

Error Detection Circuit $\left(\frac{\Sigma}{N_\Sigma} >> 1\right)$

The noise voltage has a normal distribution with a mean value of zero. The incremental probability dp of $V(t) - V_{DC}$ being in the range $V(t) - V_{DC}$ and $V(t) - V_{DC} + dV(t)$ is

$$dp = \frac{d\,V(t)}{[2\,\pi\,V_N{}^2]^{1/2}}\,e^{-\frac{[V(t) - V_{DC}]^2}{2\,V_N{}^2}} \qquad (C\text{-}14)$$

V(t) = instantaneous values of (C-12)

$$V_N = \left(\frac{2}{\pi}\right)[kT\,\beta\,Z\,(NF)]^{1/2} \qquad (C\text{-}15)$$

$$V_{DC} = \frac{2}{\pi}\sqrt{2}\;H\,(Z\,\eta_r\,\eta_t\,L_r\,L_t)^{1/2}\,KG^{1/2}\,\alpha \quad (C\text{-}16)$$

Equation (C-14) is included to emphasize the random nature of the noise voltage V_N. The noise voltage (C-15) can be specified in terms of an equivalent error angle α_N by equating (C-15) and (C-16).

$$\alpha_N = \frac{[k\,T\,\beta\,(NF)]^{1/2}}{\sqrt{2}\;H\,(\eta_r\,\eta_t\,L_r\,L_t)^{1/2}\,KG^{1/2}} \qquad \text{for } (\Delta = 0) \qquad (C\text{-}17)$$

The angle α_N then has the random characteristics of the noise described by equation (C-14). On any one measurement of α, if all the parameters of (C-17) are known, the probability of the error exceeding a certain amount can be determined by integrating (C-14). If a large number of measurements are taken, the standard deviation is given by (C-17). For a stationary target α can be determined as closely as desired by averaging a large number of measurements.

For uncorrelated errors on two axes the total angular error is

$$\xi = [\alpha_N{}^2 + \varepsilon_N{}^2]^{1/2} \qquad (C\text{-}18)$$

b. Case II—Target Off Antenna Axis—Linear Ratio Circuit—$\frac{\Sigma}{N_\Sigma} >> 1$.—If the target is off the antenna axis and the system parameters in H remain fixed, the angular errors will increase. The increased errors result from a decrease in gain, G, of the even far field patterns and the decrease in slope, K, of the odd far field function. Both K and G appear in the denominator of equation (C-17).

In order to eliminate the variables, range and target area, in the ECS voltage it is necessary to use a ratio circuit. If a linear ratio circuit is used (see equation (B-52)) the ECS is

$$ECS = \frac{\Delta + N_\Delta}{\Sigma + N_\Sigma} \qquad (C\text{-}19)$$

If a ratio circuit performs the operation, indicated by equation (C-19), at IF frequencies then no further detection is necessary. While no such ratio circuit is available, the analysis will be based upon the use of the hypothetical ratio device. This will indicate the optimum performance.

Each noise voltage, N_Σ and N_Δ, can be considered as the sum of two voltages in time quadrature as described in Appendix IIIa. The IF ratio is

$$ECS = \frac{(\sqrt{2}\;\Delta + N_{\Delta x})\sin\omega t + N_{\Delta y}\cos\omega t}{(\sqrt{2}\;\Sigma - N_{\Sigma x})\sin\omega t + N_{\Sigma y}\cos\omega t} \quad (C\text{-}20)$$

For the assumed condition $\left(\frac{\Sigma}{N_\Sigma} >> 1\right)$ the quadrature term $N_{\Sigma y}$ will have little effect upon the denominator. The mean value of the cos ωt/sin ωt term is zero so it will also be dropped. The DC ratio remaining is

$$ECS = \frac{\sqrt{2}\;\Delta + N_{\Delta x}}{\sqrt{2}\;\Sigma\left[1 + \frac{N_{\Sigma x}}{\sqrt{2}\;\Sigma}\right]} \qquad (C\text{-}21)$$

By dividing the bracketed term in the denominator into one, and then retaining only the first order term, the noise voltages appear only in the numerator. With first order terms retained the ECS is

$$ECS = \frac{\Delta}{\Sigma} + \frac{\partial\left(\frac{\Delta}{\Sigma}\right) d\alpha}{\partial\alpha} + \frac{N_{\Delta x}}{\sqrt{2}\;\Sigma}\left[1 + \left(\frac{\Delta}{\Sigma}\right)^2\right]^{1/2} \qquad (C\text{-}22)$$

The incremental value of Δ/Σ in equation (C-22) is included to allow determining the equivalent noise angle, α_N, corresponding to the noise voltage term in (C-22). The equivalent noise angle, α_N, is

$$\alpha_N = \frac{[k\,T\,\beta\,(NF)]^{1/2}\left[1 + \left(\frac{\Delta}{\Sigma}\right)^2\right]^{1/2}}{\sqrt{2}\left[\frac{\partial\Delta}{\partial\alpha} - \left(\frac{\Delta}{\Sigma}\right)\frac{\partial\Sigma}{\partial\alpha}\right]} \qquad (C\text{-}23)$$

for $\Delta \geqq 0$.

The evaluation of the partial derivatives in equation (C-23) requires that the basic definition of gain, equation (B-19) be used to determine G_r and G_t. This is a straightforward problem though it may involve some calculation.

When $\Delta \rightarrow 0$, the equivalent noise angle α_N is the same if computed by (C-23) as by (C-17).

c. Case III—Target Off Antenna Axis—Limiter Type of Ratio Circuit—$\frac{\Sigma}{N_\Sigma} >> 1$.—A method of introducing an instantaneous ratio action is to limit the vector sum of $\Sigma \pm j\,\Delta$. A block diagram of such a ratio circuit is shown in the sketch.

Consider first the input signal to one of the limiters. Since the limiter acts upon the absolute magnitude of the signal, the signal is considered as a magnitude at a phase angle.

$$[(\Sigma + N_\Sigma)^2 + (\Delta + N_\Delta)^2]^{1/2}\;\underline{/\psi + \psi_N} \qquad (C\text{-}24)$$

$$\psi + \psi_N = \tan^{-1}\frac{\Delta + N_\Delta}{\Sigma + N_\Sigma} \qquad (C\text{-}25)$$

After limiting the signal has a constant amplitude and a phase angle given by equation (C-25). If the constant amplitude factor is omitted the signal after the upper limiter is

$$\frac{(\Sigma + N_\Sigma) + j\,(\Delta + N_\Delta)}{[(\Sigma + N_\Sigma)^2 + (\Delta + N_\Delta)^2]^{1/2}} \qquad (C\text{-}26)$$

Thus the signal is unchanged by the limiter except for the normalizing term in the denominaior. equation (C-26). The output of the lower limiter differs by a negative sign on the quadrature term. The linear combination immediately following the limiters separates the in-phase and quadrature terms of the upper and the lower limiter outputs. After a 90 degree phase shift the second linear combination adds and subtracts the normalized sum and difference signals which are now in-phase. The amplifier is added to the sum channel to keep the sum signal larger than the difference signal. If the sum and difference signals are equal then one linear detector functions on noise only, and the analysis of the detector action is more complex.

As in Case II, only the in-phase components of N_Δ and N_Σ are retained at the detectors. The difference of the linear detector outputs is

$$ECS \doteq \frac{\sqrt{2}\;\Delta + N_\Delta\,\Sigma\,[\Sigma^2 + \Delta^2]^{-1/2}}{[\Sigma^2 + \Delta^2]^{1/2}}. \qquad (C\text{-}27)$$

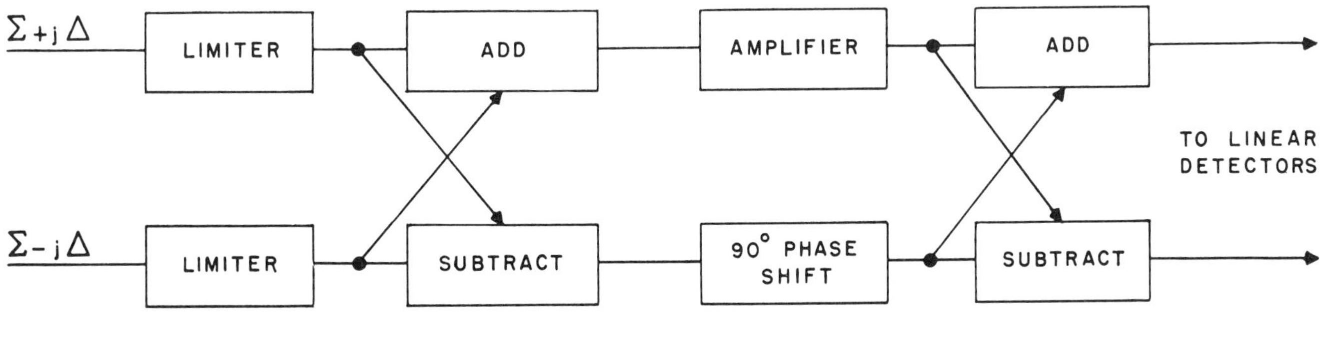

Limiter Type of Ratio Circuit

To obtain the equivalent angular error, α_N, the derivative of $\sqrt{2}\ \Delta/[\Sigma^2 + \Delta^2]^{1/2}$ is used as the slope of the ECS curve at the angle α.

The result is

$$\alpha_N = \frac{[k\,T\,\beta\,(NF)]^{1/2}\,\Sigma}{\sqrt{2}\left[(\Sigma^2+\Delta^2)^{1/2}\dfrac{\partial\Delta}{\partial\alpha} - \Delta\,\dfrac{\partial(\Delta^2+\Sigma^2)^{1/2}}{\partial\alpha}\right]} \tag{C-28}$$

for $\Delta \geqq 0$.

As in Case II if $\Delta \rightarrow 0$, then α_N is the same as equation (C-17).

The application of (C-28) to practical problems may not be as complicated as equation (C-28) might indicate. For example, with the aid of Table III, some possible functional forms of Δ and Σ can be made up as follows:

$$\Delta = k_2\, f^2\,(\alpha)\,\sin\alpha/4\,\cos\alpha/4 \tag{C-29}$$

$$\Sigma = k_2\, f^2\,(\alpha)\,\cos^2\alpha/4 \tag{C-30}$$

These linear phase Δ and Σ functions can be used in equation (C-28). The result is

$$\alpha_N = \frac{[k\,T\,\beta\,(NF)]^{1/2}\,4}{\sqrt{2}\ \ k_3\, f^2\,(\alpha)\,\cos\alpha/4} \tag{C-31}$$

for $(\alpha \geqq 0)$.

By using a desired function of α for $f^2(\alpha)$, the variation of α_N with α can be determined. Since equations (C-31) and (C-17) are equal when $\alpha \rightarrow 0$, the value of k_3 can be determined by equating (C-17) and (C-31).

In Section D the similarity of the aperture functions for approximately linear phase to overlapped illuminations for linear phase is pointed out. Thus if equations for overlapped illuminations from Table III are used instead of those for no overlap, equations (C-29) and (C-30), the results for α_N should apply to approximately linear phase aperture functions over the region of the beam where $\dfrac{\Sigma}{N_\Sigma} >> 1$.

2. Effect of Integration of Signal Pulses

Thus far the analysis of the effect of noise on angular accuracy has dealt with the linear detection of CW signals. In anticipation of this discussion of pulsed radar the equivalent noise angle, α_N, for CW radar is derived on a round trip basis in subsection Cb. For pulse operation the IF bandwidth, β, is made as wide as is commensurate with the pulse length, τ. A general rule is

$$\beta\,\tau > 2 \tag{C-32}$$

With the IF bandwidth, β, large enough to handle pulses, the equations for α_N predict the effect of thermal noise on the amplitude of the detected pulses. The distribution of the pulse amplitudes follows the normal law, equation (C-14), if $\left(\dfrac{\Sigma}{N_\Sigma} >> 1\right)$.

In a tracking radar, a range gate and a narrow filter subsequent to the second detector remove as much of the noise as is consistent with the ability of the radar to follow accelerations of the target. If the error signal pulses are used as video pulses to improve a visual display through "beam sharpening," a certain amount of averaging of the angular information can be done by the observer of a radar indicator.

Through an analysis, Appendix IV, the improvement in signal to noise ratio of the tracking error signal by a narrow filter is determined. For large signal to noise ratios at the second detector the improvement in S/N is

$$\text{maximum improvement} = \left[\frac{f_r}{2\,\beta_n}\right]^{1/2} \quad \text{(monopulse)} \tag{C-33}$$

f_r = pulse repetition frequency, pps

β_n = bandwidth of narrow filter, cps

The maximum improvement in db, is

$$\text{db} = 20\log_{10}\left[\frac{f_r}{2\,\beta_n}\right]^{1/2} \tag{C-34}$$

To illustrate the magnitude of this improvement take $f_r = 2400$ and $\beta_n = 2.5$ then

$$\text{db} = 26.9 \tag{C-35}$$

Unfortunately the full amount of this improvement is not available when it would be the most valuable. If the signal to noise ratio at the second detector is low, then the improvement decreases markedly. At the present time tracking radars are designed to operate at sum channel (Σ) signal to noise ratios well above unity at the second detector to obtain satisfactory performance. An idea of the decrease in performance at low signal to noise ratio can be determined from equation (8) in Appendix IV. The following values were calculated from this approximate equation and are probably optimistic.

Voltage $\left(\dfrac{\Sigma}{N_\Sigma}\right)$	db $\left[\text{subtract from } 10\log_{10}\left(\dfrac{f_r}{2\,\beta_n}\right)\right]$.
5	$\doteq$ 0.0
1	− 3.9
1/2	− 7.4
1/4	−12.7
1/8	−18.5

Loss in Second Detector at Low S/N

It is possible to avoid part or nearly all of the loss at signal to noise ratios greater than one-half if coherent doppler techniques can be employed. If the target being tracked is moving radially relative to the radar, then a doppler shift of the received signal frequency results. By making use of a coherent oscillator, audio frequencies are obtained which can be filtered with narrow filters before the actual second detector action. Thus more efficient filtering results. Coherent doppler circuits for a tracking radar are described in the Quarterly Engineering Reports on "Anti-Clutter Techniques," Signal Corps Contract D.A. 36-039-sc-5446.

3. Angular Accuracy Calculations for Tracking Radar

A useful concept in the calculation of angular accuracy is the "ideal" tracking radar. The "ideal" radar has an antenna with a gain G_o and a slope K_o, a unity noise figure receiver, and achieves the signal integration indicated by equation (C-33). From equation (C-17) the equivalent angular error for the azimuth axis is

$$\alpha_N = \frac{[k\,T\,\beta\,(NF)]^{1/2}}{H\,(2\,\eta_r\,\eta_t\,L_r\,L_t)^{1/2}\,KG^{1/2}} = \frac{N}{\sqrt{2}\,\frac{\partial\Delta}{\partial\alpha}} \qquad \text{(C-36)}$$

From earlier sections the values of these parameters for the "ideal" radar are

$$H = \left[\frac{P_t\,A_{eff}\,\lambda^2}{(4\,\pi)^3\,R^4}\right]^{1/2}$$

$\eta_r, \eta_t, L_r, L_t = 1$ for "ideal" radar.

$$G_o = \frac{4\,\pi\,hd}{\lambda^2} = \text{maximum value of } G.$$

$$K_o = \left(\frac{G_o}{12}\right)^{1/2} = \text{maximum value of } K, \text{ and}$$

$$\alpha_N = \frac{2\,\pi\,d}{\lambda}\,(\theta_{az})_N. \qquad \text{(C-37)}$$

The improvement from filtering is given by equation (C-33). The solution of (C-36) for $(\theta_{az})_N$, the one axis angular error is

$$(\theta_{az})_N = \left[\frac{kT\,\beta\,\beta_n\,12}{P_t\,A_{eff}\,f_r\,\pi}\right]^{1/2}\frac{\lambda^2\,R^2}{hd^2} \qquad \text{(C-38)}$$

where $\left(\frac{\Sigma}{N_\Sigma} >> 1\right)$.

Equation (C-38) gives the angular error of the "ideal" radar. For design purposes it is desirable to know the minimum transmitted power required as a function of the various radar parameters. The minimum transmitted power for the "ideal" radar for a particular RMS error, $(\theta_{az})_N$ is

$$(P_t)_{min} = \frac{kT\,\beta\,\beta_n\,12\,\lambda^4\,R^4}{(\theta_{az})^2\,A_{eff}\,\pi\,h^2\,d^4\,f_r} \qquad \text{(C-39)}$$

The minimum power must be increased to account for the actual antenna gain and slope factor, spillover losses, collimator and RF plumbing losses, and the receiver noise figure. If the sum channel signal to noise ratio is not appreciably greater than one, then the full improvement from filtering is not realized as discussed in sub-section C2.

It is convenient to make up a family of curves for tracking radars which differ by —10, —20, —30, etc. db from a particular ideal radar. Since all the parameters in equation (C-39) are in the form of quotients or products, logarithms are particularly applicable to the problem of scaling to new parameters. For a particular range (R = 89,700 units) the equation for $(P_t)_{min}$ reduces to

$$(P_t)_{min} = \left.\frac{\beta\,\lambda^4\,\beta_n}{A_{eff}\,(\theta_{az})_N^{\,2}\,h^2\,d^4\,f_r}\right|_{R\,=\,89{,}700} \qquad \text{(C-40)}$$

if $\left(\frac{\Sigma}{N_\Sigma} >> 1\right)$.

where λ, A_{eff}, h, d, and R are in a consistent set of units.

The slope of the log $(P_t)_{min}$ versus log R curve is 40 db/decade so it is necessary to know only one point to plot the function. This is illustrated in the sketch on page 29.

A curve for a particular set of parameter values is given in Figure 18. A sample calculation sheet, Figure 19, shows that for new parameter values the minimum power can be quickly determined with the aid of Figure 18.

Reference to equation (C-36) shows that equation (C-39) must be modified as follows to account for actual antenna factors.

$$(P_t)_{actual} = (P_t)_{min}\left(\frac{G_o}{G}\right)\left(\frac{K_o}{K}\right)^2\left(\frac{1}{\eta_r\,\eta_t\,L_r\,L_t}\right) \qquad \text{(C-41)}$$

a. Ratio of $(\theta_{az})_N$ to One Way Beamwidth.—The equivalent noise angle, α_N, can also be expressed as a function of the sum channel signal to noise ratio. From equation (C-8) the RMS sum signal is

$$\Sigma = Z_\Sigma^{\,1/2}\,H\,\eta_t\,L_r^{\,1/2}\,L_t^{\,1/2}\,G \qquad \text{(C-42)}$$

From (C-17) and (C-33) the equivalent noise angle is

$$\alpha_N = \frac{1}{H\,K}\left[\frac{kT\,\beta\,\beta_n\,(NF)}{G\,\eta_t\,\eta_r\,L_t\,L_r\,f_r}\right]^{1/2} \qquad \text{(C-43)}$$

if $\left(\frac{\Sigma}{N_\Sigma} >> 1\right)$.

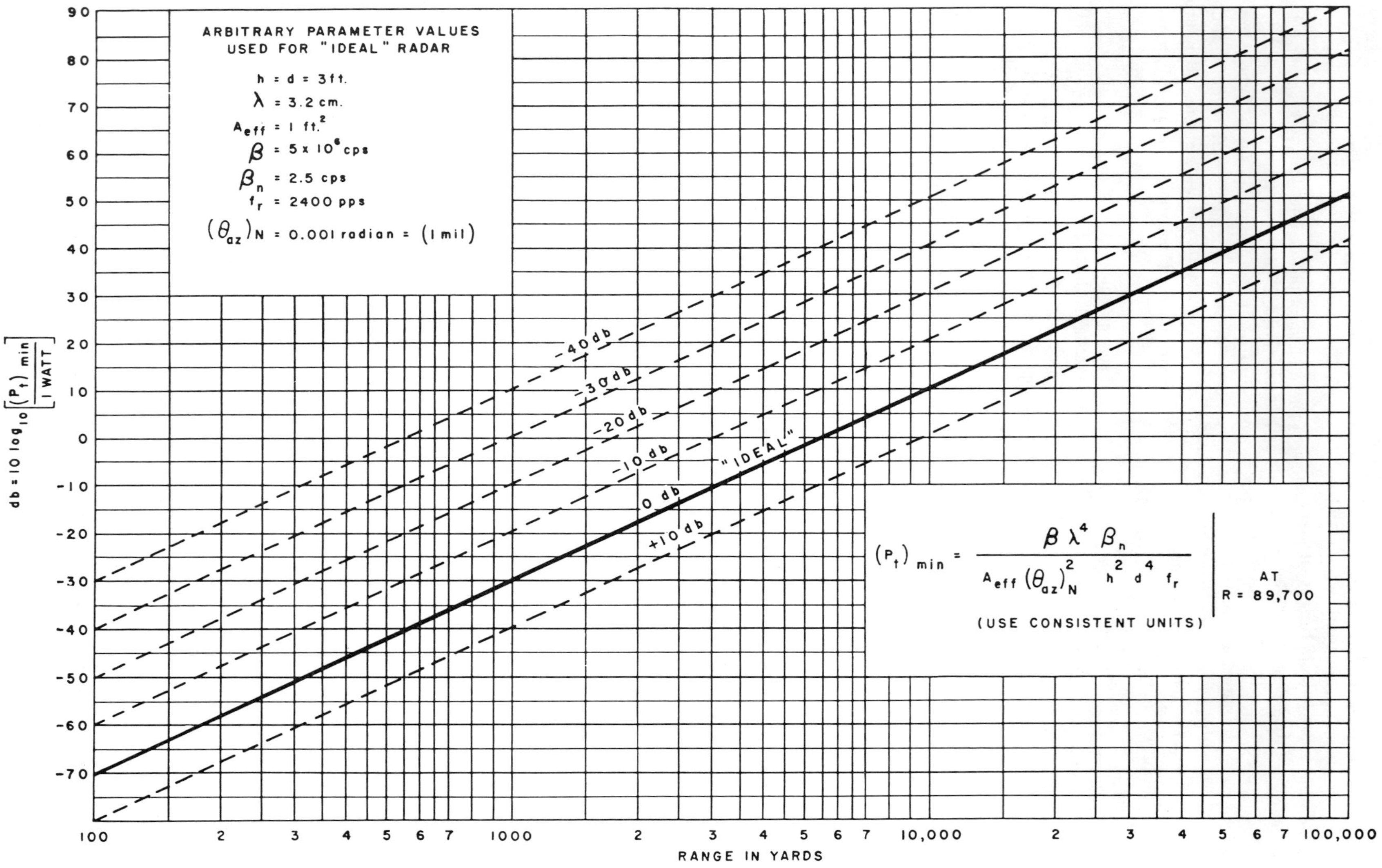

Figure 18. Minimum Peak Power for 1 Angular Mil Accuracy (RMS Error on Azimuth Axis).

Problem

Given a radar with the following parameter values, determine the minimum power required to track with 0.1 angular mil RMS error (one axis) at a range of 15,000 yards.

Method of Solution

Determine the db change of each radar parameter relative to the corresponding parameter value of the radar set described in Figure 18. Then by addition of the decibel changes determine the total change in performance and use Figure 18 to obtain the db above 1 watt of transmitter power required.

Parameter	Symbol	Figure 18 Value	New Value	db change
Angular Accuracy	$(\theta_{az})_N$	1 mil	0.1 mil	—20 db
Target Area	A_{eff}	1 ft^2	0.01 ft^2	—20
Wavelength	λ	3.2 cm.	3.2 cm.	0.0
IF Bandwidth	β	$5x10^6$ cps	$5x10^6$ cps	0.0
Aperture Height	h	3 ft.	6 ft.	+ 6.0
Aperture Width	d	3 ft.	6 ft.	+12.0
Antenna Gain	G_t	G_o	11.4 db taper	— 0.84
Antenna Slope	K	K_o	linear	— 0.42
Spillover Factor	γ_{it}	1	0.812	— 0.90
Spillover Factor	γ_{ir}	1	0.452	— 3.45
RF Plumbing Losses	L_r, L_t	1	0.945	— 0.25
Receiver Noise Figure	(NF)	1	15.81	—12.00
Repetition Frequency	f_r	2400 pps	2,400	0.0
Servo Bandwidth	β_n	2.5 cps	2.5 cps	0.0
			TOTAL	—39.86

Answer

From Figure 18, at a range of 15,000 yards, the power required for a radar with a performance of —39.9 db relative to the "ideal" curve is 57.5 db greater than 1 watt, or

$$P_{actual} = \log^{-1}\frac{57.5}{10} = 560,000 \text{ watts if } \left(\frac{\Sigma}{N_\Sigma} >> 1\right).$$

A check on the Σ/N_Σ ratio using equations (C-8) and (C-9) or equation (C-45) gives

$$\frac{\Sigma}{N_\Sigma} = 4, \text{ so } \left(\frac{\Sigma}{N_\Sigma} > 1\right).$$

Thus a peak power of 0.56 megawatts would be the minimum power with which the desired performance can be obtained. Fluctuations in target area and other unpredictable effects make it desirable to have an excess of two to four times the minimum power.

Figure 19. Sample Calculation of Peak Transmitter Power.

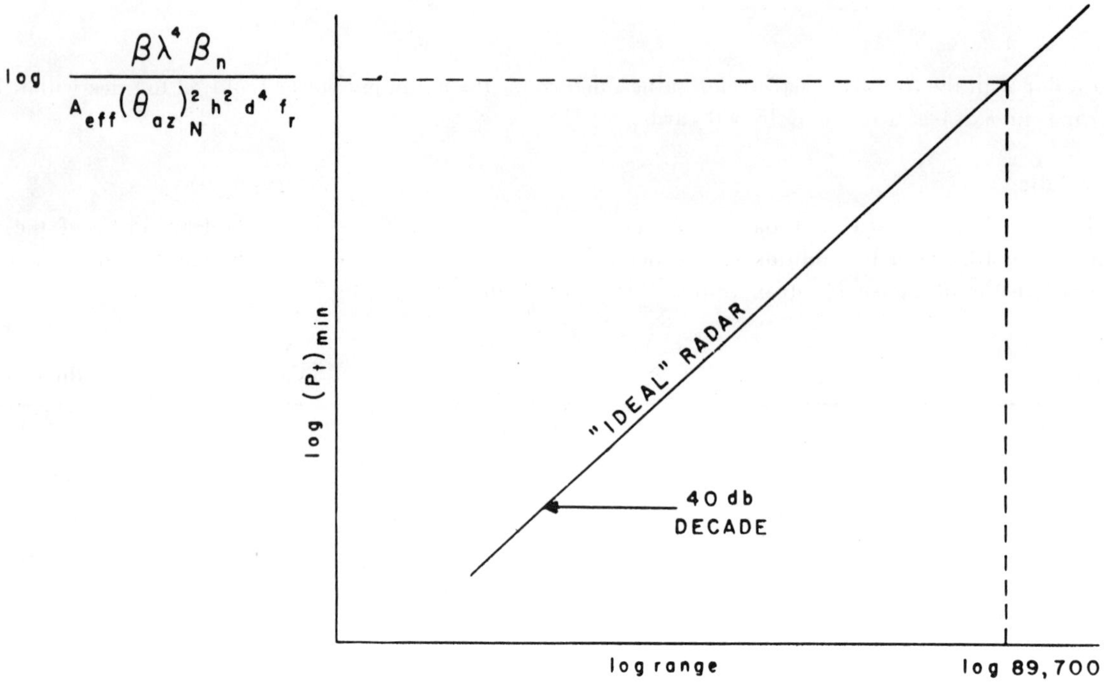

Log (P_t) Minimum Versus Log R (Round-Trip)

Certain parameters in equation (C-43) can be evaluated in terms of (C-42).

$$\alpha_N = \frac{N_\Sigma}{\Sigma}\left[\frac{\eta_{it}\, G\, \beta_n}{\eta_{ir}\, f_r}\right]^{1/2} \frac{1}{K} \qquad \text{if } \left(\frac{\Sigma}{N_\Sigma} >> 1\right). \qquad \text{(C-44)}$$

Σ = RMS sum signal at IF

N_Σ = RMS noise at IF

Since the azimuth beamwidth is a function of the ratio λ/d the angular error can be expressed as a fraction of the one-way beamwidth for $\frac{\Sigma}{N_\Sigma} >> 1$ as

$$\frac{(\theta_{az})_N}{BW} = \frac{N_\Sigma}{\Sigma} \frac{1}{2\pi K k_3}\left[\frac{\eta_{it}\, G\, \beta_n}{\eta_{ir}\, f_r}\right]^{1/2} \qquad \text{(C-45)}$$

$$BW = \frac{k_3 \lambda}{d} = \text{one way beamwidth of even far field in radians.} \qquad \text{(C-46)}$$

k_3 = a constant dependent upon A_e (x)

$k_3 = 0.88$ for $A_e(x) = 1$

$k_3 = 1.2$ for $A_e(x) = \cos \frac{\pi x}{d}$

In case the target signal is near the minimum allowable Σ/N_Σ ratio, it may be desirable to have some device attached to the radar receiver to indicate when the tracking errors are likely to exceed allowable limits. Since all the parameters in equation (C-45) except N_Σ/Σ are fixed, this ratio is a measure of tracking accuracy. A device attached to the radar which will reflect as nearly as possible the ratio of the RMS signal to RMS noise would help in evaluating tracking data. Since the noise is a statistical quantity, the ratio of the RMS quantities cannot be determined exactly except for a fixed target.

4. Angular Accuracy of Conical Scan Radar

In sub-section B 4 *c* a hypothetical sequential lobing antenna is described. This antenna has a square aperture, for which the $KG^{1/2}$ angular sensitivity factor is given in Appendix II for three aperture illuminations. The $KG^{1/2}$ factor for the sequential lobing antenna is the same as the $K_a G^{1/2}$ factor for the combination phase-amplitude antenna, Table VI. In Appendix V, the S/N ratio of a sequential lobing error signal is determined. Since the lobing frequency was quite low ($f_r >> \omega_m/2\pi$), the analysis can be considered to apply to the conical scan form of sequential lobing antenna.

The error signal to noise ratio of a conical scan radar is 3 db better than the amplitude axis of an otherwise equivalent combination phase-amplitude radar. The relative error signal to noise ratio of several radars is compared in tabular form on the following page.

Radar	Even Aperture Illumination	Relative Error Signal to Noise Ratios
"Ideal" Monopulse	Uniform	0
Symmetrical Monopulse	11.4 db taper	— 5.6 db
Conical Scan	11.4 db taper (squinted)	--8.6 db
Combination Monopulse	11.4 db taper (squinted)	—11.6 db

Relative Error Signal to Noise Ratios

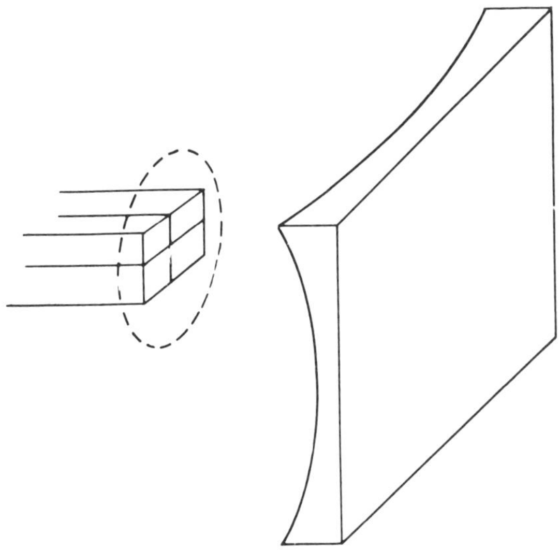

Monopulse Antenna

In making up the above tabulation, the only factors considered were K, G, and η from Tables V and VI and the analysis of detector action in Appendices IV and V.

D. Monopulse Radar Components

The design of monopulse radars requires specialized circuits which are not described in the literature. This section will discuss components and techniques which are applicable to monopulse design.

1. Antenna Feeds

Instead of attempting to discuss all possible monopulse antenna feeds, the four horn feed will be examined in some detail.

The portion of the sketch within the dotted lines is the four horn feed. RF combination devices can be added subsequent to the feed horns to obtain desired linear combinations of the RF signals. An analysis of a one axis, two horn feed will be used to illustrate how the far field produced by each horn can be separated into even and odd components.

The effect of moving a single feed in the focal plane of a collimator is to move the beam in the far field. This can be considered to be mathematically equivalent to tilting the phase front in the aperture plane of the collimator. Two adjacent feed horns set symmetrically about a line normal to the secondary aperture will produce "squinted" beams.

The aperture functions are considered to be the aperture function for the on-axis feed with a linear phase multiplier. It is assumed that mutual coupling between the feeds is negligible, though in sub-section E 1 *d* a method of analyzing mutual coupling effects is described.

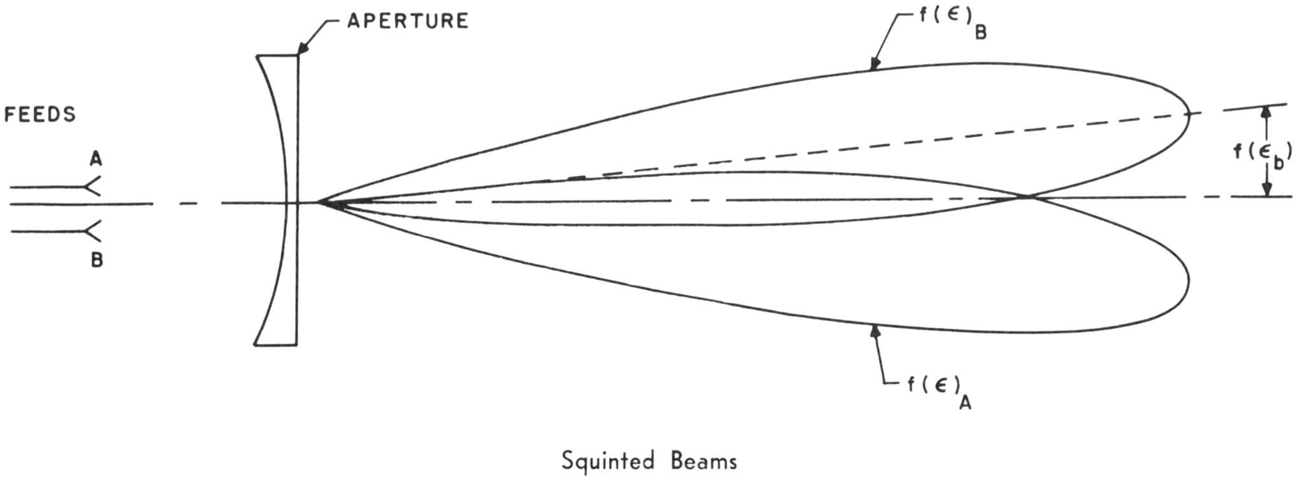

Squinted Beams

$$A(y)_A = A(y)\, e^{j\frac{\varepsilon_b y}{h}} \qquad \text{(D-1)}$$

$$A(y)_B = A(y)\, e^{-j\frac{\varepsilon_b y}{h}} \qquad \text{(D-2)}$$

$A(y)_A$, $A(y)_B$ = aperture field produced by each feed

$A(y)$ = aperture field of one feed for $\varepsilon_b = 0$

$\varepsilon_b = \dfrac{2\pi h \theta_{el}}{\lambda}$ = normalized squint angle

Exponential-trignometric relationship can be used to expand the phase front tilt term in (D-1) and (D-2).

$$A(y)_A = A(y)\left[\cos\frac{\varepsilon_b y}{h} + j\sin\frac{\varepsilon_b y}{h}\right] \qquad \text{(D-3)}$$

$$A(y)_B = A(y)\left[\cos\frac{\varepsilon_b y}{h} - j\sin\frac{\varepsilon_b y}{h}\right] \qquad \text{(D-4)}$$

$A(y)$ will usually be an even function so $A(y)\cos\frac{\varepsilon_b y}{h}$ is also even and $A(y)\sin\frac{\varepsilon_b y}{h}$ is an odd aperture function. For an example assume $A(y)$ is a uniform illumination and $\varepsilon_b = \pi$. The aperture functions are as shown in the sketch.

RF devices for giving linear combinations of the fa field signals will be described.

a. Feeds for Approximately Linear Phase.—For ap proximately linear phase the odd far field must be in time quadrature with the even far field.

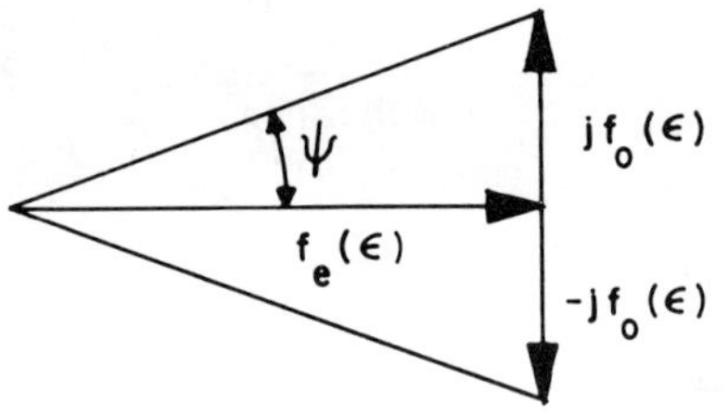

Odd and Even Functions Added in Quadrature

However, in equation (D-7) the odd terms are in-phase with the even terms. A straightforward way to produce the desired phase shift of the odd terms is to use RF sum and difference devices to separate the odd from the even terms. After a relative phase shift of 90 degrees, a second linear recombination will give the desired signals. This is illustrated by a sketch for a one-axis case (page 71).

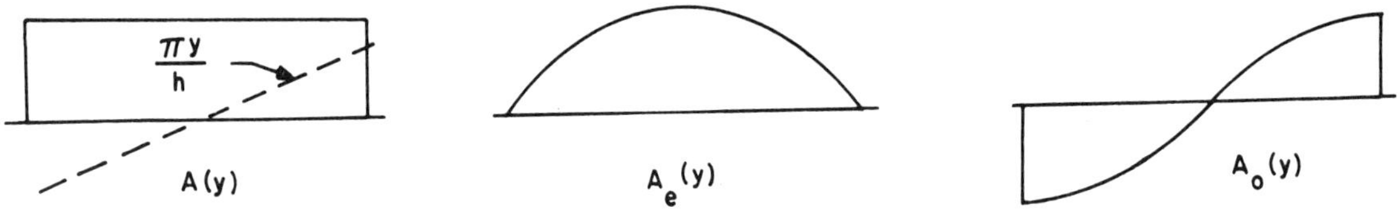

Aperture Functions For One Horn of Four Horn Feed

The Fourier transforms of (D-3) and (D-4) will produce two component far fields.

The indicated linear combinations of the signal and the 90 degree phase shift can take place at RF and/or IF.

$$\mathcal{F}\,[A(y)_A] = \mathcal{F}\,[A_e(y) + jA_o(y)] = f(\varepsilon)_A = f_e(\varepsilon) - f_o(\varepsilon) \qquad \text{(D-5)}$$

$$\mathcal{F}\,[A(y)_B] = \mathcal{F}\,[A_e(y) - jA_o(y)] = f(\varepsilon)_B = f_e(\varepsilon) + f_o(\varepsilon) \qquad \text{(D-6)}$$

The two component far fields of equations (D-5) and (D-6) can be utilized in a variety of ways. Because of the difference of sign of the odd terms in (D-5) and (D-6), linear combinations of the far field signals enable the odd far field to be separated from the even. The analysis of the two horn feed can be generalized and used for the four horn case. By analogy for the four feeds the far fields are

$$\left.\begin{aligned}
f(\varepsilon,\alpha)_A &= [f_e(\varepsilon) + f_o(\varepsilon)]\,[f_e(\alpha) + f_o(\alpha)],\\
f(\varepsilon,\alpha)_B &= [f_e(\varepsilon) + f_o(\varepsilon)]\,[f_e(\alpha) - f_o(\alpha)],\\
f(\varepsilon,\alpha)_C &= [f_e(\varepsilon) - f_o(\varepsilon)]\,[f_e(\alpha) + f_o(\alpha)], \text{ and}\\
f(\varepsilon,\alpha)_D &= [f_e(\varepsilon) - f_o(\varepsilon)]\,[f_e(\alpha) - f_o(\alpha)].
\end{aligned}\right\} \quad \text{(D-7)}$$

The same method of signal combination can be extended to two axes.

Another method of obtaining the required 90 degree phase relation between the even and odd far field components is to use three db directional couplers. In a two axis application, four directional couplers are required as shown in the sketch on page 71.

The hybrid couplers have the property of coupling one-half the energy from one guide to another, but shifting the phase of the coupled signal by 90 degrees. The signals E_A, E_B, E_C, and E_D are the far field signals given by equation (D-7).

The assumption, the mutual coupling between the feeds is negligible, allows an analysis of the hybrid coupled

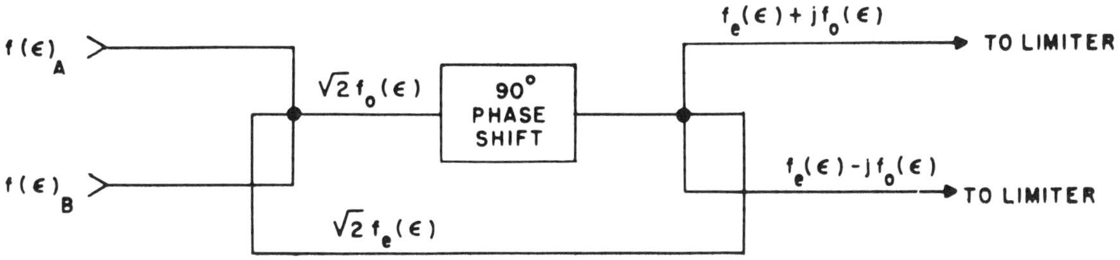

Source of Approximately Linear Phase Signals

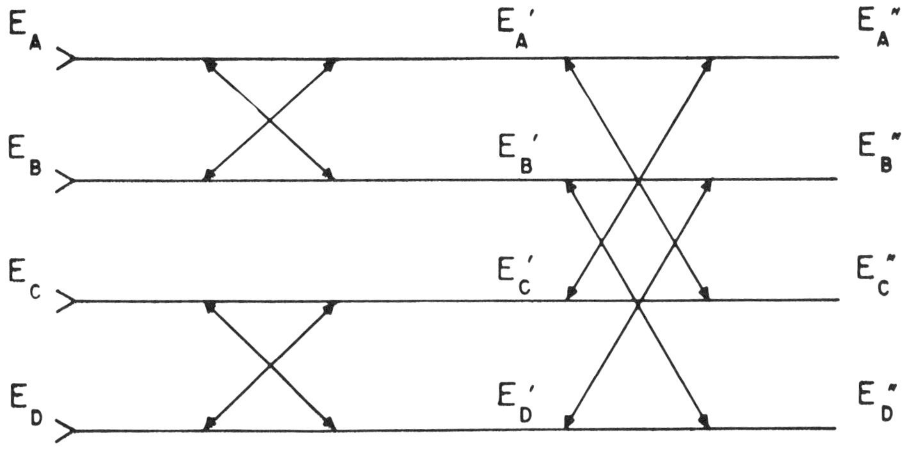

Hybrid Coupled Feed

feed. The transmit pattern is not included as it is a multiplier of both even and odd terms and will cancel from the ultimate ratio.

The far field equation (D-7) can be simplified by the substitutions

$$E_l = \frac{f_o(\varepsilon)}{f_e(\varepsilon)} \qquad \text{(D-8)}$$

$$A_z = \frac{f_o(\alpha)}{f_e(\alpha)} \qquad \text{(D-9)}$$

The signals after the first set of directional couplers adding E_A to E_B and E_C to E_D are,

$$\begin{aligned}
E_A' &\approx (1 + E_l)\ (1 + A_z) - j\ (1 + E_l)\ (1 - A_z) \\
E_B' &\approx (1 + E_l)\ (1 - A_z) - j\ (1 + E_l)\ (1 + A_z) \\
E_C' &\approx (1 - E_l)\ (1 + A_z) - j\ (1 - E_l)\ (1 - A_z) \\
E_D' &\approx (1 - E_l)\ (1 - A_z) - j\ (1 - E_l)\ (1 + A_z)
\end{aligned} \qquad \text{(D-10)}$$

The signals after the second set of directional couplers are,

$$\begin{aligned}
E_A'' &\approx (E_l + A_z) - j(1 - E_l A_z) \\
E_B'' &\approx (E_l - A_z) - j(1 + E_l A_z) \\
E_C'' &\approx (-E_l + A_z) - j(1 + E_l A_z) \\
E_D'' &\approx (-E_l - A_z) - j(1 - E_l A_z)
\end{aligned} \qquad \text{(D-11)}$$

A calculation of the magnitude of the double primed quantities shows them to be equal. Thus mixers and IF amplifiers following these hybrid couplers should be phase stable since limiters can be used to equalize the signal amplitudes. After limiting, the quantity E_l may be separated by adding E_A'' to E_B'' and E_C'' to E_D'', with the result,

$$\begin{aligned}
V_1 &\approx (E_l - j1) \\
V_2 &\approx (-E_l - j1)
\end{aligned} \qquad \text{(D-12)}$$

The quantity A_z may be separated by adding E_A'' with E_C'' and E_B'' with E_D''.

$$\begin{aligned}
V_3 &\approx (A_z - j1) \\
V_4 &\approx (-A_z - j1)
\end{aligned} \qquad \text{(D-13)}$$

The target position information is obtained from the phase comparison of these voltages. The phase angle between V_1 and V_2 is,

$$2\,\psi_{el} = \tan^{-1}\left(\frac{-1}{E_l}\right) - \tan^{-1}\left(\frac{-1}{-E_l}\right)$$

$$2\,\psi_{el} = \tan^{-1}\frac{2E_l}{1 - E_l^2} \qquad \text{(D-14)}$$

The phase angle between V_3 and V_4 similarly is,

$$2\,\psi_{az} = \tan^{-1}\frac{2A_z}{1 - A_z^2} \qquad \text{(D-15)}$$

Equations (D-14) and (D-15) may be expressed in a more familiar form using the double angle relationship,

$$2\,\psi_{el} = 2\tan^{-1} E_l = 2\tan^{-1}\frac{f_o(\varepsilon)}{f_e(\varepsilon)}$$

$$2\,\psi_{az} = 2\tan^{-1} A_z = 2\tan^{-1}\frac{f_o(\alpha)}{f_e(\alpha)} \qquad \text{(D-16)}$$

Thus from equation (D-16) it is seen that ψ_{el} and ψ_{az} are proportional to twice the inverse tangent of the delta/sum ratio in elevation and azimuth.

The effect of an unbalance in the power split in the three db directional couplers is analyzed in Appendix VI.

b. Use of Higher Order Mode Feeds.—Analysis of the monopulse antenna in terms of even and odd far field components suggests that it is possible to replace the four horn feed with a feed which utilizes higher order modes. A first step would be to use a second mode on one axis. The front view of the feeds would be altered as shown in the sketch.

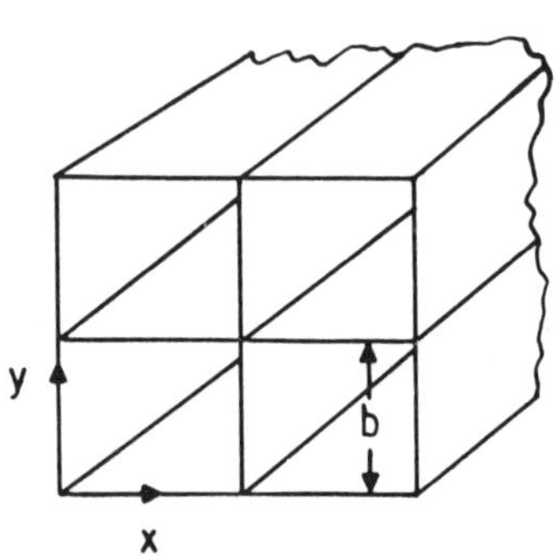

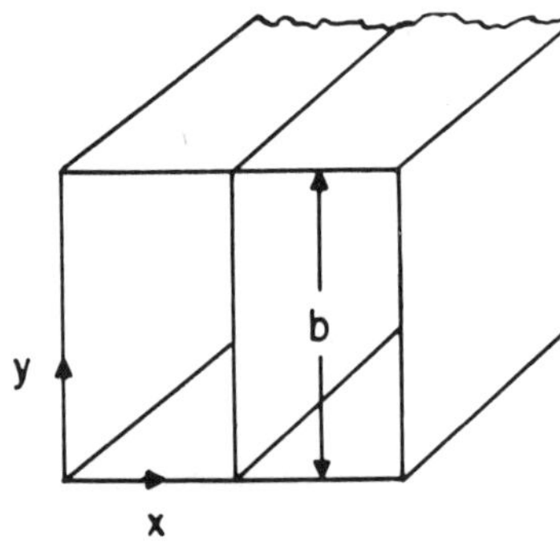

Antenna Feeds

The usual mode utilized in the four horn feed is the TM_{01}, and the electric field across the mouth of one waveguide feed is

$$E_x = k_1 \sin\frac{\pi y}{b} \qquad \text{(D-17)}$$

where E_x = x component of electric field

k_1 = proportionality constant

b = height of waveguide

By widening the waveguide in the y direction, a second mode, TM_{02}, can be used in addition to TM_{01}. The total E_x field is

$$E_x = k_1 \sin\frac{\pi y}{b} + k_2 \sin\frac{2\pi y}{b} \qquad \text{(D-18)}$$

The TM_{02} mode gives rise to an odd far field and the TM_{01} to an even far field along the y axis. The duo-mode exciter is shown in the sketch below.[20]

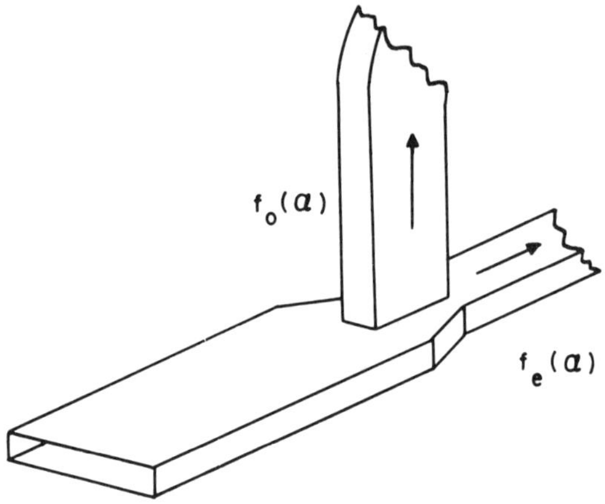

Duo-mode Exciter

[20]"Investigations of Multi-Mode Propagation in Waveguides and Microwave Optics," Northwestern University, Evanston, Illinois, Army Signal Corps Contract W36-039-sc-38240.

A duo-mode exciter enables both modes to be used with independent control of the relative phase of the two modes.

Another variation of the four horn feed is to use a horn capable of supporting dual modes in both the x and y directions.

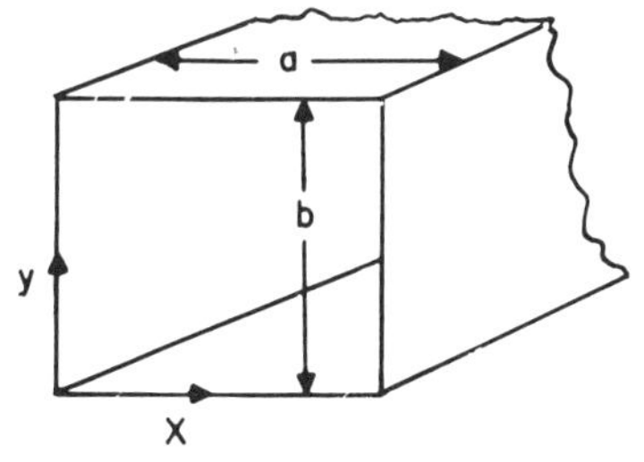

Square Horn For Higher Modes

The variation of the electric field with x and y for both TM_{mn} and TE_{mn} waves is

$$E_x = k_1 \cos \frac{m \pi x}{a} \sin \frac{n \pi y}{b} \qquad \text{(D-19)}$$

$$E_y = k_2 \sin \frac{m \pi x}{a} \cos \frac{n \pi y}{b} \qquad \text{(D-20)}$$

$$\lambda_{co} = \frac{2\,ab}{\sqrt{n^2a^2 + m^2b^2}} \qquad \text{(D-21)}$$

No modes involving zero subscripts are possible for TM_{mn} waves since this leads to $E_z = 0$. The lowest order waves which appear feasible are

Mode	Transverse Electric Field
TE_{01}	$E_x = k_1 \sin \frac{\pi y}{b}$ (D-22)
TE_{02}	$E_x = k_1 \sin \frac{2 \pi y}{b}$ (D-23)
TE_{10}	$E_y = k_1 \sin \frac{\pi x}{a}$ (D-24)
TE_{20}	$E_y = k_1 \sin \frac{2 \pi x}{a}$ (D-25)

Electric Field Components

These equations, (D-22) through (D-25), show that it is necessary to have both E_x and E_y components of electric field to obtain information in two axes with a square horn. On transmission the power must be split between the TE_{01} and TE_{10} modes. There are at least three ways in which this power split between the modes could be arranged: (a) on a sequential lobing basis, alternately producing an E_x or E_y electric field, (b) split the power equally between TE_{01} and TE_{10} modes with E_x and E_y in phase, and (c) split the power equally between TE_{01} and TE_{10} modes with E_x and E_y in time quadrature. The third method would produce a circularly polarized transmitted field. The necessity of using both E_x and E_y components in the far field implies that a 3 db decrease in the $KG^{1/2}$ angular sensitivity factor for each axis would result.

c. Comparison of Overlapped and Four Horn Aperture Illuminations.—In Table III, Aperture Functions which Give Linear Phase, the even and odd aperture functions for a "cosine overlapped" illumination are sketched. Without overlap ($r = 0$) the even far field pattern has high sidelobes (—8 db one way); with overlap the sidelobes are smaller and the phase angle, ψ, is less for a given normalized angle, α, in the far field. Since a smaller ψ angle for a given α increases the angle that can be covered in the far field without ambiguity, this is a valuable characteristic of overlapped illuminations. While overlapped illuminations are desirable, no entirely satisfactory method of producing the exact $A_e(x)$ and $A_o(x)$ are available. If interlaced line feeds are used, the same slots which must radiate most strongly on the even pattern, $f_e(\alpha)$, will be in a region where cancellation must occur on $f_o(\alpha)$. One method of partially overcoming this difficulty, through the use of a resonant line feed structure, was recently described in the literature.[21] A resonant structure is frequency sensitive.

Comparison of the even and odd aperture functions for overlapped illuminations, Table III, and the even and odd aperture functions for four horn feeds, Table IV, reveals a marked similarity. The major difference is the relative phase of the even and odd aperture and far field patterns. The even and odd far fields of the overlapped illuminations are in time quadrature, while the even and odd far fields for the four horn feed are in-phase. This difference can be altered by a 90 degree phase shift in the circuits. Sketches of the overlapped and four horn aperture functions are reproduced here to show the similarities.

---

[21] C. B. Watts, Jr., "Simultaneous Radiation of Odd and Even Patterns by a Linear Array," Proceedings of IRE, October, 1952, pp. 1236-1239.

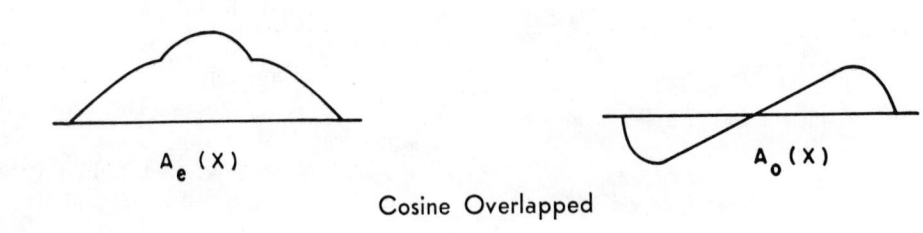

Cosine Overlapped

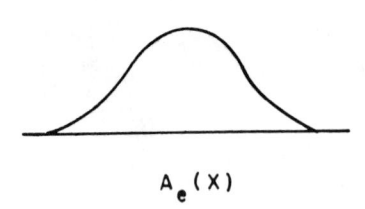

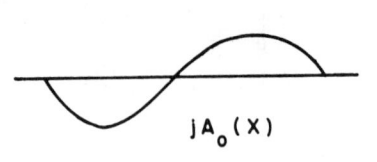

One Axis of Four Horn Feed

d. Coupling Between Horn Feeds.—While an exact solution of coupling effects is a difficult analytical problem, an AC circuit analogy to the RF circuits can be used to give a physical picture of coupled feeds. The analogy will be on a one axis basis though it could be extended to two axes. If an RF hybrid device is used to obtain the sum and difference of the feed horn signals, the RF schematic is as shown in the sketch.

A possible analogy to the RF circuit would be the following circuit.

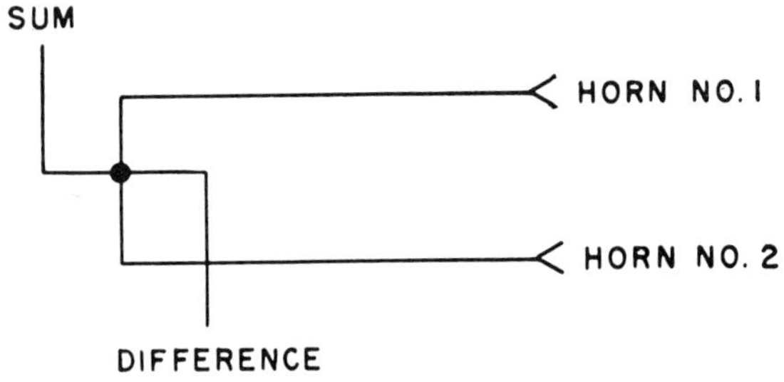

RF Sum and Difference Circuit

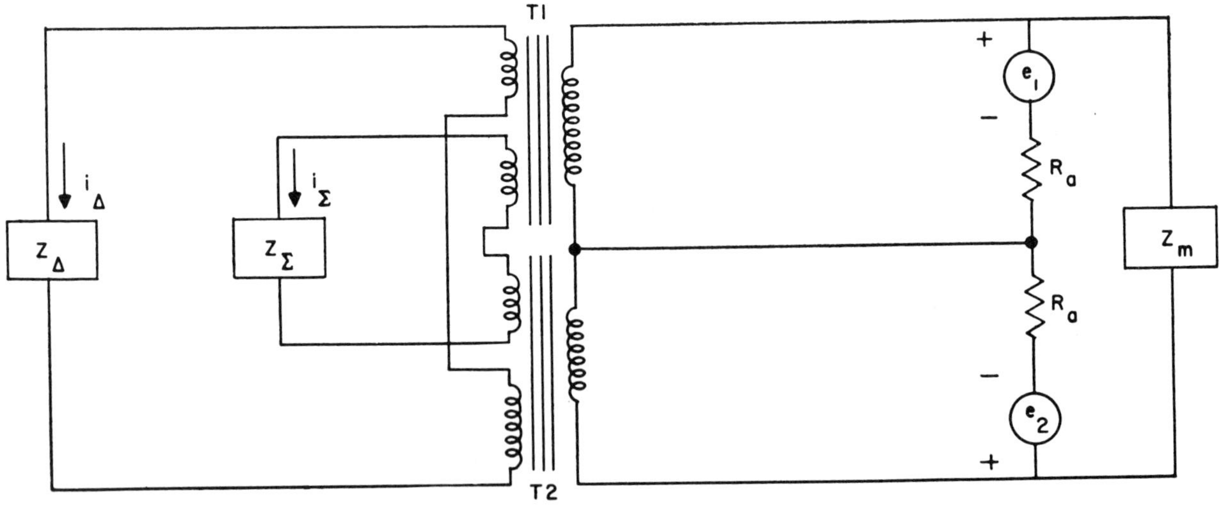

AC Circuit Analogy to Coupled Feeds

T_1, T_2 = perfect transformers (unity coupling and $X_p \rightarrow \infty$)

R_a = radiation resistance

e_1 = received signal in horn #1

e_2 = received signal in horn #2

Z_m = mutual impedance to represent coupling

By AC circuit analysis the current in the load impedances Z_Δ and Z_Σ is

$$i_\Sigma = \frac{e_1 + e_2}{Z_\Sigma + 2 R_a}, \text{ and} \qquad \text{(D-26)}$$

$$i_\Delta = \frac{e_1 - e_2}{\frac{2 R_a Z_\Delta}{Z_m} + Z_\Delta + 2 R_a}. \qquad \text{(D-27)}$$

Equation (D-26) shows that the mutual coupling has no effect on the sum current. In the difference current (D-27), however, both the phase and magnitude of the current are affected by the mutual, Z_m. The form of equation (D-27) is such that the effect of the coupling term can be nullified by making

$$Z_\Delta = \left(2 R_a - \frac{2 R_a Z_\Delta}{Z_m}\right) \qquad \text{(D-28)}$$

Solving equation (D-28) for Z_Δ gives

$$Z_\Delta = \frac{2 R_a Z_m}{Z_m + 2 R_a} \qquad \text{(D-29)}$$

The lack of dissipative elements at the feed horns would indicate that the coupling would be of a reactive nature, and that it is possible to eliminate the coupling effect by the use of a reactive load, equation (D-29). This simplified analysis indicates that (1) the use of small horns which have mutual coupling will not decrease the K factor of the feeds if the difference channel load is properly matched. (2) The presence of mutual coupling can be detected by comparing the input impedance of the sum and difference arms. (3) The sum and difference signals will not be in-phase if there is mutual coupling unless the difference channel is properly matched.

2. Ratio Circuits

A ratio circuit eliminates range and target size from the radar error signal. A ratio circuit should have the following characteristics:

a. operate over a wide dynamic range of input signals
b. maintain the "sense" of the error signals

Several ratio circuits which possess these characteristics are tabulated below.

The operation of each type of ratio circuit will be described.

a. IF Limiters.—The IF limiter gives a particular kind of ratio action. This ratio circuit is best suited for use with monopulse antennas which give linear phase far field functions. If the input signal to the limiter is

$$(\Sigma \pm j\,\Delta)\, e^{j\omega t} \qquad \text{(D-30)}$$

The output of the limiter is

$$k_1 \underline{/\omega t + \psi} \qquad \text{(D-31)}$$

where k_1 = amplitude of limiter output—assume unit amplitude unless specified and,

$$\psi = \tan^{-1}\left(\frac{\pm\Delta}{\Sigma}\right). \qquad \text{(D-32)}$$

Expressed in ratio form the output of the limiter is

$$k_1 \frac{\Sigma \pm j\,\Delta}{|\Sigma^2 + \Delta^2|^{1/2}}\, e^{j\omega t} \qquad \text{(D-33)}$$

An ideal limiter would give a constant output for all input levels. Practical single stage limiter circuits have limiting characteristics as illustrated by the sketch on page 76.

Ratio Circuit	Variations of Basic Ratio Circuit	Response Time
IF Limiters	1. cascaded 6BN6 limiters 2. cascaded pentode limiters 3. cascaded diode limiters	nearly instantaneous
Electronic Dividers	1. Fixed reference signal type a. AGC ratio circuit b. Delay Line AGC c. IAGC	 several pulses two pulses fraction of a pulse width
	2. Δ reference signal type a. FM-AM divider b. Tube characteristic	 from fraction of a pulse to several pulses depending upon design
Logarithmic	1. Log IF amplifiers	nearly instantaneous

Ratio Circuit Characteristics

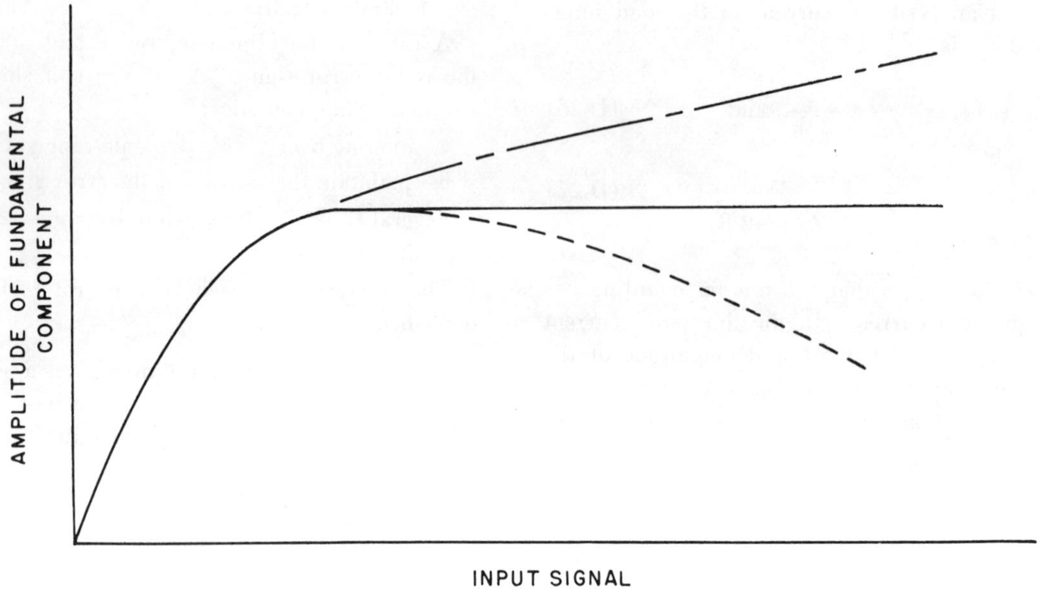

Limiter Characteristics

The solid line in the sketch shows the characteristic of a properly adjusted limiter. The dotted line is characteristic of the limiting action of an overdriven pentode. The limiting action does not begin until the input signal reaches an appreciable amplitude of perhaps 1 or 2 volts, and it is not practical to obtain an input signal greater than perhaps 15 db over the minimum limiting signal. It is necessary to use cascaded limiter stages to cover a wide dynamic range of input signals.

Since IF limiters are usually used in pairs, it is important that the relative phase shift through the pair of cascaded limiters remain constant with variations in input signal amplitude. Two possible sources of unpredictable relative phase shifts are coupling through grid to plate capacity and harmonics in the limiter outputs.

The coupling through the grid to plate capacity of a limiter stage will not introduce errors if it is balanced by a similar coupling in the corresponding stage of the other cascaded limiter. For large input signals the grid-plate capacity current can become an appreciable fraction of the total plate current. The first order output voltage from a limiter is

$$V = Z_l(i_p - ji_c) \qquad \text{(D-34)}$$

Z_l is the load impedance, i_p the electron conduction component of plate current, and i_c is the grid-plate capacity current.

If the electron current, i_p, is constant, the capacity current, i_c, will change the phase of the output signal by the angle

$$\gamma = \tan^{-1}\frac{i_c}{i_p} \qquad \text{(D-35)}$$

If the capacity current becomes a large enough fraction of the conduction current, the output voltage of the limiter stage will be altered. A rising limiter output

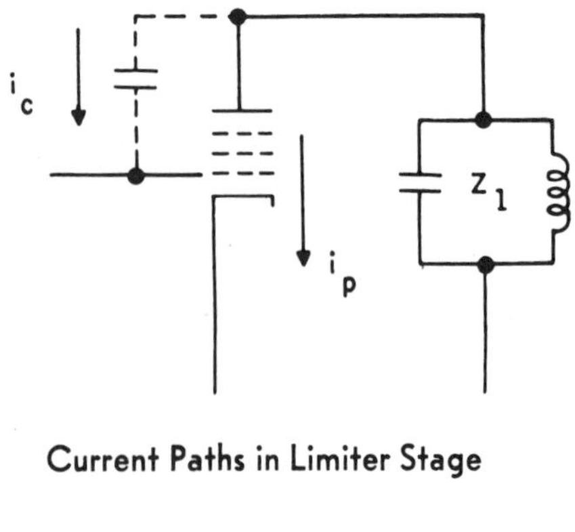

Current Paths in Limiter Stage

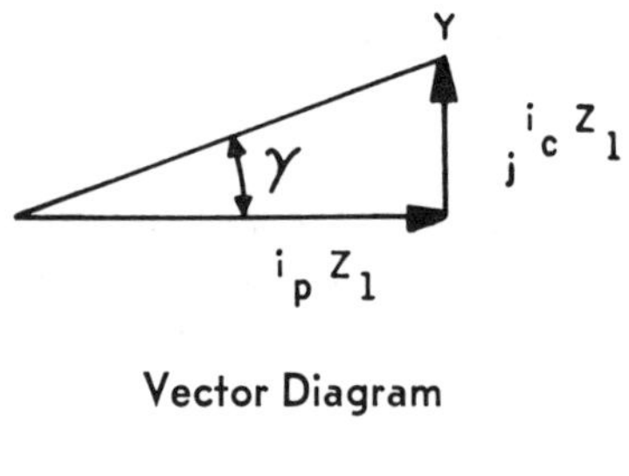

Vector Diagram

curve such as shown by the dash-dot line in the sketch of a limiter characteristic can result from excess grid-plate capacity.

A sine wave which is clipped on one or both peaks contains a considerable amount of harmonics. A symmetrically clipped wave has only odd harmonics, and the third harmonic of a square wave has an amplitude one-third of the fundamental amplitude. Unsymmetrical clipping will result in even harmonics. The second harmonic, if present, is likely to cause difficulty in cascaded limiters. The second harmonic generated by one limiter will add to the fundamental at the input to a succeeding limiter. The non-linear action of the second limiter results in the generation of an additional fundamental component of signal in the second limiter output. It is this new component at the fundamental frequency which may cause an undesired phase shift of the resultant fundamental signal.

The harmonics in the plate current of a limiter stage are minimized by tuned circuits. If the limiter load impedance is a parallel tuned circuit, 30 megacycles center frequency, a total capacity of 15 micromicrofarads, and a Q of 4 the third harmonic output voltage is small. For example, for a square plate current pulse the third harmonic is 3.1% of the fundamental output. A double tuned circuit would give an even further reduction of harmonic output.

b. Electronic Dividers.—The best known application of an electronic divider circuit is AGC, which keeps the output of a receiver constant. If the AGC voltage is applied to a second identical amplifier channel, the output of the second amplifier will be divided by the input signal to the first channel. The sketch shows the block diagram of an AGC ratio circuit.

The ratio action can be described briefly by the following equations:

$$\mu\,|\Sigma| \doteq v_s \qquad \text{(D-36)}$$

μ = gain of sum IF amplifier

v_s = reference voltage amplitude

The AGC voltage controls the gain, μ, of the identical IF amplifiers. The output of the Δ IF amplifier is

$$\mu\,\Delta \doteq v_s\,\frac{\Delta}{\Sigma} \qquad \text{(D-37)}$$

The ratio action is analyzed more exactly in Appendix II of the fifth quarterly report. The analysis of the AGC ratio circuit differs from the analysis of a linear feedback circuit. The feedback signal, the AGC voltage, is not linearly superimposed upon the input signal, rather it multiplies the input signal. The result of the analysis is subject to these assumptions:

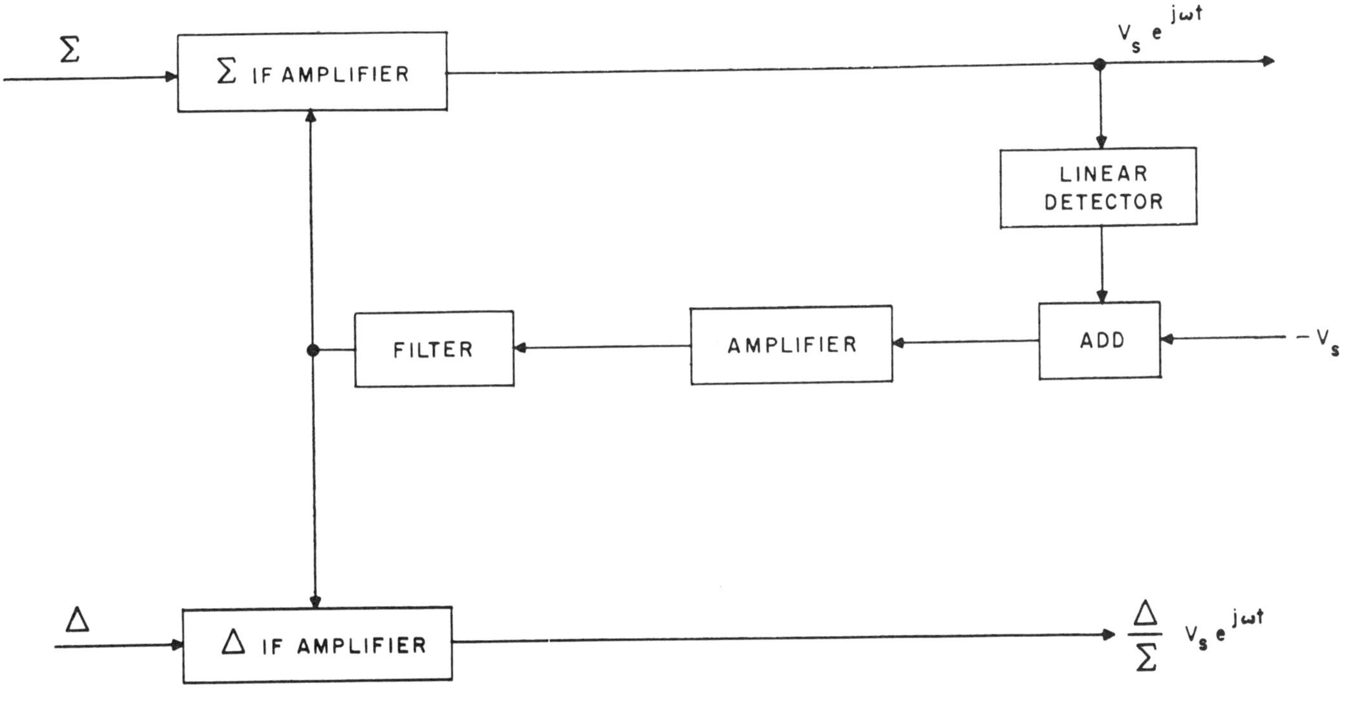

AGC Ratio Circuit

(1) Only one time constant exists in the AGC loop, and the transfer constant is $\frac{k_3}{1 + j\omega T}$.

(2) The pulse repetition frequency is much larger than the reciprocal of T, or $T >> \frac{1}{f_r}$.

(3) The sum and difference signals have the same waveform.

(4) The AC component of the input signal fluctuation is small enough so that the slope, k_2, of the IF gain control curve is constant over the region of interest.

The sum channel output waveform and transient response using the above assumptions is

$$V_\Sigma = \frac{\mu_o + k_2 k_3 v_s}{T} [k + g_i(t)] \frac{1}{1 - \exp\left[-\frac{2\pi}{T\omega}(1 + k_2 k_3 k)\right]} \times$$

$$\int_0^{\frac{2\pi}{\omega}} \exp\left[-\frac{(1 + k_2 k_3 k)v}{T} - \frac{k_2 k_3}{T}\int_{t-v}^{t} g_i(x)\,dx\right] dv. \qquad \text{(D-38)}$$

where V_Σ = amplitude of Σ channel output
k = the steady state input
$g_i(t)$ = the modulation component of the input
T = time constant of the AGC filter
v = variable of integration

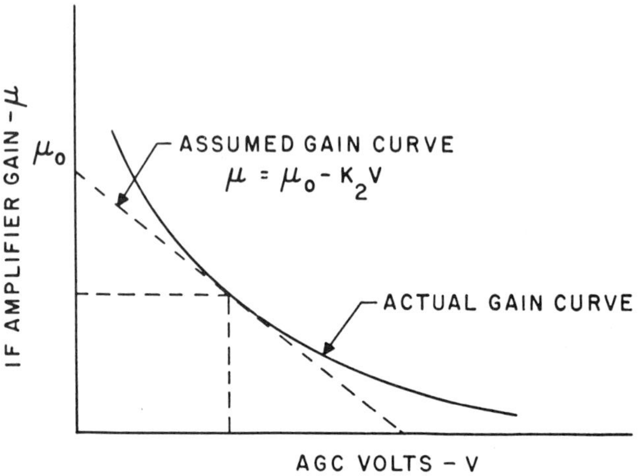

Definition of μ_o and k_2

Equation (D-38) is a general form from which the response to a particular input function can be determined. If $g_i(t) = 0$, the output voltage, equation (D-38) becomes

$$V_\Sigma = \frac{\mu_o + k_2 k_3 v_s}{1 + k_2 k_3 k} k \qquad \text{(D-39)}$$

The design objective is to make

$$k_2 k_3 k >> 1 \text{ or } \mu_o \qquad \text{(D-40)}$$

so that the output will be V_s as in the simplified analysis (D-36). The difference between a linear feedback problem and a ratio circuit analysis is brought out by equation (D-39). The input signal, k, appears in the denominator of the transfer ratio, (D-39), whereas in a linear feedback problem the return ratio is independent of the input signal. It is desirable to adjust the gain curve of the IF amplifier so that the product $k_3 k_2 k$ is nearly constant over the range of input signals.

As just described, the AGC ratio circuit is suitable for use with a tracking radar. If the filter circuit is replaced with a delay line integrator, the AGC ratio circuit can be used with a scanning radar. The action of a delay line AGC circuit is described in Appendix VII.

The delay line AGC ratio circuit has not been compared thoroughly with the instantaneous type of ratio circuit such as the limiter. However, the following observations will be made: (1) in the presence of rapid fluctuations in signal from an external source, the instantaneous circuit will give better results; (2) with internal noise the AGC ratio circuit may give more accurate results as it averages the Σ channel signal before computing the ratio.

The usual external sources of rapid signal fluctuations are rapid scanning and propeller modulation. The internal noise would be thermal noise, and differs in the two channels.

Δ *Reference Dividers.*—Examination of the AGC ratio circuit sketch indicates that replacing the fixed reference voltage V_s with the Δ signal will result in another type of ratio circuit. A block diagram of the Δ reference divider is shown on page 79.

A simplified analysis indicates that the output is

$$V_o \doteq \frac{\Delta}{\Sigma} \frac{1}{k_2}. \qquad \text{(D-41)}$$

The significant factor in equation (D-41) is k_2. k_2 was previously defined as the slope of the IF gain curve, and shown to be dependent upon the input signal Σ. For

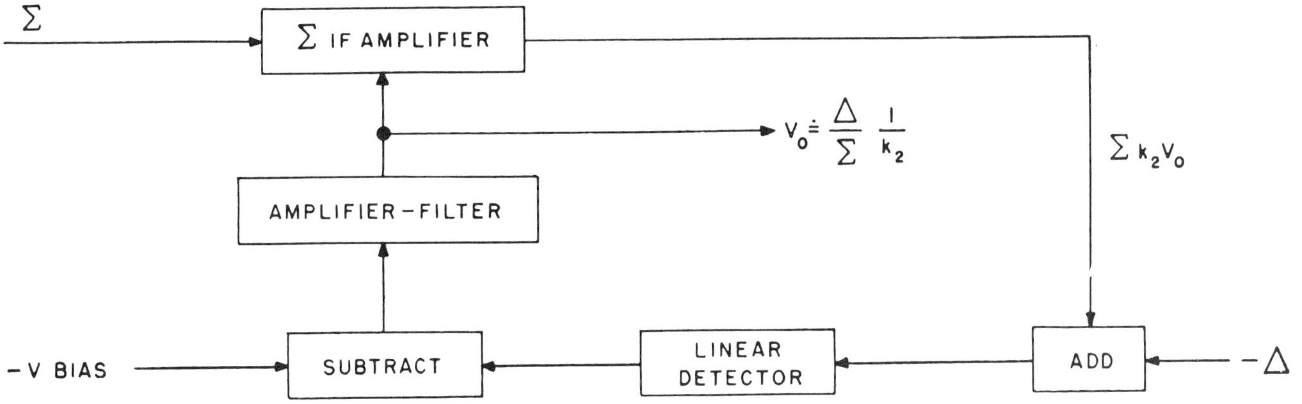

Δ Reference Divider

accurate results k_2 must be a constant for all input signal levels. This indicates that the Σ IF amplifier must be replaced by a multiplier circuit, i.e., a circuit which keeps k_2 constant over a wide operating range. There are many possible variations of this type of divider circuit though all require a multiplier capable of operating through a wide dynamic range. Multiplier circuits capable of operating over a limited dynamic range are the FM-AM multiplier[22] and the tube characteristic multiplier.

c. Logarithmic Ratio Circuits.—In subsection B2*c*(1), it is shown that the logarithm of two squinted Gaussian far field patterns is linear. A practical method of obtaining the logarithm of each antenna signal is to use an IF amplifier with a log characteristic over the required dynamic range. A log IF amplifier design with wide dynamic range is utilized in the SPS-2 multiple-beam height-finding radar.

3. Detectors

Following the ratio circuits it is customary to convert the IF error signal to DC or video pulses for tracking or indicator purposes respectively. Some form of linear recombination of the signal output of the ratio circuits may be necessary before the signals are linearly or quadratically detected. Some of the possible linear combination. detector arrangements are:

a. phase sensitive detector

b. 90 degree phase detector

c. extended range phase detector

d. extended linearity phase detector

The circuit techniques used to effect linear combinations of the signals prior to detection can vary considerably. Representative methods will be given here and others will occur to the reader.

a. Phase Sensitive Detector.—This might also be called the basic "error sensing" detector. The block diagram is self-explanatory.

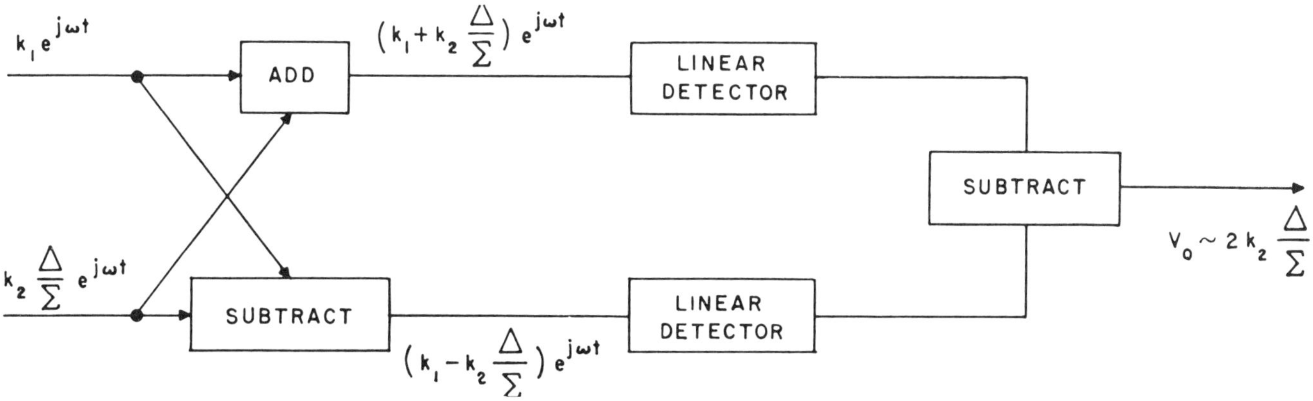

Phase Sensitive Detector

[22] "Interim Development Report for Radar Set AN/SPS-2," Contract NObsr 49018. General Electric Company.

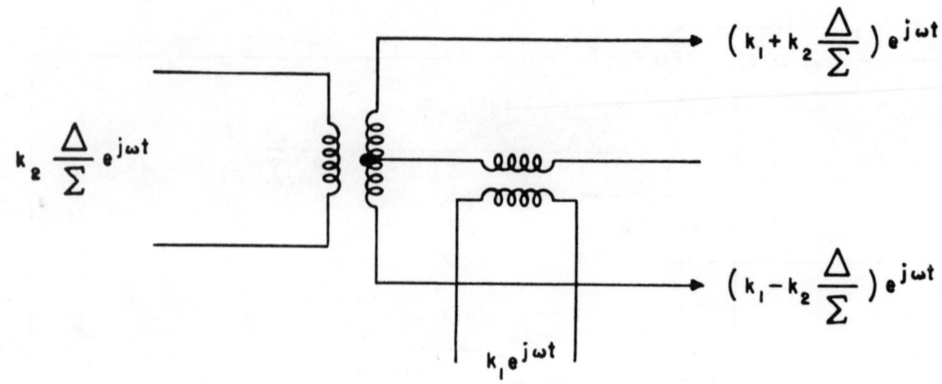

Add and Subtract Circuit

The IF add and subtract circuits must be designed with due consideration for amplitude and phase response for pulses. A possible method of obtaining the IF addition and subtraction is to use a transformer.

b. 90 Degree Phase Detector.—An ideal phase detector circuit would produce an output which is linearly proportional to the phase angle between the input signals. Also the output would be independent of the magnitude of either of the input signals. The phase detectors described here do not meet either of these requirements. If a limiter type of ratio circuit is used ahead of the phase detector, the input amplitude will remain constant.

If the ratio action has been produced by a limiter, the Σ and Δ signals are still combined in time quadrature. The Σ and Δ signals can be separated by a linear network. The output characteristic will depend upon the functional form of Δ/Σ.

From the vector diagram, page 81

$$\frac{\Delta}{D} = \frac{\Delta}{(\Sigma^2 + \Delta^2)^{1/2}} = \sin\psi \qquad \text{(D-42)}$$

From the section on ECS signals, equation (B-43), for linear phase far field signals.

$$\psi = \tan^{-1}\frac{\Delta}{\Sigma} = p\,\alpha \qquad \text{(D-43)}$$

From equations (D-42) and (D-43) the phase detector characteristic is sketched (page 42).

c. Extended Range Phase Detector.—If the gain, μ, in the sum channel is much greater than one, the 90 degree

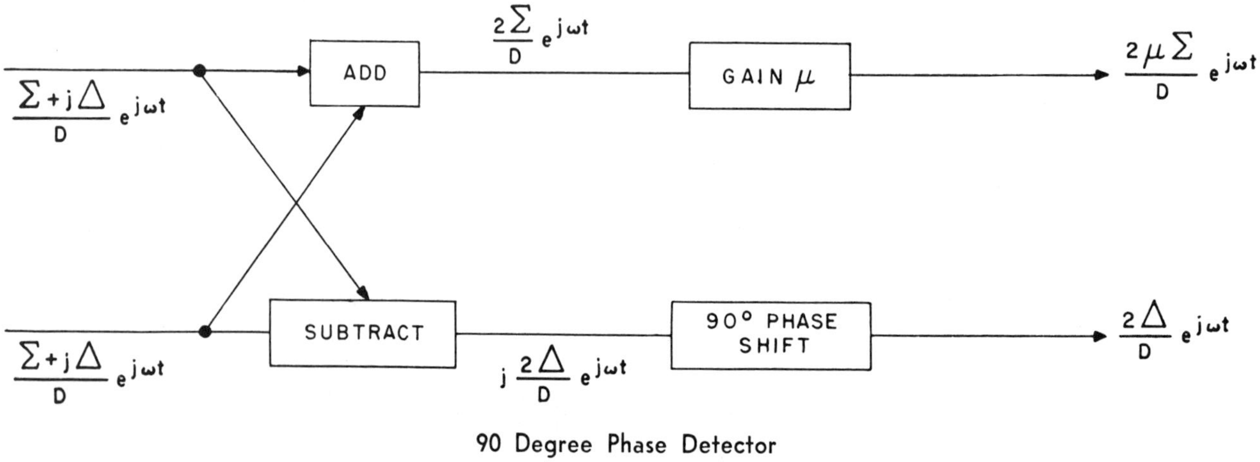

90 Degree Phase Detector

An alternative method of making a linear combination of signals simplifies the 90 degree phase detector (see page 81).

phase detector has an extended range characteristic as shown by the dotted line in the sketch. This same characteristic can be obtained by making the delay line in the

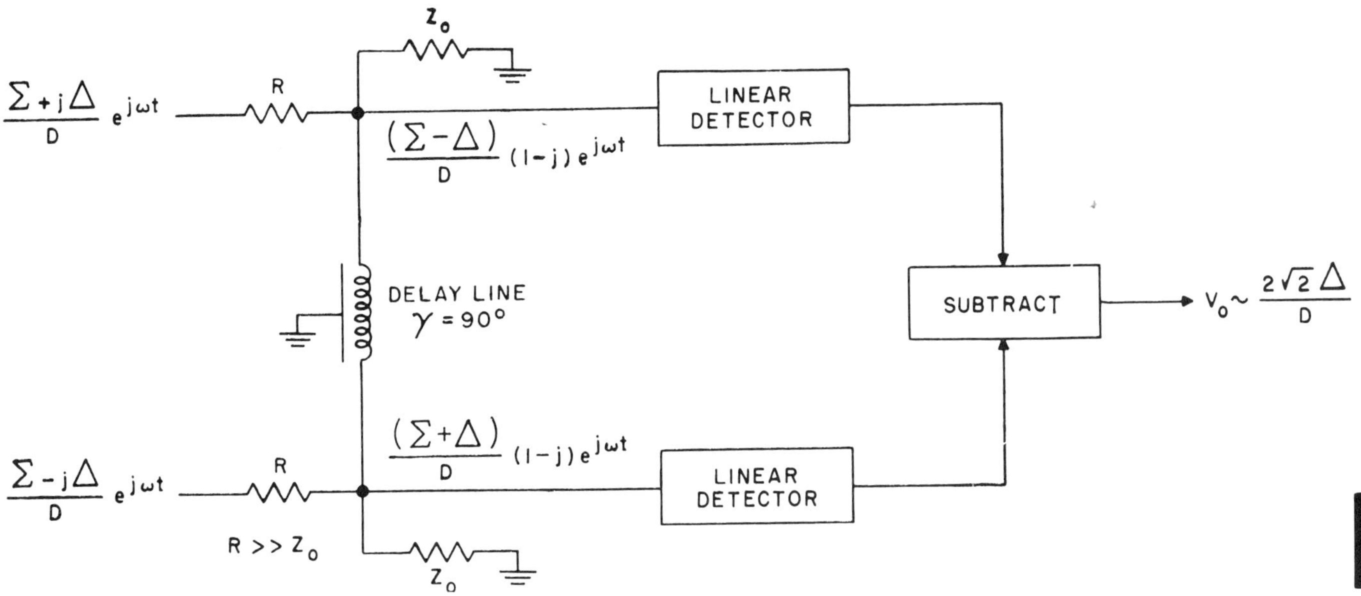

Delay Line 90 Degree Phase Detector

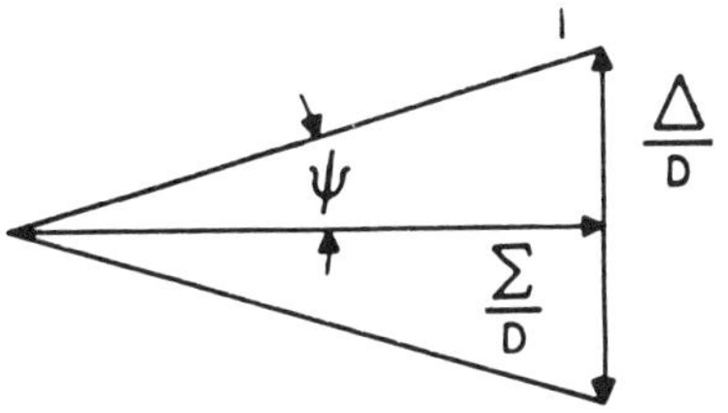

Vector Relationship

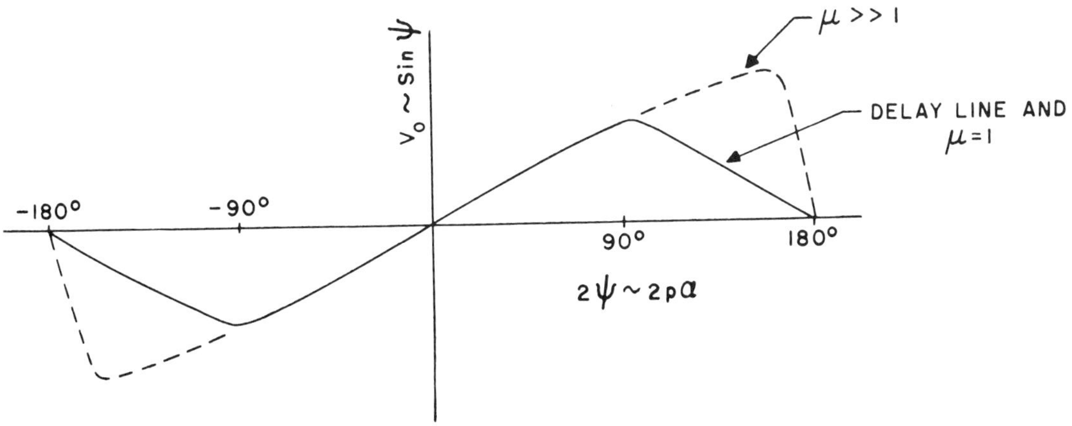

90 Degree Phase Detector Characteristics

81

delay line phase detector less than 90 degrees long. The signals at the terminals of the delay lines are proportional to

$$\frac{[(\Sigma + j\,\Delta) + e^{-j\gamma}(\Sigma - j\,\Delta)]e^{j\omega t}}{D} \quad \text{and} \qquad \text{(D-44)}$$

$$\frac{[(\Sigma - j\,\Delta) + e^{-j\gamma}(\Sigma + j\,\Delta)]e^{j\omega t}}{D} \qquad \text{(D-45)}$$

where $D = (\Sigma^2 + \Delta^2)^{1/2}$.

These equations can be rearranged to clarify the effect of making γ less than 90 degrees.

$$\frac{\left(\Sigma + \Delta \tan \frac{\gamma}{2}\right)}{D}(1 + e^{-j\gamma})\,e^{j\omega t} \quad \text{and} \qquad \text{(D-46)}$$

$$\frac{\left(\Sigma - \Delta \tan \frac{\gamma}{2}\right)}{D}(1 + e^{-j\gamma})\,e^{j\omega t} \qquad \text{(D-47)}$$

The output of the detector is proportional to the difference of equations (D-46) and (D-47).

$$V_o \sim \frac{\Delta}{(\Sigma^2 + \Delta^2)^{1/2}} \tan \gamma/2 = \sin \psi \tan \gamma/2 \qquad \text{(D-48)}$$

Equation (D-48) makes it clear that using a delay line of less than 90 degrees is equivalent to attenuating the Δ signal by the factor $\tan \gamma/2$ before detection. If γ is made quite small (perhaps 10 electrical degrees) the phase detector characteristic is extended from 90 degrees to approximately 150 degrees. The dotted curve in the sketch also applies to a short delay line, however, the factor by which the output is reduced must not be neglected.

d. Extended Linearity Phase Detector.— Another method of combining the ratio circuit outputs prior to detection gives improved linearity and also extended range. The signal combinations are shown in the sketch.

The output of a 90 degree phase detector following the addition, amplifying, and limiting circuit shown in the sketch is

$$V_o \sim \left[1 + \frac{\Delta}{D}\right]^{1/2} - \left[1 - \frac{\Delta}{D}\right]^{1/2} \qquad \text{(D-49)}$$

Equation (D-49) can also be written as

$$V_o \sim |1 + \sin \psi|^{1/2} - |1 - \sin \psi|^{1/2} \qquad \text{(D-50)}$$

The resulting output curve is more linear than the extended range phase detector characteristic. Because of the additional circuit for adding, amplifying, and limiting in one signal path, precautions must be taken to insure that the phase and amplitude characteristics of the two paths are the same if the circuit is to function on short pulses. An approximate sketch of the extended linearity characteristic is included on page **83**.

The improvement in linearity is brought out by the sketch of the characteristics.

While this may occur to the reader, it will be pointed out that if the original IF vector signals are Σ and Δ, then only the addition, $\Sigma + j\,\Delta$, is needed to create the signals used in the extended linearity detector.

Since the antenna patterns, the ratio circuit, and the phase detector functions enter into the linearity of the error signal versus far field angle, the radar components should be designed to give the best overall linearity.

e. Linearizing Circuit.—The output of a phase detector is of the form

$$V_o = p_1 \sin \psi \qquad \text{(D-51)}$$

It is desirable to convert this non-linear output to a linear one.

$$v_o = p_2\,\psi \qquad \text{(D-52)}$$

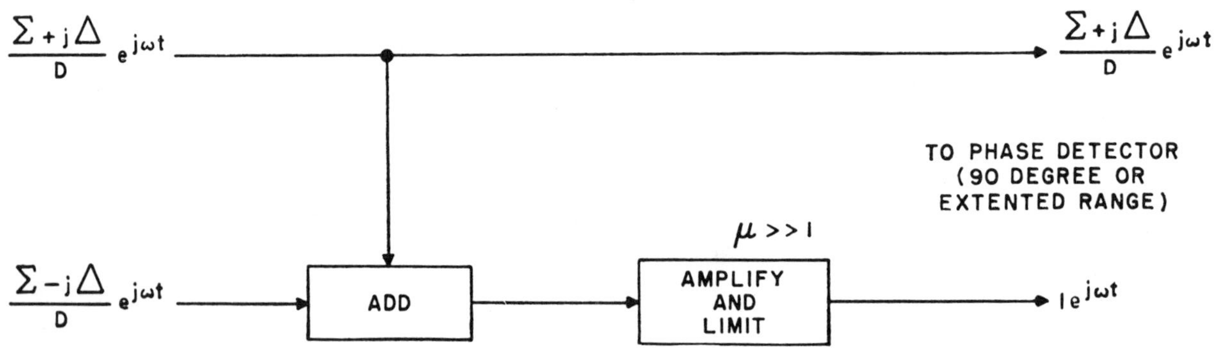

Extended Linearity Phase Detector

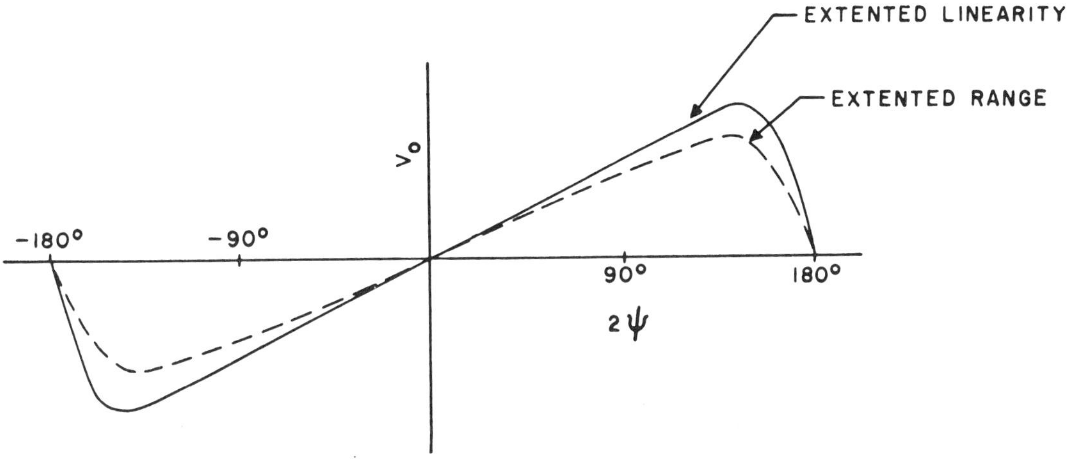

Extended Linearity Phase Detector Characteristic

A non-linear circuit element can be used to convert the non-linear output, equation (D-51), to a linear one, equation (D-52).

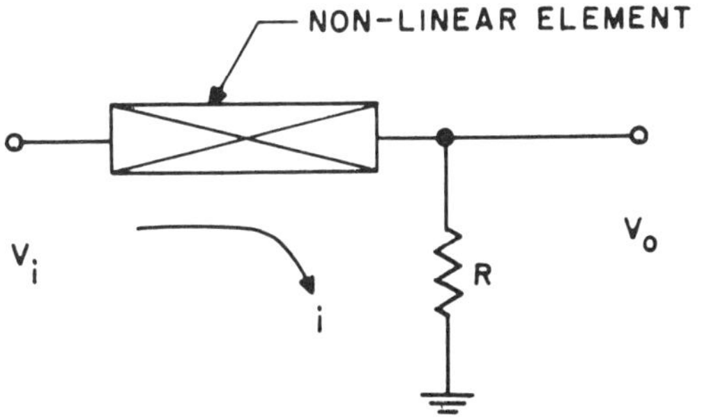

Linearizing Circuit

$$V_o = v_i = iR + v_t \qquad \text{(D-53)}$$

$$v_o = iR \qquad \text{(D-54)}$$

where

v_t = voltage drop across non-linear element

The voltage drop across the non-linear element can be approximated by a power series.

$$v_t = a\,i + b\,i^2 + c\,i^3 + \dots \qquad \text{(D-55)}$$

Substituting equations (D-53) and (D-55) for V_o in equation (D-51) gives

$$p_1 \sin\psi = i\,[R + a + b\,i + c\,i^2 + \dots]. \qquad \text{(D-56)}$$

but from (D-54) and (D-52)

$$v_o = iR = p_2\,\psi \qquad \text{(D-57)}$$

Dividing equation (D-56) by (D-57)

$$\frac{\sin\psi}{\psi} = \frac{p_2}{p_1 R}\,[1 + a + b\,i + c\,i^2 + \dots]. \qquad \text{(D-58)}$$

The sin ψ term can be expanded into a series and the series substituted in equation (D-58).

$$1 - \frac{\psi^2}{3!} + \frac{\psi^4}{5!} - \frac{\psi^6}{7!} = \frac{p_2}{p_1 R}\left[1 + a + \frac{p_2 b \psi}{R} - \left(\frac{p_2\psi}{R}\right)^2 c + \dots\right]. \qquad \text{(D-59)}$$

The above equality holds if terms on the left are equal to corresponding terms on the right. The value of the coefficients on the right should be

$$a = \frac{p_1 R}{p_2} - 1,$$

$$b = 0,$$

$$c = -\frac{1}{3!}\left(\frac{R}{p_2}\right)^3 p_1, \qquad \text{(D-60)}$$

$$d = 0,\ \text{etc.}$$

If a material with non-linear resistance characteristics, such as Thyrite, has the proper coefficients, the phase detector curve can be linearized. This has been demonstrated experimentally.

4. Stabilizing Circuits

Two methods of stabilizing the radar ECS output against errors caused by changes in relative gain or phase shift are:

(1) Use of a monitor signal

(2) Signal commutation

Either of these methods can be applied to stabilize only certain critical components or to the complete radar. For example, a monitor signal could be used to stabilize the relative gain and phase shift of the IF amplifiers, or the monitor signal can be introduced at the antenna and monitor all subsequent components. Monitoring circuits of the latter type were described in quarterly engineering report number four.

The signal commutation can be done in a number of ways which are equivalent in the method of operation.

(1) RF commutation

(2) IF commutation

(3) Double conversion at IF

(4) Audio displaced LO signals

The use of double conversion at IF was reported in the fifth quarterly report. Because only one main IF amplifier was used, this circuit is also called "One Channel Monopulse Tracking Radar." Subsequently considerable work has been done on the AN/TPQ-5 Project under Contract DA-36-039-sc-5408 to explore more fully the potentialities of RF and IF commutation. The reader is referred to the quarterly engineering reports on the AN/TPQ-5 Project for a comprehensive treatment of these topics. In particular the use of a gyromagnetic ferrite switch for RF commutation appears promising. Figure 20, which illustrates RF commutation, was taken from Figure 14 of the Sixth Quarterly Report on Radar Set AN/TPQ-5(XE-1). The double conversion method of stabilizing relative gain and phase shift is subsequently described in a qualitative manner.

The requirements for stability of gain and phase shift in the IF amplifiers of a monopulse radar are difficult to satisfy with separate IF amplifier channels. A method of minimizing the stability requirements of both the IF amplifiers and error detector circuits makes use of a double conversion receiver. At present this technique appears useful only for tracking radars where the desired information is concentrated in one pulse length (range element).

The second conversion changes the signals in the sum and difference IF channels to new IF frequencies which differ by an audio frequency. The audio frequency is preferably a fraction of the pulse repetition frequency. After the second conversion, the sum and difference IF signals can be added together and amplified in one IF strip.

The operation of the circuit will be explained in more detail with the aid of a block diagram, Figure 21. The circuits from the antenna to the second mixers are the conventional monopulse circuits of a combination radar, or one axis of a radar with separate delta outputs for each axis. After pre-amplification, the pulse IF signals are mixed for a second time to convert to two new IF frequencies labeled f_6 and f_7. These two frequencies differ by an audio frequency f_8 which is less than the pulse repetition frequency f_r. After addition, the sum and difference signals are amplified simultaneously in a single IF amplifier. With the two IF frequencies only a few hundred cycles apart, there is no chance for differential phase shift to occur if operation is on a linear part of the phase curve of the IF amplifier. After IF amplification, an amplitude demodulator gives video pulses modulated by the delta signal at an audio rate. With some video amplification first, a boxcar circuit can be used to remove the repetition rate components.

If the unwanted cross products are filtered from the boxcar circuit output, then an audio frequency, f_8, error signal remains. By mixing the second oscillator outputs, at frequencies f_4 and f_5, an audio reference signal is obtained which can be used to resolve the amplified error signal into azimuth and elevation ECS voltages.

It is necessary to filter the DC output of the boxcar circuit sufficiently to remove the audio modulation before using the voltage for AGC control. This will reduce the response speed of the AGC over that of a radar which does not use the double conversion. However, most of the amplitude noise energy[23] is contained in a narrow spectrum close to zero frequency, and the tracking performance will not be materially affected. If a fast response AGC is necessary, the slow AGC can be supplemented by a fast AGC channel. This fast AGC channel would derive its signal from the sum pre-amplifier before the second mixing action. A band-pass audio filter would derive components of pulse amplitude fluctuation below $f_r/2$ and above the cutoff of the low pass AGC circuit. A combination of the low pass and band-pass AGC voltage would be used for gain control. The use of the voltage from the low pass filter of the common IF amplifier channel preserves the accuracy of the ratio of delta to sum.

[23]J. E. Meade, "Noise in Tracking Radars," Naval Research Laboratory Report 3759.

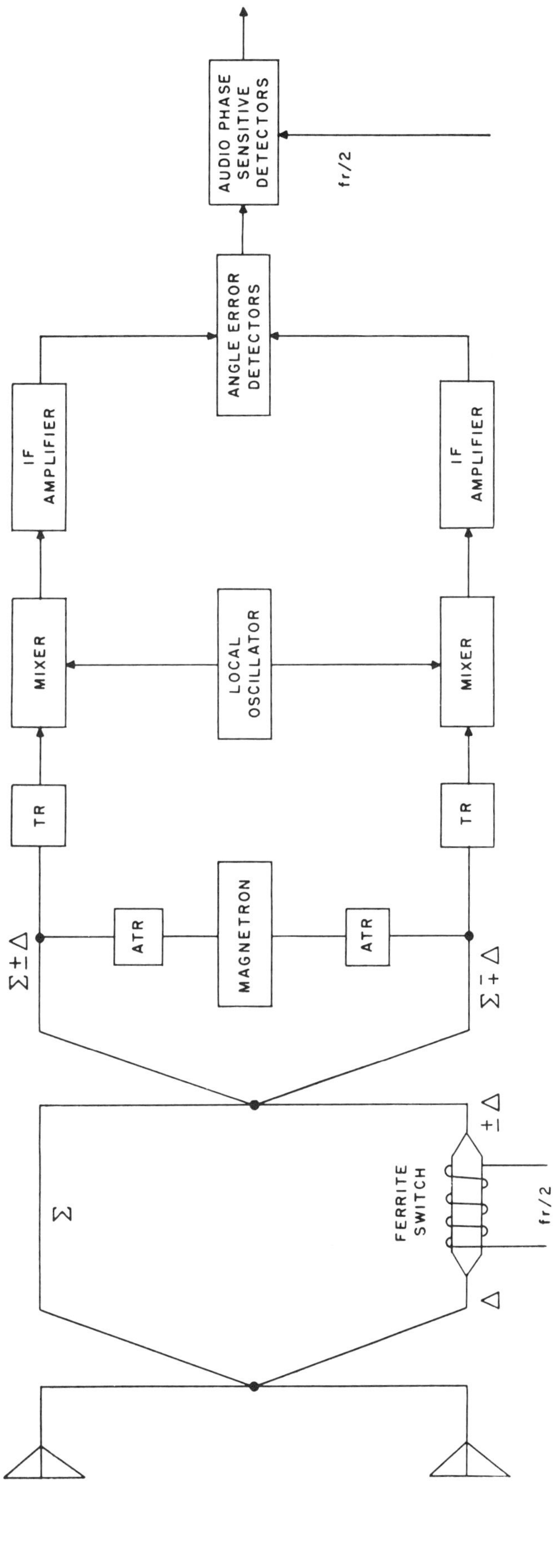

Figure 20. RF Commutation.

85

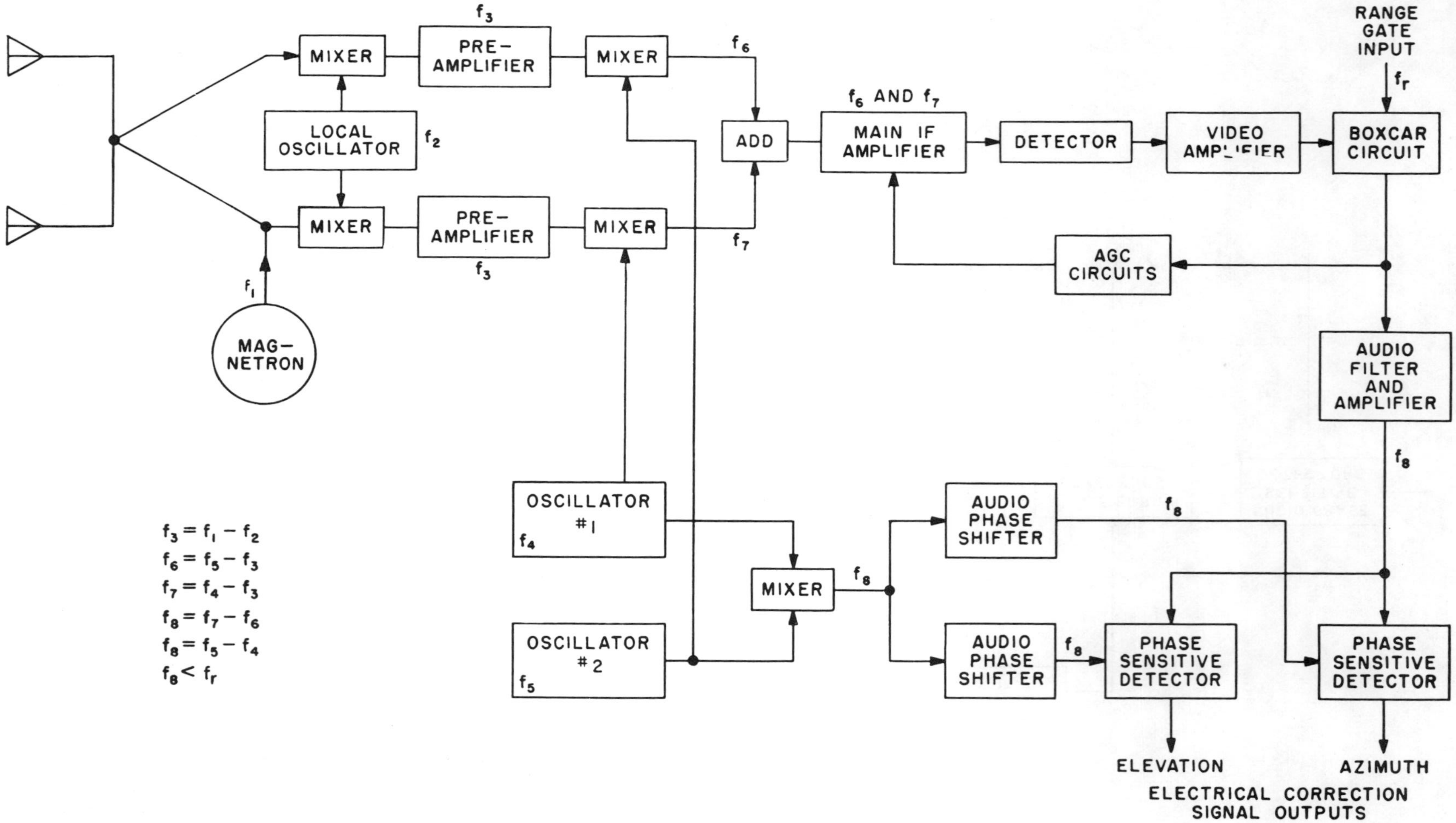

$f_3 = f_1 - f_2$
$f_6 = f_5 - f_3$
$f_7 = f_4 - f_3$
$f_8 = f_7 - f_6$
$f_8 = f_5 - f_4$
$f_8 < f_r$

Figure 21. One Channel Monopulse Tracking Radar.

The conversion of the error to an audio voltage at the output of the IF will minimize the DC drift problem usually encountered at this point in a tracking radar. The audio voltage can be amplified to a high level before it is converted to a DC, ECS output. If a 60 or 400 cycle error voltage is desired for an AC servo system, the difference frequency, f_8, could be made equal to the desired AC error frequency. The frequency of one of the oscillators could be controlled by a reactance tube and a phase discriminator, or single side-band techniques could be used to create the second frequency at the oscillator frequency, f_4, plus the AC error frequency.

The use of the single amplifier channel following the second mixers will not eliminate the possibility of phase shift in the first mixers and pre-amplifiers. The use of two local oscillator signals differing by an audio frequency, or a rotary frequency changer in the RF difference channel, are possible methods of further improving the gain and phase stability.

V. CONCLUSIONS

1. ECS Principle

The conclusions on the ECS principle will be divided into two parts, analytical and experimental.

a. Analytical.—Within the limitations of the assumptions used in analysis, it is possible to obtain linear, independent error signals (ECS) for two axes. Three methods, phase, ratio, and log were investigated analytically, and are believed suitable for practical applications.

b. Experimental.—The lens used for the experimental work has a different focal point for the elevation and azimuth planes at a wave length of 3.33 cm. The horn feed was adjusted to the optimum focus for the azimuth plane and all subsequent ECS data were taken in azimuth. The results indicate:

(1) Over the central portion of the beam (±1/2 beamwidth, one way sum pattern) the accuracy of the azimuth ECS is better than the measuring system. (The electrical angles were measured to ±3 electrical degrees. This is equivalent to ±0.15 angular mils in the far field. The position of the antenna was measured to ±0.4 angular mil.)

(2) Beyond the central portion of the beam (beyond ±1/2 beamwidth) the azimuth ECS is fairly linear (±0.3 angular mils in the far field) to the first null of the one way sum pattern.

(3) The azimuth ECS is relatively independent of the elevation angle over elevation angles of plus and minus one beamwidth of the one way sum pattern.

2. Angular Sensitivity Factor $KG^{1/2}$ for Antennas

It is concluded that the factor, $KG^{1/2}$, provides a suitable basis for comparing the angular sensitivity of monopulse antennas. The maximum value of K, the slope factor, for a given aperture is determined in the report. This maximum value of K is designated K_0 and has the same significance for direction finding as G_0, the maximum aperture gain, has for search radar. Tables of odd and even aperture functions and the efficiencies relative to the optimum, K_0 and G_0, enable proposed monopulse antenna designs to be readily compared.

3. Limitation of Thermal Noise on Angular Accuracy

It is concluded that the effect of receiver thermal noise on the angular accuracy of monopulse radar can be predicted. The factor α_N, the equivalent noise angle, is determined from the angular factor, $KG^{1/2}$; the receiver noise figure, NF; and other radar parameters. While the performance of existing monopulse radars is 18 db or more below theoretical for a particular aperture and receiver at X band, very intensive effort on antenna and receiver development will be required to materially decrease the 18 db gap between existing designs and theoretical. More experimental work is necessary to verify certain assumptions made in the noise analysis.

4. Stabilization Against DC Drift and Relative Gain or Phase Shift Errors

In the fifth quarterly report a method of stabilization against DC drift and relative gain or phase shift in IF amplifiers was described. The feasibility of this method, called "One Channel Monopulse Tracking Radar," was demonstrated in the laboratory with a minimum of equipment. It is believed that extensions of this method, especially RF commutation, offer a practical method of stabilizing monopulse tracking radars.

5. Application of ECS Principle

Several applications of the ECS principle are suggested in the report. For convenience these applications are summarized here. From the experimental work it is concluded that these ideas are feasible, though additional investigations are necessary to design an optimum system for each application.

a. Since the ECS output is derived ahead of the servo amplifier in a tracking radar, the ECS can be used to correct the antenna position data to the true target position. Thus as long as the antenna structure has sufficient mechanical rigidity, errors arising from antenna servo backlash or jitter can be eliminated. The ECS signal can be used to correct the electrical data takeoff from elevation and azimuth potentiometers, or to drive a small, fast, accurate servo operating on a differential gear adder.

b. When tracking overhead targets, the angular rates in polar coordinates may become excessive, momentarily, for a conventional antenna drive servo. By using ECS signals to correct the antenna data, it is possible to obtain accurate data on overhead targets. Since overhead targets generally return a strong signal, it should be possible to utilize the ECS out to the first null of the one way sum pattern.

c. The ability to determine the relative angular positions of two targets at different ranges in the beam could be utilized for a firing error indicator.

d. The use of "beam sharpening," as described in the introduction, allows the angular position of a point target to be determined more accurately and quickly on a visual presentation.

e. By stopping the main beam before it strikes the ground it should be possible to track to lower angles using the ECS signal without excessive "low angle" tracking errors.

f. It may be possible to use "offset" elevation boresighting procedures if the ECS signal at angles of 1/2 beamwidth off the antenna axis can be made accurate enough.

g. The use of the ECS principle offers a method of determining the target angle in a track-while-scan radar. The ECS signal can be added to the antenna position signal to determine the target position. The average of the target angles when the target is within the half power points of the beam should give the best target position. By using a coincidence principle, which would require both the presence of a target and the target to be within a certain beam angle, extraneous data can be excluded. The presence of a target would be indicated by the sum signal, and the target position by the ECS.

6. Use of Coherent Integration

It is concluded that the use of coherent integration can provide adequate tracking accuracy in certain cases which would otherwise require increased transmitter power or antenna size. The analysis for the equivalent noise angle, α_N, may predict an allowable RMS error, however, if the sum channel IF signal to noise ratio is less than one, then the predicted error will be exceeded unless coherent integration is used. Narrow band coherent doppler techniques are promising for this application as well as for minimizing tracking clutter.

Coherent integration will not improve the angular accuracy unless the radar parameters are within certain limits. While this has not been verified experimentally, it is believed that the following inequality indicates the IF S/N region where coherent integration is useful.

$$2 > \frac{\Sigma}{N_\Sigma} > \frac{\lambda}{\underbrace{(\theta_{az})_N}_{\text{(maximum allowable RMS value of noise error)}} K\, 2\pi d} \left[\frac{\eta_t\, G\, \beta_n}{\eta_r\, f_r} \right]^{1/2}$$

There are undoubtedly additional restrictions on target acceleration and doppler spectrum.

VIII. APPENDICES

Appendix I—Conditions for Linear Ratio

The proof for the conditions on the aperture illumination for a linear ratio is taken directly from the Data Folder 51-E-234, "Relation Between the Far Field and the Illumination of Antenna Apertures" by W. L. Murdock and J. S. Kerr. This is a General Electric Company Publication for use within the Company, and therefore the material is not readily available. The desired result is

$$\frac{f_o(\alpha)}{f_e(\alpha)} = p\,\alpha \quad \text{or} \tag{1}$$

$$f_o(\alpha) = p\,\alpha\, f_e(\alpha). \tag{2}$$

Equation (2) can be written

$$\int_{-d/2}^{d/2} A_o(x) \sin\frac{\alpha x}{d}\, dx = p\,\alpha \int_{-d/2}^{d/2} A_e(x) \cos\frac{\alpha x}{d}\, dx \tag{3}$$

The right-hand side of equation (3) can be integrated by parts

$$= pd\, A_e(x) \sin\frac{\alpha x}{d}\Big|_{-d/2}^{d/2} - pd \int_{-d/2}^{d/2} A_e'(x) \sin\frac{\alpha x}{d}\, dx \tag{4}$$

Using the integrated value for the right side of (4), this equation can be rearranged.

$$\int_{-d/2}^{d/2} |A_o(x) + pd\, A_e'(x)| \sin\frac{\alpha x}{d}\, dx = 2\, pd\, A_e\left(\frac{d}{2}\right) \sin\frac{\alpha}{2} \tag{5}$$

where $A_e'(x) = \frac{d}{dx}(A_e)$.

Equation (5) is true for all values of α. However, the limit of the left-hand side as $\alpha \to \infty$ is zero (Riemann-Lebesque Theorem). Therefore, $A_e(d/2)$ must be zero. The integral on the left-hand side can be identically zero for all values of α only if

$$A_o(x) = -pd\, A_e'(x) \tag{6}$$

Thus the necessary and sufficient conditions for a linear ratio, equation (1), are

$$A_e\left(\frac{d}{2}\right) = 0$$
$$A_o(x) = -pd\, A_e'(x) \tag{7}$$

Appendix II—K and G of Combination and Conical Scan Antennas

The formulas for G and K were derived using equations (B-20) and (B-34) respectively from the text.

1. Squinted Beam (or Split Aperture Squinted Beams of Combination Antenna)

$$G^{1/2} = G_o^{1/2} \frac{\left[\dfrac{\sin\left(\dfrac{\pi r - \varepsilon_b}{2}\right)}{\left(\dfrac{\pi r - \varepsilon_b}{2}\right)} + \dfrac{\sin\left(\dfrac{\pi r + \varepsilon_b}{2}\right)}{\left(\dfrac{\pi r + \varepsilon_b}{2}\right)} \right] \dfrac{\sin \pi r/2}{\dfrac{\pi r}{2}}}{1 + \dfrac{\sin \pi r}{\pi r}} \tag{1}$$

$$K_a = G_o^{1/2} \frac{\left[\dfrac{\sin\left(\dfrac{\pi r + \varepsilon_b}{2}\right) - \left(\dfrac{\pi r + \varepsilon_b}{2}\right)\cos\left(\dfrac{\pi r + \varepsilon_b}{2}\right)}{\left(\dfrac{\pi r + \varepsilon_b}{2}\right)^2} - \dfrac{\sin\left(\dfrac{\pi r - \varepsilon_b}{2}\right) - \left(\dfrac{\pi r - \varepsilon_b}{2}\right)\cos\left(\dfrac{\pi r - \varepsilon_b}{2}\right)}{\left(\dfrac{\pi r - \varepsilon_b}{2}\right)^2} \right] \dfrac{\sin \dfrac{\pi r}{2}}{\pi r}}{1 - \dfrac{\sin \pi r}{\pi r}} \tag{2}$$

where $\pi r \neq \varepsilon_b$

r = taper factor of illumination (both x and y axes)

ε_b = normalized squint angle $= \dfrac{2\pi\, h\, \theta_{el}}{\lambda}$

a. Uniform Illumination $(r = 0)$

If r is zero then the illumination is uniform along both the x and y axes, and equations (1) and (2) reduce to

$$G^{1/2} = G_o^{1/2} \frac{\sin \varepsilon_b/2}{\varepsilon_b/2}, \text{ and} \tag{3}$$

$$K_a = \frac{G_o^{1/2}}{\varepsilon_b^2}\left[\sin \frac{\varepsilon_b}{2} - \frac{\varepsilon_b}{2}\cos\frac{\varepsilon_b}{2}\right]. \tag{4}$$

b. Cosine Illumination $(r = 1)$

If $\pi r/2$ is equal to $\pi/2$ then the illumination is cosine along both the x and y axes, and equations (1) and (2) reduce to

$$G^{1/2} = \frac{G_o^{1/2}\, 2 \cos \dfrac{\varepsilon_b}{2}}{\left(\dfrac{\pi}{2}\right)^2 - \left(\dfrac{\varepsilon_b}{2}\right)^2}, \text{ and} \tag{5}$$

$$K_a = G_o^{1/2} \frac{\left[\left(\dfrac{\pi}{2}\right)^2 - \left(\dfrac{\varepsilon_b}{2}\right)^2\right] \sin\dfrac{\varepsilon_b}{2} - \varepsilon_o \cos\dfrac{\varepsilon_b}{2}}{\left[\left(\dfrac{\pi}{2}\right)^2 - \left(\dfrac{\varepsilon_b}{2}\right)^2\right]^2} \tag{6}$$

The equations for K, $G^{1/2}$, and the product $KG^{1/2}$ are plotted on curve sheets Figures II-1, II-2, and II-3. The plot of the product is particularly useful to determine the optimum squint angle. The maximum value of the product was used for the calculation of the angular sensitivity for the amplitude comparison plane in Table VI.

2. Phase Comparison Plane of Combination Antenna

It is necessary to determine a value for K for the phase comparison plane of the combination antenna separately. The result is

$$K_p = G_o^{1/2} \frac{1}{4} \frac{\sin\dfrac{\pi r}{2}}{\dfrac{\pi r}{2}} \frac{\dfrac{\sin \dfrac{\pi r - \varepsilon_b}{2}}{\dfrac{\pi r - \varepsilon_b}{2}} + \dfrac{\sin\left(\dfrac{\pi r + \varepsilon_b}{2}\right)}{\dfrac{\pi r + \varepsilon_b}{2}}}{1 + \dfrac{\sin \pi r}{\pi r}} \tag{7}$$

where r and ε are defined in equation (2). It is useful to note that equation (7) can be rewritten

$$K_p = \frac{1}{4} G^{1/2} \tag{8}$$

where $G^{1/2}$ is given by equation (1). Thus K_p can be determined from Figure II-2 and equation (8).

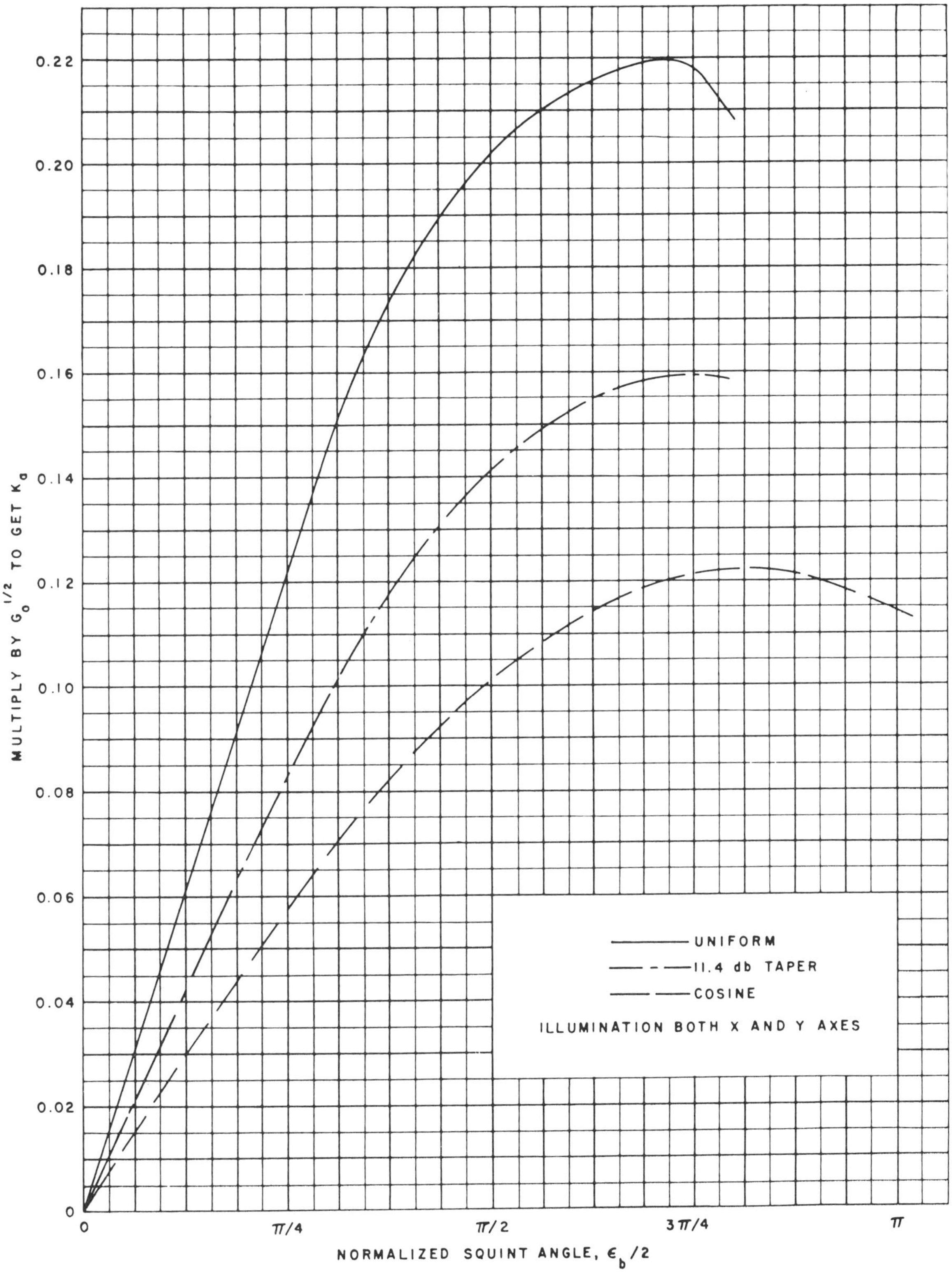

Figure II-1. K_a, Angular Sensitivity.

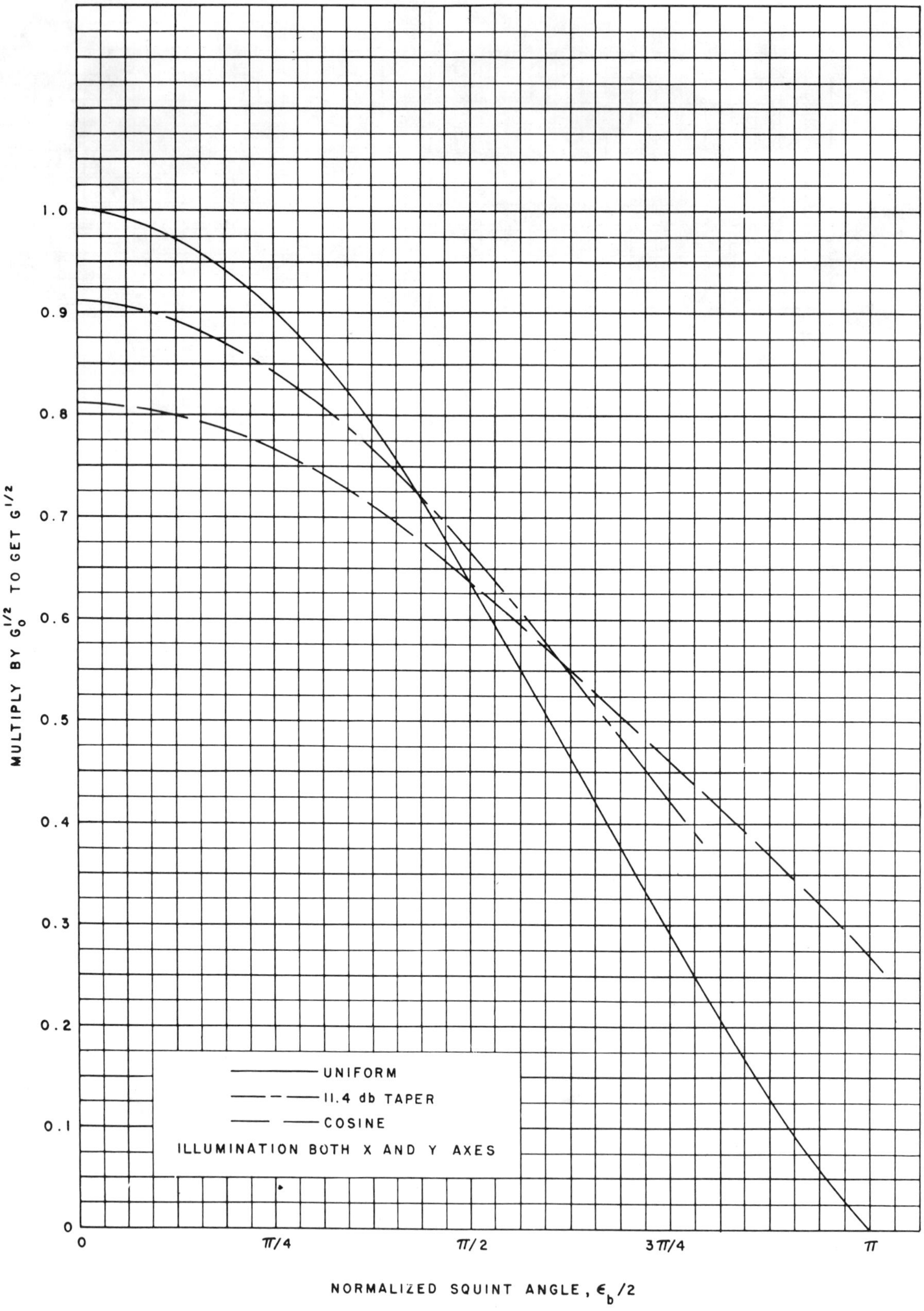

Figure II-2. $G^{1/2}$, Antenna Gain.

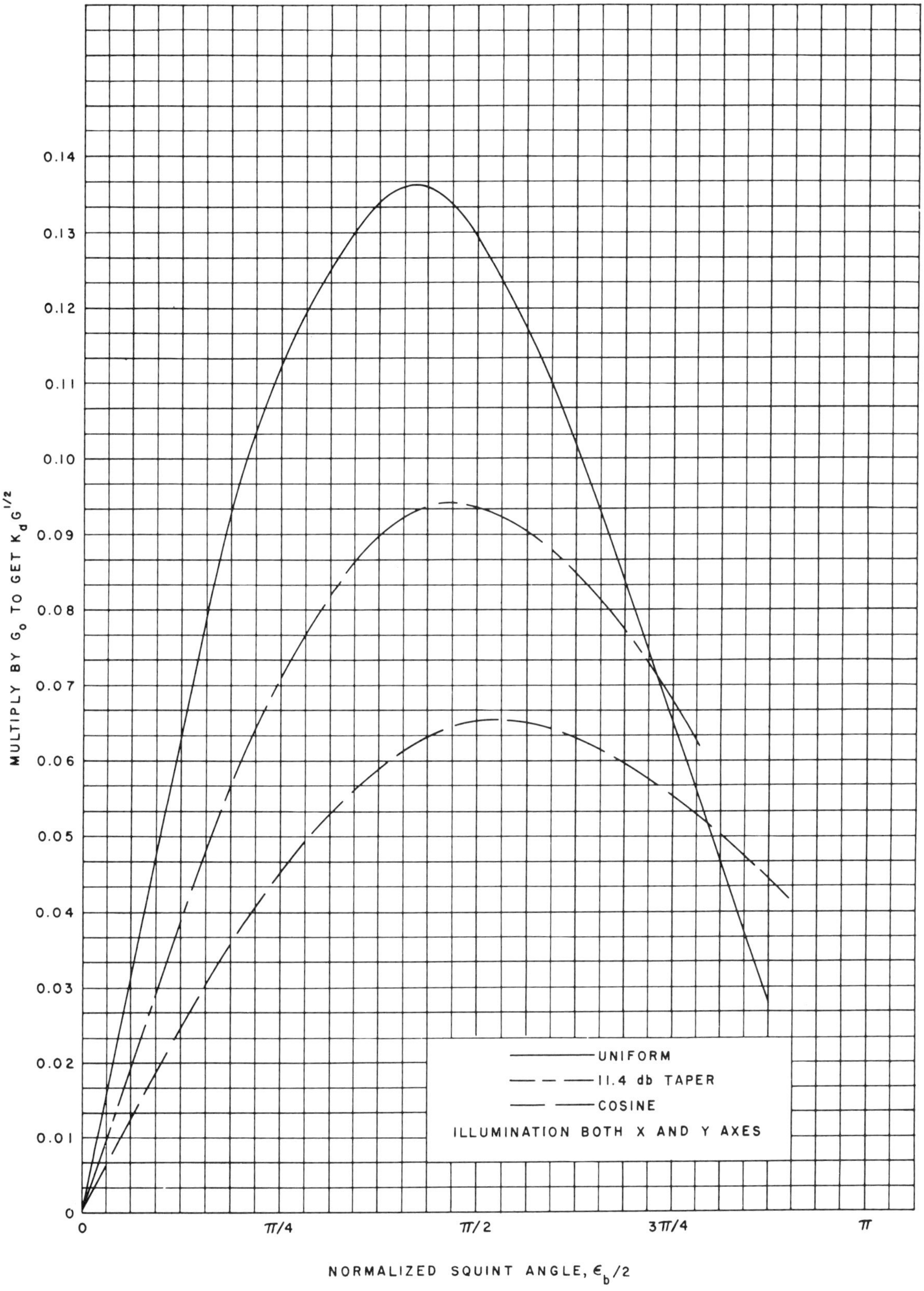

Figure II-3. $K_a G^{1/2}$, Round Trip Angular Sensitivity.

Appendix III—Linear Detection of Signal Plus Noise

A linear detector is assumed to have the following characteristic

$$I(t) = 0 \qquad \text{for } S(t) < 0 \qquad (1)$$

$$I(t) = g\,S(t) \qquad \text{for } S(t) > 0 \qquad (2)$$

The rectification of a sine wave signal will give a repetitive half sine wave signal.

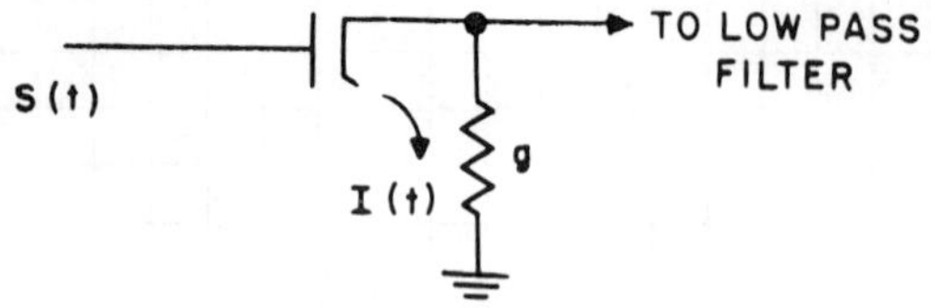

Linear Detector

The current wave I(t) can be expanded into a Fourier Series

$$I(t) = \frac{I\ \text{peak}}{\pi}\left[1 + \frac{\pi}{2}\cos\omega t + \sum_{n=1}^{\infty} a_n \cos(2n\omega)t\right] \qquad (3)$$

where

$$a_n = \frac{(-1)^{n+1}}{4n^2 - 1}$$

A low pass filter which cuts off at a frequency below the sine wave frequency, ω, will pass only the DC term of (3).

1. Modulated CW Signal Plus Noise

$$S(t) = \sqrt{2}\ S\,|1 + \sigma\cos\omega_m t|\sin\omega t + N(t) \qquad (4)$$

where

$S(t)$ = instantaneous signal voltage

S = RMS value of the CW Signal

σ = modulation index, $\dfrac{\text{RMS Modulating Signal}}{\text{RMS CW Signal}}$

ω_m = angular modulating frequency

ω = CW angular frequency

$N(t)$ = instantaneous noise voltage

When the noise has a small bandwidth compared to ω, the noise can be represented as

$$N(t) = N_x \sin\omega t + N_y \cos\omega t \qquad (5)$$

where N_x, N_y = uncorrelated noise voltages with normal distributions.

The mean squared value of N(t) is

$$\overline{N(t)^2} = \overline{N^2} = \overline{(N_x \sin\omega t + N_y \cos\omega t)^2} \qquad (6)$$

N = RMS noise voltage

$$\overline{N^2} = \frac{\overline{N_x^2}}{2} + \frac{\overline{N_y^2}}{2} \qquad (7)$$

Therefore

$$\overline{N^2} = \overline{N_x^2} = \overline{N_y^2} \qquad (8)$$

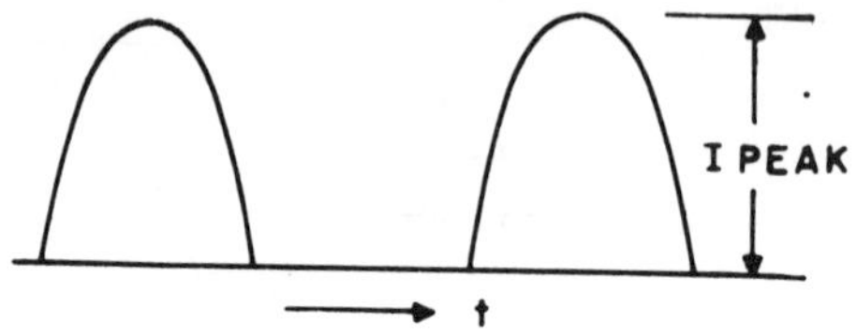

I (t)

Using the form for N(t) given in equation (5) the voltage S(t) is

$$S(t) = [\sqrt{2}\ S(1 + \sigma\cos\omega_m t) + N_x]\sin\omega t + N_y\cos\omega t \qquad (9)$$

If $S >> S\sigma$, N_x, and N_y then the N_y component will have little effect upon the instantaneous voltage S(t). Therefore the low frequency output current is

$$I(t) = \frac{g}{\pi}\left|\sqrt{2}\ S(1 + \sigma\cos\omega_m t) + N_x\right|, \qquad (10)$$

and the mean squared output current is

$$\overline{I(t)^2} = \left(\frac{g}{\pi}\right)^2 \overline{|\sqrt{2}\ S(1 + \sigma\cos\omega_m t) + N_x|^2} \qquad (11)$$

With the condition $S >> S\sigma$ and N_x, the absolute magnitude sign can be ignored and the mean square of (11) reduces to

$$\overline{I(t)^2} = \left(\frac{g}{\pi}\right)^2 [2S^2 + (\sigma S)^2 + \overline{N_x^2}] \qquad (12)$$

However, in equation (8), $\overline{N_x^2}$ is shown to equal $\overline{N^2}$. Thus the final form for the mean square current is

$$\overline{I(t)^2} = \left(\frac{g}{\pi}\right)^2 [2S^2 + (\sigma S)^2 + \overline{N^2}] \qquad (13)$$

Equation (13) is limited to cases where the S/N ratio is large. It is useful for limiting case checking against such references as "Mathematical Analysis of Random Noise," by S. O. Rice in January, 1945, B.S.T.J. The diagram on page 148 of this article agrees with equation (13) in the limiting case for S/N large.

In the book, "Threshold Signals," by J. L. Lawson and G. E. Uhlenbeck, equation 30b on page 160 of the book does not agree with equation (13). Equation 30b of the book is

$$\overline{I(t)^2} \doteq \frac{S^2}{2} + \frac{S_o^2}{2} + \sigma^2 B \qquad (14)$$

where $N^2 = \sigma^2 B = kT\,(NF)\,B$

The reason for this disagreement is not clear. In a limiting case the equations for boxcar action taken from the same book agree with equation (13). [See equation (21) this appendix.]

2. Boxcar Spectrum of Modulated Signal Plus Noise

If a linear detector is followed by a boxcar circuit the low frequency components of the signal are enhanced. The spectrum for a boxcar output is given in the book, "Threshold Signals," by J. L. Lawson and G. E. Uhlenbeck. In particular, the spectrum for a linear detector followed by a "short gate" boxcar circuit is given on page 280. With certain symbols altered for convenience the spectrum is

$$W_{S+N}(f) = \frac{g^2N^2f_r \sin^2(\pi f/f_r)}{\pi^3 f^2}\left\{\frac{4}{\pi}(1+z^2) - F^2 + f_r F^2 \delta(f) + f_r \sigma^2 z^4 F'^2\left[f - \left(sf_r \pm \frac{\omega_m}{2\pi}\right)\right]\right\}. \qquad (15)$$

where

$$z = \frac{S}{N} = \frac{\text{RMS signal voltage}}{\text{RMS noise voltage}}$$

$z =$ S/N ratio of the undetected IF signal

$W_{S+N}(f) =$ power per cycle in one ohm

$g =$ conductance of detector load

$f_r =$ pulse repetition frequency

$\delta(f) = \delta$ function $\quad \int_0^\infty \delta(f)\,df = 1/2$

$\sigma =$ index of modulation

$F =$ confluent hypergeometric function $F(a, b, z^2)$

$n =$ integers 0, 1, 2, 3, etc.

$F' = \partial F/\partial z^2$

F can be expressed as a series for large and small z. From page 174 of "Threshold Signals" for large z,

$$F\left(-\frac{1}{2}, 1, z^2\right) = \frac{2}{\sqrt{\pi}}\,z\left[1 + \frac{1}{4}\frac{1}{z^2} + \frac{1}{16}\frac{1}{z^4} + +\right]_{(z>>1)} \qquad (16)$$

then

$$F^2 = \frac{1}{\pi}\left[z^2 + \frac{1}{2} + \frac{3}{16}\frac{1}{z^2} + +\right]_{(z>>1)} \qquad (17)$$

$$F' = \frac{1}{\pi}\frac{1}{z} + + \qquad (z >> 1) \qquad (18)$$

$$F = 1 + \frac{1}{2}z^2 - \frac{1}{16}z^4 + + \qquad (z < 1) \qquad (19)$$

Equation (15) can be checked against equation (13) by integrating equation (15) and comparing results.

$$\int_0^\infty W(f)\,df = \left(\frac{g}{\pi}\right)^2\left[\frac{N^2\,2}{\pi}\int_0^\infty \frac{\sin^2 x}{x^2}\,dx + 2S^2 + \sigma^2 S^2\right] \qquad (20)$$

for $z >> 1$.

$$\overline{I(t)^2} = \int_0^\infty W(f)\,df = \left(\frac{g}{\pi}\right)^2 [N^2 + 2S^2 + (\sigma S)^2] \qquad (21)$$

Equations (21) and (13) are the same.

For noise alone

$$\overline{I(t)} = I_{DC} = \frac{gN}{\sqrt{2\pi}} \qquad (22)$$

$$\overline{I(t)^2} = I_{DC}^2 + \overline{I_{AC}^2} = \left(\frac{g}{\pi}\right)^2 2N^2 \qquad (23)$$

$$\overline{I_{AC}^2} = \left(\frac{g}{\pi}\right)^2 N^2\,\frac{(4-\pi)}{2} \qquad (24)$$

Appendix IV—Analysis of Narrow Filter Action on Monopulse Error Signal

The boxcar spectrum of a signal plus noise is given in Appendix III. By integrating the spectrum over the pass-band of the narrow filter the effect of the narrow filter can be determined.

$$\overline{(S+N)^2} = \left(\frac{1}{g}\right)^2 \int_0^{\beta_n} W_{S+N}(f)\, df \qquad (1)$$

For $f_r >> \beta_n$ and $z >> 1$ the integration is easily performed and the result is

$$\overline{(S+N)^2} = \frac{2}{\pi^2}\left[\frac{N^2}{f_r}\beta_n + S^2\right] \qquad (2)$$

Equation (2) gives the RMS output of one linear detector-boxcar-filter circuit. The output from the other circuit for the usual balanced detector will have the same RMS value only the two noise voltages are not correlated as pointed out in equation (C-10). The two detector outputs are

$$V_1 = \frac{\sqrt{2}}{\pi}\left[N_1\left(\frac{\beta_n}{f_r}\right)^{1/2} + \Sigma + \Delta\right] \text{ and} \qquad (3)$$

$$V_2 = \frac{\sqrt{2}}{\pi}\left[N_2\left(\frac{\beta_n}{f_r}\right)^{1/2} + \Sigma - \Delta\right] \qquad (4)$$

where

$$\overline{N_1 N_2} = 0 \qquad (5)$$

$$N_1 = N_2 = \sqrt{2}\, N_\Delta = \sqrt{2}\, N_\Sigma \qquad (6)$$

The difference of (3) and (4) is

$$V_1 - V_2 = \frac{2\sqrt{2}}{\pi}\left[N_\Delta\left(\frac{\beta_n}{f_r}\right)^{1/2} + \Delta\right] \qquad (7)$$

From equation (C-13) the error signal to noise ratio for CW conditions is

$$\frac{\sqrt{2}\,\Delta}{N_\Delta} \qquad \text{(CW radar)} \qquad (8)$$

From equation (7) the error signal to noise ratio for linear detectors-boxcar-narrow filters is

$$\frac{\Delta}{N_\Delta}\left(\frac{f_r}{\beta_n}\right)^{1/2} \text{(monopulse)} \qquad (9)$$

The improvement in error signal to noise ratio by the narrow filter is the ratio, equation (9) divided by equation (8), or

$$\text{improvement} = \left[\frac{f_r}{2\beta_n}\right]^{1/2} \text{for } (z >> 1). \qquad (10)$$

For $f_r >> \beta_n$ and $z < 1$ the integration, equation (1), can be performed with the series for $z < 1$ in Appendix III. The error signal to noise ratio is

$$\frac{\Delta}{N_\Delta}\left[\frac{f_r}{\beta_n}\right]^{1/2} \frac{1}{\left[\frac{2}{z} + z\right]\left[\frac{4}{\pi} - 1\right]^{1/2}} \quad \text{for } (z < 1) \qquad (11)$$

The improvement in this case is a function of the sum channel signal to noise ratio.

$$\text{improvement} = \left[\frac{f_r}{2\beta_n}\right]^{1/2} \frac{1}{\left[\frac{2}{z} + z\right]\left[\frac{4}{\pi} - 1\right]^{1/2}} \quad \text{for } (z < 1) \qquad (12)$$

Appendix V—S/N of Sequential Lobing Error Signal

The desired error signal is present as a modulation of the RF antenna signal pulses. After conversion of the RF pulses to IF pulses a linear detector, boxcar circuit, and narrow audio filter can recover the desired modulation component. In block diagram form the circuits are:

The total RF signal for a one-axis error is

$$S\left(1+\frac{1}{S}\frac{\partial S}{\partial \alpha}d\alpha \sin \omega_m t\right) \tag{4}$$

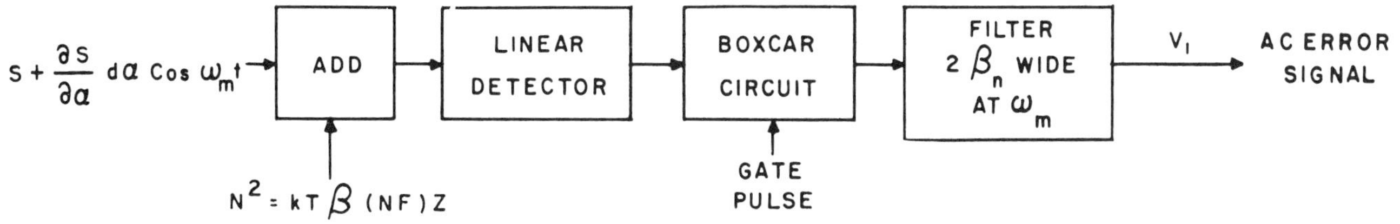

IF Detector and Audio Filter

The received signal, S, can be calculated from equation (C-8).

$$S = [Z\,\eta_t\,\eta_r\,L_r\,L_t\,G_r\,G_t]^{1/2}\,H \tag{1}$$

H is defined by equation (C-5). A small change in α close to the antenna axis will produce a change in S of

$$\frac{\partial S}{\partial \alpha}d\alpha = H\,[Z\,\eta_t\,\eta_r\,L_r\,L_t]^{1/2}\left[\frac{\partial G_r^{1/2}}{\partial \alpha}G_t^{1/2}+\frac{\partial G_t^{1/2}}{\partial \alpha}G_r^{1/2}\right] \tag{2}$$

For a small azimuth error, $d\alpha$, this signal will modulate the carrier at the conical scan frequency. Since G_t and G_r are equal for a conical scan antenna, equation (2) can be reduced to a single term. Also the derivative of $G^{1/2}$ with respect to α is defined as K, the slope factor by equation (B-34).

$$\frac{\partial S}{\partial \alpha}d\alpha = 2\,H\,\eta\,(Z\,L_r\,L_t)^{1/2}\,K G^{1/2}\,d\alpha \tag{3}$$

The spectrum of a modulated boxcar output following a linear detector is given in Appendix III, equation (15). The bandwidth of the narrow filter at frequency $\omega_m/2\pi$ should be at least 2 β_n to give the same performance as the monopulse radar investigated in Appendix IV.

With $f_r >> \omega_m/2\pi$

$$V_1 = \left(\frac{1}{g}\right)^2 \int_{\frac{\omega_m}{2\pi}-\beta_n}^{\frac{\omega_m}{2\pi}+\beta_n} W_{(S+N)}(f)\,df \tag{5}$$

$$\overline{V_1^2} = \left(\frac{1}{\pi}\right)^2\left[N^2\frac{4\beta_n}{f_r}+\left(\frac{\partial S}{\partial \alpha}d\alpha\right)^2\right] \tag{6}$$

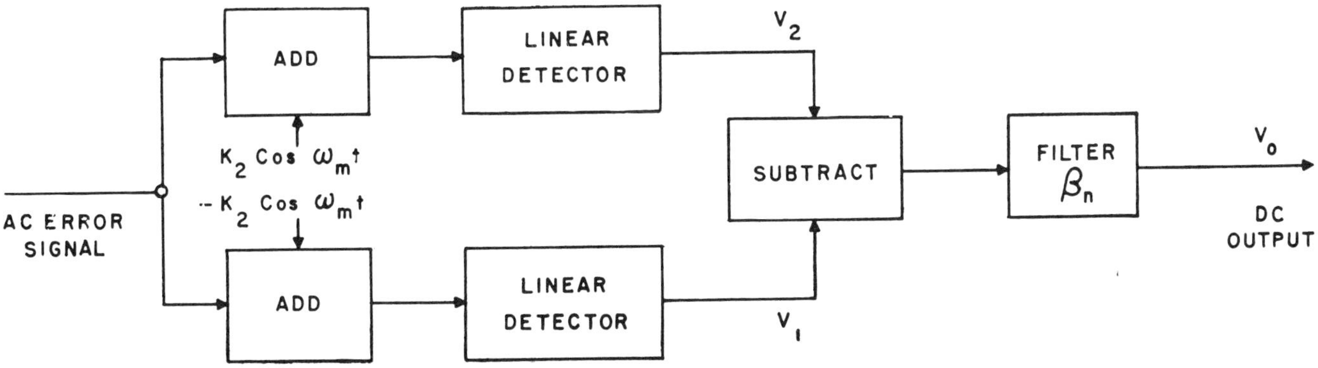

Synchronous Detector

The error signal term in equation (6) is an AC error signal. It is customary to convert to a DC signal with a synchronous detector. A noise free reference signal is supplied at the lobing frequency $\omega_m/2\pi$.

The detection of a CW signal plus random noise of narrow bandwidth is analyzed in Appendix IIIa. The output of one linear detector is

$$\overline{V_2^2} = \left(\frac{1}{\pi}\right)^4 \left[\frac{N^2\, 4\, \beta_n}{f_r} + 2\left(\frac{\partial S}{\partial \alpha}\, d\alpha\right)^2\right]. \quad (7)$$

The output of the narrow filter is

$$V_1 - V_2 = \frac{2}{\pi^2}\left[2\, N\left(\frac{\beta_n}{f_r}\right) + \sqrt{2}\,\frac{\partial S}{\partial \alpha}\, d\alpha\right] \quad (8)$$

From equation (8) the ratio of the DC error signal to noise is

$$\frac{\frac{\partial S}{\partial \alpha}\, d\alpha}{\sqrt{2}\; N\left(\frac{\beta_n}{f_r}\right)} \quad (9)$$

The partial derivative, equation (3) can be substituted into (9)

$$\frac{H\, \eta\, K}{N}\left[\frac{2\, Z\, L_r\, L_t\, G\, f_r}{\beta_n}\right]^{1/2} d\alpha \quad (10)$$

The equation for Δ, the RMS monopulse error signal is given by equation (C-6). Making the substitution of Δ the ratio is

$$\frac{\Delta}{N_\Delta}\left[\frac{2\, f_r}{\beta_n}\right]^{1/2} \text{(conical scan)} \quad (11)$$

In Appendix IV the error signal to noise ratio for linear detectors-boxcars-narrow filters is given by equation (9) as

$$\frac{\Delta}{N_\Delta}\left[\frac{f_r}{\beta_n}\right]^{1/2} \text{(monopulse)} \quad (12)$$

Conical scan, equation (11), gives a 3 db better error signal to noise ratio than monopulse, equation (12), for antennas with equal $KG^{1/2}$ factors. The value of $KG^{1/2}$ for a sequential lobing antenna is the same as $K_aG^{1/2}$ for the combination phase-amplitude antenna, Appendix II.

Appendix VI—Unbalance in Directional Coupler Power Split

The effect of an unbalance in the power split of the three db directional couplers is obtained by extending the analysis from the fifth quarterly report. Assume the signals at the antenna feeds to be in phase and of amplitude E_A, E_B, E_C, and E_D, and the phase of the signal through the coupler to be retarded by exactly 90 degrees. Let the power division of each coupler be less than three db by the amount δ in voltage. Using the notation shown in the Figure VI-1 for convenience, the signals after the first pair of directional couplers are:

These signals are added and the azimuth ECS, $2\,\psi_{az}$, obtained in the following manner:

$$\psi_1 \approx \tan^{-1} \frac{\mathcal{J}(E_A'' - E_C'')}{\mathcal{R}(E_A'' - E_C'')}$$

$$\psi_2 \approx \tan^{-1} \frac{\mathcal{J}(E_B'' - E_D'')}{\mathcal{R}(E_B'' - E_D'')} \qquad (3)$$

$$2\,\psi = \psi_1 - \psi_2$$

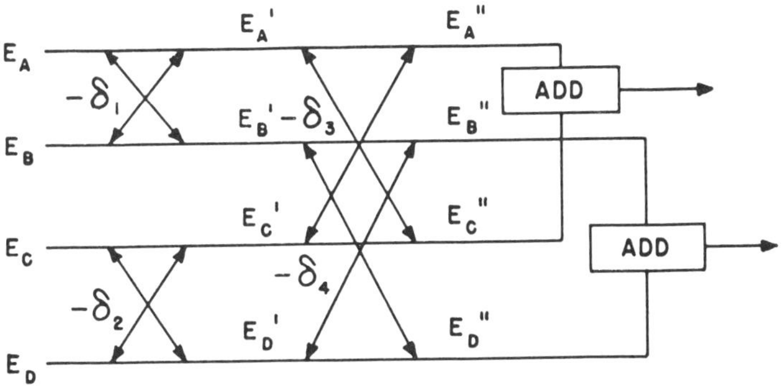

Figure VI-1. Power Unbalance In The Directional Couplers.

$$\begin{aligned}
E_A' &\approx E_A(1+\delta_1) - jE_B(1-\delta_1)\\
E_B' &\approx E_B(1+\delta_1) - jE_A(1-\delta_1)\\
E_C' &\approx E_C(1+\delta_2) - jE_D(1-\delta_2)\\
E_D' &\approx E_D(1+\delta_2) - jE_C(1-\delta_2)
\end{aligned} \qquad (1)$$

The signals after the second set of directional couplers are:

Some simplification of these equations is obtained by assuming the target to be in three different positions in the antenna beam and assigning values to the directional coupler variation in power split. On the boresight axis $(E_A = E_B = E_C = E_D)$ if all the δ values are assumed equal, the error in measurement becomes,

$$2\,\psi_{az} = \tan^{-1} \frac{-(1-\delta^2)}{2\,\delta} - \tan^{-1} \frac{-(1-\delta^2)}{2\,\delta} = 0 \qquad (4)$$

$$\begin{aligned}
E_A'' &\approx [E_A(1+\delta_1)(1+\delta_3) - E_D(1-\delta_2)(1-\delta_3)] - j[E_B(1-\delta_1)(1+\delta_3) + E_C(1+\delta_2)(1-\delta_3)]\\
E_B'' &\approx [E_B(1+\delta_1)(1+\delta_4) - E_C(1-\delta_2)(1-\delta_4)] - j[E_A(1-\delta_1)(1+\delta_4) + E_D(1+\delta_2)(1-\delta_4)]\\
E_C'' &\approx [E_C(1+\delta_2)(1+\delta_3) - E_B(1-\delta_1)(1-\delta_3)] - j[E_D(1-\delta_2)(1+\delta_3) + E_A(1+\delta_1)(1-\delta_3)]\\
E_D'' &\approx [E_D(1+\delta_2)(1+\delta_4) - E_A(1-\delta_1)(1-\delta_4)] - j[E_C(1-\delta_2)(1+\delta_4) + E_B(1+\delta_1)(1-\delta_4)]
\end{aligned} \qquad (2)$$

This is not always true on the boresight axis. If the power unbalance is such that one directional coupler couples too much energy (has a positive δ) there are several possible outputs. Reversing the values δ_1 and δ_3 in sign, the output expression is,

$$2\,\psi_{az} = \tan^{-1}\frac{-1}{-\delta} - \tan^{-1}\frac{-1}{\delta} \qquad (5)$$

By letting the δ values all equal 0.1 in voltage (0.84 db) the output angle becomes $2\,\psi_{az} = 11.4$ degrees. If only δ_1 and δ_2 are reversed in sign, $2\,\psi_{az} = 0$. If just one of the δ's, δ_1, is reversed, $2\,\psi_{az} = 0$ while if δ_3 is reversed in sign, $2\,\psi_{az} = 11.4$ degrees.

Now the target is taken on the elevation axis at a point 0.50 degrees from the boresight axis. Assuming $E_A = E_C$, $E_B = E_D$, and all δ's equal, the output equation is,

$$2\,\psi_{az} = \tan^{-1}\frac{-(1-\delta^2)(E_A+E_B)}{(1+\delta)^2E_A-(1-\delta)^2E_B} - \tan^{-1}\frac{(1-\delta^2)(E_A+E_B)}{(1+\delta)^2E_B-(1-\delta)^2E_A} \qquad (6)$$

From the antenna patterns $E_A = E_C = 0.237$ and $E_B = E_D = 0.803$. Assuming all δ's $= 0$, the angle measured is $2\,\psi_{az} = 57.2$ degrees. Other power unbalances in the couplers were found to give angular errors of the same order of magnitude.

The target is now taken at a point on the plane 45 degrees with the azimuth and elevation axis and 0.7 degrees from the boresight axis. Under these conditions, $E_A = E_D = 0.289$, $E_B = 1$, and $E_C = 0.0543$. Simplified equations can no longer be obtained. Resorting to equations (2) and (3) and by assuming all δ's $= 0.1$, the output angle is $2\,\psi_{az} = 59.6$ degrees. This is the same value obtained with the δ's equal to zero. If δ_1 and δ_3 are reversed in sign, $2\,\psi_{az} = 69.8$ degrees. This seems to be a large error, but considering the shift of the boresight null (11.4 degrees) the target measurement is 58.4 degrees. In another condition calculated the δ's $= 0.1$ with just δ_1 reversed in sign. In this case the boresight angle is $2\,\psi_{az} = 11.4$ degrees, and the target angle is $2\,\psi_{az} = 71.5$ degrees to give a target measurement angle of 60.1 degrees.

Using assumed δ values, the amplitude variation of the E_A'', E_B'', E_C'', and E_D'' signals is calculated for the target on the boresight axis. If they are of the same sign, the relative output signals are equal though the amplitudes change with the value of δ (0.1 db if the δ's $= 0.1$). If the signs of δ_1 and δ_3 are reversed the relative amplitudes will vary by 1.76 db (each output varying 0.88 db.).

The amplitude variation of the E_A'', E_B'', E_C'', and E_D'' signals can be used to determine the coupler unbalance and is therefore of interest. Solving for the four outputs of the coupler unit using a single input signal, two coupler δ's values are found from the expressions:

$$\begin{aligned} E_A'' &= E_A(1+\delta_1)(1+\delta_3) \\ E_B'' &= jE_A(1-\delta_1)(1+\delta_4) \\ E_C'' &= -jE_A(1+\delta_1)(1-\delta_3) \\ E_D'' &= -E_A(1-\delta_1)(1-\delta_4) \end{aligned} \qquad (7)$$

A single input signal to the E_B, E_C, and E_D inputs gives similar equations for separating the δ_3 and δ_4 values. To solve for δ_1 and δ_2 the coupler unit must be turned around, the signal fed into the double prime terminals and measured at the lettered terminals.

Appendix VII—Delay Line AGC Ratio Circuit Operation

Figure VII-1 shows in block diagram form a method of using AGC on a scanning radar. The ultra-sonic delay line allows the AGC to act on all range elements during each repetition period. The action during each range element is the same as described by equations (D-36) and (D-37). A simplified block diagram of the AGC ratio circuit is shown below.

Figure VII-I. Simplified Delay Line AGC Ratio Circuit.

The delay line is used in an integrator loop to improve the accuracy of the ratio circuit. The expression for the ratio output, equation (1), shows the importance of the integrator action in improving the accuracy and still maintaining stability.

ratio output =

$$\frac{k_2 k_3 v_s}{\mu_{oc}} [k - f_i(t)] \left[\frac{1 - \mu_{oc}^q [1 - k_2(k - f_i(t))]^q}{1 - \mu_{oc} [1 - k_2(k - f_i(t))]} + \frac{\mu_{oc}^q [1 - k_2(k - f_i(t))]^q}{1 - \mu_{oc}(1 - k_2 k)} \right]$$
(1)

where k = the steady state input of each range element
$f_i(t)$ = step input applied at zero time
k_2 = slope of IF gain curve in volts/volt/volt
k_3 = ratio of odd to even input
v_s = reference voltage for AGC circuit
μ_{oc} = open loop gain of integrator circuit
q = number of repetition periods, integers
f_r = repetition period and reciprocal of delay time

As is customary in the solution of a difference equation, equation (1) applies only for time intervals, $1/f_r$ long, corresponding to particular values of q. From (1) for $q = 0$

$$\text{ratio output} = V_s k_3 \left[1 + \frac{f_i(t)}{k} \right] \text{ for } 0 < t < 1/f_r$$
(2)

Equation (2) shows that there is no correction of the error in the ratio during the first repetition period. This is the expected result since no correction can reach the IF amplifier until the delay period $1/f_r$ elapses. By inspection, it is apparent that the quantities which are raised to the power q must be less than one if the output is to be finite as q approaches a large integer, or

$$|\mu_{oc} [1 - k_2(k - f_i(t))]| < 1$$
(3)

The above inequality can also be expressed in a slightly different manner.

$$1 + \frac{1}{\mu_{oc}} > k_2 [k - f_i(t)] > 1 - \frac{1}{\mu_{oc}}$$
(4)

If the integrator loop is to be stable μ_{oc} must be less than one, and it is desirable for μ_{oc} to approach one for ratio action,

$$k_2 [k - f_i(t)] >> \frac{1}{\mu_{oc}} - 1$$
(5)

When $\mu_{oc} \to 1$, to maintain stability

$$2 > k_2 [k - f_i(t)] > 0$$
(6)

If the two inequalities (5) and (6) are satisfied, then as q becomes large the ratio output reduces to

$$\text{ratio output} \cong k_3 v_s \cong \frac{\Delta}{\Sigma} v_s$$

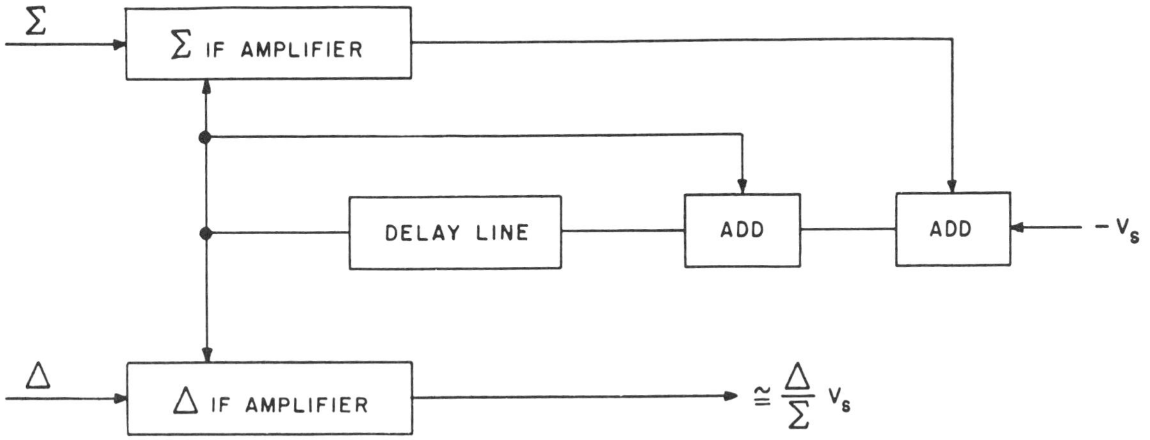

the desired result. The transient response will depend upon the DC level k, and can be any of the following.

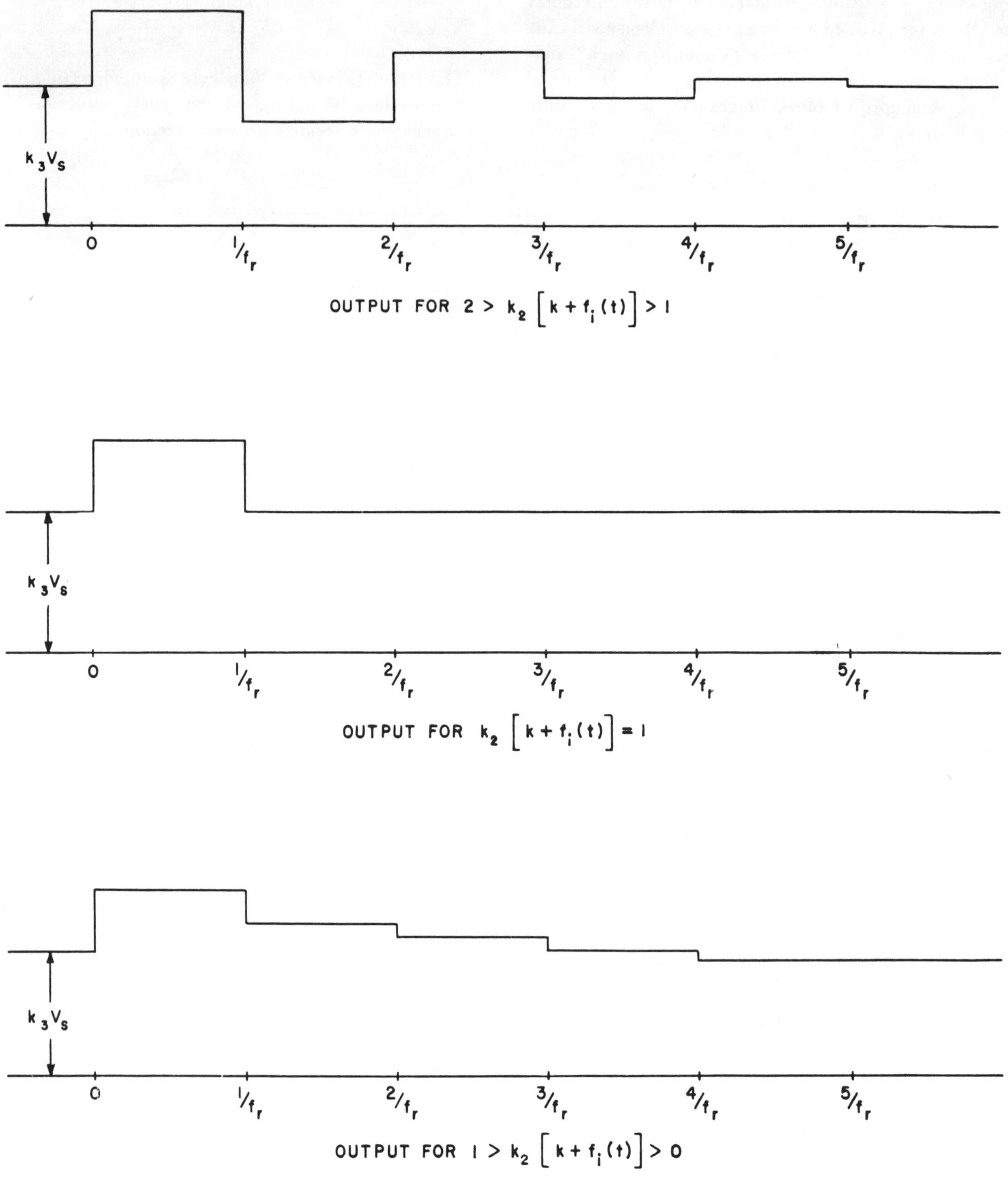

Figure VII-2. Delay Line AGC Ratio Circuit Transient Response.

The three outputs shown corespond to the transient solutions as follows: oscillatory for large signals, critically damped for a medium signal and over damped for a small signal. While the solution given applies to DC signals, it also applies to pulse signals of a repetitive nature. Thus, the circuit can handle β/f_r repetitive pulse signals of different levels simultaneously, where β is the IF amplifier bandwidth. The bandwidth required of the delay line will depend upon μ_{oc} although it will in general be much larger than β. The action of the circuit can best be described as providing AGC for each range element of a repetition period $1/f_r$.

The same delay line feedback arrangement can be used to monitor the gain or phase shift of an amplifier with the aid of an auxiliary monitoring signal. For phase monitoring, the usual AM detector would be replaced with a phase detector, and a suitable reactance tube phase shifter employed for phase control.

If this circuit is used for gain control of the IF of a scanning radar, the output variations are reduced to prevent overload of the IF amplifier or the indicator tube.

In Figure 2 of the fourth quarterly report, the use of a short video delay line was indicated to compensate for the delay in the IF amplifier. If more than one stage is gain controlled, then taps can be used on the short delay line to allow each stage to receive the control voltage at the instant of arrival of a particular repetitive signal. Enough hits per beamwidth must be available to allow the circuits to approach steady state when used with a scanning radar.

DAVID K. BARTON received his A.B. Degree in Physics from Harvard University in June 1949. Upon graduation he returned to White Sands Proving Ground, New Mexico, to resume, under Civil Service, the development work on guided missiles range instrumentation started during his Army service in 1947-48. Four years' experience in tracking radar and computer system engineering at White Sands was followed by two years in the Radar Division of the Signal Corps Engineering Laboratories at Fort Monmouth. Mr. Barton joined the Missile and Surface Radar Section at Moorestown, New Jersey early in 1955 and has since worked in the Systems Engineering Group on Instrumentation Radar and advanced systems problems.

IRVING STOKES, Manager of Information Handling and Radar Projects Engineering, received a BS in EE degree in 1938 from Newark College of Engineering. In 1939 he completed the graduate course in Thermionic Tube Theory at the Stevens Institute of Technology. He has also successfully completed courses in Management and Application of Statistical Analysis to Design of Experiments.

In 1938 Mr. Stokes joined Tung Sol Lamp Works, where he was connected with analysis and control of radio tube quality. He joined the U. S. Army Signal Corps as a civilian design and development engineer in 1940. Mr. Stokes was engaged in the design and development of radar applications for searchlight and anti-aircraft control; recognition and identification; paratrooper and advance airport homing beacons; ground navigation systems; electronic surveying and missile trajectory determination and guidance. During this period he was awarded four patents.

Prior to joining RCA in 1955, as Manager of the Precision Tracking Radar Systems Projects Group, Missile and Surface Radar Engineering, Mr. Stokes was Deputy Director of the Radar Division of the Signal Corps Engineering Laboratories. Mr. Stokes is a member of Tau Beta Pi and a former member of industry panels and committees in the field of radar and guided missiles.

GUIDED MISSILE INSTRUMENTATION RADAR

By

I. STOKES and D. K. BARTON

Missile and Surface Radar Engineering
Defense Electronic Products
Moorestown, N. J.

In 1946 THE FIRST German V-2 rocket to be fired in this country sat on the launcher at White Sands Proving Ground. After final count-down began, the men at the radar desert site, 1 mile south of the launcher tensely awaited the rocket launcher signal. The historical significance of the imminent event was of secondary importance in the minds of the participants to the desire to achieve a successful conclusion to nine or more months of intensive design, development and planning effort. At zero time, a cloud of smoke, flame and sand rose beneath the missile as it slowly lifted itself from the base and hovered precariously a few feet off the ground, seeming to drop back a little just prior to the start of a determined acceleration skyward. At launch a strong radar reply from the missile beacon appeared on the radar scope and automatic tracking began soon thereafter. At approximately 9,000 feet above the desert floor, the V-2 began to wobble noticeably so that observers at all points of the compass received the illusion that the rocket was about to fall in his direction. At this point, one of the fins tore off and fluttered towards the ground as the missile laid over on its side and headed for Texas. The fuel cut-off device was activated and impact 18,000 yards from the launcher resulted in a crater 30 feet deep and 30 feet across. The radar had tracked and plotted the complete trajectory, in spite of the fact that the lost fin had contained the beacon receiving antenna. This rather gratifying radar performance established the radar as a key device for guided missile instrumentation.

EARLY U.S. SYSTEMS

The development of high-altitude research rockets and guided missiles in the United States brought together instrumentation methods from many fields. From the aircraft test field came internal instrumentation, radio telemetry and recording devices. Anti-aircraft artillery evaluation has made extensive use of theodolites, and these were brought to bear on missiles in the earliest stages of development. Various fixed camera methods were derived from other ordnance projects, as well as from the field of Astronomy, which also contributed the long focal length telescope. Electronic instrumentation came from the fields of navigation, direction finding, fire control, close-support bombing, and from specialized methods such as the doppler radio techniques used by the Germans in the V-2 program. Since the program got under way toward the end of World War II, it was logical that military equipment, surplus and captured, would play a significant role, and this was certainly true in the case of radar equipment.

THE SCR-584 RADAR

One radar in particular, the SCR-584, proved particularly valuable to the test ranges, and has maintained its usefulness to this day. The primary reasons for this were its availability (some fifteen hundred sets had been built by the end of the war), reliability (a tremendous production and field engineering effort had followed the original MIT Radiation Laboratory development work), and flexibility. Designed for control of 90mm AA guns and similar weapons, the SCR-584 was equipped with alternate forms of data output, an oversized modulator and power supply, and ample space for additions and modifications that proved necessary. Furthermore, development of instantaneous electronic plotting boards (operating in rectangular coordinates) had already been carried out for use in close-support bombing and in mortar location. Transponder beacons were also available from the close-support bombing program, as was a standard radar modification for long-range tracking of beacons. Thus, the SCR-584 was ready for the earliest V-2 and Wac Corporal firings at White Sands Proving Ground, and was in use at other rocket ranges. Along with the radar, of course, there were used the famous German Askania theodolite, Bowen-

106

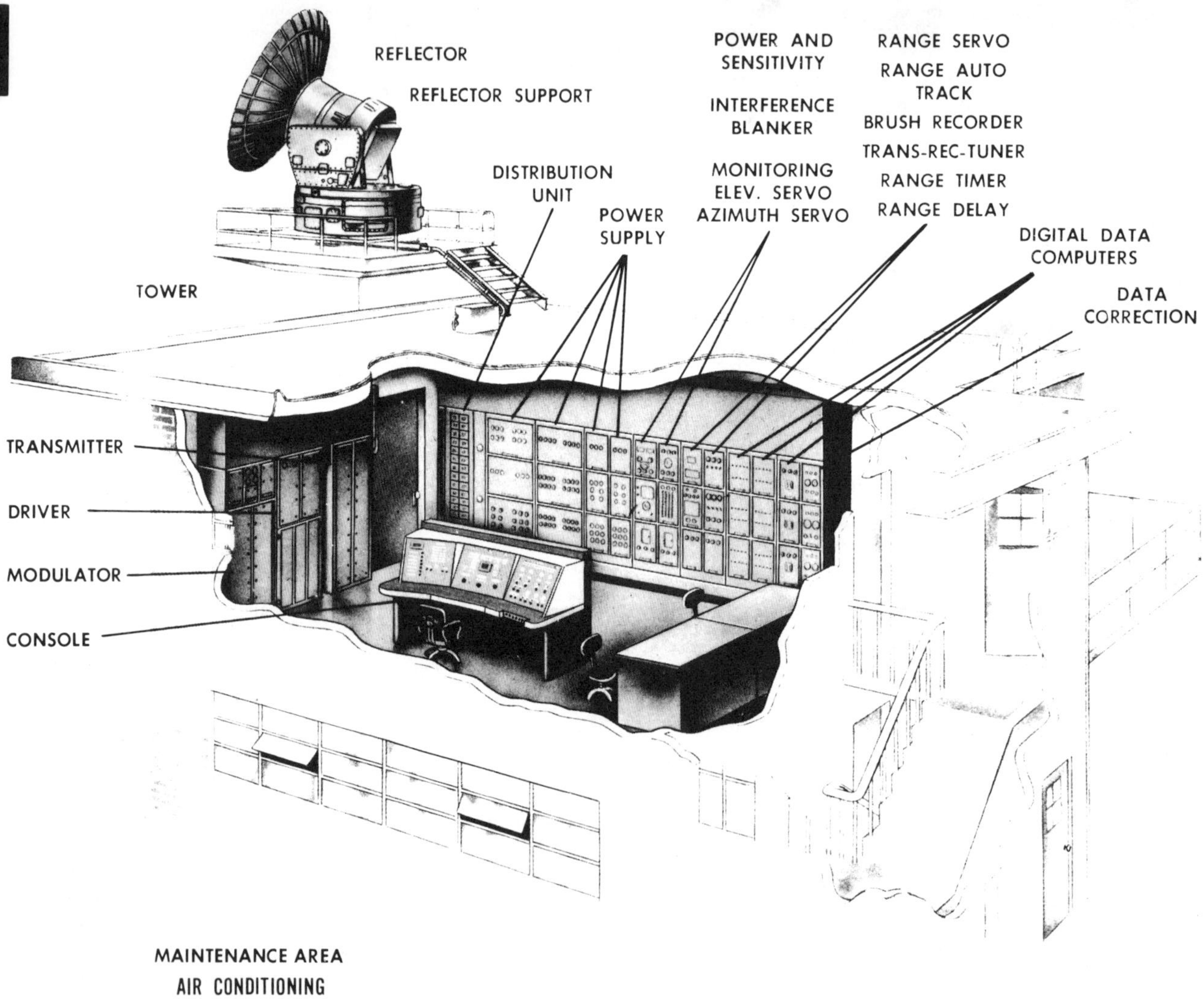

Fig. 1—Artist's conception of typical Instrumentation Radar equipment setup.

Knapp fixed cameras, and doppler radio system.

MODIFICATIONS AND REFINEMENTS

The flexibility of the original SCR-584 has already been mentioned. Dozens of important field modifications were made in order to incorporate design advances to this radar at various times in the range instrumentation programs. These were concerned primarily with extending range of tracking and range rates, synchronizing adjacent radars for reliability of operation without interference, recording of position and signal strength data, and providing control links to the missile through modulation of radar pulse rate or pulse code. Ultimately these and other improvements were incorporated through factory modification in the AN/MPQ-12 radar, which first became available in 1949, and in the improved AN/MPQ-18, which followed the MPQ-12 in design and has been operational since late 1954. These radars have been used primarily at WSPG, but similar modifications have been applied to the SCR-584 for use at NAMTC (Pt. Mugu), NOTS (China Lake), AFMTC (Cocoa), Edwards Air Force Base, and Eglin Field. Other nomenclature applied to modified SCR-584 radars includes AN/MPG-2 (refers to X-band set using MC607 kit on standard SCR-584), AFMTC tracking radar Mod II (similar to the MPQ-18) and AFMTC tracking radar Mod III (C-band version of Mod II). Some of these sets are still in the process of factory modification, with delivery extending through 1956. In each case, the basic SCR-584 pedestal, modulator and power supply, and 30-cycle conical scan is employed, while most other major components have been replaced or extensively modified. In some cases, the cost of modification has run to two or three times the original cost of the radar.

Parallel to evolution of improved radars has been a great effort in system design, involving chain radar operation, real-time computing, plotting and data transmission, precision data recorders, communications systems, transponder beacons, control and telemetry equipment. The effect of this effort, which will not be described in detail here, has been to provide further uses for radar data, and to keep a continuing pressure towards development of radars with greater precision and range, so that significant data can be supplied to the instrumentation system as a whole.

PRECISION RADARS

The inherent advantages of the use of radar for military purposes have encouraged great efforts in development of precision radars, which have proceeded continuously at RCA since the end of World War II. The chief results of this effort have been four major radar equipments culminating in the Instrumentation Radar, AN/FPS-16 (See Cover Illustration and Fig. 1).

The first of these major radar programs is known as the Bumblebee radar which was developed by RCA as an associate contractor of the Applied Physics Lab., Johns Hopkins University. The Bumblebee radar has been tested extensively and has been in continuous operation at Applied Physics Lab. since May 1953.

The second precision radar built by RCA is known as the Terrier radar, and was designed for use in the Army Application of the Navy Terrier Guided Missile System. This radar has been in continuous operation since early 1954, at the Naval Ordnance Test Station, Inyokern, California, where it was tested extensively by the Marine Corps.

The third member of this family of precision radars is the Instrumentation Radar AN/FPS-16 (XN-1) which was developed for the Navy Bureau of Aeronautics. The radar has been completed and has been evaluated by the Naval Research Laboratory. The results achieved with this radar have shown that its tracking performance is superior to any radar tested to date.

Two AN/FPS-16 (XN-2) radars are presently being built at RCA's Moorestown Plant. These radars are production prototypes and are expected to exceed the performance results achieved with the XN-1 model. (See equipment photos of Figs. 2, 3, 4 and 5.) A production contract for twenty-two AN/FPS-16 radars has been awarded to RCA with delivery commencing in December 1957. The proven high accuracy of the AN/FPS-16 design has resulted in the adoption

Fig. 2—David K. Barton, one of the authors, is shown at the control console of the Instrumentation Radar.

of this precision instrumentation radar for use at the Navy, Army and Air Force guided missile test ranges at Pt. Mugu, California; White Sands Proving Ground; and Patrick Air Force Base, Florida.

The net result of these developments, occurring over the last twelve years has been to make available a radar capable of precision and instrumental accuracy to a factor of ten greater than that obtained with modified SCR-584's.

BRIEF DESCRIPTION OF THE AN/FPS-16 INSTRUMENTATION RADAR

The Military Performance or operational characteristics of the AN/FPS-16 are classified and cannot be discussed here.

The radar is designed for installation as a fixed station and will be housed in a building similar to the artist's conception shown in Fig. 1. The antenna pedestal is mounted on a tower which is isolated from the building proper in order to minimize vibrations transmitted to the tower. The electronic equipment is housed in specially designed rack-type cabinets with ease of maintenance as a primary consideration. (See Figs. 3 and 5.) The equipment is cooled by a closed circulating air-conditioning system. A regulator is provided at the bottom of each cabinet to control the amount of cold air entering, with the return through a duct at the top of the cabinet. The operating console of modern design was "human engineered" for optimum operator efficiency. Only one operator is required even though the radar is one of the most complex built to date.

The obvious importance of reliable performance during operations involving the expenditure of thousands of dollars has dictated the use of the latest reliability techniques in the equipment design. In addition built-in checkout equipment has been included to insure adequate monitoring of component performance.

The antenna pedestal shown on the front cover was built to give high mechanical accuracy and smooth tracking, and was made rugged to give reliable operation under severe environmental conditions. Its performance has been such that it represents a major advancement in the state of the art.

Since Instrumentation Radar installations represent a considerable investment of money, effort and training, a high degree of flexibility has been achieved. This makes feasible the later inclusion of new advances in radar and thus forestalls early system obsolescence. In addition, the flexibility provides an instrument adaptable to a multiplicity of missions required by the guided missile test ranges. Some examples of this flexibility include the ability to:

a. Change frequency readily.
b. Provide synchro, potentiometer and digital data individually or simultaneously.
c. Track skin reflections or beacons.
d. Accept two or three coordinate designation data.
e. Be readily modified by substitution of chassis or racks in a generously spaced equipment layout.

In spite of all the sterling characteristics achieved in this radar design, RCA is currently engaged in the development of some significant improvements to enhance the system usefulness and maintain the enviable lead over competitive equipments.

Fig. 3—Merritt Sheeder, Project Engineer, is making a circuit check at the readily accessible underside of an Instrumentation Radar Chassis.

THE RADAR AND ITS TASK

A tracking radar is essentially a device whereby a highly directive antenna is positioned by a servomechanism onto the selected target, from which range and angular error signals are derived by reflection of the radar's transmitted wave or by use of a transponder in the missile or aircraft being tracked. The Instrumentation Radar operates in the microwave frequency. region and uses pulse transmission. The position of the target is given by two angles of the antenna (azimuth and elevation) and by slant range from the radar. The principal advantages of employing the radar for guided missile instrumentation, as compared with the best of competitive systems, are its ability to track with high accuracy in darkness and through clouds, as well as atmospherically clear conditions to long ranges, and to yield data which can be readily and instantaneously reduced to its final form. These advantages are possessed by no other means of range instrumentation. The Instrumentation Radar will therefore be used to track missile and aircraft targets (with or without beacon transponders) and to produce spherical coordinate data outputs of high accuracy from which an instantaneous record of the target trajectory can be obtained.

REAL-TIME PLOTTING AND RANGE CONTROL

Test ranges have for some time exploited the obvious advantages of the radar method where high accuracy has not been required. One such application has been instantaneous display of target data on plotting boards, for quick evaluation of flight operations and for range control and safety. In this application, the presence of all data in electrical form at one station, with reduction to rectangular form within the capability of simple analog computers, has hastened the growth of large installations for plotting of multiple targets in various projections, using data from several radars simultaneously. The ability of radars to track aircraft by reflection, as well as the abilities to conduct several tracks from a single station without interference, and to transmit commands to the target through special beacons, have made radar a primary means of control for drone aircraft and a potential means of guidance for various missile tests.

Existing modifications of the SCR-584 have proven sufficiently accurate for real-time plotting of position on a full-scale basis, but difficulties arise when small regions in space are expanded for detailed plots, or where velocity data is required. Limitations of these radars have prevented full exploitation of the radar-plotting system for impact prediction of high altitude missiles and similar tasks. With the availability of radars such as the AN/FPS-16, the limiting errors will be shifted back to the computing or plotting equipment in most cases, and predictions from velocity data will be greatly improved.

CHAIN RADAR SYSTEMS

As long as missile tests were conducted at short range or on near-vertical trajectories, a single radar station sufficed for tracking. However, as soon as long range horizontal flights or other trajectories of military interest were undertaken, the need for two or more coordinated tracking stations became apparent, and "chain radar" systems were developed. To this day radar systems with the ready availability of electrical data in a form for simple reduction, have been essentially the only source of real-time data for exchange between stations. Once available for transmission, radar data is used to position other types of instrument as well as succeeding radars in the chain. Accuracy of modified SCR-584 radars has proved barely sufficient for applications involving transmission of real-time data, and considerable system work has been done to preserve even this accuracy through the various conversions of coordinates and electrical form required. Present systems use analog data outputs from the radar to feed coordinate conversion analog computers located at each radar site. Analog data in rectangular form is then converted to digital form for transmission over communication links, and the reverse process followed after reception at the remote point. With the high accuracy of the AN/FPS-16 radar, the analog processes used in chain operation can be eliminated entirely. Designation of a target, remote plotting of position on an expanded scale, or recording of data for trajectory analysis purposes will thus be possible to the full accuracy of the radar instrument.

RADAR DATA PROCESSING

In addition to real-time use of analog data from the radar, most test ranges require some more accurate permanent record of radar tracking data. With the arrival of precise instrumentation radars such as the AN/FPS-16, this facet of radar instrumentation will assume increased importance. There is presented the prospect of precise tracking data, available under all-weather operations, with negligible reduction time required, and limited in accuracy chiefly by uncertainties of tropospheric refraction. Where most present installations depend upon photographic recording of synchro data, range scopes, mechanical scales and (when possible) boresight telescopes (Fig. 4), the new radars will provide direct digital read-out of angle and range data, as precise as (or more precise than) present theodolite data and requiring only solution of two simple triangles for reduction to rectangular form. Refraction correction-to-elevation angle data will also be needed to obtain highest accuracies, and in general some modification from rectangular coordinates is needed to put the data in form for final use. None of these processes need take longer than a few seconds per data point reduced, even in such machines as present card-programmed calculators, and in many cases real-time reduction should be possible. Direct electrical transfer (possibly through magnetic tape recording) will eliminate all manual reading operations. The method has already been demonstrated at the White Sands Proving Ground and elsewhere, limited by the accuracy of modified SCR-584 radars. Of particular importance in the new radar, is the elimination or correction within the radar of the several sources of significant error which have formerly required calibration and correction in the reduction process. In the AN/FPS-16, for instance, these errors will not merely be calibrated and known, but will be reduced below the threshold of significance for data reduction. Means of applying the refractive correction for average atmospheric conditions automatically have also been investigated and are feasible within the confines of the radar. Output data to the computers will thus be capable of immediate reduction by simple processes, without manual operations of any sort, and without any film processing.

COMMUNICATIONS FOR RADAR INSTRUMENTATION SYSTEMS

One of the major advantages of the radar method is that each station is self-sufficient as far as trajectory measurement is concerned. Communications channels are used for data transmission in the chain radar system assuring target acquisition both for radars and for optical and other instrumentation remote from the launcher. Other than this, communications requirements for a radar instrumentation system are chiefly for voice communications, signalling and control circuits, and timing. Even in the timing circuits there do not exist the extreme requirements for precision that apply to baseline instrumentation systems—it is only necessary to match the complete trajectory data obtained at the radar to some range time reference within a few tenths of a millisecond to achieve full system accuracy. Present digital data transmission equipment (such as AN/TSQ-1) is designed to operate reliably within the poorest voice channels, and thus an entire radar instrumentation sys-

Fig. 4—Defense Electronic Product engineers (left to right: Merritt Sheeder, Lawrence Carapellotti, and Simpson Adler) are examining an Instrumentation Radar schematic with the radar boresite tower in the background.

tem can be operated using standard voice communications links.

RADAR BEACONS

Transponder beacons developed for navigation and close-support bombing were used in the first V-2 and Wac Corporal in 1946. In the succeeding ten years, intensive development has made available a series of radar beacons expressly designed for range instrumentation. Small size, light weight and low-power drain are the distinguishing characteristics of most of these, and ruggedized construction far beyond that required in guided missiles has been achieved.

The use of a beacon improves radar tracking in three respects, as compared to reflection tracking. First it extends the range by an amount depending chiefly on the ability of the beacon receiver to sense the radar interrogation. Secondly, a beacon provides a point source for the radar to track, eliminating target scintillation. A comparison of reflection and beacon tracking results for similar flight paths shows that improvement is obtained at relatively short ranges. Thirdly, use of a beacon provides positive identification of the desired target and rejection of ground clutter. With the beacon transmitter offset several megacycles (or any desired amount within the radar band) reflective targets and clutter are attenuated 30 to 40 db. Special codings in the beacon receiver or transmitter can distinguish one beacon from another when several are used at once in the same area.

A further advantage of using a beacon is that it provides a two-way communication path between the target and the ground radar. Equipment is available to utilize this channel for guidance commands, emergency cut-off or detonation commands, and for telemetry from the target. The high gain and reliability in the radar-beacon path is available for these functions, while frequency selectivity, pulse coding, and the narrow radar beam provide protection from interference. Channel bandwidth is limited, of course, since the radar repetition rate provides the "carrier" for all information exchanged. Depending on the method of modulation, bandwidths in the order of a hundred cycles are available for each pulse in a code group used for information. Five or more such pulses may be used simultaneously in each direction of transmission, meeting most requirements for simple telemetry.

CONCLUSIONS

The Instrumentation Radar AN/FPS-16, based on extensive research and development effort carried on continuously since the last war, has been demonstrated to be capable of instrumental accuracy and precision well beyond the known uncertainties of tropospheric propagation. Combined results of field tracking tests, mechanical and electrical lab tests, and error analysis indicates extremely high overall accuracy.

The radar can maintain the above accuracy and precision under all normal weather conditions, target maneuvers and target locations (within the observed hemisphere of coverage) except at zenith and below 2 to 3 degrees in elevation angle.

Systems for using radar data in real-time and for later analysis are far advanced, and can provide complete trajectory data without appreciable time lags.

No other instrumentation system can attain higher accuracy over extended paths (10 miles or greater) under average weather conditions. Only plate camera systems (against a star field at night) offer any substantial margin of accuracy at any time.

ACKNOWLEDGEMENT

The list of technical contributors to the ten-year program leading to this major advance in the radar field would be too extensive to reproduce here. The authors have merely recorded the biographical sketch of the growth of an important new equipment. The credit for engineering advances of the system belong to the Missile and Surface Radar Department collectively and to greater and lesser degrees with individual engineers in the organization.

Fig. 5—Lawrence T. Carapellotti, Project Leader of System Evaluation is applying the test prods to one of the rack chasses, while Simpson B. Adler, Project Leader of Precision Tracking Radar Development, reads the meter.

REPRINTED FROM 1959 CONFERENCE PROCEEDINGS, 3RD NATIONAL CONVENTION ON MILITARY ELECTRONICS, INSTITUTE OF RADIO ENGINEERS, PROFESSIONAL GROUP ON MILITARY ELECTRONICS

ACCURACY OF A MONOPULSE RADAR

By David K. Barton
Radio Corporation of America
Missile and Surface Radar Department
Moorestown, New Jersey

1. Background

The precision monopulse tracking radar is the result of twenty years of intensive research and development in radar tracking devices. Beginning with the earliest British and American designs of lobe-switched beams for angle measurement, requirements for higher accuracy and more rapid measurement have led to a succession of improved radars. The "sequential scan" types, including lobe-switched and conical scanning radars widely used in World War II, remain in use in most airborne systems and in ground systems of moderate accuracy. Present designs for high accuracy, however, are primarily "simultaneous lobe comparison" or monopulse radars. A historical view of the development of this technique is contained in Donald Rhodes recent book "Introduction to Monopulse" (Ref. 1), along with an analysis of the design problems encountered.

Current research and development objectives aim at obtaining accuracies in both target position and velocity measurement which are limited only by factors external to the radar: target characteristics and propagation anomalies. The following discussion presents a method of quantitatively evaluating bias and noise errors in a refined monopulse tracker, and of predicting the accuracy of both position and velocity data for a variety of tracking conditions through analysis of spectral characteristics of the error. The Instrumentation Radar AN/FPS-16 is used to illustrate the application of the method, and actual test results are given. Characteristics of this radar are given in Appendix A.

2. Inventory of Radar Errors

Radar errors can be classified in several different ways:

a. By source, into radar-dependent, target-dependent propagation and apparent errors (those arising in the instruments used for radar testing);

b. By frequency distribution, into noise or AC component, and bias or DC component;

c. By coordinate, into range, azimuth and elevation errors; and

d. By point of entry into the system, into tracking errors (causing the boresight axis or range gate to depart from the target), and translation errors (causing the output data to misrepresent the position of the boresight axis or range gate).

For a monopulse tracker of the type discussed here, Figures 1 and 2 list the error components which require evaluation before the total error can be predicted. Time and space do not permit a detailed discussion here of each error component, the theory describing it and the practical problems in controlling and measuring it. Reference 2 contains detailed data on many of these considerations. It is entirely possible, however, to analyze and measure each error, and to design, maintain and operate the radar so that total error is within 0.1 mil in angle and 5 yards in range. Furthermore, as will be shown, the precision (noise level) of the data is frequently as good as 0.03 mil and 2 yards. Before proceeding

	Bias	Noise
Radar-dependent tracking errors	Boresight axis collimation Axis shift with: RF and IF tuning Receiver phase shift Target amplitude Temperature Wind force Antenna unbalance Servo unbalance	Receiver thermal noise Multipath (elevation only) Wind gusts Servo electrical noise Servo mechanical noise
Radar-dependent translation errors	Levelling of pedestal North alignment Static flexure of pedestal and antenna Orthogonality of axes Solar heating	Dynamic deflection of pedestal and antenna Bearing wobble Data gear nonlinearity and backlash Data take-off nonlinearity and granularity
Target-dependent tracking errors	Dynamic lag	Glint Dynamic lag variation Scintillation Beacon modulation
Propagation errors	Average refraction of troposphere Average refraction of ionosphere	Irregularities in tropospheric refraction Irregularities in ionospheric refraction
Apparent or instrumentation errors (for optical reference)	Telescope or reference instrument stability Film emulsion and base stability Optical parallax	Telescope, camera or reference instrument vibration Film transparent jitter Reading error Granularity error Variation in optical parallax

Fig. 1 Inventory of angle error components.

further, we should note that the various error components are essentially uncorrelated, from which it follows that total error is the rms sum of the components present during the period to be considered.

3. Angle Tracking Noise Equations

Referring to the noise components listed in Figure 1, it can be shown that thermal noise, multipath error, glint, lag and tropospheric propagation contribute the major portion of the total noise error in a precise tracking radar. Formulae for evaluating these components will be given below. Thermal noise in an efficiently-designed monopulse tracker is:

(1) $$\sigma_t = \frac{\theta}{\sqrt{2 \frac{S}{N} \frac{f_r}{\beta_n}}}$$ mils rms (measured normal to the line-of-sight),

112

where θ is half-power antenna beamwidth in mils, S/N is single-pulse signal-to-noise ratio, f_r is repetition rate in pps and β_n is equivalent noise bandwidth in cps of the servo at whose output σ_t is measured (this and other relationships used are derived in Ref. 2). The error due to multipath or ground reflection is:

(2) $$\sigma_m = \frac{\theta \rho}{\sqrt{8A_s}}$$ mils rms (in elevation only),

where ρ is ground reflectivity, and A_s is the power ratio of mainlobe gain to gain in the side-lobe receiving reflections. Normally, the reflected energy is received at an off-axis angle equal to twice the elevation angle of the target. This error is cyclic over an interval:

(3) $$\Delta E = \frac{\lambda}{2h}$$ radians (in elevation),

where λ is radar wavelength and h is antenna height above ground (λ and h are in similar units). The glint error is approximately equal to a constant fraction (say 1/4) of the target span L_t measured normal to the beam, and its angular amplitude can be approximated by:

(4) $$\sigma_g = \frac{250\, L_t}{R}$$ mils,

where R is range measured in the same units as L_t. Lag error for a continuous track (after elimination of lock-on transients) is given by:

(5) $$\xi = \frac{V_t}{K_v} + \frac{A_t}{K_a}$$

where V_t and A_t are target velocity and acceleration, respectively; K_v and K_a are radar velocity and acceleration error constants, and ξ is in the same length or angle units that are used in expressing velocity and acceleration. If radar angular lag is to be calculated, V_t must be in units of angle per second, A_t in angle per second^2, and both must include any "geometrical" terms resulting from transformation of straight-line target courses into the radar's spherical coordinates (see, for instance, Ref. 3). The servo error constants depend upon servo design, but for any servo the acceleration error constant is limited to:

(6) $$K_a \leq 2.5\, \beta_n^2$$

For estimating the variable component of lag, the rate of change of (5) may be used, and the variation over whatever period is appropriate to the problem found:

(7) $$\text{Variation} = \dot{\xi}\, \Delta t = \left(\frac{A_t}{K_v} + \frac{\dot{A}_t}{K_a} \right) \Delta t$$

Lastly, the error due to tropospheric anomalies can be estimated as:

(8) $$\sigma_p \cong 2\, \Delta N \times 10^{-3} \sqrt{L/l_o}$$ mils rms (measured normal to the line-of-sight)

where ΔN is the magnitude of the anomaly in refractive index modulus ($N = [n-1] \times 10^6$), L is the length of the path traversing the region of anomalies, and l_o is the scale length of the anomalies (this equation is derived in Ref. 4, with a factor $\sqrt{\pi}$ instead of 2).

	Bias	Noise
Radar-dependent tracking errors	Zero range setting Range discriminator shift Servo unbalance Receiver delay	Receiver thermal noise Multipath Servo electrical noise Servo mechanical noise
Radar-dependent translation errors	Range oscillator frequency Data take-off zero setting	Range resolver error Internal jitter Variation in receiver delay Data gear nonlinearity and backlash Data take-off nonlinearity and granularity Range oscillator instability
Target-dependent tracking errors	Dynamic lag Beacon delay	Dynamic lag Glint Scintillation Beacon jitter
Propagation error	Average tropospheric refraction Average ionospheric refraction	Irregularities in tropospheric refraction Irregularities in ionospheric refraction

Fig. 2 Inventory of range error components.

Noise due to wind gusts, scintillation, beacon modulation and ionospheric irregularities can also be evaluated theoretically, but these components are so small for the case considered here that the formulae are omitted. The values of these components are included, however, in the overall error analysis below.

4. Experimental Techniques

Noise due to servo electrical and mechanical features, pedestal design, and the data take-off components can best be evaluated experimentally for the specific radar concerned. In performing such an evaluation on the AN/FPS-16, a series of new test techniques were devised in attempts to measure the extremely small errors encountered. A brief description of some of these tests is given below.

To measure servo noise under actual tracking conditions, especially at low rates where "cogging" and other mechanical problems are severe, free balloons carrying reflective spheres were tracked by the radar. Boresight telescope photographs were carefully read to determine noise amplitude over various periods of the track, and power spectra were obtained to assist in separating the various error components. Figure 3 presents results of one such

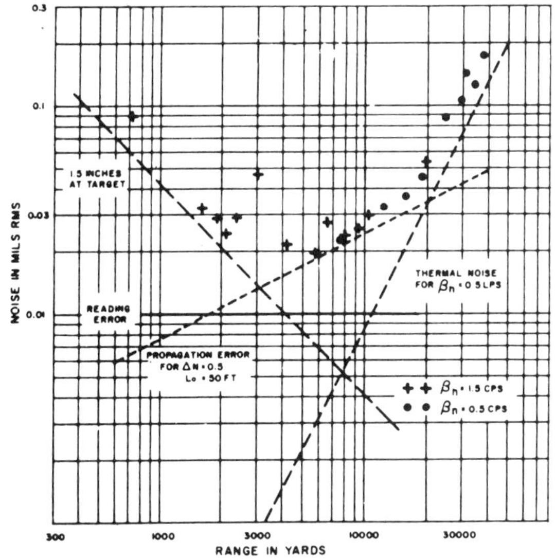

Fig. 3 Azimuth tracking noise vs. range, sphere track No. 7.

test. The calculated values of thermal noise from Equation (1), propagation error from (8), reading error of the film, and uncertainty in sphere location relative to the balloon are shown as straight lines. Multipath error is absent from the azimuth data, glint was eliminated by using a six-inch diameter sphere, and lag was minimized (at least at the longer ranges) by use of the free balloon in relatively smooth air. It can be concluded that servo noise did not exceed about 0.01 mil in this case. Other tracks indicated values up to 0.03 mil, depending upon the servo gain used. In order to obtain this data, it was necessary to attach the sphere in the center of the balloon, presenting a stable configuration and avoiding the rapid bouncing of the sphere with respect to the visible balloon which was encountered when the sphere was suspended below the balloon. On the first test run, this bounce amounted to some twelve inches, contributing errors greater than 0.03 mil at all ranges less than 10,000 yards. Although the sphere could not be held any closer than one or two inches to the balloon center, the error from this was relatively small beyond 5000 yards.

Measurements of pedestal and data take-off errors, which show up in translation of the tracking axis to azimuth and elevation readings, were carried out using optical autocollimators and a set of precisely calibrated polygonal mirrors and flats. Static values obtained were:

Bearing wobble	0.005 mil rms
Data gear nonlinearity	0.025 mil rms
Data gear backlash	0.004 mil rms
Pedestal total 0.026 mil	
Data take-off nonlinearity	0.018 mil rms
Data take-off granularity	0.015 mil rms
Encoder total 0.024 mil	

Dynamic deflections, measured by attaching a photocell to the autocollimator, did not exceed the 0.05 mil threshold sensitivity of the sensor, even during peaks exceeding one radian/sec^2 acceleration. As a result, they can be assumed entirely negligible compared to lag error.

5. Total Angle Noise Analysis

Since the angle noise depends upon conditions of the track, a specific example will be chosen for analysis:

Target	Earth satellite, velocity 4 mi/sec, altitude 115 miles, crossing 40 miles from radar zenith.
S/N Ratio	+15 to +30 db
Repetition Rate	285 pps
Servo Bandwidth	5 cps
Beamwidth	20 mils
Sidelobe Ratio	45 db at all angles beyond 15°
K_v	300 per second
Weather	Clear, $\Delta N = 0.5$, $l_o = 50$ ft.

Calculations will be made at minimum range R_1 of about 125 miles, elevation 70°; and at maximum range R_2 of 250 miles, elevation 25°. Figure 4 shows the results in tabular form. The actual radar noise error, measured as standard deviation from a smooth curve passed through

Component	Noise Error in RMS Mils			
	Azimuth		Elevation	
	At R_1	At R_2	At R_1	At R_2
Thermal Noise	0.180	0.360	0.060	0.330
Multipath	0.000	0.000	0.020	0.020
Servo Noise	0.010	0.010	0.010	0.010
Pedestal	0.026	0.026	0.026	0.026
Data Take-off	0.024	0.024	0.024	0.024
Glint	0.010	0.002	0.010	0.002
Lag Variation	0.025	0.000	0.020	0.000
Tropospheric	0.060	0.030	0.020	0.030
Wind Gusts	0.002	0.002	0.002	0.002
Scintillation	0.010	0.000	0.003	0.000
Ionospheric	0.000	0.001	0.000	0.001
Total RMS Error	0.185	0.360	0.076	0.330
Total Error for S/N=40db	0.080	0.050	0.050	0.054

NOTES: Effective target span of 10 feet was assumed.
Lag variation calculated over a one-second period.
Tropospheric effects assumed to 20,000 ft. altitude.
Winds up to 30 mph with 2 mph gusts assumed.
Scintillation calculated for ± 3 db echo variation.

Fig. 4 Angle noise analysis for satellite track.

the recorded data, is shown in Figure 5 along with the actual S/N ratio plot for the target, which was the Soviet Sputnik II. The theoretical curve on Figure 5 is that for thermal noise only, which is the predominating component. The azimuth error from Figure 4 has been transformed back to traverse error (measured normal to the line-of-sight, rather than in a hori-

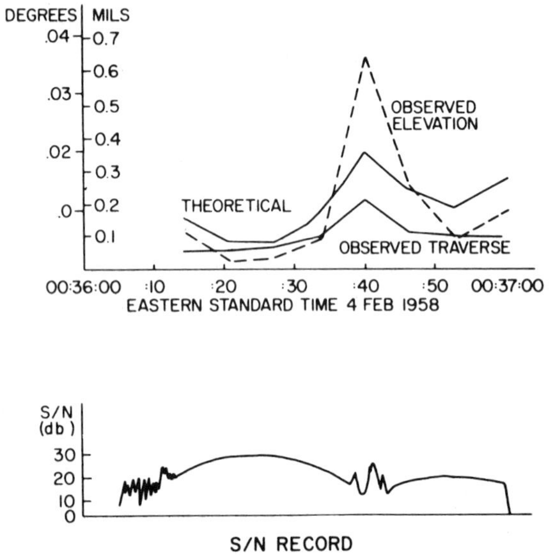

Fig. 5 Angle precision,
Feb. 4 track of Sputnik II.

zontal plane) for plotting on Figure 5. It should be noted that the effect of azimuth error on target position data is proportional to the cosine of elevation angle, and hence the traverse error is a good measure of radar accuracy.

The satellite track exemplifies data in which thermal noise governs overall error. In the many cases where a beacon is available to assist the radar, the overall tracking error commonly drops to the 0.05 to 0.08 mil level shown on the last line of Figure 4 for an assumed S/N = 40 db. In this case, the controlling errors are those due to pedestal, data take-off, multipath and tropospheric terms, and increase in S/N ratio beyond +40 db proves to have little value. Figure 6 shows the dispersion measured from the recorded digital data on a typical beacon track, which extended to a range in excess of 500 miles. The break in data between 280 seconds and 380 seconds represents the time required to reacquire the target "second time around" on the range system.

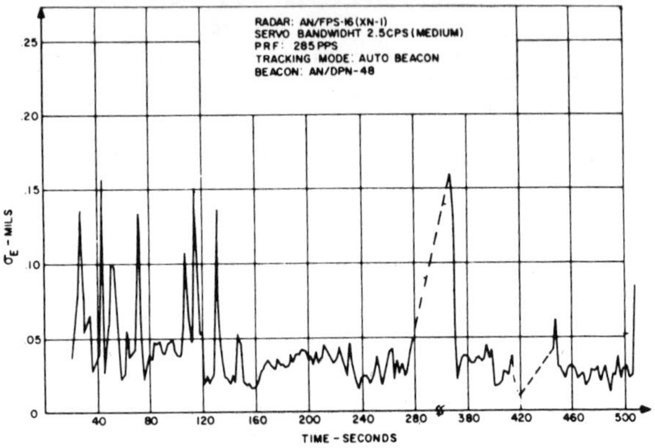

Fig. 6.

6. Angle Bias Error

Although handicapped by lack of good absolute reference data, it has been possible to evaluate separately each component of bias error, and to form an estimate of total bias which checks with field observations. The radar contributions are approximately constant for a given level of maintenance and calibration, with values as given in Figure 7.

In addition, there will be more or less constant lag errors over some periods of tracking, which may be evaluated according to Equation (5) for any given tracking situation and servo bandwidth setting. Finally, the propagation term (entirely due to the troposphere) must be considered. This will be large for elevation data, but may be estimated and corrected for standard atmospheric conditions or for measured conditions. The correction ΔE may be calculated approximately using the following equation (see Ref. 2, Appendix J):

$$(9)\quad \Delta E = \frac{N_s x 10^{-3}}{\tan E} \times \frac{H}{H + 35000} \text{ mils} \quad (E > 3°)$$

where N_s is the surface refractive index modulus (ranging from 220 to 380 for various U.S. locations at various times), E is elevation angle, and H is target altitude in feet above sea level. In Reference 5, Pearson and his associates at White Sands have arrived at a similar relationship, using H/tan E = ground range of target (for the flat earth case) and adjusting the value of the constant over the range from 16,500 to 26,000 feet, depending upon the N_s value, instead of the 35,000 feet used above. Using this or similar

Component	Bias Error in RMS Mils	
	Azimuth	Elevation
Boresight Axis Collimation	0.02	0.02
Electrical Axis Shift	0.03	0.03
Wind Force	0.02	0.01
Antenna Unbalance	0.00	0.01
Servo Unbalance	0.01	0.01
Pedestal Levelling	0.00	0.01
North Adjustment	0.02	0.00
Static Flexure	0.00	0.03
Orthogonality of Axis	0.01	0.00
Solar Heating	0.01	0.02
Total Radar Bias	0.05	0.055

Fig. 7 Radar-dependent angle bias errors.

correction formulae, the tropospheric refraction error can be reduced to 0.05 mil for elevation angles above 3°. Overall system errors within 0.1 mil are then possible when lag errors fall within that value.

7. Range Error

The ability of pulsed radar to measure range with great accuracy is well-known, and the treatment of range errors here will be brief. Receiver thermal noise contributes a component:

$$(10) \quad \sigma_{rt} = \frac{\tau}{\sqrt{\frac{S}{N}\frac{f_r}{\beta_n}}} \text{ feet rms}$$

where τ is the equivalent radar pulse length in feet (1 μsec = 500 feet), and the other symbols are as used in (1) above. The multipath error has a value:

$$(11) \quad \sigma_{rm} = \frac{\rho h \sin E}{\sqrt{8A_s}} \text{ feet rms,}$$

where ρ, h (in feet), E and A_s are as defined in (2) and (3) above. Dynamic lag and glint terms can be calculated according to (4), (5), (6) and (7) above, converting target rates and span into radial components. The tropospheric bias equation is (from Ref. 2):

$$(12) \quad \Delta R = \frac{0.0235N_s}{\tan E} \times \frac{(H+10^{-4}H^2)}{(50,000+H+10^{-7}H^2)} (E > 3°)$$

where ΔR is the excess range seen by the radar and E, N_s and H are defined in Equation (9). The maximum value of excess range is in the order of 150 feet, and corrections using (12) reduce error to 1 or 2 feet. Anomalies have such a small effect on range noise that they can be neglected, as can ionospheric effects at C-band. Values of error in Figure 8 are calculated for the example used in part 5 above, using a K_v of 3000, a servo bandwidth of 7 cps, and pulse width of 1 μsec. Experimentally determined values are used for components not discussed above. Figures 9 and 10 illustrate the measured noise errors in recorded digital data for the satellite track and for the beacon track. Errors of 4 to 10 feet are the rule. No independent check of bias error is possible due to lack of a reference standard.

8. Velocity Error

A frequently used relationship for estimating the accuracy of velocity data obtained over a period T seconds by differentiation of position data is:

$$(13) \quad \sigma_v = \frac{\sigma_p \sqrt{2}}{T}$$

where σ_v and σ_p are standard deviations in velocity and position data respectively. The relationship is exact for a two-point difference method where the points are T seconds apart and have uncorrelated errors. As shown in Ref. 2, the relationship is valid for a wide variety of noise errors, and can be made exact for any noise component by introducing a

Component	Error in RMS Feet	
	Bias	Noise
Zero Setting	5	-
Discriminator Shift	3	-
Servo Unbalance & Noise	2	1
Receiver Delay	1	0
Thermal Noise (S/N=30db)	-	2
Multipath	-	
Range Resolver	-	5
Range Oscillator	2	2
Data take-off	1	1
Data Gears	1	0
Internal Jitter	-	4
Dynamic Lag	8	2
Glint	-	2
Scintillation	-	1
Tropospheric Refraction (corrected)	1	0
Ionospheric Refraction	1	0
Total Error	11	8

NOTE: A momentary lag error of 20 feet appears at the point of closest approach. The 8-foot error appears over the remainder of the track.

Fig. 8 Range error analysis for satellite track.

115

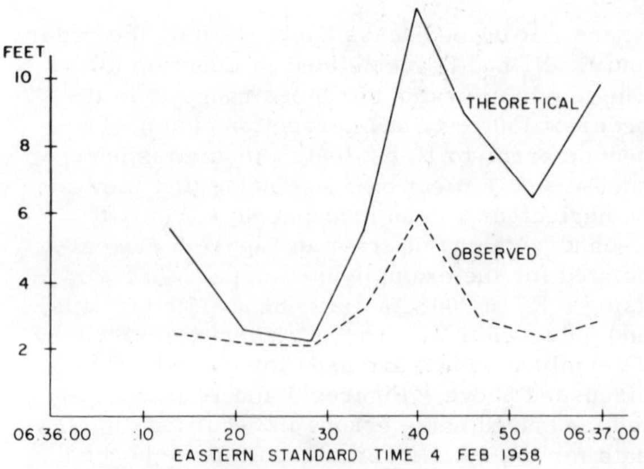

Fig. 9 Range precision,
Feb. 4 track of Sputnik II.

correction factor $F(\omega T)$ or $A(aT)$, depending upon the frequency ω of cyclic noise or the half-power frequency a of Markoffian noise. Figure 11 is a plot of the factors $F(\omega T)$ and $A(aT)$ for a typical differentiating network response. It is noted that both factors are greatest when ωT or aT is equal to π; that is, when one-half cycle of the cyclic noise or of the half-power frequency of Markoffian noise occupies the smoothing period T. At values of ω or a above or below this critical frequency, equation (13) gives higher velocity error estimates than are necessary. An inspection of each position noise component is required to determine its spectral distribution before (13) can be applied. In general, the total velocity error is then found by summing the squares of the various components, each modified by its correction factor:

$$(14)\quad \sigma_v^2 = \frac{2}{T}\left[\sigma_{p_1}^2 F^2(\omega_1 T) + \sigma_{p_2}^2 F^2(\omega_2 T) + \cdots + \sigma_{p_n}^2 A^2(a_n T) + \sigma_{p_{n+1}}^2 A^2(a_{n+1} T) + \cdots\right]$$

where σ_{p_1} and $F(\omega_1 T)$, etc., describe a series of cyclic error components and σ_{p_n} and $A(a_n T)$, etc., describe the Markoffian components. Each Markoffian component is characterized by a frequency spectrum (for power density):

$$(15)\quad S(\omega) = K_o \frac{a^2}{a^2 + \omega^2} \quad (-\infty < \omega < \infty)$$

and a position error (variance):

$$(16)\quad \sigma_p^2 = \int_{-\infty}^{\infty} S(\omega)\, d\omega = \pi K_o a$$

where K_o is zero frequency noise density in (position units)^2 per radian/sec, and a is the half-power frequency in radians/sec.

As an example, assume that the velocity of the Sputnik II at maximum radar range (250 miles) is to be measured with a smoothing time of ten seconds. (The analysis will be valid provided the acceleration due to gravity is compensated in the computer). For simplicity, only errors due to thermal noise, pedestal or resolver error, and the troposphere will be considered, and spectral characteristics will be assumed for the position noise as given in Figure 12. Target velocities in the directions measured by radar angles are seen to be accurate to 15 ft/sec, while radial velocity is accurate to 0.24 ft/sec. Calculations for T = 3 sec produce total angle velocity error of 0.055 mil/sec (80 ft/sec), range velocity error of 1.5 ft/sec. Calculation for S/N = 30 db and T = 3 sec. reduces thermal noise and results in angle velocity error of 0.018 mil/sec (27 ft/sec), range velocity error of 0.5 ft/sec. A similar

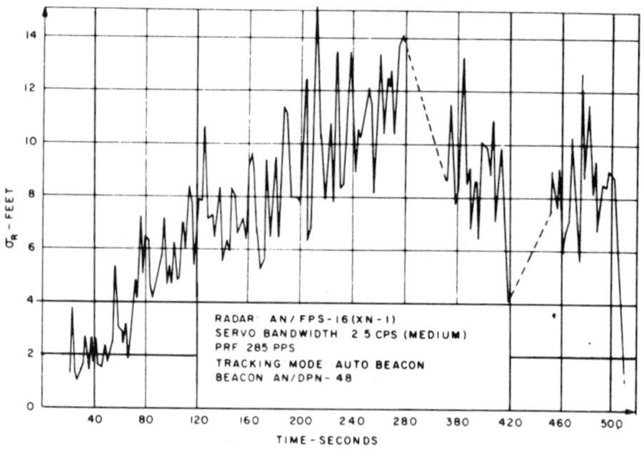

Fig. 10.

analysis can be extended to all error components and modified to cover any desired smoothing time. A complete power spectrum for tracking error can be obtained in cases where a reference such as boresight telescope data is available. Figure 13 represents such a spectrum from the sphere track of Figure 3. The high amplitude below ω = 1 rad/sec is presumed to be propagation noise, Markoffian in nature with σ_p = 0.035 mil and a = 1 rad/sec, close to the value used in Figure 12. In Figure 13, the spectrum has been plotted for positive frequencies only, using densities just double those represented by K_o in Equations (15) and (16). Using 10-second smoothing time on the balloon data, all components of wind could be measured to 0.25 ft/sec at 50,000 feet range from the radar.

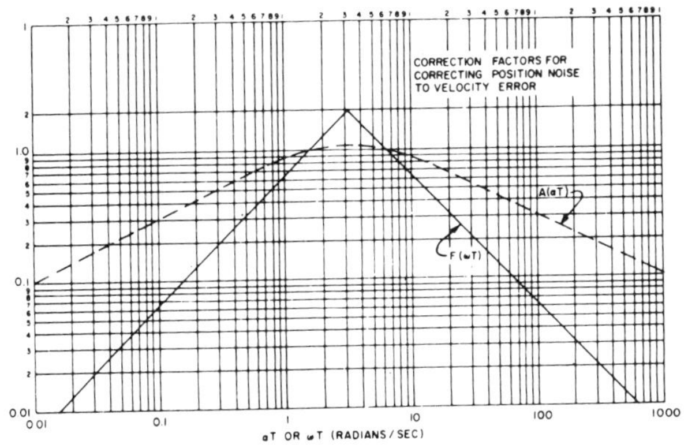

Fig. 11.

9. Conclusions

Sources of radar error have been analyzed by theoretical and experimental means, and broken down into components whose amplitudes and effects can be estimated under given tracking conditions. Test data taken at field locations show that modern monopulse radar can achieve accuracies heretofore reserved to laboratory equipment and complex optical networks. The data are available in electrical form during tracking and require only correction for tropospheric refraction, according to simple correction formulae, to obtain absolute accuracy of 0.1 mil in angle and 5 yards in range. This full accuracy is maintained without dilution from geometry at elevation angles down to a few degrees, and with moderate reduction in accuracy to about two degrees above the horizon. Precision or smoothness of tracking permits derivation of velocity data on cooperative targets to about 25 ft/sec at 250 miles range with 3-second smoothing, with the radial component 50 times better. Echo tracking results deteriorate, of course, as range approaches the limit set by the radar equation, but actual measurement on a Soviet Sputnik has given data of higher accuracy than achievable by other means.

Component	Spectrum	Frequency Parameter (rad/sec)	σ_p	A(aT) or F(ωT)	σ_v
Thermal (angle)	Markoffian	a = 31	0.36	0.18	0.009
Thermal (range)	Markoffian	a = 44	11	0.15	0.24
Pedestal (angle)	Cyclic	ω = 0.13	0.026	0.8	0.003
Resolver (range)	Cyclic	ω = 13	5	0.05	0.035
Troposphere (angle)	Markoffian	a = 1	0.030	0.9	0.004

Total angle velocity error (mils/sec) 0.01
Total angle velocity error (ft/sec) 15
Total range velocity error (ft/sec) 0.24

NOTES: σ_p is in mils or feet

σ_v is in mils/sec or ft/sec

Smoothing time T = 10 sec

Fig. 12 Conversion of position noise to velocity error.

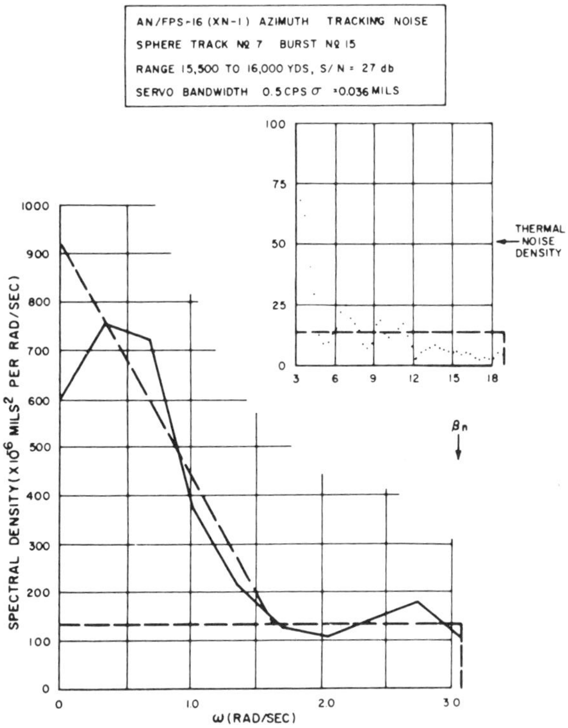

Fig. 13 Spectrum of azimuth tracking noise.

Appendix A

AN/FPS-16 INSTRUMENTATION RADAR CHARACTERISTICS

ANTENNA PEDESTAL
- Axes - Azimuth and Elevation
- Weight - 12,000 Pounds
- Plunging Capability - Yes
- Azimuth Coverage - Continuous 360°
- Elevation Coverage - 10° to +85° (200° Movement Available)

ANTENNA SYSTEM
- Size - 12-Foot Diameter Parabola
- Feed - 4-Horn Monopulse
- Gain - 45 db
- Beamwidth - 1.1°
- Polarization - Vertical
- Rotary Joints and Waveguide - 3 Megawatt Capability

TRANSMITTER

Type - Magnetron

Peak Power	Frequency	Max. Duty Cycle
1 MW Fixed Tuned	5480 ± 35 MC	.001
250 KW Tunable	5450 - 5825 MC	.0016

PRF - Internal: 12 Steps between 160 to 1707 PPS
External: Any PRF between 160 to 1707 PPS
Pulse Width - 0.25 μs, 0.5 μs and 0.1 μs
Coding - Up to 5 pulses within Duty Cycle Limitations

RECEIVER SYSTEM
- System Noise Figure - <11 db
- I. F. Center Frequency - 30 MC
- Bandwidths - Wide 8.0 MC - Narrow 1.6 MC
- Local Oscillators - Two - Skin and Beacon
- AFC - Skin and Beacon
- Non-Tracking I. F. - Non-gated manually gain controlled receiver; video added to tracking video for display only.

RANGE TRACKING SYSTEM

Gate Widths	-	Pulse Width	0.25 µs	0.50 µs	1.0 µs
		Acquisition	1.0 µs	1.25 µs	1.75 µs
		Tracking	0.5 µs	0.75 µs	1.25 µs
Aided Tracking	-	Yes			
Maximum Slew Rate	-	40,000 yds. per second			
Maximum Tracking Rate	-	8,000 yds. per second			
Automatic Lock-On	-	Search ± 1000 yds. and auto lock			
Servo Bandwidth	-	Continuously adjustable manually or automatically between 0.5 cps (K_V = 2000) and 6.0 cps (K_V = 3000)			

ANGLE TRACKING SYSTEM

Aided Tracking	-	Yes
Maximum Slew Rate	-	Azimuth 48° per second Elevation 37° per second
Maximum Track Rate	-	Azimuth 40° per second Elevation 30° per second
Servo Bandwidth	-	Continuously adjustable manually or automatically between 0.5 cps (K_V = 150) and 4.0 cps (K_V = 300)
Scan	-	Circle scan of adjustable radius and rate. Also sector scan in azimuth and elevation.

DATA OUTPUTS

Potentiometer		
Synchro	-	Army or Navy Speeds
Digital	-	Serial straight binary Range - 20 Bits (0.5 yd. quanta) Angle - 17 Bits (<.05 mil quanta)

DISPLAYS

Range	-	Dual "A" Scope Dials Digital - Octal numeral
Angle	-	Dials Digital - Octal numeral

MONITORING

Noise Figure Measurement
Transmitter Power
Range and Angle Servo Performance
Significant Waveforms
Significant Voltages and Currents
Strip Chart Recorder and Signal Patch Panel

CABINETS

Transmitter	-	2
Console	-	1
Equipment Racks	-	27

PRIMARY POWER REQUIREMENTS - 75 KVA, 208/115V
3 Ø, 60 cycles

Acknowledgements

The author is indebted to Mr. John Dunn of the Naval Research Laboratory and Mr. A. E. Hoffman-Heyden of the RCA Service Company, Patrick AFB, for the data on observed tracking precision of the radar on Sputnik II and the typical beacon track. Development and evaluation of the radar has been carried out by RCA under contracts NOas55-869c and NOas56-1026d, under cognizance of the Bureau of Aeronautics, U. S. Navy.

References

1. Rhodes, Donald R., "Introduction to Monopulse," McGraw-Hill Book Co., New York, 1959.

2. Barton, David K., "Final Report, Instrumentation Radar AN/FPS-16(XN-1)," RCA Moorestown, Contract DA-36-034-ORD-151.

3. Locke, Arthur S., "Guidance," D. Van Nostrand Company Inc., Princeton, N. J., 1955, pp 246-263.

4. Muchmore, R. B. and Wheelon, A. D., "Line-of-Sight Propagation Phenomena," Proc. IRE Vol. 43 Vol. 43 No. 10, October 1955.

5. Pearson, K. E; Kaspark, D. D; and Tarrant, L. N; "The Refraction Correction Developed for the AN/FPS-16 Radar at White Sands Missile Range," U. S. Army Signal Missile Support Agency Technical Memorandum No. 577, November 1958.

6. Barton, D. K., "Sputnik II as Observed by C-Band Radar," presented at National IRE Convention, New York, March 24, 1959.

MULTIPLE TARGET RESOLUTION OF MONOPULSE VERSUS SCANNING RADARS

SAMUEL F. GEORGE and ARTHUR S. ZAMANAKOS

U. S. Naval Research Laboratory, Washington, D. C.

Abstract.—The lack of information on the multiple target resolution characteristics of operational radars prompted a study of the inherent angle resolving powers and angle error sensitivities of the most prevalent tracking systems: amplitude comparison monopulse, conical scan, and two forms of sequential lobing.

The measure of multiple target resolution used is the minimum angular target separation which will permit the radar to determine the presence of more than a single target and allow the radar to locate accurately each target present.

The results of this study show that the monopulse radar has the greatest resolution capability for a given antenna aperture and beamwidth. It also has greatest angle-error sensitivity for a given antenna gain. However, the advantages of monopulse in these respects are relatively small.

I. INTRODUCTION

The problem of multiple target resolution in tracking radars has always been of importance, but the requirements of space surveillance are so demanding that it becomes mandatory to examine very critically the inherent angle resolving powers of the basic methods employed for angle tracking. It is the purpose of this paper to report the results of a study of both the angular resolutions and the angle-error sensitivities of our most prevalent tracking systems: monopulse, sequential lobing and conical scan.

To represent monopulse, the amplitude comparison system has been selected because it is perhaps the most widely used for high-accuracy tracking. Two forms of sequential lobing are used: (1) the so-called "conventional" method, which transmits on a sum pattern and receives on four separate horns sampled sequentially for comparing the up-down and left-right signals, and (2) the "paired" sequential, which also transmits on an unlobed sum pattern and receives on added pairs from a square array for comparison. Hereafter, these will be referred to as sequential and paired lobing, respectively.

II. ANTENNA PATTERN CONSIDERATIONS

In order to provide as unbiased a comparison as possible, it is most important that the same antenna aperture and beamwidth be used for all four systems. The basic transmitted antenna pattern which is used throughout was constructed from experimental data from a modern monopulse radar. An approximate equation for the main lobe of this sum pattern is found to be $S = \cos^2 56.25$ where θ is in degrees. Rather than the actual data, this theoretical pattern is used in the analysis. It has a beamwidth of 1.2° measured at the half-power points. This same basic monopulse sum pattern is used as the main lobe for the conical scan and sequential systems and as the transmitted pattern for paired lobing.

The difference pattern used for monopulse was also constructed from actual data and turned out to be $D = 0.707 \sin 112.5\,\theta$ where again θ is in degrees. Again this theoretical pattern rather than the raw data is used in the analysis.

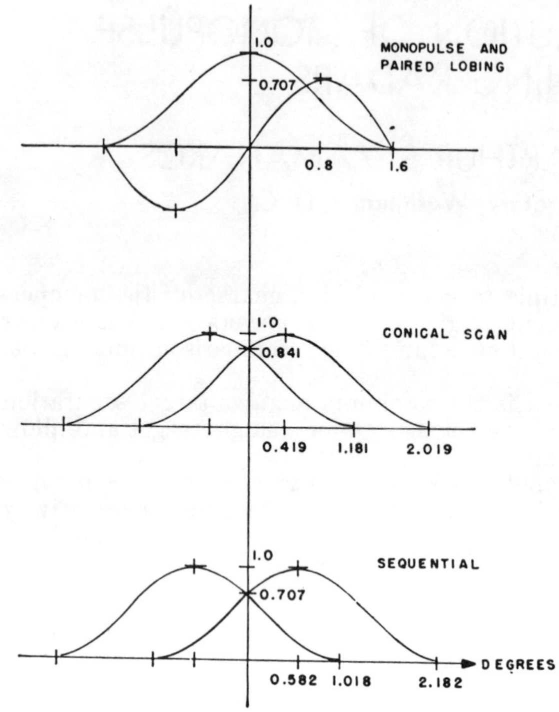

Fig. 1—Basic antenna patterns.

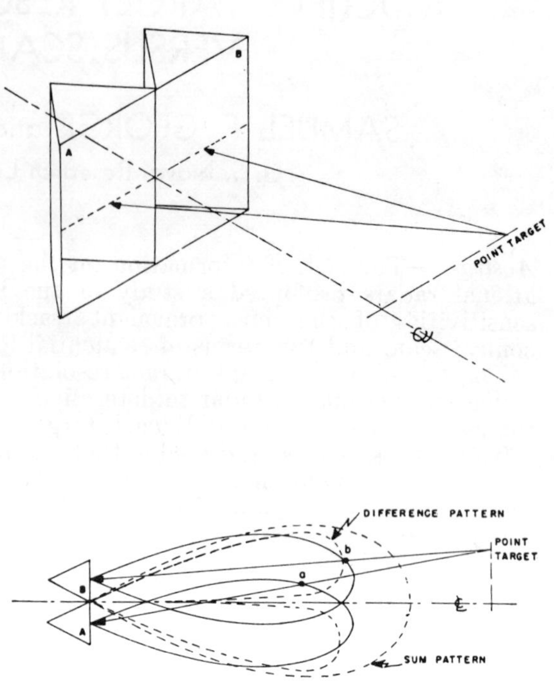

Fig. 2—Monopulse radar — azimuth mode.

Crossover for the conical scan one-way pattern is down 1.5 db and for sequential is down 3 db. The paired lobing system uses the same sum and difference pattern as monopulse. The antenna patterns used in the study of the four systems are shown in Fig. 1.

III. DEVELOPMENT OF ANGLE-ERROR SIGNALS

Since both angular resolution and angle-error sensitivity may be determined from the angle-error signals, it is desirable to examine the manner in which these signals are developed in the various systems. Figure 2 may be thought of as a pictorial representation of the physical relationships of a monopulse antenna operating in the azimuth mode on a single point target. Although two lobes are shown emanating from horns A and B in the lower part of Fig. 2, modern techniques permit tighter sum and difference beams as indicated. In practice, then, the amplitude comparison monopulse system transmits on the basic sum pattern and receives on both the sum and difference patterns. If the sum pattern is called S and the difference pattern D, both functions of the scan angle θ, then monopulse determines the angle-error voltage by forming the product S^3D. In Fig. 3 is depicted the situation existing for two targets displaced in both azimuth and elevation. For the purposes of this analysis only a single channel will be considered, say azimuth, and there will be no discussion of the effects of any possible interaction between the two angle channels. When two targets are present, the angle-error voltage is once again proportional to S^3D, but now S and D contain voltages returned from both targets. The phase relationship between the r-f returns from these two targets thus enters the picture. A brief analysis of the forms taken by the angle-error signals as functions of this phase relationship is presented for all systems in Appendix I. The result for monopulse is

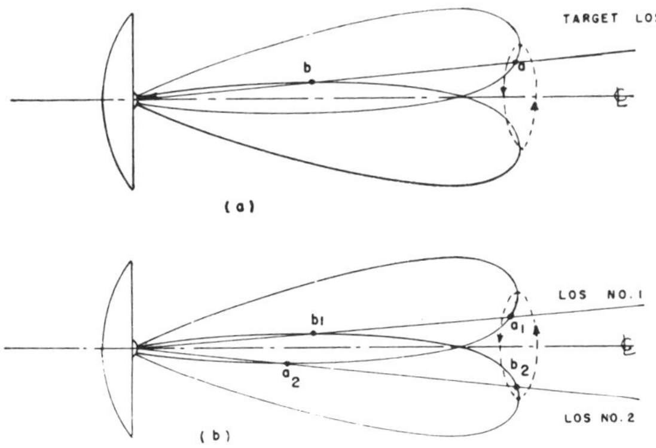

Fig. 4—Conical scan radar.

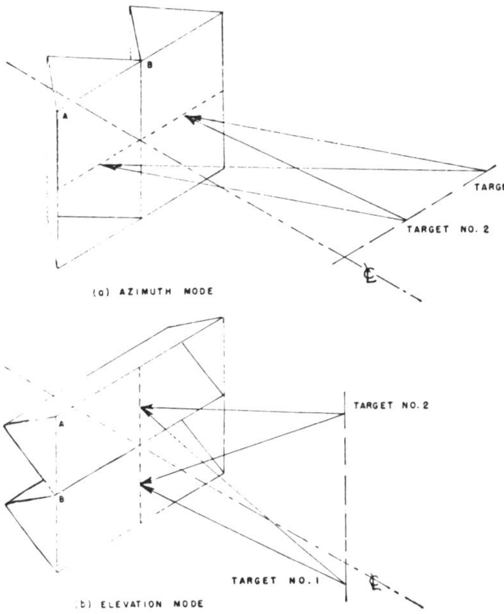

Fig. 3—Monopulse radar — two targets.

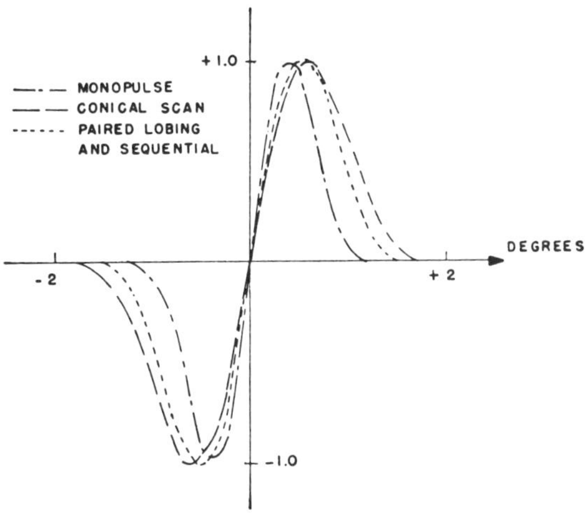

Fig. 5—Single target angle-error signal.

(1) $$E_{v(m)} \propto S_1{}^3D_1 + S_2{}^3D_2 + S_1S_2(S_2D_1 + S_1D_2)\cos\varphi$$

where φ represents the coherence factor or r-f phase difference between the returned voltages from the two targets. The subscripts identify the two targets; e.g., S_1 means the sum pattern voltage amplitude due to target No. 1. Discussion of the significance of (1) will be deferred until after all of the systems have been considered.

In Fig. 4 the operation of the conical scan antenna system is portrayed. Here a single lobe, offset by a squint angle, is rotated or nutated about the radar axis as shown. The amount of modulation at the nutating frequency is a measure of the angle error. It is seen that there is zero modulation on axis. It is also noted that both transmission and reception take place on the same pattern in contrast to monopulse. A simple analysis shows that the error voltage is proportional to the difference in the squares of the voltages which would obtain on the two lobes formed by an intersection of a theoretical plane, containing the radar axis, normal to the antenna. This is illustrated by part (a) of Fig. 4 and shows the voltages as a and b. The angle-error voltage would then be proportional to $a^2 - b^2$ or $(a + b) \times (a - b)$. If $S \cong a + b$ and $D \cong a - b$ it is seen that the error voltage for a conical scan system is SD. It must be remembered that S and D are symbolic of the processes and are not numerically the same as in monopulse. The two-target case is represented in Fig. 4(b) and the analysis in Appendix I yields the following result for conical scan

(2) $$E_{v(c)} \propto (a_1{}^2 - b_1{}^2) + (a_2{}^2 - b_2{}^2) + 2\,(a_1a_2 - b_1b_2)\cos\varphi$$

where the a's and b's are defined in the figure.

The angle-error signals for both forms of sequential lobing are obtained by forming the product SD. From Appendix I the result for sequential, using the antenna structure of the last sketch of Fig. 1, is

$$(3) \qquad E_{v(s)} \propto S_1(a_1 - b_1) + S_2(a_2 - b_2) + [S_1(a_2 - b_2) + S_2(a_1 - b_1)] \cos \varphi$$

and for paired lobing, using the same sum and difference patterns as monopulse, is

$$(4) \qquad E_{v(p)} \propto S_1D_1 + S_2D_2 + (S_1D_2 + S_2D_1) \cos \varphi$$

The same coherence factor φ is used in all systems.

IV. ANGLE-ERROR SENSITIVITY

The angle-error sensitivity is defined as the rate of change of angle-error voltage for a single target with respect to the angular displacement of the radar antenna at the zero error voltage point. An alternate way of expressing this is that the sensitivity is the slope of the angle-error voltage curve as it passes thru zero. A little reflection on the subject reveals that the sensitivity is a function not only of the error voltages just defined but also of the error servo loop gain and of the signal-to-noise ratio. In fact, so long as the echo strength is high and system noise is not a limiting factor, there should be no essential difference in the angle-error sensitivity between any of the systems under discussion. In all cases the servo system should provide the maximum gain possible without becoming unstable. Furthermore, the automatic gain control (AGC) on a strong signal would tend to equalize the slopes of the error-signals about the null. However, when the thermal noise begins to limit system performance, there appears a difference in angular sensitivity. In order to evaluate this difference we must determine a fair method of normalizing the error voltages. The method we have adopted is one of making all of the maximums of the error voltages the same. Figure 5 shows the normalized angle-error signals from a single point target. The following line of reasoning is used in choosing this method of normalization. In order to afford an equitable comparison of angular sensitivity in the presence of noise, one must first be certain that the same signal-to-noise ratio exists. Assuming the same thermal noise level for all systems, one way to do this is to insure the same maximum signal level in all systems. The slopes of the curves in Fig. 5 then determine the angle-error sensitivity. The slopes at crossover were determined from the angle-error voltages to be as shown in Table I which shows that the monopulse system enjoys a superiority in angular sensitivity when the systems are operating near their noise threshold.

TABLE I

ANGLE-ERROR SLOPES AT CROSSOVER

System	*Slope*	*Slope Normalized to Monopulse*
monopulse	4.6	1
sequential	3.3	.72
paired lobing	3.1	.67
conical scan	2.7	.59

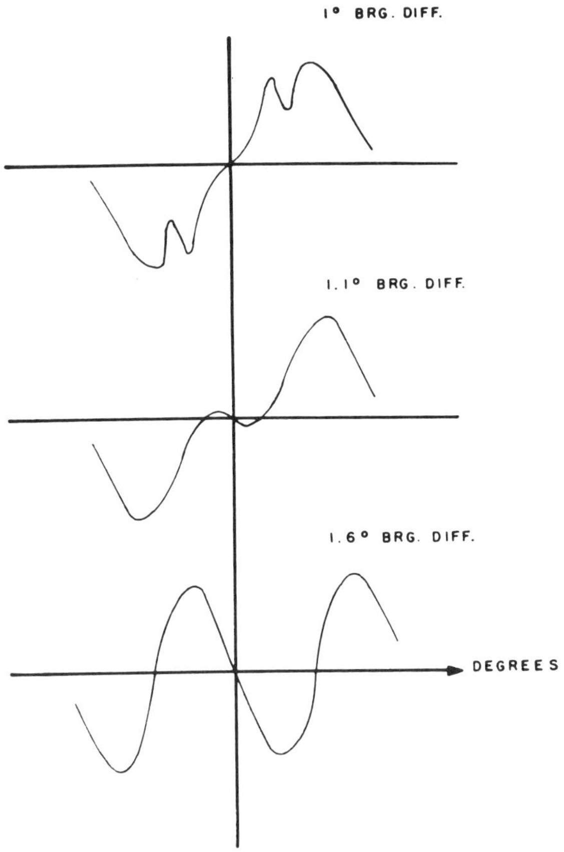

Fig. 6—Behavior of monopulse near resolution.

123

V. ANGULAR RESOLUTION

In considering the capability of a tracking system for resolving multiple targets, one starts with the premise that the system must first be able to resolve two adjacent targets. Hence a study of how well the system can resolve two targets indicates its multiple target discrimination capacity. In this paper only two targets will be discussed. Angular resolution is defined as the minimum target separation which will: (1) permit the radar to determine the presence of more than a single target and (2) allow the radar to locate and track accurately each target present, in the absence of noise. This is a somewhat arbitrary definition and is perhaps more severe than the Rayleigh criterion, but it is felt that when the noise problem must be considered, it offers a fair basis of system comparison.

A graphical procedure was used in determining the error signal for two targets as given in equation form by (1) through (4). The value which should be assigned to the coherence factor φ raises an interesting problem. The situation in practice, of course, is that φ is a random variable with time giving rise to angle noise and glint. As a first step in determining the influence of φ on resolution it was felt that probably coherent targets, $\varphi = 0$, would offer the worst case. Thus, the angle-error signals for $\varphi = 0$ were plotted for a varying bearing difference between the two targets, taken with equal amplitude echo returns. The results for monopulse are shown in Fig. 6. From the top curve it is seen that for a bearing difference of 1.0° there is no resolution. Instead, the system indicates the presence of a single target midway between the two targets (as would be expected). When the angular separation is increased to 1.1°, as shown in the middle curve of Fig. 6, the two targets are just separated by the monopulse radar and there is no longer a false target indication at the midpoint. However,

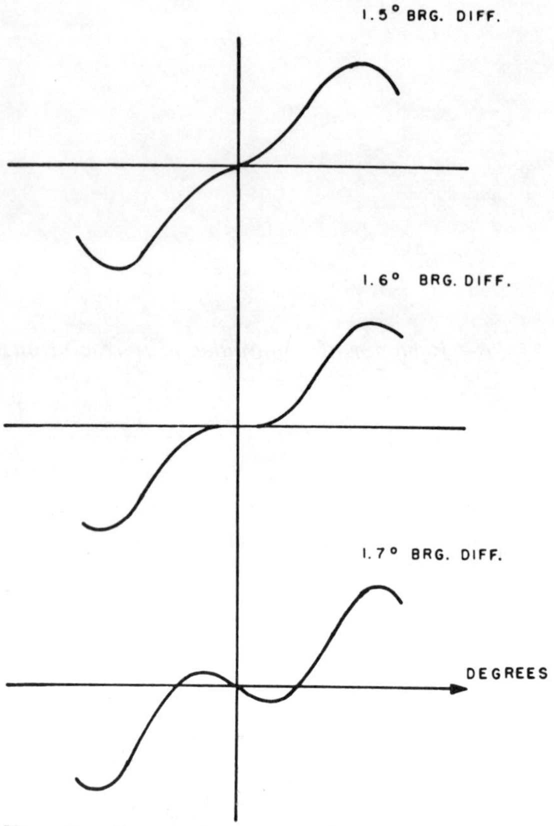

Fig. 7—Composite curve for paired lobing, conical scan and sequential radar.

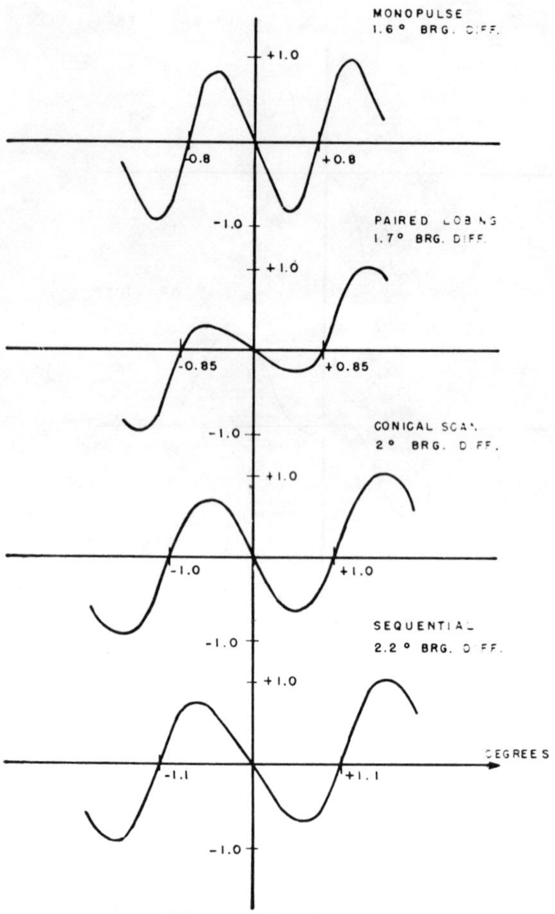

Fig. 8—Resolution.

resolution has not yet occurred according to the definition since the correct bearings of the two targets cannot be determined. The final curve in Fig. 6 shows that for a bearing difference of 1.6° the monopulse system can resolve two in-phase point targets of equal echo amplitude. The null at zero would not appear as a target because of the reversed sense of the error voltage. In accordance with the manner in which the error-voltage curves have been set up here, a positive voltage would drive the antenna to the left, hence the positive voltage shown on the lowest curve in Fig. 6, to the left of the null at zero, would drive the system away from zero.

A similar procedure was followed to determine the angular resolution for the conical scan, paired lobing and sequential radars. The results are shown in Fig. 7. The three systems exhibited such similarity of behavior that a composite curve was drawn. For a 1.5° bearing difference, as illustrated in the top curve, none of these three systems obtains resolution but rather they all indicate the presence of a single target at midpoint. As the separation is increased to 1.6° all three systems show a broad null at midpoint as seen in the middle curve of Fig. 7. The null is so broad that an operator would undoubtedly recognize an unusual situation without realizing the cause. By the time separation reaches 1.7° all systems will indicate the presence of more than one target as revealed by the final curve in Fig. 7, but not all systems will resolve. The bearing separation at which each of the four systems under study will resolve two targets is summarized in Fig. 8. The results are given in Table II,

TABLE II

COMPARISON OF THE FOUR SYSTEMS

System	*Bearing Diff.*	*Normalized Bearing Diff.*	*Slope on Target*
monopulse	1.6°	1.3	3.9
paired lobing	1.7°	1.4	2.6
conical scan	2.0°	1.7	2.7
sequential	2.2°	1.8	3.3

where the second column represents the bearing difference at resolution normalized to the half-power beamwidth (1.2°) of the basic transmitted antenna pattern. It is seen that there is very little difference in the resolving capability of the monopulse and paired lobing systems both of which are superior (about 30%) to the conical scan and sequential systems. It is interesting to note that the angular sensitivity decreases in monopulse and paired lobing but remains the same in sequential and conical scan when the second target enters the scene. However, monopulse still retains a slight edge in sensitivity.

In Fig. 9 are resolution curves for $\varphi = 90°$ and $\varphi = 180°$. Although the shape of the curves changes slightly with φ, it is noted that the separation required for resolving two targets remains independent of the coherence factor. This is very significant for it indicates that for the practical case where φ is a random variable, resolution should still obtain as in Figs. 8 and 9. A brief but still incomplete analysis has shown that this appears in fact to be true; i.e., the nulls on the average separate two targets the same as for $\varphi = 0$. But there exists a dispersion about the mean for random φ which has not been fully investigated.

The effect of targets which return unequal echo amplitudes or are fading has also been very briefly studied. Since only range-gated systems are being considered, these unequal returns could arise from difference in target size, aspect angle, etc. It appears that amplitude differences do not change the theoretical noise free two-target resolution of any of the systems. However, the depth of the null is less for the weaker target, and thus when noise becomes a factor the system naturally will suffer from loss in sensitivity. Also, the angular sensitivity decreases for the weaker target.

The effect of antenna sidelobes on multiple target resolution is beyond the scope of the present work, although it is certain that such an effect can be serious. There is no doubt that sidelobes do deteriorate angular resolution and can produce false targets, but a comprehensive study must be deferred to the future. As an extrapolation from the analysis done for this paper, it would appear that the method of processing the angle-error signal employed by the monopulse technique would render monopulse less receptive to the deleterious effects of sidelobes than the other systems insofar as angular resolution is concerned.

VI. DISCUSSION

The angle-error sensitivity and angular resolution of multiple targets by monopulse, sequential lobing and conical scan type radars have been studied. Every

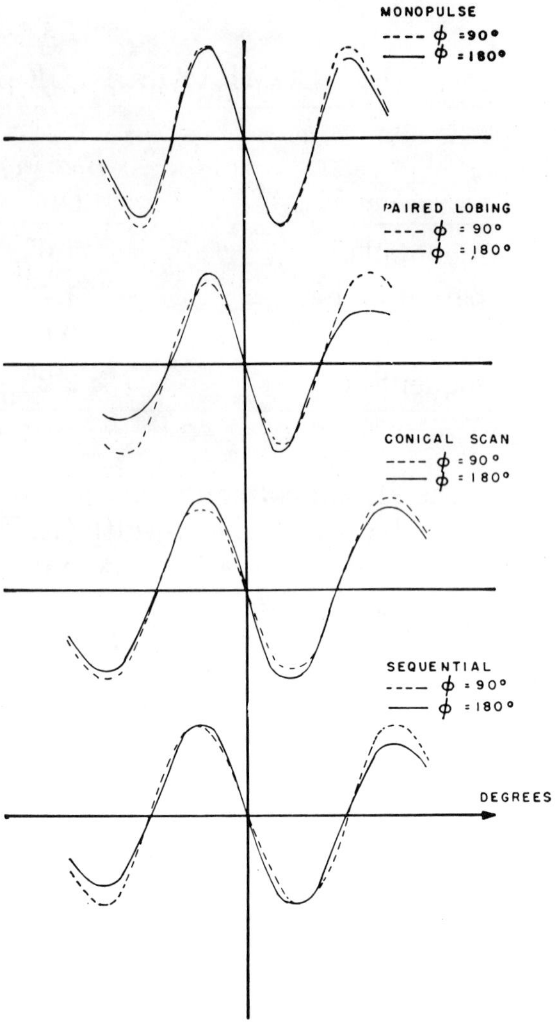

Fig. 9—Affect of varying coherence factor.

126

attempt has been made to compare all of the systems on a fair and unbiased basis by using the same antenna aperture size, antenna gain, and the same basic antenna beamwidth for every radar considered. When the signal-to-noise ratio is adequate for good AGC action, that is, under strong signal conditions, there is no essential difference in the angle-error sensitivity between the monopulse, paired lobing, conical scan and sequential lobing radar systems. However, near the threshold, when there is little AGC action, the monopulse system exhibits an angle-error sensitivity greater than the other systems, ranging from about 40% greater than sequential to nearly 70% greater than conical scan.

The relative merits of radar systems, regarding angular resolution, depend somewhat on the definition of resolution adopted. A very severe require ment has been imposed here in that accurate tracking, for a noise-free stationary point target has been demanded. From Figs. 6-8 one sees that the separation of two targets occurs, in general, long before resolution, using this stringent definition. For example, separation occurs for monopulse at a 1.1° bearing difference and for the other systems at a 1.7° bearing difference, at about 55% greater difference. However, resolution occurs at 1.6° for monopulse, 1.7° for paired lobing (only 6% greater), 2.0° for conical scan (25% greater than monopulse) and 2.2° for sequential (about 40% greater than monopulse). Thus, monopulse does show superiority in resolving targets, although compared to paired lobing this is relatively small.

VII. CONCLUSION

In conclusion it should be stated that the differences in both angle-error sensitivity and angular resolution determined for the four systems in this paper may not be considered sufficient in themselves to warrant the selection of any one system over another. But, upon adding these slight differences to others such as scan loss, range sensitivity, system complexity, jamming immunity, and many others, a relative overall figure of merit might be evolved depending upon the operational requirements.

ACKNOWLEDGMENT

The authors wish to express their appreciation to the members of the Tracking Branch, Radar Division, of the U. S. Naval Research Laboratory, whose encouragement and very helpful discussions made this paper possible.

APPENDIX I

An outline is given here for the method used in obtaining the four equations found in the text. The monopulse will be analyzed first. Denote the magnitude of the transmitted (sum) signal for Target No. 1 by S_1, then upon reception in the sum channel the magniude (two-way) is proportional to S_1^2. Likewise the sum channel return for Target No. 2 would be proportional to S_2^2. Now, if it is assumed that a relative r-f phase difference, φ, exists between the two target returns, the two-way return voltage received in the sum channel, S^2, may be written as

$$S^2 = S_1^2 \cos \omega + S_2^2 \cos (\omega t + \varphi) \tag{5}$$

where ω represents the angular r-f carrier. Similarly, if D stands for the difference pattern, the two-way return voltage in the difference channel, SD, may be written as

$$SD = S_1 D_1 \cos \omega t + S_2 D_2 \cos (\omega t + \varphi) \tag{6}$$

where the S pattern has illuminated the target and the D pattern has been used on reception only. Now, monopulse determines the error signal by forming the product *sum channel* $\times$ *difference channel* or S^3D, retaining all but the r-f terms. Hence, multiplying (5) by (6) and filtering yields

$$E_{v(m)} \propto \langle S^3 D \rangle = \tfrac{1}{2} \{ S_1^3 D_1 + S_2^3 D_2 + S_1 S_2 (S_2 D_1 + S_1 D_2) \cos \varphi \} \tag{7}$$

where the symbol $\langle\ \rangle$ stands for r-f filtering. This is the same as (1) in the text.

A more elegant, and perhaps simpler, way of handling the problem is to use the vector notation. For monopulse this would be

$$\Sigma = \mathbf{S}_1 + \mathbf{S}_2 \tag{8}$$

for the sum channel and

$$\Delta = \mathbf{D}_1 + \mathbf{D}_2 \tag{9}$$

for the difference channel received signals, where the subscript-1 vectors and subscript-2 vectors differ in direction by φ. Here $|\mathbf{S}_1| = S_1^2$, $|\mathbf{D}_1| = S_1 D_1$, etc. The monopulse system then forms the error voltage by taking the dot product, $\Sigma \cdot \Delta$ so

$$\Sigma \cdot \Delta = \mathbf{S}_1 \cdot \mathbf{D}_1 + \mathbf{S}_2 \cdot \mathbf{D}_1 + \mathbf{S}_1 \cdot \mathbf{D}_2 + \mathbf{S}_2 \cdot \mathbf{D}_2 \tag{10}$$

and thus

$$E_{v(m)} \propto |\mathbf{S}_1|\,|\mathbf{D}_1| + |\mathbf{S}_2|\,|\mathbf{D}_1| \cos\varphi + |\mathbf{S}_1|\,|\mathbf{D}_2| \cos\varphi + |\mathbf{S}_2|\,|\mathbf{D}_2| \tag{11}$$

which is easily seen to reduce to (7) and (1).

Using the vector notation, the conical scan system also forms the dot product $\mathbf{S} \cdot \mathbf{D}$ where $|\mathbf{S}_1| = a_1 + b_1$, $|\mathbf{D}_1| = a_1 - b_1$, etc., and again the angle φ separates the vectors with different subscripts, hence

$$\mathbf{S} \cdot \mathbf{D} = (\mathbf{S}_1 + \mathbf{S}_2) \cdot (\mathbf{D}_1 + \mathbf{D}_2) \tag{12}$$

which yields the same result as (10) except that the symbols have different meanings. Substitution into (12) yields, therefore,

$$\begin{aligned} E_{v(c)} \propto\ & (a_1 + b_1)(a_1 - b_1) + (a_2 + b_2)(a_1 - b_1)\cos\varphi \\ & + (a_1 + b_1)(a_2 - b_2)\cos\varphi + (a_2 + b_2)(a_2 - b_2) \end{aligned} \tag{13}$$

which upon simplification is seen to be equivalent to (2).

The same formula (10) holds also for both forms of sequential lobing with the appropriate absolute values. For sequential $|\mathbf{S}_1| = S_1$, $|\mathbf{D}_1| = a_1 - b_1$, etc., and

$$\begin{aligned} E_{v(s)} \propto\ & S_1 (a_1 - b_1) + S_2 (a_1 - b_1)\cos\varphi \\ & + S_1 (a_2 - b_2)\cos\varphi + S_2 (a_2 - b_2) \end{aligned} \tag{14}$$

which is the same as (3) and for paired lobing $|\mathbf{S}_1| = S_1$, $|\mathbf{D}_1| = D_1$, etc., and we obtain (4) by inspection.

For $\varphi = 0$ the four system formulas reduce to:

monopulse:	$(S_1^2 + S_2^2)(S_1 D_1 + S_2 D_2)$
conical scan:	$(a_1 + a_2)^2 - (b_1 + b_2)^2$
sequential:	$(S_1 + S_2)[(a_1 + a_2) - (b_1 + b_2)]$
paired lobing:	$(S_1 + S_2)(D_1 + D_2)$

Reprinted from Volume XV
PROCEEDINGS OF THE NATIONAL ELECTRONICS CONFERENCE
Hotel Sherman, Chicago, Illinois, October 12, 13, 14, 1959.

AMPLITUDE—AND PHASE—SENSING MONOPULSE SYSTEM PARAMETERS

An analysis of amplitude- and phase-sensing sum-and-difference monopulse systems is made in an attempt to specify the effects certain parameters have on system performance. The effects of pre-comparator voltage unbalance, phase shifts and post-comparator phase shifts are shown and compared for each type of sensing system. Since the basic difference between a phase-sensing and amplitude- sensing monopulse system lies in the antenna configuration, major attention is given to the antenna and the microwave assembly which produces the monopulse additions and subtractions. To complete the evaluation, the parameters which affect the receiving equipment performance have also been developed. Equations describing the performance of the systems have been derived and system evaluation curves have been plotted using these equations.

WILLIAM COHEN

and

C. MARTIN STEINMETZ

STAVID ENGINEERING, INC.
PLAINFIELD, NEW JERSEY

PART I

Introduction

The intent of this paper is to give the reader an understanding of the operation of amplitude- and phase-sensing sum-and-difference monopulse systems. This understanding may be enhanced by examining the most important parameters and the manner in which they affect system performance. To establish these parameters and their operational significance in the clearest form, certain simplifying assumptions are made to reduce the mathematical complexity. In such cases where the assumed functions closely resemble the actual radar characteristics, the curves plotted herein may be used to predict its system performance. Where the assumptions are not valid, the reader may make a similar analysis by substituting the appropriate radar characteristics.

Monopulse Versus Conventional Radar

The transmitting function of a sum-and-difference monopulse system is the same as for a conventional radar; that is, a short burst of energy is radiated for every pulse cycle—illuminating the area included within the antenna beamwidth. In a monopulse system the transmitted beamwidth is that of the sum antenna pattern; and due to reciprocity, this pattern is obtained by inserting the transmitter output into the r.f. sum channel. Physically this is accomplished using a duplexer which may consist of two short slot hybrids and a dual TR tube. During the transmit period the TR tubes are fired and the hybrid arrangement causes all the transmitter energy to flow in the port feeding the sum channel. After the transmitted pulse, the TR tubes deionize, and the hybrid arrangement causes all the received energy to flow in the port feeding the receiver.

The major difference, however, between the two radar systems lies in their receiving functions. The received signal in a basic monopulse system is utilized by two channels — a sum and a difference channel, whereas one

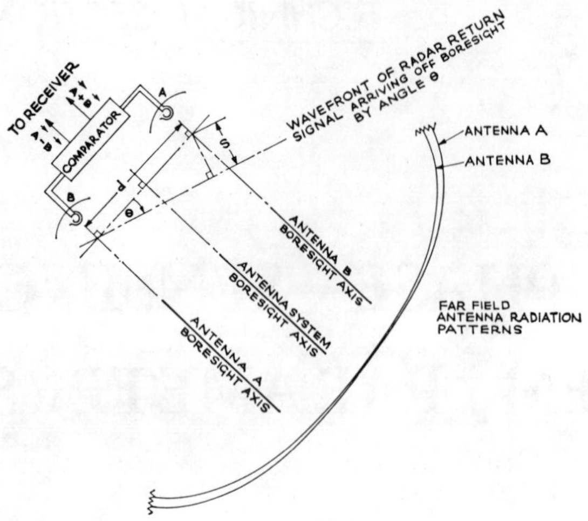

Figure 1 — Phase-sensing antenna configuration.

channel is used in the conventional radar system. The sum and the difference channels described herein may be derived from either the azimuth or elevation plane; the analysis is the same. The sum channel is similar in operation to a conventional radar receiver channel where a maximum signal level is obtained for radar return signals arriving along the boresight axis of the antenna; and, the degree of target resolution is determined by the antenna beamwidth. However, an opposite condition exists in the difference channel where a minimum signal level is obtained for radar return signals arriving along the boresight axis of the antenna. The antenna characteristics and the r.f. comparison circuitry shape the r.f. signal envelope in the difference channel such that this minimum signal level defines the lowest point of a deep, sharp null. Using this sharp null indication to precisely define the boresight axis of the antenna, permits the monopulse system to measure target angles much more accurately than conventional radar systems having the same antenna beamwidth.

An additional improvement, of lesser degree, obtained by using monopulse techniques is increased target resolution. This resolution improvement is achieved by subtracting the video envelope of the sum and difference channels in what is termed, "Monopulse Resolution Improvement Circuits." This subtraction produces a sharp target pulse having, generally, the inverted shape of the difference channel envelope. Since this pulse is much narrower than that obtained from the sum channel alone (or the conventional radar channel), greater target resolution is obtained in the sum-and-difference monopulse system over a conventional radar having the same antenna beamwidth.

A Comparison of Amplitude- and Phase-Sensing Antenna Systems

A phase-sensing antenna system having the required amplitude and phase characteristics is constructed using two or more antenna apertures separated by several wavelengths, see Figure 1. The antenna apertures, each having a single feed, produce identical radiation patterns which are symmetrical about their individual boresight axes. Since the individual boresight axes are parallel, the far field radiation patterns from such an antenna system are almost overlapping; giving essentially equal amplitude signals in each antenna for targets included in their beamwidths. However, there will be a time or phase delay between the signals arriving at the feeds of each antenna due to the spacing between them for all return signals which are off of the boresight axis ($\theta \neq 0°$). Thus, due to the phase characteristic of the received signals, the addition and subtraction of the separated feed signals in the comparator results in establishing a well defined system boresight axis; which is midway between and parallel to the individual antenna aperture boresight axes.

An amplitude-sensing antenna system having the required amplitude and phase characteristics is constructed using a single antenna aperture having two or more closely spaced feeds, see Figure 2. However, in an amplitude-sensing antenna system each of the closely spaced feeds produces a radiation pattern which is displaced from the antenna boresight axis. This displacement angle is a function of the separation of the feedhorn phase centers from the focal point of the antenna aperture. Although the antenna patterns are identical they do not overlap as in the phase-sensing system. The displaced patterns do, however, intersect on the boresight axis of the antenna. Thus, for all radar return signals within the beamwidth of the antenna, except those arriving along the boresight axis, unequal signal amplitudes will be induced in the feeds. When the two signals are subtracted a null will be produced for only those return signals arriving along the boresight axis — since the feedhorns have equal signal voltages only at $\theta = 0$. This subtraction of the feed signals in the comparator results in establishing a well defined system boresight axis along the antenna aperture boresight axis due to the amplitude characteristics of these signals. Thus, both sensing systems produce a null in the difference channel for signals arriving along their boresight axis.

Comparator and Receiver

The sum and difference signal envelopes are produced

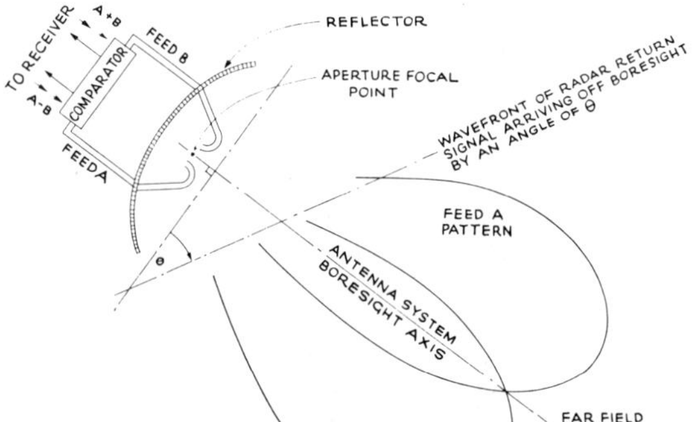

Figure 2 — Amplitude-sensing antenna configuration.

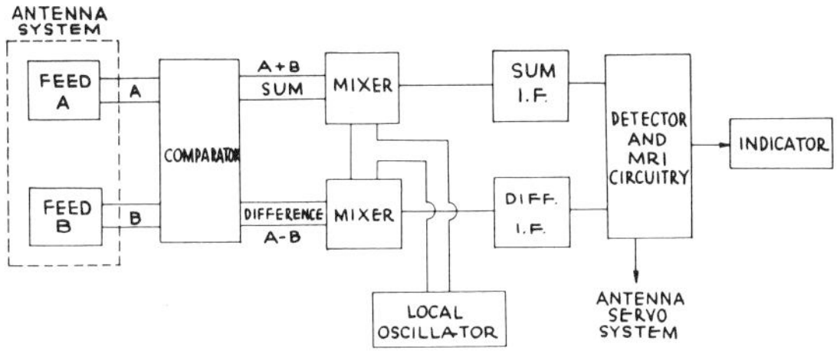

Figure 3 — Block diagram of a basic sum-and-difference monopulse system.

in what shall be termed a "comparator." The comparator for a two channel basic monopulse system could consist in its simplest form, of a folded "T" hybrid. The sum channel envelope is obtained by a vectorial addition, in the hybrid sum arm, of the received r.f. signals arriving at the individual antenna system feeds. The difference channel envelope is produced by a vectorial subtraction of these same r.f. signals in the difference or orthogonal arm of the hybrid. A monopulse system for improving the resolution in both azimuth and elevation would use a more elaborate comparator, but its basic function would be essentially the same as described above.

A block diagram of a basic sum-and-difference monopulse system is shown in Figure 3. The r.f. signals and their associated envelopes coming from the sum and difference outputs of the comparator are heterodyned down, in the mixer, to an IF frequency where they are amplified in their individual IF channels and processed through appropriate detectors and monopulse resolution improvement circuitry.

Thus, it is seen that the two sensing systems differ mainly in the antenna configuration. Physically the comparator and the receiving equipment can be the same, but the exact function the comparator performs and the errors incurred by its use are quite different for each sensing system.

It should be noted here that in the forthcoming analysis only the received signal is being considered. It is assumed that the transmitted pulse will effectively "spray" the area where the antenna is pointed with sufficient energy to get a large return signal. In order to derive usable and easily understood equations many parameters, such as noise, have been neglected, (see reference 5) but the following results may be extended further to include these effects.

AMPLITUDE-SENSING MONOPULSE SYSTEM

Antenna Parameters

It is assumed that the antenna voltage radiation pattern, produced by the displaced phase centers of the feedhorns are described; for Lobe A by

$$\sin \frac{\pi(\theta - \theta_{NA})}{\epsilon} \qquad \theta_{NA} \leqslant \theta \leqslant \theta_{NA} + \epsilon$$

and, for Lobe B by

$$\sin \frac{\pi(\theta - \theta_{NB})}{\epsilon} \qquad \theta_{NB} \leqslant \theta \leqslant \theta_{NB} + \epsilon$$

where θ = angle of the return signal wavefront with respect to the plane of the antenna aperture, see Figure 2.

θ_{NA} = angle at which the first null of lobe A occurs, in the domain $0° > \theta > -90°$.

ϵ = the beamwidth of each lobe (assumed to be the same for the two lobes) measured between the null points of the lobe.

θ_{NB} = angle at which the first null of lobe B occurs in the domain $0° > \theta > -90°$.

These expressions have been plotted in Figure 4 for a typical beamwidth and beam separation. A sine function was chosen to describe the amplitude radiation patterns instead of the more common $\frac{\sin X}{X}$ expression, since a good approximation of the actual patterns are obtained and it simplifies the expressions that will be derived.

The separation between the two lobes A and B can be defined as $\psi = \theta_{NB} - \theta_{NA}$. If the value of ψ is small, it can be computed approximately from the antenna parameters by the following expression:

$$\psi = \theta_{NB} - \theta_{NA} = K_p \tan^{-1} \frac{n}{f} \text{ (See Reference No. 2)}$$

where n is the distance between the feedhorn centers and f is the focal length of the dish in the desired plane. K_p is a constant which is a function of the size and shape of the paraboloid and the type of illumination used across the aperture. $K_p = .6$ for a small dish is a reasonable value.

If the individual radiation patterns of each horn are considered to be proportional to a sine function, the resultant sum channel and difference channel envelope patterns can be found. The sum envelope is the addition

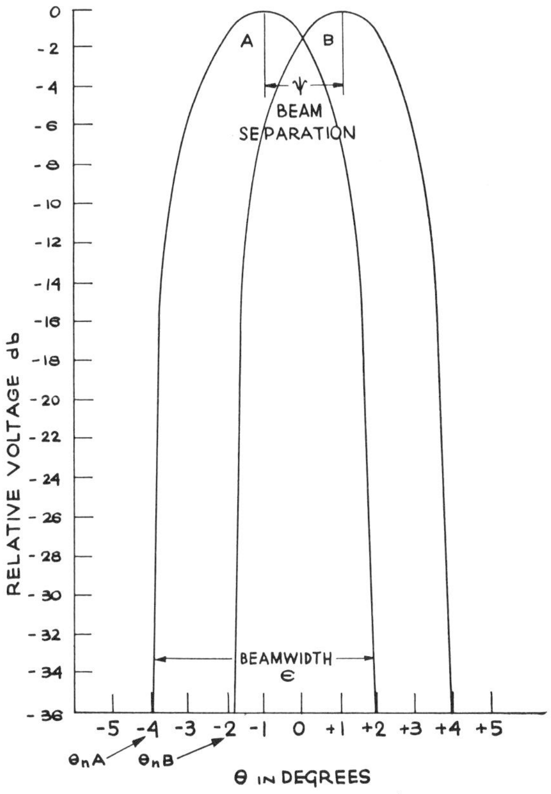

Figure 4 — Antenna lobes of an amplitude-sensing system assuming sine radiation patterns.

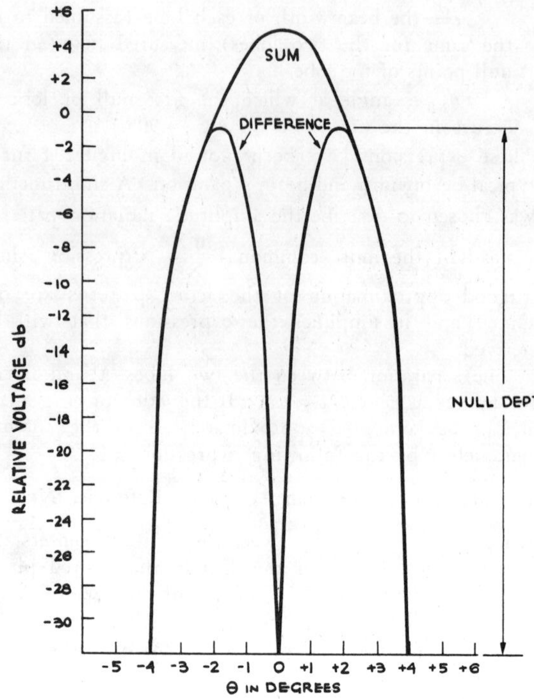

Figure 5 — Sum and difference envelope patterns for sine antenna patterns.

of lobe A and lobe B and the difference envelope is the magnitude of lobe A minus lobe B. These have been plotted in Figure 5.

The equations for the two lobes can be rewritten in a form which shows explicitly the antenna parameters of beamwidth, ϵ, and beam separation, ψ.

It can be seen from Figure 4 that

$\theta_{NA} = -\frac{1}{2}(\epsilon + \psi)$ and

$\theta_{NB} = -\frac{1}{2}(\epsilon - \psi)$

The expression for lobe A becomes

$$\sin\left[\frac{\pi\theta}{\epsilon} + \frac{\pi}{2}\left(1 + \frac{\psi}{\epsilon}\right)\right]$$

$$-\tfrac{1}{2}(\epsilon + \psi) \leqslant \theta \leqslant \tfrac{1}{2}(\epsilon - \psi) \qquad (1)$$

and for lobe B

$$\sin\left[\frac{\pi\theta}{\epsilon} + \frac{\pi}{2}\left(1 - \frac{\psi}{\epsilon}\right)\right]$$

$$-\tfrac{1}{2}(\epsilon - \psi) \leqslant \theta \leqslant \tfrac{1}{2}(\epsilon + \psi) \qquad (2)$$

Phase Shift Effects

It has been assumed that an equal phase signal is being received by both horns near the position of the null ($\theta = 0°$). However, as a result of some phase shift, γ, in the reflector, feedhorn and comparator, (which will be termed pre-comparator phase shifts) the voltages from the two lobes will not completely cancel at the boresight and the null level of the difference pattern will not be zero. For signals arriving on the boresight axis ($\theta = 0$) the magnitude of lobe A equals the magnitude of lobe B.

$$\sin\frac{\pi}{2}\left(1 + \frac{\psi}{\epsilon}\right) = \sin\frac{\pi}{2}\left(1 - \frac{\psi}{\epsilon}\right) = \cos\frac{\pi}{2}\frac{\psi}{\epsilon}$$

However, if there is a pre-comparator phase shift, γ, these signals are not 180° out-of-phase, and will produce a resultant output whose magnitude is:

$$\sin\frac{\pi}{2}\left(1 + \frac{\psi}{\epsilon}\right)\sqrt{2\,(1 - \cos\gamma)}$$

$$= 2\cos\left(\frac{\pi}{2}\frac{\psi}{\epsilon}\right)\sin\tfrac{1}{2}\gamma \qquad (3)$$

The null depth is defined as the ratio of the level in the null to the level at the peak of the difference pattern as shown in Figure 5.

The peak occurs at

$$\frac{d}{d\theta}\sin\left[\frac{\pi\theta}{\epsilon} + \frac{\pi}{2}\left(1 + \frac{\psi}{\epsilon}\right)\right] = 0$$

$$\frac{\epsilon}{2} \leqslant \psi < \epsilon$$

which is found to be $\theta = -\frac{\psi}{2}$. Substituting back into the lobe equation, a value of one is obtained which is the peak of the difference pattern when $\frac{\epsilon}{2} \leqslant \psi < \epsilon$. Therefore, the null depth in db for large beam separatitons is $20\log_{10}$

$$\left(\frac{1}{2\cos\frac{\pi}{2}\frac{\psi}{\epsilon}\sin\frac{1}{2}\gamma}\right) \qquad (4)$$

in the domain $\frac{\epsilon}{2} \leqslant \psi < \epsilon$. However, when the beam separation is less than $\frac{\epsilon}{2}$ the null depth ratio changes because the peak amplitude is no longer one. It can be seen from Figure 4 that the peak amplitude will always occur at θ_{NB} for the smaller beam separations. Hence, the null depth in db for the domain, $0 < \psi \leqslant \frac{\epsilon}{2}$ is $20\log_{10}$

$$\left(\frac{\sin\frac{\pi}{2}\frac{\psi}{\epsilon}}{\sin\frac{1}{2}\gamma}\right) \qquad (5)$$

A family of curves showing the effects of phase shift, γ, in the feed versus null depth as a function of the ratio of beam separation and beamwidth in the two domains are given in Figure 6.

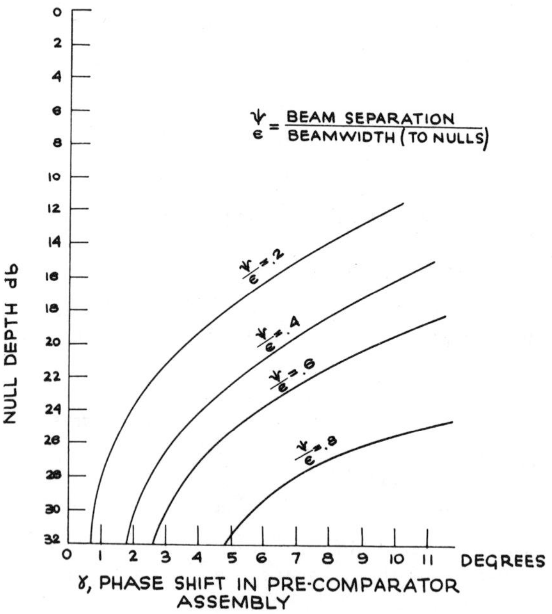

Figure 6 — Null depth vs. pre-comparator phase shift for an amplitude-sensing monopulse system.

Since the null depth will affect the resolution of the radar system and its angle measurement precision, it is most desirable to keep the null depth as deep and sharp as possible. For a fixed tuned radar set, it is fairly easy

sultant outputs of the sum and difference signals are found vectorially. The magnitude of the difference channel was found to determine null depth, shown in equation (3). Its phase angle is found to be

$$\alpha_D = \tan^{-1} \frac{\sin\gamma \sin\left[\frac{\pi\theta}{\epsilon} + \frac{\pi}{2}\left(1 + \frac{\psi}{\epsilon}\right)\right]}{\sin\left[\frac{\pi\theta}{\epsilon} + \frac{\pi}{2}\left(1 - \frac{\psi}{\epsilon}\right)\right] - \sin\left[\frac{\pi\theta}{\epsilon} + \frac{\pi}{2}\left(1 + \frac{\psi}{\epsilon}\right)\right]\cos\gamma} \qquad (6)$$

Similarly for the sum channel resultant signal output, its phase angle is found to be

$$\alpha_S = \tan^{-1} \frac{-\sin\gamma \sin\left[\frac{\pi\theta}{\epsilon} + \frac{\pi}{2}\left(1 + \frac{\psi}{\epsilon}\right)\right]}{\sin\left[\frac{\pi\theta}{\epsilon} + \frac{\pi}{2}\left(1 - \frac{\psi}{\epsilon}\right)\right] + \sin\left[\frac{\pi\theta}{\epsilon} + \frac{\pi}{2}\left(1 + \frac{\psi}{\epsilon}\right)\right]\cos\gamma} \qquad (7)$$

to compensate for pre-comparator phase shifts, but in a tunable radar system the phase shifts have to be kept small over the frequency range. Figure 6 indicates that for good null depths the pre-comparator phase shifts must be kept small, and for a given amount of phase shift in the feed, better null depths are obtained for the largest possible $\frac{\psi}{\epsilon}$ ratios. There is, however, a limit on how large the $\frac{\psi}{\epsilon}$ ratio can be made without appreciably affecting the sum pattern. Using narrow beamwidths and increasing beam separation will cause the sum pattern to flatten until eventually an undesirable dip occurs in the pattern. Other factors such as the allowable phase center separation of the feedhorns and the dish size will also influence the engineering compromise required to obtain optimum system performance.

In addition to "filling-in" the null, pre-comparator phase shifts also result in a second order effect which is a shifting of the phase crossover; or in other words, the antenna boresight indication. To demonstrate the degree of boresight shift, the phase angle of the re-

The phases of these two signals are compared in the receivers' amplitude-sensitive phase detector, which will be discussed later. The phase reversal point of the detector's output voltage is an indication of the antenna boresight. Any shift in this phase reversal point corresponds to an error in the pointing accuracy of the antenna.

Mathematically we can take the difference between the two phase angles to determine the response of the detector. Figure 7a and 7b show $(\alpha_S - \alpha_D)$, the phase comparison of the difference and sum channel resultant signals for selected values of γ and ϵ, with varying $\frac{\psi}{\epsilon}$ ratios; and, selected values of ϵ and $\frac{\psi}{\epsilon}$ with a varying γ. It can be seen that the phase of the compared signals reverses sharply when there is no pre-comparator phase shift, and becomes more gradual with increasing amounts of pre-comparator phase shifts or smaller $\frac{\psi}{\epsilon}$ ratios. But it should be noted that in all cases with no post-comparator phase shifts the reversal occurs on the boresight. However, as can be seen from Figure 8, post-comparator phase shifts change the point at which the curve intersects the detector phase reversal

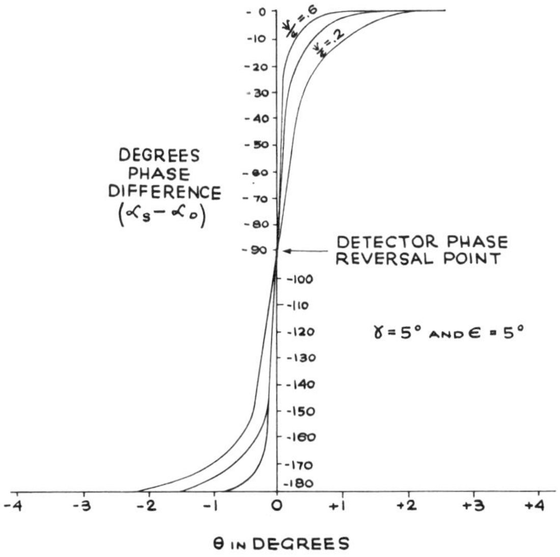

Figure 7(a) — Phase difference, $(\alpha_S - \alpha_D)$, vs. angle θ for various ratios of ψ/ϵ (amplitude-sensing monopulse).

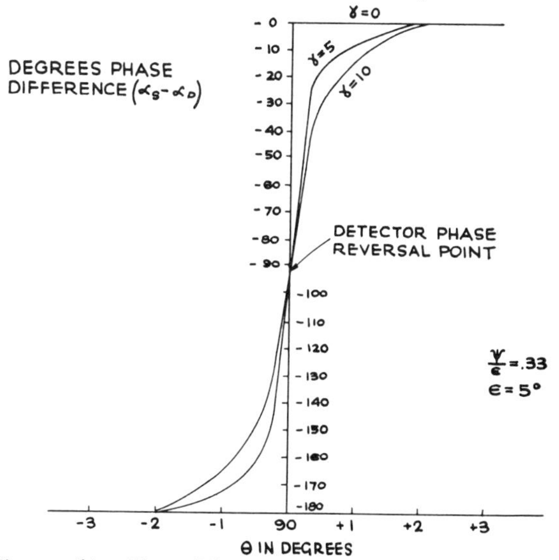

Figure 7(b) — Phase difference $(\alpha_S - \alpha_D)$, vs. angle θ for various pre-comparator phase shifts. (Amplitude sensing monopulse).

point. Hence, the combination of pre-comparator and post-comparator phase shifts will result in a change of boresight indication.

Voltage Unbalance Effects

As previously stated, the null point should occur at the antenna boresight ($\theta = 0°$) for symmetrical antenna patterns, when both lobes are of equal amplitude.

$$\sin \frac{\pi}{2}\left(1 + \frac{\psi}{\epsilon}\right) = \sin \frac{\pi}{2}\left(1 - \frac{\psi}{\epsilon}\right)$$

This equality, however, is contingent on a perfect voltage balance between the two channels, A and B. Should the signal levels be unequal, due to more attenuation in one channel than the other, the equation can be rewritten as

$$\sin\left[\frac{\pi\theta}{\epsilon} + \frac{\pi}{2}\left(1 + \frac{\psi}{\epsilon}\right)\right] = K \sin\left[\frac{\pi\theta}{\epsilon} + \frac{\pi}{2}\left(1 - \frac{\psi}{\epsilon}\right)\right]$$

where $K < 1$ represents the loss in relative voltage of the weaker signal channel. In order to solve this equation for θ_S which will now indicate the boresight shift, let

$$A = \frac{\pi \theta_S}{\epsilon}, \quad B = \frac{\pi}{2}\left(1 + \frac{\psi}{\epsilon}\right)$$

and $C = \frac{\pi}{2}\left(1 - \frac{\psi}{\epsilon}\right)$.

Thus the above equation may be rewritten as

$$\sin(A + B) = K \sin(A + C)$$

By grouping terms

$$A = \tan^{-1} \frac{K \sin C - \sin B}{\cos B - K \cos C}$$

The boresight shift as a function of voltage unbalance is

$$\theta_S = \frac{\epsilon}{\pi} \tan^{-1} \frac{K \sin\left[\frac{\pi}{2}\left(1 - \frac{\psi}{\epsilon}\right)\right] - \sin\left[\frac{\pi}{2}\left(1 + \frac{\psi}{\epsilon}\right)\right]}{\cos\left[\frac{\pi}{2}\left(1 + \frac{\psi}{\epsilon}\right)\right] - K \cos\left[\frac{\pi}{2}\left(1 - \frac{\psi}{\epsilon}\right)\right]} = \frac{\epsilon}{\pi} \tan^{-1} \frac{1 - K}{1 + K} \cot \frac{\pi}{2} \frac{\psi}{\epsilon} \quad (8)$$

The above equation divided by ϵ is plotted in Figure 9 as a function of various $\frac{\psi}{\epsilon}$ ratios. Multiplying the factor obtained from the curve, by ϵ, results in the boresight shift in degrees. It can be seen that the greater values of the ratio $\frac{\psi}{\epsilon}$, the smaller the boresight shift will be for a given voltage unbalance. However, for any particular ratio $\frac{\psi}{\epsilon}$ and voltage unbalance K, smaller boresight shift will result when using the narrowest beamwidth.

If the sine function assumed to describe the antenna lobes is not a reasonable approximation of the actual antenna patterns, the amount of null shift as a function of voltage unbalance can still be computed from the slopes of the two antenna patterns around the crossover.

Figure 10a shows the antenna patterns with and without a voltage unbalance. Lobe B_1 indicates the equal voltage balance condition and Lobe B_2 the unbalanced condition. Since the boresight point is at the intersection of Lobe A and Lobe B, it can be seen that the angle at which the intersection occurs is shifted by an angle, θ_S, in changing from Lobe B_1 to B_2. Let us assume that the

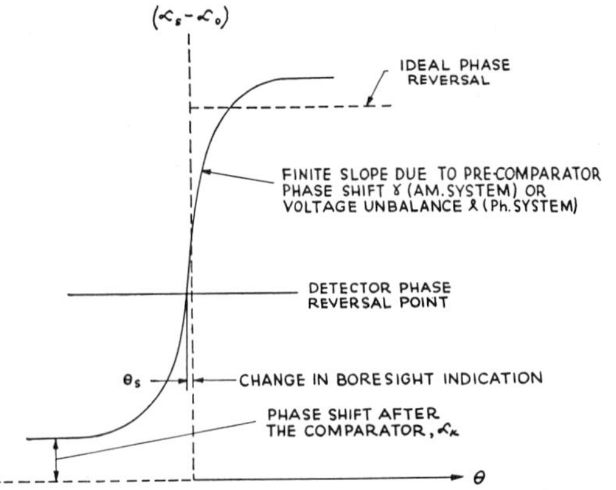

Figure 8 — Boresight shift due to pre-comparator and post-comparator phase shifts. (Amplitude-sensing monopulse).

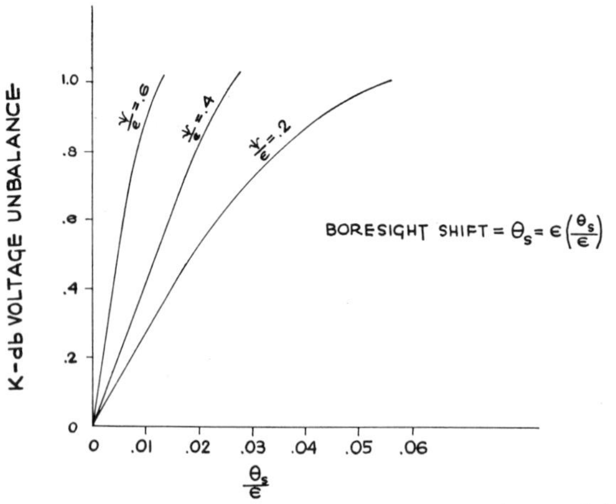

Figure 9 — Boresight shift vs. pre-comparator voltage unbalance for an amplitude-sensing monopulse system.

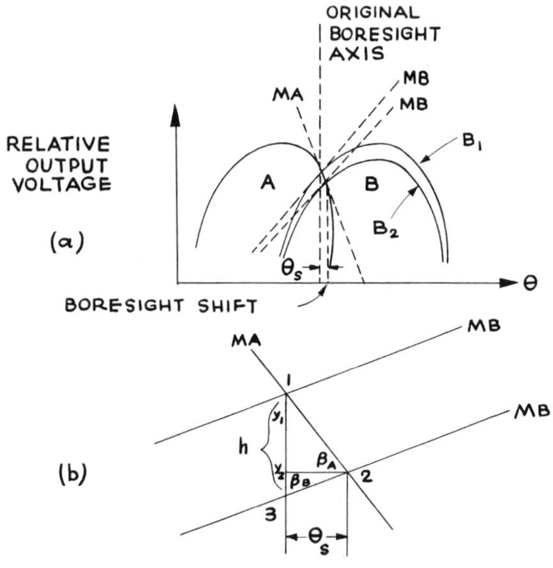

Figure 10 — Boresight shift in amplitude-sensing monopulse.

slope of Lobe A is constant near the intersection of the two lobes and equal to M_A. This is generally true for small losses in the reception of Lobe B. Let the slope of Lobe $B = M_B$ around the intersection point. Let $h =$ the loss in channel B and $\theta_S =$ the shift of the boresight indication due to this loss. Referring to Figure 10b, a vertical line (h) is drawn from points 1 to 3 and the line θ_S drawn perpendicular to h and passing through point 2.

We can deduce by trigonometry that

$$\tan \beta_A = M_A \text{ and } \tan \beta_B = M_B$$

From the diagram one can see that

$$\tan \beta_A = \frac{y_1}{\theta_S} \qquad \tan \beta_B = \frac{y_2}{\theta_S}$$

and $y_1 = h - y_2$

substituting $|M_A| = \frac{y_1}{\theta_S}$ and $|M_B| = \frac{y_2}{\theta_S}$

and solving for θ_S

$$\theta_S = \frac{y_2}{|M_B|} = \frac{y_1}{|M_A|}$$

substituting

$$\frac{|M_B|}{|M_A|}(h - y_2) = y_2$$

and transposing

$$\frac{|M_B|\,h}{|M_A|} = y_2 + \frac{|M_B|\,y_2}{|M_A|}$$

therefore

$$y_2 = \frac{|M_B|\,h}{|M_A|\left(1 + \frac{|M_B|}{|M_A|}\right)} = \frac{|M_B|\,h}{|M_A| + |M_B|}$$

so

$$\theta_S = \frac{h}{|M_A| + |M_B|} \qquad (9)$$

Since it is desirable to keep θ_S as small as possible and assuming h is a given quantity, it then becomes obvious that M_A and M_B, the slopes, must be as large as possible. For example, if the actual antenna patterns were of the general shape as shown in Figure 4, it would be best to have the beamwidth, ϵ, as narrow as possible and the beam separation, ψ, as large as possible. Of course, as was mentioned before, there are many other parameters which would actually limit the values of ϵ and ψ.

END PART I

Bibliography

1. "Monopulse Radar," R. N. Page, *IRE Convention Record,* 1955, Pages 132-134, Part I.
2. *Aerials for Centimetre Wave-Lengths,* Fry & Coward, Cambridge University Press — 1950.
3. *Reference Data for Radio Engineers,* I.T. and T., 1956.
4. *Introduction to Monopulse,* Donald R. Rhodes, McGraw Hill Book Company — 1959.
5. "Monopulse Automatic Tracking and the Thermal Bound," H. T. Budenbom, *IRE National Convention on Military Electronics,* June 1957.

List of Symbols

a, a_2, a_3 Coefficients in Taylor Series
d Distance between Feedhorn Centers
e_D Difference I.F. Signal Voltage
e_S Sum I.F. Signal Voltage
E_D Signal Voltage in Difference Channel
E_{D2} Signal Voltage in Difference Channel with Voltage Unbalance
E_S Signal Voltage in Sum Channel
E_{S1} Signal Voltage in Sum Channel with Phase Shift
E_{S2} Signal Voltage in Sum Channel with Voltage Unbalance
E_o Output Voltage from Phase Detector
f Focal Length of Paraboloid Antenna
h Loss in Channel B
i_c Mixer Crystal Current
K Ratio of Signal Voltage Loss in Channel B to Channel A
K_1 Received Signal Level in a Single Feedhorn
K_D Amplitude of Difference Signal
K_{D1} Amplitude of Difference Signal with Phase Shift
K_{D2} Amplitude of Difference Signal with Voltage Unbalance
K_S Amplitude of Sum Signal
K_{S1} Amplitude of Sum Signal with Phase Shift
K_{S2} Amplitude of Sum Signal with Voltage Unbalance
K_{OD} Amplitude of Local Oscillator Signal at Difference Mixer
K_{OS} Amplitude of Local Oscillator Signal at Sum Mixer
l Ratio of Signal Voltage in Channel B to that in Channel A
K_p Parabolic Antenna Constant (.6)
M_A Slope of Lobe A Near the Boresight Axis
M_B Slope of Lobe B Near the Boresight Axis
n Distance between Phase Center of Closely Spaced Feedhorns
S Extra Path Delay in Antenna B
α_D Phase Angle of Difference Signal Voltage
α_{D1} Phase Angle of Difference Signal Voltage with Phase Delay γ
α_{D2} Phase Angle of Difference Signal Voltage with Voltage Unbalance
α_K Phase Shift Between Sum and Difference Channel (After the Comparator)
α_S Phase Angle of Sum Signal Voltage
α_{S1} Phase Angle of Sum Signal Voltage with Phase Delay γ
α_{S2} Phase Angle of Sum Signal Voltage with Voltage Unbalance l
α_{OD} Phase Angle of Local Oscillator Voltage at Difference Mixer
α_{OS} Phase Angle of Local Oscillator Voltage at Sum Mixer
γ Spurious Phase Shift
ϵ Beamwidth of Each Lobe
θ Angle of Return Signal Wavefront with Respect to the Plane of the Antenna Apertures (See Figures 1 & 2)
θ_{NA} Angle at which the First Null of Lobe A Occurs $0° > \theta > -90°$
θ_{NB} Angle at which the First Null of Lobe B Occurs $0° > \theta > -90°$
θ_S Boresight Shift
λ Wavelength of Carrier
π 3.14
ϕ Phase Delay Due to Distance S
ψ Separation Between Lobes
ω_c 2π Carrier Frequency
ω_o 2π Local Oscillator Frequency

AMPLITUDE—AND PHASE—SENSING MONOPULSE SYSTEM PARAMETERS

An analysis of amplitude- and phase-sensing sum-and-difference monopulse systems is made in an attempt to specify the effects certain parameters have on system performance. The effects of pre-comparator voltage unbalance, phase shifts and post-comparator phase shifts are shown and compared for each type of sensing system. Since the basic difference between a phase-sensing and amplitude- sensing monopulse system lies in the antenna configuration, major attention is given to the antenna and the microwave assembly which produces the monopulse additions and subtractions. To complete the evaluation, the parameters which affect the receiving equipment performance have also been developed. Equations describing the performance of the systems have been derived and system evaluation curves have been plotted using these equations.

WILLIAM COHEN

and

C. MARTIN STEINMETZ

STAVID ENGINEERING, INC.
PLAINFIELD, NEW JERSEY

PART II

PHASE-SENSING MONOPULSE SYSTEM

Antenna Parameters

Figure 1 shows the antenna configuration of a phase-sensing system and the general situation of the reflected radar energy returning at an angle θ with respect to the plane of the antenna apertures. It is assumed here that the phase centers of the antennae are at the same point as the geometrical centers. A signal arriving at antenna A will be delayed from that arriving at antenna B due to the distance S. From the diagram one can see that $S = d \sin \theta$.

Letting ϕ = phase delay due to this distance S we have

$$\phi = \frac{2\pi s}{\lambda} = \frac{-2\pi d \sin\theta}{\lambda} \tag{10}$$

where λ is the wavelength of the carrier frequency.

The antenna voltages may be given as follows:

$$A = K_1 \sin(\omega_c t + \phi)$$

and

$$B = K_1 \sin(\omega_c t) \tag{11}$$

where ω_c is the angular velocity of the carrier and K_1 an arbitrary value which is related to the transmitted power, range, backscattering conditions and antenna parameters.

For the difference channel (A-B) we have:

$$E_D = K_1 [\sin(\omega_c t + \phi) - \sin(\omega_c t)] \tag{12}$$

The resultant output, E_D, is then composed of the subtraction of two voltages of the same frequency but of changing phase. This expression can be rewritten as a single resultant voltage and phase angle as follows:

$$E_D = K_D \sin(\omega_c t + \alpha_D)$$

where

$$K_D = K_1 \sqrt{(\cos\phi - 1)^2 + (\sin\phi)^2} = 2K_1 \sin \tfrac{1}{2}\phi \tag{13}$$

and

$$\alpha_D = T_{AN}^{-1}\left[\frac{\sin\phi}{\cos\phi - 1}\right]$$

The phase angle expression can be simplified further and may be expressed in two ways:

$$\alpha_D = \frac{\pi}{2} + \tfrac{1}{2}\phi \qquad 0^\circ < \phi < 360^\circ$$

or

$$\alpha_D = \frac{\pi}{2} + \tfrac{1}{2}\phi \qquad 0^\circ < \phi < +180^\circ$$

$$\alpha_D = \frac{3\pi}{2} + \tfrac{1}{2}\phi \qquad 0^\circ > \phi > -180^\circ$$

The second expression will become useful when this phase angle is compared with the sum signal later on.

Similarly for the *sum channel* we have

$$E_S = K_1 [\sin(\omega_c t) + \sin(\omega_c t + \phi)] \text{ or} \tag{14}$$

$$E_S = K_S \sin(\omega_c t + \alpha_S)$$

where $$K_S = K_1 \sqrt{(1 + \cos\phi)^2 + (\sin\phi)^2} \tag{15}$$

$$= 2K_1 \cos \tfrac{1}{2}\phi$$

and $$\alpha_S = T_{AN}^{-1}\left(\frac{\sin\phi}{1 + \cos\phi}\right)$$

The sum channel phase expression can also be simplified and may be expressed as

$$\alpha_S = \tfrac{1}{2}\phi \qquad 180^\circ > \phi > -180^\circ$$

Using the above derived equations, the relative amplitude of the sum and difference channels, as a function of ϕ, are plotted in Figure 11.

If the magnitude of ϕ is allowed to become larger than $\pm 180^\circ$, the even order functions shown will repeat giving rise to ambiguous nulls. In view of the fact that the null is used to define the antenna bore-

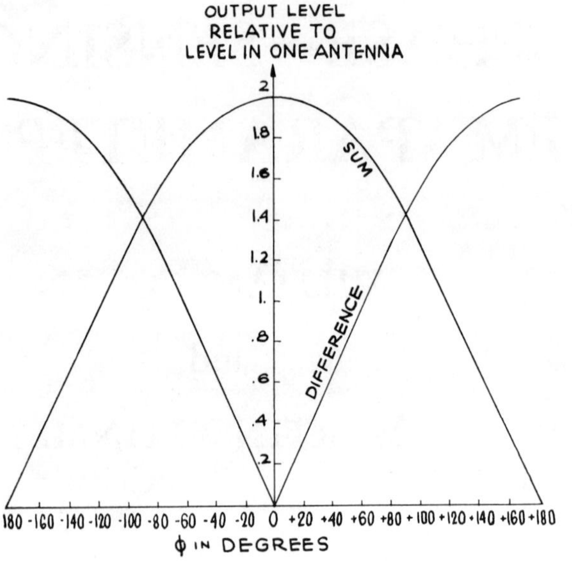

Figure 11. Signal Amplitude of sum and difference channels in a phase-sensing monopulse system.

sight, the system would then be inoperative unless these ambiguities are resolved.

The method for eliminating the ambiguities in a phase-sensing monopulse system is by limiting the angles of signal arrival θ. This is accomplished by restricting the antenna beamwidth so that $\phi \left(\frac{2\pi d}{\lambda} \sin\theta\right)$ is within $\pm 180°$. In both sensing systems similarly shaped sum and difference patterns result from the comparator.

Comparison of the phase angles of the sum and difference signals yields the following results:

$$\alpha_S - \alpha_D = \tfrac{1}{2}\phi - \left[\frac{\pi}{2} + \tfrac{1}{2}\phi\right] = -\frac{\pi}{2} \qquad 0° < \phi < +180°$$

$$\alpha_S - \alpha_D = \tfrac{1}{2}\phi - \left[\frac{3}{2}\pi + \tfrac{1}{2}\phi\right] = -\frac{3}{2}\pi \qquad 0° > \phi > -180° \qquad (16)$$

These equations have been plotted in Figure 12, where it can be seen that for $\gamma \doteq 0°$, an instantaneous phase reversal occurs at the boresight or null, $\theta \doteq 0°$ or $\phi = 0°$. It will be remembered that this was also true for the amplitude-sensing system.

Phase Shift Effects

However, if we now assume that a phase delay, γ, occurs in the antenna feed, B, the corresponding equations change as follows:

$$E_{S1} = K_{S1} \sin\left[\omega_c t + T_{AN}^{-1} \frac{\sin\phi - \sin\gamma}{\cos\phi + \cos\gamma}\right]$$

One change lies in the value of α_S which is now

$$\alpha_{S1} = T_{AN}^{-1}\left[\frac{\sin\phi - \sin\gamma}{\cos\phi + \cos\gamma}\right]$$

or $\alpha_{S1} = \frac{1}{2}(\phi - \gamma) \qquad 180° > \phi + \gamma > -180°$

Similarly for the difference channel,

$$\alpha_{D1} = T_{AN}^{-1} \frac{\sin\phi + \sin\gamma}{\cos\phi - \cos\gamma} \text{ or}$$

$$\alpha_{D1} = \frac{\pi}{2} - \tfrac{1}{2}(\gamma - \phi) \qquad 0° < \gamma + \phi < +180°$$

$$\alpha_{D1} = \frac{3\pi}{2} - \tfrac{1}{2}(\gamma - \phi) \qquad 0° > \gamma + \phi > -180° \qquad (17)$$

The phase difference between the sum and difference channels $\alpha_{S1} - \alpha_{D1}$ will equal the following expressions:

$$\alpha_{S1} - \alpha_{D1} = -\frac{\pi}{2} \qquad 0° < \gamma + \phi < +180°$$

$$\alpha_{S1} - \alpha_{D1} = -\frac{3}{2}\pi \qquad 0° > \gamma + \phi > -180° \qquad (18)$$

The two conditions, before and after the null, are similar to those calculated before, except that the phase reversal occurs when $\phi = -\gamma$ and not when $\phi = 0$. This would indicate a change of boresight if a true phase detector were used.

Now, returning to the general sum and difference equations and solving for the amplitude expression we find for the sum amplitude

$$K_{S1} = K_1 \sqrt{(\cos\gamma + \cos\phi)^2 + (\sin\phi - \sin\gamma)^2} = 2 K_1 \cos \tfrac{1}{2}(\gamma + \phi) \qquad (19)$$

and for the difference amplitude

$$K_{D1} = K_1 \sqrt{(\sin\phi + \sin\gamma)^2 + (\cos\phi - \cos\gamma)^2} = 2 K_1 \sin \tfrac{1}{2}(\gamma + \phi) \qquad (20)$$

In order to find when $K_{D1} = 0$, i.e. a null, we can set equation 20 equal to zero and solve for ϕ.

$$2 K_1 \sin \tfrac{1}{2}(\gamma + \phi) = 0$$

$$\sin \tfrac{1}{2}(\gamma + \phi) = 0$$

$$\phi = -\gamma \qquad (21)$$

Since $\phi = -\gamma$ at the null, it can be seen that the amplitude null advances the same amount as the phase reversal point advances for the same phase delay γ.

The advance of an amplitude null has been shown in Figure 13 for $\gamma = +10°$. Likewise, it can be seen in Figure 12, how the phase reversal point advances because of a pre-comparator phase shift of $\gamma = +10°$.

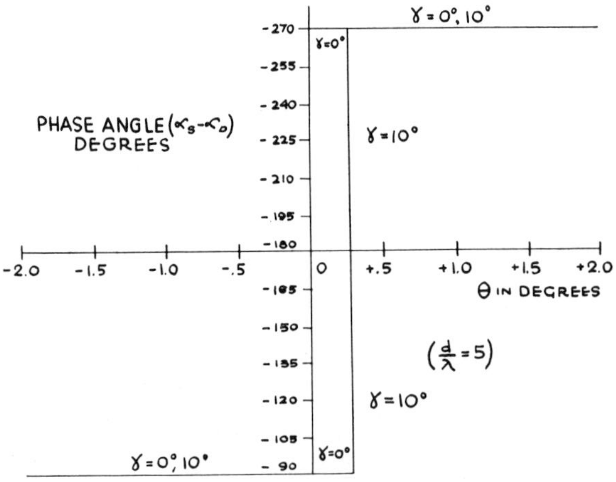

Figure 12. Phase difference vs. angle (phase-sensing system.)

Thus it may be seen that the amount of null or phase reversal shift for a given phase shift depends on the value. $\frac{2\pi d}{\lambda}$.

To indicate the effect the $\frac{d}{\lambda}$ ratio has on the amount the null and phase reversal will shift, a plot for various values of $\frac{d}{\lambda}$ is shown in Figure 14. Observe that the greater the value of $\frac{d}{\lambda}$ the less the shift. This demonstrates that in designing a phase-sensing system the spacing of the antennas and the wavelength of the carrier are important parameters in determining the errors that will be encountered in the operation of the system.

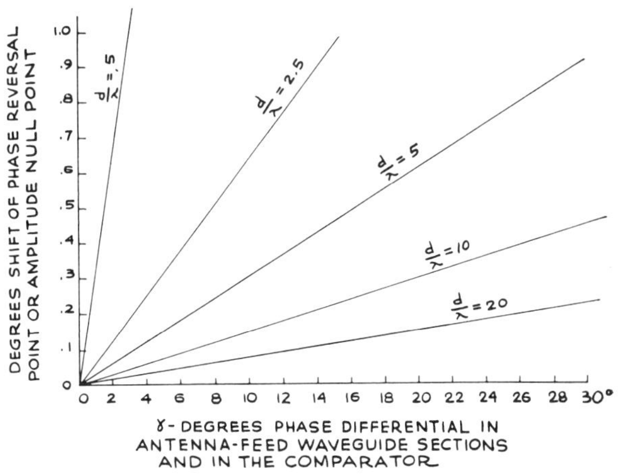

Figure 14. Null or phase reversal shift vs. pre-comparator phase shift (phase-sensing monopulse system.)

Voltage Unbalance Effects

Again consider the case of some attenuation in one channel, A or B. For the sake of argument, let the voltage in the A channel be reduced by a factor $l < 1$. The corresponding equations change as follows: For the difference channel

$$E_{D2} = K_{D2} \sin\left[\omega_c t + T_{AN}^{-1}\left(\frac{\sin\phi}{\cos\phi - l}\right)\right] \tag{22}$$

where

$$K_{D2} = K_1 \sqrt{(\cos\phi - l)^2 + (\sin\phi)^2}$$

and

$$\alpha_{D2} = T_{AN}^{-1}\left(\frac{\sin\phi}{\cos\phi - l}\right)$$

Similarly for the sum channel voltage under these conditions we find

$$E_{S2} = K_{S2} \sin\left[\omega_c t + T_{AN}^{-1}\left(\frac{\sin\phi}{l + \cos\phi}\right)\right]$$

where

$$K_{S2} = K_1 \sqrt{(l + \cos\phi)^2 + (\sin\phi)^2}$$

and

$$\alpha_{S2} = T_{AN}^{-1}\frac{\sin\phi}{l + \cos\phi} \tag{23}$$

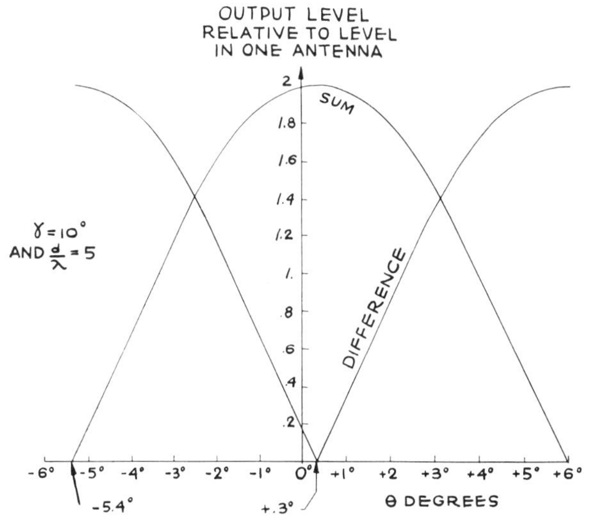

Figure 13. Signal amplitude of sum and difference channels in a phase-sensing monopulse system with pre-comparator phase shift.

A plot of these equations for amplitude effects are shown in Figure 15 for voltage unbalances of 1.0 db, 0.4 db and 0.1 db. The sum channel amplitude does not change over this range, but it can be seen how the difference null "fills in."

When these nulls "fill in," the resolution of the radar system declines. A representative plot of null depth vs voltage unbalance is shown in Figure 16. Since the actual null depth varies with the antenna shaping parameters, this curve is only indicative of the change in null depth with a change in voltage unbalance.

The resultant phase angle ($\alpha_{S2} - \alpha_{D2}$) has been plotted in Figure 17 for various voltage unbalances. Instead of a sharp phase transition at $\phi = 0°$, it becomes more gradual as the voltage unbalance is increased. This is similar to the amplitude-sensing system which had a pre-comparator phase shift. Here also post-comparator phase shifts will cause a reversal in phase at a point other than $\phi = 0°$ which would indicate an error in boresight. This results in the same effect shown in Figure 8 for the amplitude sensing system.

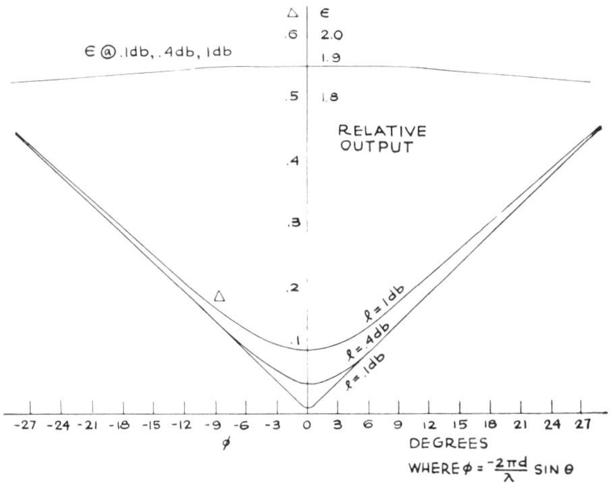

Figure 15. Amplitude of sum and difference channels with voltage unbalance (phase-sensing monopulse system.)

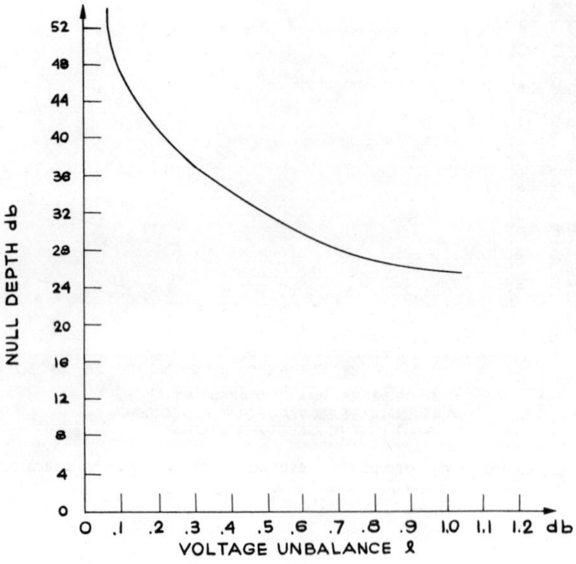

Figure 16. Null depth vs. pre-comparator voltage unbalance (phase-sensing system).

RECEIVING EQUIPMENT

Phase Detector

Both amplitude and phase information from either monopulse system can be utilized in a phase-sensitive amplitude detector shown in Figure 18. The detector's output voltage can be described by the following equation:

$E_O = \sqrt{|E_S|\,|E_D|}\cos(\alpha_S - \alpha_D + \alpha_K)$ where α_K is any constant phase shift after the comparator, E_O is the output voltage of the detector, $|E_S|$ is the magnitude of the sum channel signal and $|E_D|$ is the magnitude of the difference channel signal. It was shown in the amplitude section that $(\alpha_S - \alpha_D)$ is either 0 or $-\pi$ and if $\alpha_K = 0$, the term $\cos(\alpha_S - \alpha_D + \alpha_K)$ is either $+1$ or -1. This also holds true in a phase system if a constant phase shift of $\frac{\pi}{2}$ is added to the expression, $(\alpha_S - \alpha_D)$. The magnitude of the detector output depends on $|E_S|$ and $|E_D|$ and changes sense when the phase of the two signals reverse. Usually $|E_S|$ is made larger than $|E_D|$ and supplies the necessary phase reference. Thus the output of the phase-sensitive amplitude detector will be mainly a function of the amplitude characteristic of $|E_D|$. This output is commonly referred to as the "S" curve.

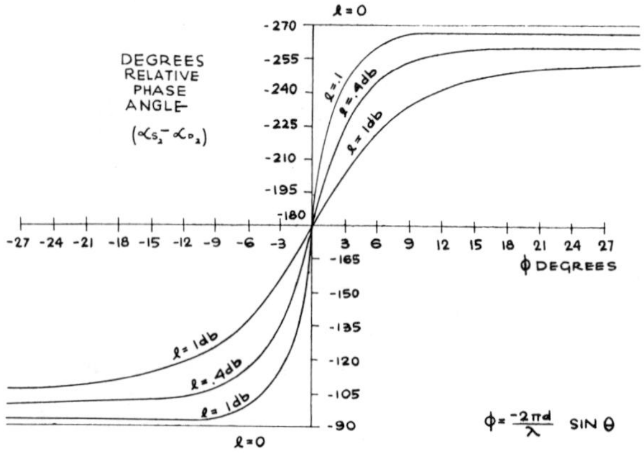

Figure 17. Phase difference vs. angle for phase system with voltage unbalance.

The slope of the "S" curve around the crossover point depends upon many factors.

1. The range
2. The receiver gain
3. The backscattering area of the target or ground
4. STC effectiveness
5. Other receiver and transmitter constants

If $(\alpha_S - \alpha_D + \alpha_K)$ is not 0° or $-$ 180° due to an additional phase constant, $\alpha_K \neq 0$, added to it, the cosine term will not equal ± 1 but plus or minus some smaller value and the sensitivity of the detector will be reduced. This will reduce the signal to noise ratio of the receiver, and will cause the second order change in the crossover point of the "S" curve as described in Figure 8. Calculations of receiver desensitization as a function of post comparator phase shift, α_K, have been plotted in Figure 19.

A point worth noting here is the output of the phase sensitive amplitude detector

$$\sqrt{|E_S|\,|E_D|}\cos\left(\alpha_S - \alpha_D + \frac{\pi}{2}\right) \text{ for the}$$

case of a voltage unbalance in the phase-sensing system. The output of the phase detector for a .4 db unbalance has been plotted in Figure 20 where it can be seen that

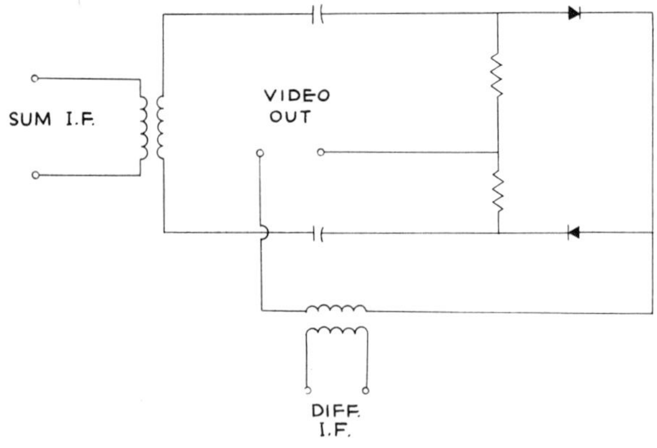

Figure 18. Phase sensitive amplitude detector.

the output is the same as for no voltage unbalance. The presence of the gradual phase reversal (Figure 17) compensates for the "filling in" of the null (Figure 15). This clearly shows the advantage of using a phase-sensitive amplitude detector in an application where precomparator voltage unbalances are anticipated. A similar advantage is derived in the amplitude-sensing system having a pre-comparator phase shift.

Local Oscillator

The following calculations are performed to investigate how the phase angle of the I.F. voltage change as a function of the phase angles entering the mixers from the local oscillator.

The mixer crystal current can be expressed by a Taylor Series of the general form

$$i_c = a_1 e + a_2 e^2 + a_3 e^3 + \cdots a_n e^n$$

The local oscillator voltage at the difference mixer is:

$$K_{OD} \sin(\omega_o t + \alpha_{OD})$$

ω_o = local oscillator angular velocity

α_{OD} = local oscillator phase angle

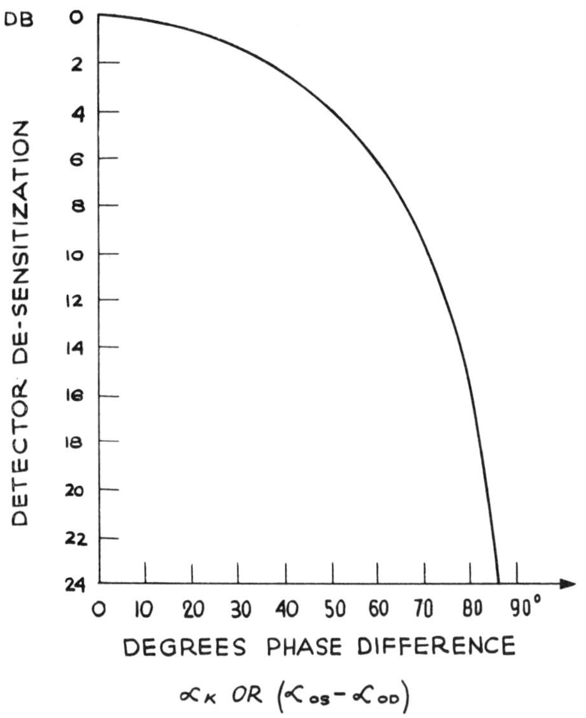

Figure 19. Phase detector desensitization vs. post comparator phase shift.

K_{OD} = arbitrary amplitude factor

The entire signal presented to the difference channel mixer is

$$e_D = K_D \sin(\omega_c t + \alpha_D) + K_{OD} \sin(\omega_o t + \alpha_{OD})$$

After substituting this value in the Taylor Series and selecting the term which represents the signal which would pass through the I.F. amplifier, we obtain for the difference channel:

$$e_{D\,(IF)} = a_D K_D K_{OD} \cos[(\omega_c - \omega_o)\,t + (\alpha_D - \alpha_{OD})]$$

and similarly one can obtain for the sum channel

$$e_{S\,(IF)} = a_S K_S K_{OS} \cos[(\omega_c - \omega_o)\,t + (\alpha_S - \alpha_{OS})]$$

Where α_{OS} is the phase angle of the local oscillator signal presented to the sum channel and K_{OS} the amplitude factor associated with the local oscillator.

If the phase of the two signals are compared, $(\alpha_S - \alpha_D) - (\alpha_{OS} - \alpha_{OD})$, it can be seen that the term $(\alpha_{OS} - \alpha_{OD})$ introduces a constant phase shift, but does not change the point of phase reversal by itself. However, this phase shift $(\alpha_K = \alpha_{OS} - \alpha_{OD})$ should be compensated for in the I.F amplifier because it would tend to desensitize the detector as shown previously in Figure 19. Moreover if the radar set were tunable and $(\alpha_{OS} - \alpha_{OD})$ changed as a function of frequency, it would be difficult to compensate for in the I.F. amplifiers. In such a case, the various lengths of local oscillator plumbing should be nearly equal so that α_{OS} and α_{OD} would remain constant over the frequency range.

It is also important in such a radar system to maintain the amplitudes of the local oscillator signals (K_{OD} and K_{OS}) at a constant value to optimize noise figure and maximum conversion gain.

Conclusions

Some analogies may be drawn between the amplitude-sensing and phase-sensing monopulse systems. In the amplitude-sensing system any voltage unbalance (before and in the comparator) will cause a boresight shift; while in the phase system any phase shift before and in the comparator causes a boresight shift. On the other hand any pre-comparator phase shift in the amplitude system will cause a reduction in the null depth while any voltage unbalance before and in the comparator in the phase system will cause null depth reduction. Also, a pre-comparator phase shift in the amplitude system and a pre-comparator voltage unbalance in the phase system would each cause a more gradual phase reversal at boresight. Under these conditions post-comparator phase shifts will produce second order boresight shifts.

In the amplitude system the shape of the antenna patterns, the beamwidth ϵ and beam separation ψ, affect the magnitude of null depth and boresight shift. Within limits it is desirable to have a large $\frac{\psi}{\epsilon}$ ratio and a small ϵ.

In the phase-sensing system the separation between antennas, d, and the operating wavelength, λ, affect the magnitude of the null depth and the boresight shift. Within limits it is desirable to have a large $\frac{d}{\lambda}$ ratio and a beamwidth such that

$$\phi\left(\frac{2\pi d}{\lambda}\sin\theta\right) < \pm 180°.$$

The sum and difference patterns from both systems are similar and the processing of the information can be the same for both systems. After the comparator any phase shifts or voltage unbalances which occur do not directly affect the position or the null depth, but will influence the sensitivity of the detector.

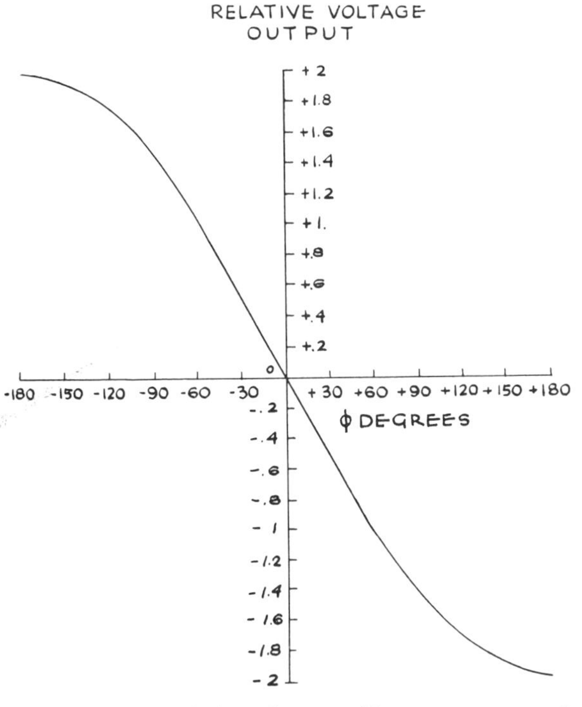

Figure 20. Output of phase detector with pre-comparator voltage unbalance (phase-sensing system.)

Bibliography

1. "Monopulse Radar," R. N. Page, *IRE Convention Record,* 1955, Pages 132-134, Part 1.
2. *Aerials for Centimetre Wave-Lengths,* Fry & Coward, Cambridge University Press — 1950.
3. *Reference Data for Radio Engineers,* I.T. and T., 1956.
4. *Introduction to Monopulse,* Donald R. Rhodes, McGraw Hill Book Company — 1959.
5. "Monopulse Automatic Tracking and the Thermal Bound," H. T. Budenbom, *IRE National Convention on Military Electronics,* June 1957.

List of Symbols

a, a_2, a_3 Coefficients in Taylor Series
d Distance between Feedhorn Centers
e_D Difference I.F. Signal Voltage
e_S Sum I.F. Signal Voltage
E_D Signal Voltage in Difference Channel
E_{D2} Signal Voltage in Difference Channel with Voltage Unbalance
E_S Signal Voltage in Sum Channel
E_{S1} Signal Voltage in Sum Channel with Phase Shift
E_{S2} Signal Voltage in Sum Channel with Voltage Unbalance
E_o Output Voltage from Phase Detector
f Focal Length of Paraboloid Antenna
h Loss in Channel B
i_c Mixer Crystal Current
K Ratio of Signal Voltage Loss in Channel B to Channel A
K_1 Received Signal Level in a Single Feedhorn
K_D Amplitude of Difference Signal
K_{D1} Amplitude of Difference Signal with Phase Shift
K_{D2} Amplitude of Difference Signal with Voltage Unbalance
K_S Amplitude of Sum Signal
K_{S1} Amplitude of Sum Signal with Phase Shift
K_{S2} Amplitude of Sum Signal with Voltage Unbalance
K_{OD} Amplitude of Local Oscillator Signal at Difference Mixer
K_{OS} Amplitude of Local Oscillator Signal at Sum Mixer
l Ratio of Signal Voltage in Channel B to that in Channel A
K_p Parabolic Antenna Constant (.6)
M_A Slope of Lobe A Near the Boresight Axis
M_B Slope of Lobe B Near the Boresight Axis
n Distance between Phase Center of Closely Spaced Feedhorns
S Extra Path Delay in Antenna E
α_D Phase Angle of Difference Signal Voltage
α_{D1} Phase Angle of Difference Signal Voltage with Phase Delay γ
α_{D2} Phase Angle of Difference Signal Voltage with Voltage Unbalance
α_K Phase Shift Between Sum and Difference Channel (After the Comparator)
α_S Phase Angle of Sum Signal Voltage
α_{S1} Phase Angle of Sum Signal Voltage with Phase Delay γ
α_{S2} Phase Angle of Sum Signal Voltage with Voltage Unbalance l
α_{OD} Phase Angle of Local Oscillator Voltage at Difference Mixer
α_{OS} Phase Angle of Local Oscillator Voltage at Sum Mixer
γ Spurious Phase Shift
ϵ Beamwidth of Each Lobe
θ Angle of Return Signal Wavefront with Respect to the Plane of the Antenna Apertures (See Figures 1 & 2)
θ_{NA} Angle at which the First Null of Lobe A Occurs $0° > \theta > -90°$
θ_{NB} Angle at which the First Null of Lobe B Occurs $0° > \theta > -90°$
θ_S Boresight Shift
λ Wavelength of Carrier
π 3.14
ϕ Phase Delay Due to Distance S
ψ Separation Between Lobes
ω_c 2π Carrier Frequency
ω_o 2π Local Oscillator Frequency

Editor's Note: For the readers' convenience, two previously published figures, to which the authors have referred in this part, are reprinted.

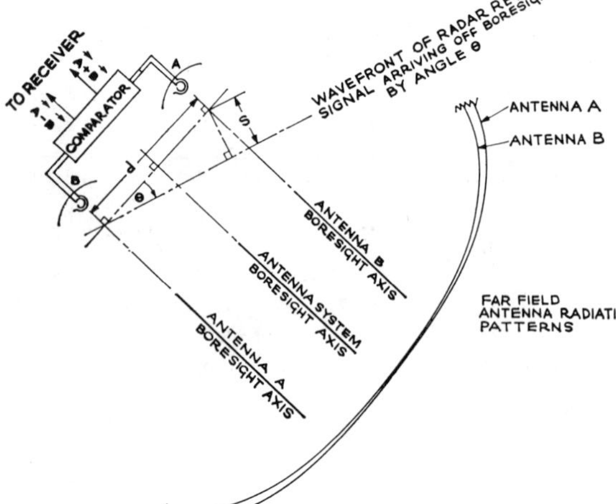

Figure 1 — Phase-sensing antenna configuration.

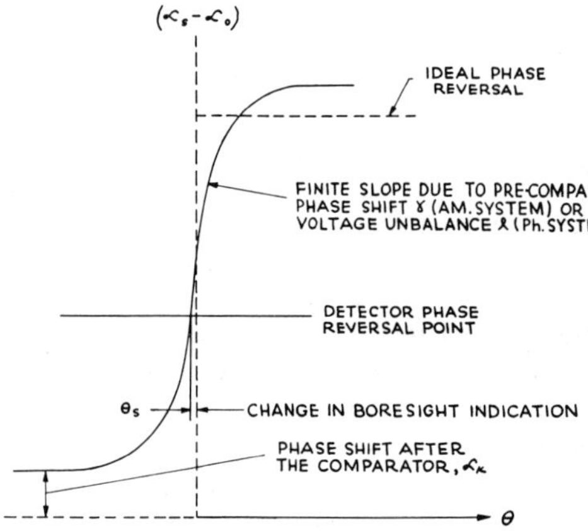

Figure 8 — Boresight shift due to pre-comparator and post-comparator phase shifts. (Amplitude-sensing monopulse.)

8.

A MONOPULSE CASSEGRAINIAN ANTENNA

R. W. Martin and L. Schwartzman
Surface Armament Division
Sperry Gyroscope Company
Division of Sperry Rand Corporation
Great Neck, L. I., New York

143

Summary

The diversity, simple configuration, and mechanical and electrical advantages of the Cassegrainian antenna have made it increasingly popular as a microwave director for radar systems. In detailed analysis of this type of antenna considered for use with a precise monopulse tracking radar, unique electrical characteristics are evidenced. These characteristics must be taken into account in the system design to insure optimum monopulse performance. The space-attenuation and monopulse-sensitivity functions are developed, and the special feed-to-subreflector near-field problem is explored with a discussion of proved solutions. Periodic design comparisons are made between the opaque-subreflector-type Cassegrainian antenna and the focal-point-feed antenna system where the single aperture, be it reflector or lens, is illuminated by a feed situated at the focal point. There is an often-used technique to circumvent the inherent aperture blocking introduced by the opaque subreflector, but it is mechanically complex and is costly. The performance limitations of the mechanically simpler opaque-subreflector-type Cassegrainian must be evaluated by the design engineer before going into these costly techniques.

Theory and Operation

The use of the Cassegrainian-type microwave antenna as a radar director has become well established. Though it is a two-reflector system, the geometry is relatively simple and fabrication is straightforward. Its electrical advantages include: image-feed reduction, advantageous positioning of the microwave-feed package, and flexibility of parameters. This paper summarizes basic design equations with specific emphasis upon unique problems evidenced when this antenna is considered for application with a monopulse system. The basic design equations may be translated into design curves, some of which are included here, to provide a tool in the initial investigation and development of such systems.

A descriptive sketch of the basic Cassegrainian design is shown in Figure 1. A spherical wave front emanating from the prime focal point F_p of the hyperboloid will be reflected as a set of spherical waves centered at the conjugate focal point F_c of the hyperboloid. In the majority of applications, F_c corresponds to the focal point of the paraboloidal main reflector, and F_p is the location of the actual feed. Consider now that an image feed exists at F_c. The solid angle θ_A subtended by the hyperboloid at F_p differs from the solid angle θ_E subtended by the paraboloid at F_c; therefore, it is a logical conclusion that the diffration aperture of the image feed varies inversely to the diffration aperture of the actual feed at F_p. A magnification factor, M = (Image-Feed Aperture)/(Actual-Feed Aperture), will therefore exist. The magnitude of the magnification factor, $M > 1$ or $M < 1$, is dependent upon the position of the hyperbolic vertex V_h with respect to F_c and F_p. It can be shown that $M = (\overline{V_h F_c})/\overline{V_h F_p})$. A more detailed form for this function is given later.

For a chosen sum amplitude illumination on the paraboloid, the dimension of the feed horn aperture is dependent upon hyperbolic contour (eccentricity), position V_h, diameter d, and the feed position KF. The parameter d most often determines the amount of aperture blocking existing in the system. It is possible for the effective aperture blocking introduced by the actual feed to exceed that due to the hyperboloid, but this is encountered in relatively few designs. A common technique for eliminating the aperture blocking by the hyperboloid is to incorporate a polarization-rotation surface on the paraboloid and a polarization-sensitive grating as the hyperboloid surface. Energy reflected from the subreflector upon the parabolic surface has its polarization rotated 90 degrees at this latter surface. Because the polarization-sensitive surface of the subreflector is transparent to the resultant orthogonal polarization, it presents essentially no aperture blocking. Although this technique affords an obvious electrical advantage at the expense of complexity and cost, it is not of prime interest here. Rather, it is the purpose of this analysis to investigate the extent to which the opaque-subreflector-type system may be utilized and the techniques possible to circumvent associated special problems.

Referring again to Figure 1, it can be seen that several variables must be considered in the design of an optimum system. Compromises must be made in the choice of such parameters as the hyperboloid size, contour, and position; feed horn size and position; and the F/D ratio of the parabola. From the basic geometry of the system, relationships linking the above parameters can be determined, but these relationships are implicit functions and ajudicious choice must therefore be made for an optimum design. The four basic design equations are given below:

1. The angle θ_A subtended at the feed position F_p by the hyperbolic reflector is given by:

$$\theta_A = 2 \arctan \frac{N}{2\frac{F}{D}(1-K) - N \cot\frac{\theta_E}{2}} \qquad (1)$$

2. The magnification, which is defined as the ratio of the image-horn size to the actual-horn size, is given by:

$$M = \frac{2(1-K) - \frac{N}{F/D}\left[\csc\frac{\theta_A}{2} - \csc\frac{\theta_E}{2}\right]}{2(1-K) + \frac{N}{F/D}\left[\csc\frac{\theta_A}{2} - \csc\frac{\theta_E}{2}\right]} \qquad (2)$$

3. The position of the hyperbolic vertex is described by:

$$X = K_1 F = \left[\frac{N}{4\frac{F}{D}}\left[\csc\frac{\theta_A}{2} - \csc\frac{\theta_E}{2}\right] + \frac{1+K}{2}\right] F \qquad (3)$$

4. The eccentricity of the hyperbola is:

$$e = \frac{1-K}{2K_1 - 1 - K} \qquad (4)$$

where:

F = Focal length of the parabola.

θ_E = Angle subtended by the parabola at its focal point.

D = Paraboloid diameter.

d = Hyperboloid diameter.

$N = d/D$.

KF = Fractional part of focal length separating the horn aperture and paraboloid vertex.

K_1F = Fractional part of focal length separating the hyperboloid vertex and the paraboloid vertex.

Since the dependent variables in Equations 1, 2, and 3 are functions of the three variables N, F/D, and K, a series of design curves that interrelate them may be generated.

An important electrical characteristic that influences the aperture amplitude distribution is the space-attenuation characteristic of the antenna system. Space attenuation is defined as the divergent path length effects between the central rays and the edge rays within the antenna system. For a parabolic reflector with the feed at the focal point, this effect reduces simply to an inverse square function of the path length of the two rays; but for a Cassegrainian antenna, it is not as simple. Referring to Figure 2, one can develop the following:

$$\Delta\phi = \frac{\ell_1 \Delta\theta}{\ell_1'} \qquad (5)$$

$$\Delta\alpha = \frac{\ell_2 \Delta\theta}{\ell_2'} \qquad (6)$$

The resulting projected parabolic area illuminated by the unit edge bundle of rays $\Delta\theta$ is:

$$A_E = \frac{\pi}{4}\left[\frac{\ell_3 \ell_2 \Delta\theta}{\ell_2'}\right]^2 \qquad (7)$$

and the resulting projected parabolic area illuminated by a similar unit central bundle of rays $\Delta\theta$ is given by:

$$A_C = \frac{\pi}{4}\left[\frac{F \ell_1 \Delta\theta}{\ell_1'}\right]^2 \qquad (8)$$

Since:

$$\ell_3 = \frac{D}{2}\csc\frac{\theta_E}{2} \qquad (9)$$

$$\ell_1 = (K_1 - K)F \qquad (10)$$

$$\ell_1' = (1 - K_1)F \quad (11)$$

$$\ell_2' = \frac{d}{2} \csc \frac{\theta_E}{2} \quad (12)$$

$$\ell_2 = \frac{d}{2} \csc \frac{\theta_A}{2} \quad (13)$$

and the space attenuation (S. A.) in db is:

$$S.A. = 10 \log \frac{A_C}{A_E} \quad (14)$$

then utilizing Equations 9 - 13, substituting Equations 7 and 8 into Equation 14, rearranging terms, and simplifying, one obtains:

$$S.A. = 20 \log \left[\frac{2 \frac{F}{D} \sin \frac{\theta_A}{2}}{\text{MAGNIFICATION}} \right] \quad (15)$$

Equation 15 reduces to the familiar expression for a focal-point-feed system when the magnification is unity, i. e. , the subreflector is a flat plate positioned halfway between F_c and F_p. The space attenuation for this special case is:

$$S.A. = 20 \log \left[\frac{F}{\rho} \right] \quad (16)$$

where ρ is the length of the edge ray (F_p to hyperbola plus hyperbola to parabola).

For a Cassegrainian system with an opaque subreflector (and designed with practical limits on the parameters), it will be found that the space attenuation is less than one db and may be neglected. This, of course, is not the case for a focal-point-feed system, in which space attenuation has an appreciable effect upon aperture illumination, particularly if the F/D ratio is small. This result has been verified experimentally and is shown in Figure 3. The feed horn pattern was measured on an arc, the radius of which was equal to the horn-to-hyperbola separation $\overline{V_h F_p}$. The feed-horn-hyperbola combination pattern was measured on an arc of parabolic contour. It is seen that these two normalized patterns are very nearly identical, indicating that (except in the vicinity of the special case of unity magnification) the Cassegrainian antenna space-attenuation function is not merely the inverse square of the ray path-lengths.

In the design of a monopulse antenna, the range- and angular-sensitivity functions, which are dependent upon the feed horn aperture and the F/D ratio, are of prime importance. The range sensitivity is defined as the product of the system on-axis sum voltage E_S on transmit and the on-axis sum voltage (proportional to E_S) on receive. The angular sensitivity is defined as the product of the on-axis sum voltage on transmit and the derivative with respect to bearing angle of the received difference-voltage on axis (E_d').

The range- and angular-sensitivity functions have been developed in existing literature for conventional focal-point-feed systems, but these functions are different for the Cassegrainian system. Conventional monopulse antenna systems normally require feed horn apertures below cut-off for F/D ratios below 0. 5. Therefore, one of the prime advantages of a Cassegrainian antenna for monopulse operation is realized for F/D ratios below this value. The sensitivity functions were therefore studied for F/D ratios ranging between 0. 3 and 0. 6. In making the analysis, it was assumed that the far-field patterns of an antenna are Guassian between the half-power points and thus of the form:

$$E = E_o \exp \left[\frac{-(h\xi)^2}{2} \right]$$

where

$$h^2 = .6932 \left(\frac{2}{BW} \right)^2$$

The difference pattern in voltage is given by:

$$E_d = \frac{E_1 - E_2}{E_o} = \exp \left[\frac{-h^2 (\xi + \xi_{1i})^2}{2} \right] - \exp \left[\frac{-h^2 (\xi - \xi_{1i})^2}{2} \right] \quad (17)$$

where ξ_{1i} is the individual beam offset angle (see Figure 4). The slope of the difference pattern on axis ($\xi = o$) is then:

$$E_d' = 2h^2 \xi_{1i} \exp \left[- \frac{2\xi_{1i}^2}{} \right] \quad (18)$$

The crossover level on axis can be shown to be:

$$|DB|_{\xi = o} = 4.34 h^2 \xi_{1i}^2 \quad (19)$$

If Equation 19 is solved for ξ_i and is substituted into Equation 18, the difference pattern slope is found to be given by:

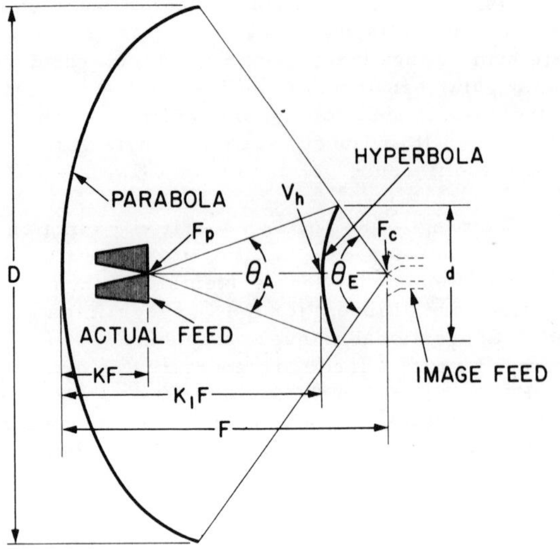

FIG. 1 CASSEGRAINIAN ANTENNA GEOMETRY

$\xi_i = \beta\phi = \beta \frac{a}{2F} \frac{180}{\pi}$ DEGREES

$E_2 = E_0 \text{ EXP}\left[\frac{-h^2(\xi+\xi_i)^2}{2}\right]$

$E_1 = E_0 \text{ EXP}\left[\frac{-h^2(\xi-\xi_i)^2}{2}\right]$

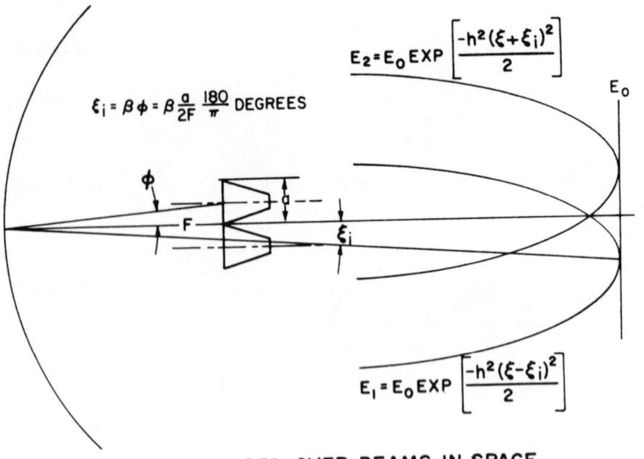

FIG. 4 CROSSED-OVER BEAMS IN SPACE

$\text{S.A.} = 20 \text{ LOG}\left[\frac{2\frac{F}{D} \text{SIN}\frac{\theta_A}{2}}{\text{MAGNIFICATION}}\right]$

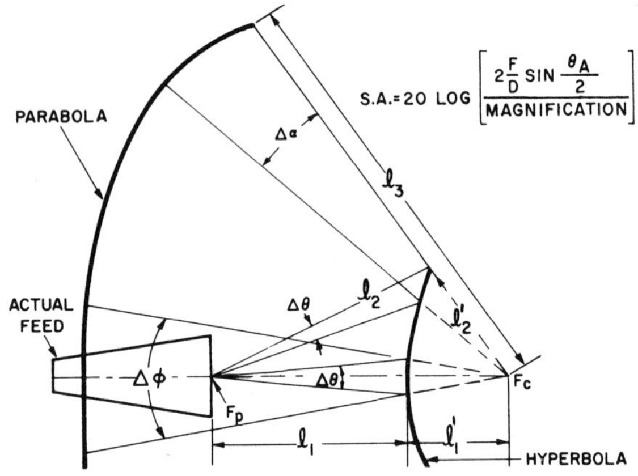

FIG. 2 SPACE ATTENUATION GEOMETRY

$E_d' \text{ (NORMALIZED)} = \frac{\sqrt{\frac{|DB|_{\xi=0}}{4.34}} \text{EXP}\left(\frac{-|DB|_{\xi=0}}{8.68}\right)}{\text{EXP}(-1/2)}$

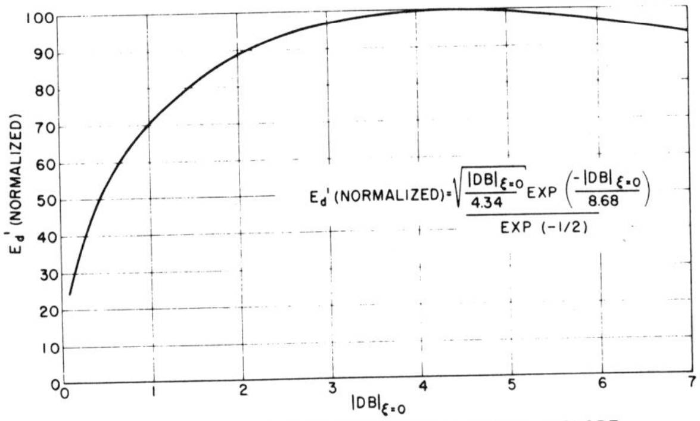

FIG. 5 NORMALIZED DIFFERENCE-PATTERN SLOPE

○ HORN-HYPERBOLA COMBINATION PATTERN MEASURED ON A PARABOLIC CONTOUR
□ HORN PATTERN MEASURED ON ARC HAVING A RADIUS EQUAL TO $\overline{V_h F_p}$

θ_A TO EDGE OF HYPERBOLA
θ_E TO EDGE OF PARABOLA

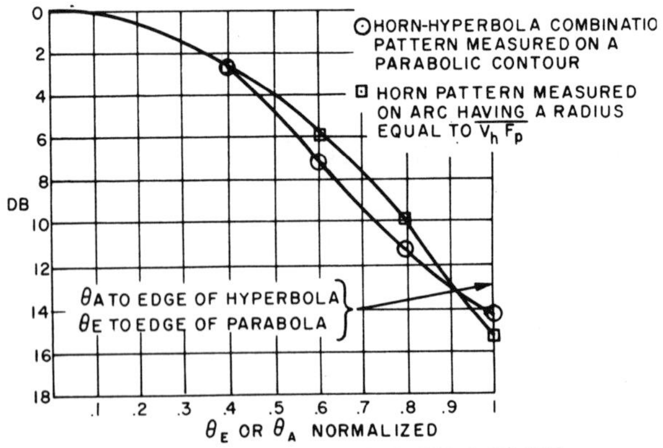

FIG. 3 COMPARISON OF DISTRIBUTIONS

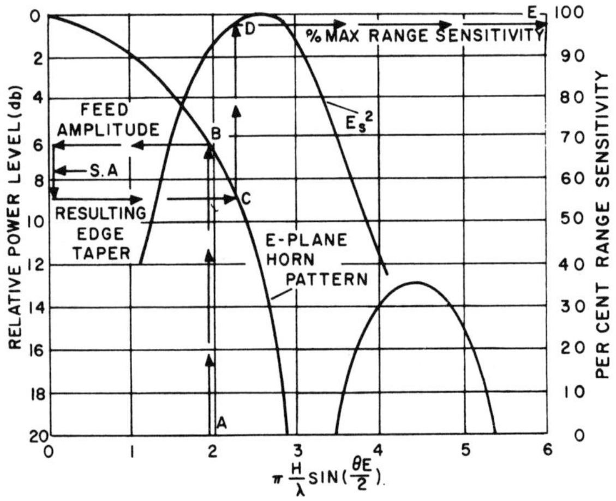

FIG. 6 E-PLANE HORN PATTERN AND RANGE SENSITIVITY

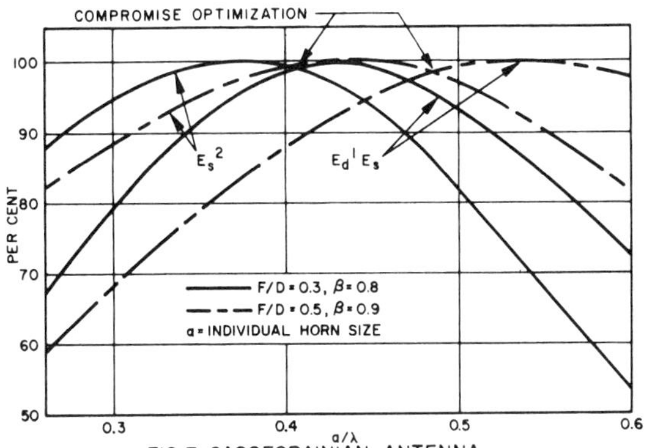

FIG. 7 CASSEGRAINIAN ANTENNA RANGE AND ANGULAR SENSITIVITY

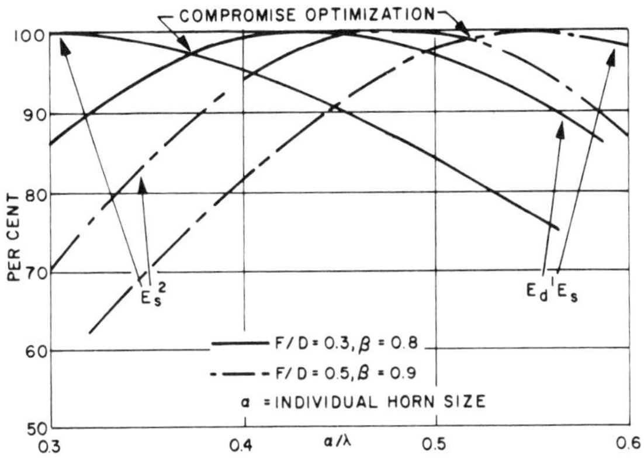

FIG. 8 PARABOLIC SYSTEM RANGE AND ANGULAR SENSITIVITY

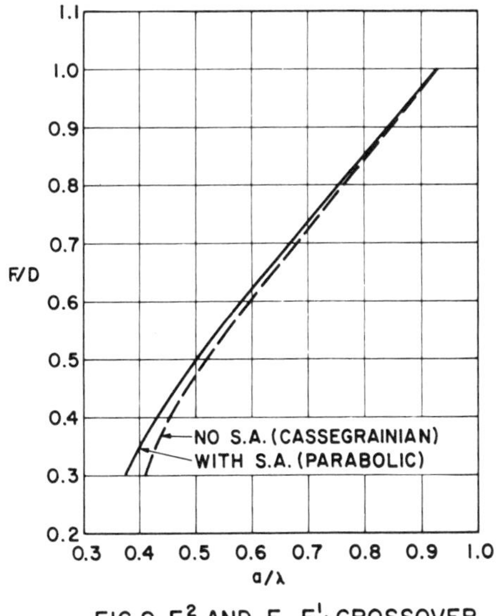

FIG. 9 E_s^2 AND $E_s E_d^1$ CROSSOVER

147

$$E_d' = 2h \sqrt{\frac{|DB|_{\xi_1 = o}}{4.34}} \exp\left[\frac{-|DB|_{\xi_1 = o}}{8.68}\right] \quad (20)$$

The normalized form of Equation 20 is plotted in Figure 5.

Differentiating Equation 20 with respect to $|DB|_{\xi_1 = o}$ and evaluating for a maximum slope, one obtains:

$$|DB|_{\xi_1 = o} = 4.34 \quad (21)$$

This equation indicates that the maximum on-axis slope of the difference pattern is obtained when the Guassian distributions cross over in space at 4.34 db down. The angle ξ_i used in the above expression is given by (Figure 4):

$$\xi_i = \beta \phi = \beta \frac{a}{2F} \frac{180}{\pi} \text{ degrees} \quad (22)$$

for $\tan \phi \cong \phi$ and where β is the beam deviation factor.

If Equation 22 is substituted into equation 19, and if it is assumed that the individual beams have a beam width given by (65) (λ/D), then Equation 19 becomes:

$$|DB|_{\xi_1 = o} = \left[1.53 \frac{a}{\lambda} \beta \frac{1}{F/D}\right]^2 \quad (23)$$

The crossover level in db can now be obtained as a function of a/λ for a fixed F/D ratio because β is primarily a function of the F/D ratio. The determination of the relationship of β to the F/D ratio in a Cassegrainian antenna is a lengthy development by itself and can be found by a geometrical analysis of the aperture phase front resulting from various configurations of the Cassegrainian antenna parameters.

Since the angular sensitivity has been defined as the product of the on-axis transmit sum signal E_s and the slope of the difference signal E_d' received on axis, it remains necessary only to determine E_s. For a circular aperture illuminated by a horn, the range sensitivity has been calculated and is plotted in Figure 6 as a function of $(\pi H/\lambda) \sin(\theta_E/2)$ where λ is the free-space wavelength and H (where H is equal to 2a) is the feed sum aperture. The maximum range-sensitivity occurs for an edge taper of 12 db. This is to be expected since the range-sensitivity and the gain of an antenna are proportional. The range sensitivity may be plotted from Figure 6 as a function of a/λ for a fixed F/D ratio. The value for the Cassegrainian antenna space-attenuation is that which is obtained from Equation 15.

The product of E_s (obtained from the E_s^2 function) and E_d' (for a constant F/D ratio) yields the angular-sensitivity function. As both the angular- and range-sensitivity functions are now known, the image feed horn may be selected for an optimum monopulse design. The variation of the two functions for F/D ratios of 0.3 and 0.5 for a Cassegrainian antenna with negligible space attenuation are depicted in Figure 7. In Figure 8, the two functions are again plotted for F/D ratios of 0.3 and 0.5; but in this case for a conventional focal-point-feed system. From Figure 7 or 8 it is seen that the two functions do not maximize for the same a/λ, and a compromise is usually chosen at the point where the two functions cross over. Figure 9 is a plot of the compromise a/λ as a function of F/D ratio. It is seen that the image horn aperture required for the Cassegrainian antenna is larger than that required if the antenna system has the feed at the parabolic focus. This difference in size increases as the F/D decreases. In conjunction with the work presented in this paper concerning the range- and angular-sensitivity functions, S. Sandler performed a rigorous theoretical and experimental analysis which substantiated the above conclusions.[1]

In the design of a Cassegrainian antenna for optimum monopulse performance, a near-field problem is encountered due to proximity of the feed horn and the hyperbola when the hyperbola is opaque and the secondary reflector is less than 75λ. The near-field effect produces a square-law phase variation across the secondary aperture and also alters the amplitude distribution of the secondary aperture from that which would be expected in the Fraunhofer region based upon feed aperture diffraction theory. A method of correcting the phase variation is to displace the subreflector and feed as a unit to introduce a compensatory square-law phase variation. However, complete phase correction in this manner is not possible because the two square-law phase functions are not identical. If the phase of the feed horn aperture is not uniform, then greater square-law phase variations will exist across the secondary aperture, and even the method of partial correction may not be sufficient for good monopulse performance. To obtain a uniform phase across the primary aperture, the usual method is to keep the flare angle of the horn small. However, since the feed horn for a Cassegrainian antenna is usually 3λ or greater, the horn length becomes excessive. A way to keep the horn length within practical limits and yet obtain a near-uniform aperture phase is to design the monopulse feed structure so that each port has an axis of symmetry such as that in the feed shown in Figure 1. This type of feed cluster allows a large reduction in horn length, because the resultant phase variation from uniform phase (as a function of horn

148

flare angle) is much less with this type of horn. The feed type as shown in Figure 2 is usually the one first considered, because the waveguide inputs to the microwave circuits are adjacent. The alternate type described above has the obvious disadvantage of separated inputs.

In a refined technique for eliminating near-field effects while still maintaining the advantage of adjacent feed outputs, the horn is focused at the hyperbola by means of a metal plate or dielectric lens in the horn. It is well known from optics that through the use of focusing techniques the Fraunhofer phase and amplitude can be obtained in the Fresnel region of an aperture along the spherical plane of focus. To achieve a focus at the hyperboloid vertex V_h, the lens must produce a spherical phase front across the horn aperture having a radius of curvature of $\overline{V_h F_p}$.

Good Cassegrainian antenna performance is based upon the premise that the wavefront impingent upon the hyperboloid is spherical, with its center at the feed point F_p. Any deviation from this is undesirable and will deteriorate the beam in space. It has been found experimentally that the feed type shown in Figure 2, having an aperture of 5λ and spaced 25λ from the hyperboloid, produces a deviation from the desired impingent spherical phase front of 60 degrees at the outer edges of the hyperboloid. With the use of a dielectric focusing lens, this phase variation has been reduced to less than five degrees.

The aperture blocking introduced by the hyperbola is the prime drawback of the type of Cassegrainian antenna under discussion. Usually, it is possible to limit this blocking to less than 3 percent. Unless specifications call for transmit sidelobes greater than 18 db down, this degree of aperture blocking is acceptable and does not affect the main lobe or monopulse performance.

Conclusions

The Cassegrainian antenna design permits the mounting of microwave circuitry behind the electrically active surfaces of the antenna. This microwave package is in an advantageous position for rigid mounting, and further is in proximity to the feeds, eliminating the need for long waveguide runs that can introduce differential phase shifts and losses that can become very critical in monopulse systems.

For a chosen illumination of the secondary aperture, greater latitude in the choice of the dimensions of the actual feed horn is afforded by a Cassegrainian antenna. The magnification between the image and actual feed horn size eliminates the problems which arise in focal-point-feed monopulse systems having F/D ratios less than 0.5. The actual feeds, being larger, are well above cutoff size. The dielectrical loading of the feed apertures is not necessary, mutual coupling between adjacent ports of the feed is reduced, and power-handling capacity is increased.

From a fabrication and cost standpoint, the advantages of the Cassegrainian antenna are evident. The basic and largest component, the parabola, can usually be selected from standard stock. The hyperboloid is a relatively small and light structure and any support structure such as a quadrapod is of a straightforward structural design. The over-all volume is reduced because the antenna depth is now less than the focal length.

The differences between the electrical properties of the opaque-subreflector-type Cassegrainian monopulse antenna and the conventional system are characterized by a different space attenuation function and modified range- and angular-sensitivity functions. The inherent aperture blocking presented by the hyperboloid must be considered as the necessary price for the advantages of simplicity, cost, and diversity.

References

[1] S. Sandler, "Monopulse Far Field Patterns," Lincoln Laboratory Group Report No. 315-1, July 23, 1959.

Acknowledgment

The research described in this paper was conducted under subcontract #98 of AF 19(604)-5200, Massachusetts Institute of Technology, Lincoln Laboratory.

The authors wish to thank the personnel of Lincoln Laboratory, particularly Mr. Leon J. Ricardi, for cooperation and encouragement during the research and development of the antenna system described above.

149

electronics
APRIL 22, 1960

FIG. 1—Two typical comparison monopulse tracking radars. Original Mk 50 monopulse radar (left) operates in X-band using 6-foot lens system. Modern monopulse radar AN/FPS-6 (right) uses 12-foot diameter parabolic reflector and operates in C-band

Precision Tracking With Monopulse Radar

Modern monopulse tracking radars offer many advantages over other basic tracking techniques such as sequential lobing and conical scan. In this article, the principles of monopulse operation are explained and its performance compared against typical radars of other types

By JOHN H. DUNN and DEAN D. HOWARD, U. S. Naval Research Laboratory, Washington, D. C.

TRACKING RADARS are classified in three major categories whose names describe the technique used to obtain information about the position of the target in relation to the direction of the center of the radar beam. These techniques are conical scan, sequentially lobed and monopulse.

One of the principal difficulties in conical scan systems was the interaction between the lobing rate and the noise-modulation components at the same frequency on the return signal. Noise modulation on the return caused by propeller rotation was particularly trouble-some since the errors caused by such high-amplitude noise caused radar to lose the target.

In the study of possible corrective measures, one of the most apparent solutions was the use of a lobing or scanning rate which was of a higher frequency than the maximum propeller rate of the aircraft. This solution required scan rates of such a high frequency that mechanical problems arose. It was also desirable to have a scan rate that could be varied so that modulation noise peaks on the return could be avoided. This need resulted in the sequential lobing system using gas tubes as switches for turning the lobes off and on. In such systems, a four-horn feed was used with either transmission on all four and reception on each of the four in sequence or a five-horn feed using a central horn for transmitting.

The ultimate in such a system would be to obtain complete information about the position of the target on each pulse. Such a system would be equivalent to an infinite lobing rate. Practical realization of such a system requires obtaining information at the r-f frequency instead of the video frequency. This became possible through the development of a device called a rat-race. The use of this device made possible the combining of the signals from all four horns to provide a signal at r-f proportional to the error in pointing direction of the antenna, and in addition, a signal which is the sum of the energy received in all four horns. Thus three signals are provided: an elevation difference signal, an azimuth difference signal and a sum signal. Since this system obtains complete information about target position on each single pulse it is called monopulse[1] (Fig. 1).

TRACKING TECHNIQUES—Most tracking radars make use of microwave optics as lenses or parabolic reflectors for convenient beam shaping and for simplicity in mechanizing target-locating functions. Usually, the focusing lens or parabola is considered in terms of a point source located in the focal plane on the antenna axis which is focused by the lens or parabola to give a beam of desired shape as indicated in Fig. 2A. However, in describing the techniques used in tracking systems it is frequently more convenient to examine the received E-field distribution in the image plane caused by a point source of echo.

This field is described in Fig. 2B which shows the $(\sin x)/x$ shaped field distribution in the image plane of a lens[2]. The lens is used in the figures because of simplicity of illustration. The center peak of this function is what is frequently called the spot to which the radiated E-field of a distant on axis point source of echo is focused in the focal plane. The spot size even for a point source is always finite, for a finite size aperture, where d (Fig. 2B), the spot width between zeros, is $d = 2f\lambda/b$, f is the focal length of the lens or parabola, b is its diameter, and λ is the wavelength of the signal.

Figure 2C demonstrates how the spot in the focal plane moves corresponding to displacement of the point source from the tracking axis. The displacement is caused by the tilt of the arriving phase front with respect to the lens[3] such that there is a continuous variation of phase of the illumination across the lens. This phenomenon is the basis upon which tracking systems can track a target. All tracking systems, using a lens or parabola, provide some means for generating a voltage corresponding to the displacement, from the antenna axis, of the spot in the focal plane. The voltage is then used to drive the antenna in a direction which will recenter the spot, thus performing closed-loop tracking by reorienting the antenna toward the echo source.

Basic difference in the three tracking systems is the means by which it senses the displacement of the spot from the antenna axis. Figure 3A gives a three dimensional plot of the spot or E-field intensity in the focal plane to aid visualization of the three dimensional problem of sensing of the spot displacement.

The conical scan tracking system senses the spot location by moving its feed center such that it describes a circular path in the focal plane around the antenna axis. Thus, as a function of time it scans around the circular cross-section of the spot. This operation is indicated in Fig. 3B which shows the path described by the center of the nutating feed for a target on axis.

If the spot is exactly centered there will be essentially a constant receive signal throughout the scan. However, when the spot in Fig. 3B is slightly displaced, the received signal during the nearer portion of the scan will be less than that received at the far portion of the scan. This effect will result in a modulation of the carrier at the lobing rate. The phase of this modulation with respect to the feed position will depend on the direction of spot shift. Therefore, the depth of the modulation is proportional to error and the phase with respect to the lobing cycle gives the direction of error. By appropriate circuits, a corresponding error voltage is generated for driving the azimuth or elevation servos accordingly.

Most common conical-scan radars also transmit from the rotating feed. Thus, the depth of the modulation for a given squint angle, resulting from an off-axis target, is increased because the target receives a similar modulation by the illuminating signal.

The sequentially-lobed radar is similar to the conical scan radar because it does sample the E-field around the spot in the focal plane in time sequence. Yet it is similar to monopulse because it uses the four-horn aperture for its sampling process. Figure 3C shows how the four horn feed of the sequentially-lobed radar is located with its aperture in the focal plane (the phase center of the feed horn is placed in the focal plane) and centered on the antenna axis. During transmission the four horns are fed in phase such that they illuminate the lens or parabola essentially the same as a conventional single horn to obtain an optimum beam pattern for illumination of the target.

Means of sensing spot displacement in the sequentially-lobed radar is provided by the four horns since they divide a centered spot into four equal quadrants such that equal energy is received in each of the four horns. However, any movement of the signal source from the antenna axis causing a displacement of the spot will unbalance the energy received in the four horns.

By sampling the energy in the output of each horn in sequence there will be a modulation of the received energy whose depth is proportional to antenna pointing error and phase with respect to the sampling sequence is a function of the direction of error. Thus, the means is provided for generating a voltage of proper magnitude and polarity to drive servo systems in the antenna to correct the error.

The monopulse radar uses the same four-horn feed as the sequentially-lobed radar as shown in Fig. 3C. In addition, both radars transmit on all four horns

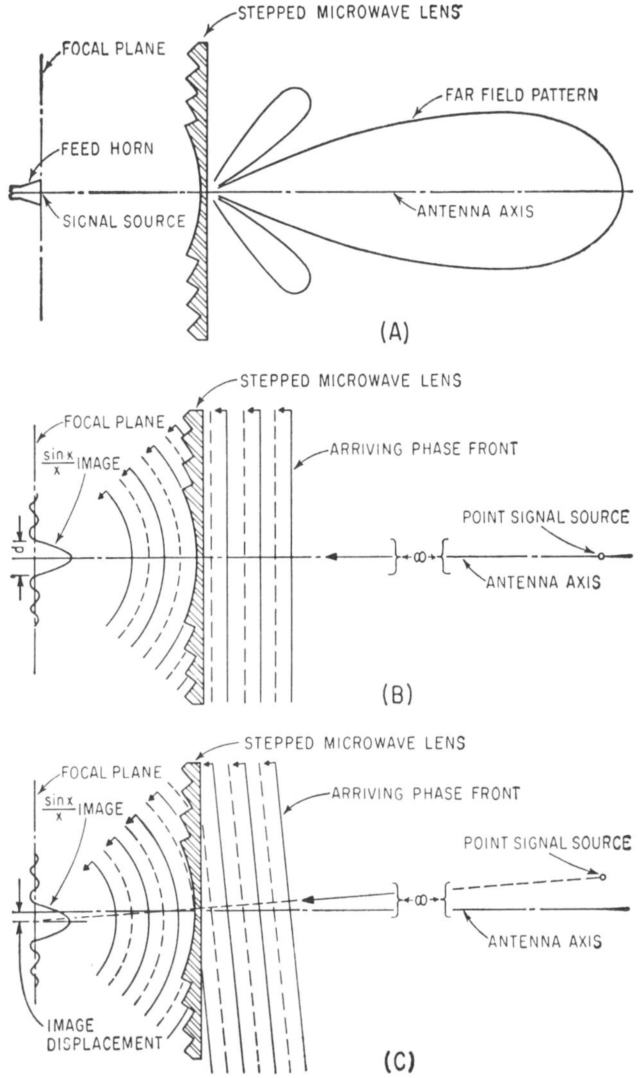

FIG. 2—Function of microwave parabolas and lenses by the conventional point source at the focal point and resultant far field pattern (A) compared with the image plane concept with a point source in the far field and resultant E-field amplitude in the focal plane for condition of on-axis source (B) and off-axis (C)

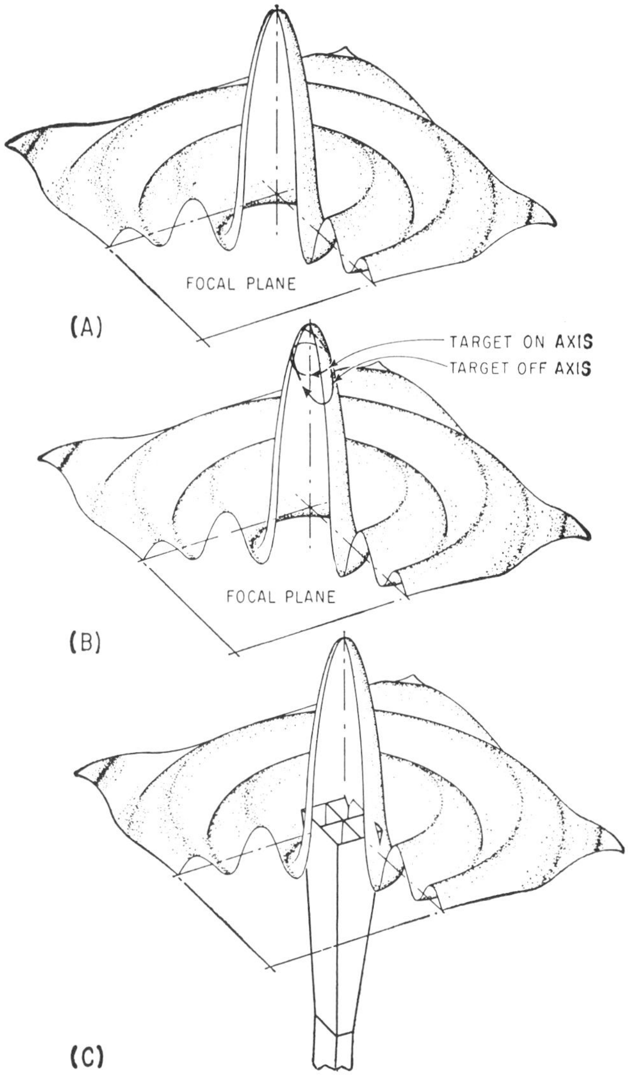

FIG. 3—Three dimensional plot of E-field amplitude in the focal plane. In (A), the phase is constant in the lobes of the pattern; (B) shows the patch described by a nutating conical scan feed for an on-axis and off-axis target; and (C) shows the three dimensional E-field in relation to the four horn feed location

simultaneously. The major difference between these radars is that where the sequentiall-lobed radar compares video from a time sequence, the monopulse radar compares r-f signals instantaneously. Through passive microwave circuits the monopulse radar gives three outputs; the sum of all four horns for range tracking and a reference to angle tracking channels, the difference of the top and bottom horn pairs for tracking in elevation and the difference of the left and right horn pairs for tracking in azimuth. The difference signals determine the magnitude of error, and the boresight line or antenna axis is the antenna pointing angle at which the difference signals are zero. The direction of error is later determined after i-f amplification by phase comparison with the sum signal.

The E-field spot in the focal plane is essentially constant phase over the main lobe of the $(\sin x)/x$ image[4] and the sum and difference signals will always have either a 0 deg. or 180 deg. phase relation. Therefore, the phase comparison is only to determine which of the two phases exists between the sum and difference signal in order to produce the error voltage of the correct polarity. Phase comparison has no significant effect on the boresight or electrical axis under normal condition.

MONOPULSE TECHNIQUE—To simplify the explanation of the operation of a monopulse radar, consider the diffraction pattern in the focal plane of the secondary aperture. This is shown in Fig. 3. The four feed horns are placed at the center of the main lobe of the diffraction pattern. On transmission, all four feed horns are excited in phase and the feed is designed and positioned to give the optimum beam from the lens or reflector. The maximum of the beam formed in this manner coincides with the tracking axis of the radar.

On receive, the signals in the four horns A, B, C and D are combined (Fig. 4) to provide a sum or reference signal and two difference signals. The difference signals are obtained by comparing the energy

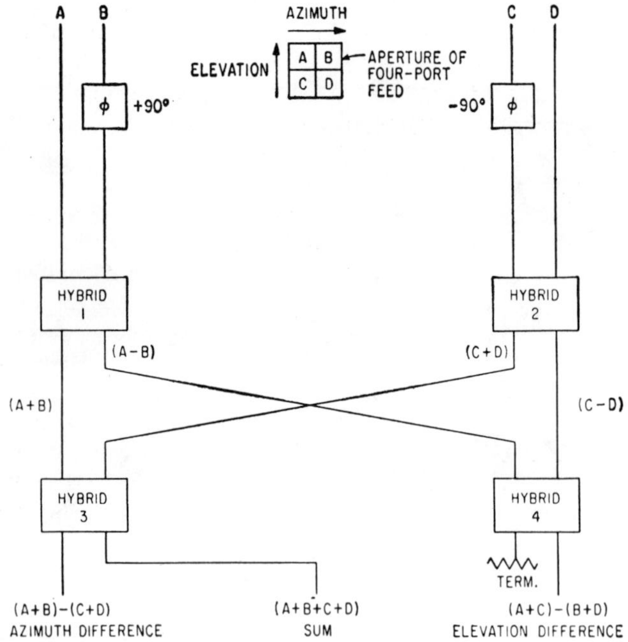

FIG. 4—Hybrids determine the sum or range track signal, and the azimuth and elevation difference signal. The difference signals, after conversion to i-f, determine the angle error signals when combined with converted sum signals in product detectors

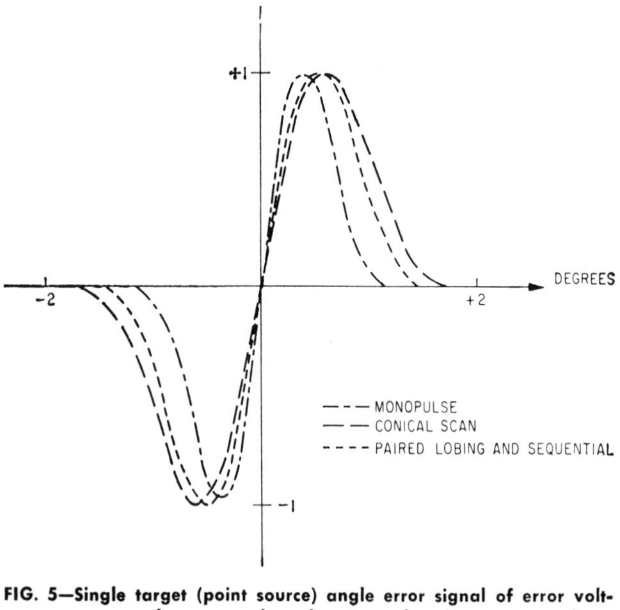

FIG. 5—Single target (point source) angle error signal of error voltage versus angular error when the target location corresponds to zero on the horizontal axis

received in the horn pairs. Thus the amplitude of the difference signal is compared with the phase of the sum signal to determine the direction of the pointing error. Figure 4 shows only the r-f comparison circuits of the system.

Earliest types of monopulse radar systems, such as the Mk 50 in Fig. 1, used a lens-type secondary aperture. Much smaller primary apertures are possible and blockage is no longer a problem. The combining of signals is accomplished within the primary feed structure through the use of short slot hybrids[5] in place of the rat race and phase stability has been greatly enhanced by the symmetrical structures made possible[6].

The most important consideration in the design of a monopulse radar is the establishment of the proper phase relationship and the elimination of phase errors in the r-f section. If proper care is used in the design of the r-f section, the boresight of the system will not be affected by phase errors or changes in phase of the i-f amplifiers.

To give a more complete picture of the effects of phase errors introduced in different points in the tracking system on the boresight of the radar, the comparison of the electrical axis in angle to the antenna axis is analyzed briefly in the appendix as a function of these phase errors. In reference to Fig. 4 and considering the azimuth plane only (for convenience), the electrical axis is determined by the sum and difference of the signals at the input to hybrid 3. These inputs are the sums of the signals from the left and right pairs of horns. When the radar is used, as previously described, with a lens or parabola, the two signals will always be in phase when the antenna feed is properly aligned.

The azimuth difference signal gives tracking error information because of the change in relative amplitude of the two signals as a function of antenna pointing angle, such that it is zero when the antenna is aligned with the target. When there are no phase errors or amplitude unbalances in the system the error detector output e_o which is the product of the sum and difference signals is simply $e_o = A^2 k \theta$ where A is the on axis amplitude in the signals from the horn pairs, k is the antenna gain constant and θ_o is the pointing angle of the antenna assuming the target is at zero angle for simplicity. This equation shows the desired condition where the electrical axis ($e_o = 0$) corresponds identically to zero antenna pointing error or $\theta_o = 0$.

If a phase error ϕ is introduced at any point after hybrid 3 either at r-f or at i-f, the error detector output (appendix) becomes $e_o = A^2 k\ \theta_o \cos \phi$. This equation shows no boresight shift since $e_o = 0$ when $\theta_o = 0$, but it does indicate a loss in sensitivity for small values of ϕ. This is not serious for a few degree phase error but if ϕ reaches 90 deg, all sensitivity is lost and if ϕ exceeded 90 deg, but is less than 270 deg, the error sense is reversed. There are possibilities that this might occur from malfunction of the equipment.

A more serious phase error is that which might occur between the two signals arriving at hybrid 3. This phase error, τ, is considered in the appendix and the error voltage derived. Analysis of e_o shows that there is a boresight shift such that when $e_o = 0$ the value of θ_o is not zero but $\theta_o \cong (\tau \tan \phi) / 2k$. Thus, it is observed that in a precision tracking radar τ must be kept essentially at zero.

The need for close proximity of the comparison circuit brings up a major disadvantage of phase comparison systems which has boresight shifts that can be even more sensitive to phase error. In conventional phase comparison systems there is a large separation of feeds such that significant boresight errors can readily arise.

It is observed that r-f phase errors which contribute only to ϕ can be corrected by an equal and opposite phase shift at i-f between sum and difference signal. However, this means of correction is not recommended since r-f phase errors are generally frequency sensitive while, for a properly tuned local oscillator, the i-f errors will be independent of frequency, thus, ϕ would be zero at only certain points in frequency.

Characteristics typical of a precision monopulse instrumentation radar have been given.[7] For a 20 mil antenna beamwidth (1.25 deg), overall absolute rms tracking accuracies of 0.1 mil (0.0057 deg) in angle are obtained. This figure includes all system biases and rms errors. The precision or angular difference in positions of a target that can be measured by the radar is less than 0.01 mil. As an example, in the tracking of a balloon-borne 6-inch sphere, a 1.5-inch wobble of the sphere inside the supporting balloon gives the major source of tracking deviation about the balloon center (measured with a boresight camera) out to approximately 2 miles where the variation of index of refraction of the atmosphere becomes the most significant source of error.

In precision radar tracking it is occasionally desired to make real time correction of angle tracking data. The amount of lag that is present with any tracking antenna, when there are angular rates, is always accompanied by a corresponding voltage output from the angle error detector circuits. Therefore, if the radar is calibrated in volts-per-unit angle error, the angle error detector output can be read in terms of angle lag providing data for real time correction of target position information from the antenna mount.

Calibration can be accomplished in monopulse with a beacon by measuring the open loop angle error voltage for given values of angle error. A conical scan radar whose sensitivity depends in part upon its scan on transmit may require boresight balloon runs with optical measurements for given bias errors introduced into the servo system.

One further point on the capability of monopulse tracking radars is that they can passively track a noise source such as the radiation from the sun.

ADVANTAGES OF MONOPULSE—The monopulse radar is the only tracking radar which utilizes the full gain capability of its secondary aperture. With present microwave techniques the amplitude of the sum signal of the monopulse radar, obtained by using all four horns in phase, is as great as if a single feed horn of optimum dimensions were used. The difference signals do not reduce the normal sum signal energy because the energy received in the difference channels is only that energy which would normally be reflected back into space if an optimum single horn feed were used.

The conical scan tracking radar has an average beamwidth (averaged over the scan cycle) which is considerably spread with respect to the optimum by the lobing process (Fig. 4). The resultant two-way antenna gain or received signal for range tracking, when using optimum squint angle for tracking, is about 3 db below that obtainable with a monopulse. In addition, the rotary joints involved in lobing may cause additional losses which, although small in reduced signal power, can seriously limit the minimum system noise temperature for use with low-noise receivers.

Sequential-lobing radars transmit an optimum beam pattern with the four horns fed in phase, but on reception the energy output from the horns which are not being observed is dumped into a load. Since the energy is divided equally between the four horns (when the target is on axis) the power in each horn will be one fourth or 6 db down from the total. Therefore, if one horn is sampled at a time the total on axis signal will be 6 db down from optimum or if it is sampled in pairs it will be 3 db down. Thus, the superiority of the monopulse is demonstrated in terms of antenna gain.

The relative capability of tracking radars to resolve multiple target and the associated angular sensitivity has been discussed in some detail[8] where the monopulse is shown to have both superior angular sensitivity and resolving capability. This is indicated (Fig. 5) by the d-c error voltages out of the angle error detectors as a function of pointing angle, for the three types of tracking radar. The sensitivity or the slopes of curves of Fig. 5 taken on axis are: (normalized to monopulse) monopulse—1.00, sequential lobing—0.72, paired lobing—0.67 (paired lobing is a form of sequential lobing where the horns are sampled in pairs), conical scan—0.59, thus indicating the superior angle sensitivity of monopulse.

The resolving capability of the radar systems is summarized[8] by the figures on the separation of two point echo sources required for resolution of the sources normalized to the separation expresses as a fraction of one way antenna beamwidth: monopulse —1.3, paired lobing—1.4, conical scan—1.7 and sequential lobing—1.8. Thus, the monopulse radar demonstrates superiority in resolution capability.

The monopulse is the only tracking radar which has no tracking noise caused by amplitude noise or amplitude scintillation[9]. This immunity to amplitude noise results because complete tracking information is obtained from each pulse on the basis of ratios of the portions of the energy of the pulse which is received in each of the horns of the feed. The echo amplitude does not affect these ratios or the electrical axis of the radar.

The sequential lobing and conical scan radar also take ratios, in a sense, but the ratios are of signals received at different times. Thus, any echo amplitude change which occurs at the rate at which the sampling or scanning takes place causes false information in the output of the angle error detector. For example, when a constant level point echo source is on axis of a conical scan radar there is no modulation of the received r-f signal. However, if the echo amplitude fluctuates up and down during a full scan, this modulation will appear to the radar as if it had a tracking error. In closed-loop tracking the radar would move off the true target position as a result of the false information. The rms tracking noise, σ_{amp}, caused in a scanning radar by amplitude scintillation[9] is $\sigma_{amp} = (0.0085)\ B\ A_{amp} \sqrt{\beta}$, where B is the two-way antenna beamwidth (the units of B determine the units of σ_{amp}), A_{amp} is the percent-noise-modulation density in the vicinity of the lobing rate in units of percent modulation per $\sqrt{\text{cps}}$, and β is the servo bandwidth in cps. The constant (0.0085) was computed assuming a parabolic-shaped pattern and takes into account the frequencies both above and below the lobing rate. For significantly different antenna patterns a new constant must be derived.

The amplitude scintillation has its highest density in the frequency range below 50 cps although there is background modulation and propeller modulation up to a few hundred cps. Therefore, since the tracking noise in scanning radars is a function of the amplitude scintillation density, it was desired to raise the lobing rate as high as possible where the amplitude scintillation density is the least. This was one of the major reasons for development of the sequential lobing radar which can be lobed electronically where the limit is the pulse repetition rate pulse-to-pulse lobing.

One further point of interest is that very-low-frequency amplitude fluctuations in the vicinity of the servo passband, although not a direct source of tracking noise, do affect the contribution target angle scintillation to tracking noise, dependent on the agc characteristics[10]. However, this effect contributes equally to all tracking systems.

A further advantage of monopulse, which is also an advantage of sequential lobing radars, is the complete elimination of mechanical moving parts from the error sensing process. This advantage is becoming increasing significant with the present and future demands for high precision tracking. The mechanical scanning not only places a limit on boresight accuracy but adds rotary joints which can seriously limit the minimum noise temperature of a low-noise system.

Reference

(1) R. M. Page, Monopulse Radar, *IRE Conv Rec*, **8**, p 132, 1955.

(2) F. A. Jenkins and H. E. White, "Fundamentals of Optics," McGraw-Hill Book Co., Inc., New York, N. Y., p 288, Third Edition, 1957.

(3) D. D. Howard, Radar Target Angular Scintillation in Tracking and Guidance Systems Based on Echo Signal Phase Front Distortion, *Proc NEC*, **15**, p 840, 1959.

(4) G. W. Farnell, Phase Distribution in the Image Space of a Microwave Lens System, Eaton Electronics Research Laboratory, McGill University, Montreal Canada, Tech Rep No. 39; October 1958 (Sponsored by A. F. Cambridge Res. Center).

(5) W. A. Tyrrell, "Hybrid Circuits for Microwaves," **35**, p 1294, Nov, 1947.

(6) P. J. Allen, A New Approach to Antenna Feeds for Precision Tracking Radars, *Rec* of the 1st *Nat Conv* of the IRE PGME, June 1957.

(7) D. K. Barton, Accuracy of a Monopulse Radar, *Proc.* 3 rd *Nat Conv* on *Mil Electronics* by *PGME* of *IRE*, 1959.

(8) S. F. George and A. S. Zamanakos, Multiple Target Resolution of Monopulse Versus Scanning Radars, *Proc NEC*, **15**, p 814, 1959.

(9) J. H. Dunn, D. D. Howard, and A. M. King, Phenomena of Scintillation Noise in Radar-Tracking Systems, *Proc IRE*, **47**, p 855, May, 1959.

(10) J. H. Dunn and D. D. Howard, The Effects of Automatic Gain Control Performance on the Tracking Accuracy of Monopulse Radar Systems, *Proc IRE*, **47**, p 430, March 1959.

APPENDIX

Derivation of the Angle Error Detector Output As A Function of R-F and I-F Phase errors.

Analysis of the effect of phase errors on the tracking boresight will be made in the azimuth plane only for simplicity. The error detector output is the smoothed output of a product detector which multiplies the sum and difference signals. These signals are derived in azimuth by hybrid 3 of Fig. 4 which takes the sum and difference of the signals from the left and right pairs of horns of the monopulse feed.

As described in the text, when using a lens or parabola, the signals in these feeds will be in-phase but their amplitudes will vary as the antenna azimuth pointing angle θ_o moves through the target such that they are equal when θ_o is equal to the target angle θ_t. Assuming the antenna patterns are linear in the region where $(\theta_o - \theta_t)$ is small, the signal is left pair of horns, e_l, and the right pair of horns, e_r, will have the functions

$$e_l = A\ 1 - k\ (\theta_o - \theta_t) \cos(\omega t)$$
$$e_r = A\ 1 + k\ (\theta_o - \theta_t) \cos(\omega t)$$

where A is the signal amplitude of the horn pairs when the antenna is on target, ω is the angular r-f frequency, and k is an antenna gain constant.

First, e_l will have a phase error τ added to the argument of the cos to simulate a phase error occurring before the difference network. Also, for simplicity, θ_t will be set to zero reference angle such that when θ_o is zero the antenna is pointing at the target. Hybrid 3 has a sum output e_s and a difference output e_d which are the sum and difference of e_l and e_r such that

$$e_s = A\ (1 + k\,\theta_o) \cos \omega t + A\ (1 - k\,\theta_o) \cos(\omega t + \tau) \text{ and}$$
$$e_d = A\ (1 + k\,\theta_o) \cos \omega t - A\ (1 - k\,\theta_o) \cos(\omega t + \tau)$$

by taking the vector sum and difference of the two vectors in e_s and e_d,

$$e_s = \sqrt{2}\,A\ (1 + k^2\theta_o^2 + (1 - k^2\theta_o^2) \cos \tau)^{1/2}$$
$$\cos\left[\omega t + \tan^{-1} \frac{(1 - k\theta_o) \sin \tau}{(1 + k\theta_o) + (1 - k\theta_o) \cos \tau}\right]$$

$$e_d = \sqrt{2}\,A\ (1 + k^2\theta_o^2 - (1 - k^2\theta_o^2) \cos \tau)^{1/2}$$
$$\cos\left[\omega t + \tan^{-1} \frac{-(1 - k\theta_o) \sin \tau}{(1 + k\theta_o) - (1 - k\theta_o) \cos \tau}\right]$$

The next source of phase error can occur at r-f or i-f between hybrid 3 and the product detector. This phase error ϕ is accounted for by addition to the argument of the cos of e_d. The product detector output smoothed, e_o, is the d-c voltage

$$e_o = A^2\ (1 + 2k^2\theta_o^2 + k^4\theta_o^4 - (1 - k^2\theta_o^2)^2 \cos^2 \tau)^{1/2}$$
$$\cos\left[-\phi + \tan^{-1} \frac{(1 + k\theta_o) \sin \tau}{(1 + k\theta_o) + (1 - k\theta_o) \cos \tau}\right.$$
$$\left. + \tan^{-1} \frac{(1 - k\rho_o) \sin \tau}{(1 + k\theta_o) - (1 - k\theta_o) \cos \tau}\right]$$

One condition to be examined is where there is no phase errors ($\phi = \tau = 0$) giving

$$e_o = A^2\,k\,\theta_o.$$

Here there is no boresight error since $e_o = 0$ when $\theta_o = 0$ and $\phi = \tau = 0$.

The next condition is where $\tau = 0$ and ϕ is finite giving

$$e_o = A^2\,k\,\theta_o \cos \phi. \quad (\text{where } \phi = 0)$$

This condition shows no boresight error but loss of sensitivity for finite values of ϕ. (see discussion in text)

When both ϕ and τ are finite, the error voltage is e_o. By analysis of the magnitude of the cos function it is found that there is no real zero. The electrical zero, $e_o = 0$, occurs only when the argument of the cos function is $\pm \pi/2$. Solving for θ_o

$$\theta_o = \frac{1}{k}\left[\frac{-1 + \sqrt{\sin^2 \tau \tan^2 \phi + 1}}{\sin \tau \tan \phi}\right]$$

and where τ is small as in the usual case

$$\theta_o \cong \frac{1}{k}\ \frac{\tau \tan \phi}{2}$$

Thus, there is a boresight shift when τ and ϕ are both finite which is the value of θ_o required for $e_o = 0$

Distribution Functions for Monopulse Antenna Difference Patterns*

O. R. PRICE†, SENIOR MEMBER, IRE, AND R. F. HYNEMAN‡, MEMBER, IRE

Summary—**Theoretical considerations of monopulse antisymmetrical or difference radiation patterns which have optimum properties in the Dolph-Tchebycheff sense are discussed. An approximate technique for synthesizing such patterns from linear arrays or continuous line sources is given. A highly desirable feature of the method is that the excitation function may be written as a simple modification of a (known) conventional Dolph-Tchebycheff excitation function. In addition, the method is also applicable to the synthesis of other specialized antisymmetrical patterns.**

I. Introduction

Design methods for the sum patterns of monopulse antennas (which are directly related to conventional antennas) are well established. If the ultimate in narrow beamwidth and low sidelobes is desired, then a pattern with all sidelobes equal must be obtained. Application of the Tchebycheff polynomial [1]–[3] yields equal sidelobes provided one considers only the space factor of an antenna. However, operational requirements sometimes dictate that the designer sacrifice some beamwidth or sidelobe level near the main lobe in order to obtain very low sidelobes far from the main beam. Design criteria for the conventional (sum pattern) antenna have also been derived for this case [4].

Many monopulse antennas in operation are modifications of parabolic-dish-reflector antennas. Laboratory prototypes of two- and three-dimensional slotted waveguide antennas have been constructed and tested. In the near future monopulse antenna arrays with independent power sources, and accompanying independent phase and amplitude control for each radiating element, will be built. Especially in view of the last two

* Received by the PGAP, March 12, 1960.
† Res. Labs., Hughes Aircraft Co., Culver City, Calif.
‡ Ground Systems Group, Hughes Aircraft Co., Fullerton, Calif.

developments it is apparent that there is a need for difference (antisymmetrical) pattern design methods.

In this paper, design methods for obtaining difference patterns satisfying various operational requirements will be derived. Because the element pattern in most antennas lacks much directivity, and in order to maintain wide applicability of the methods to be derived here, only the space factor will be considered.

II. Review of Dolph-Tchebycheff Optimum Sum Patterns

In preparation for later consideration it will be advantageous to review briefly methods for obtaining optimum patterns of arrays of isotropic sources. Considering a linear broadside array of equispaced, isotropic, in-phase sources, Dolph [1] has show that a pattern with minimum main lobe beamwidth for a given ratio of main lobe amplitude to sidelobe amplitude will have sidelobes of equal amplitude.

Fig. 1 shows the configurations under consideration. Since the amplitude distribution is symmetrical ($A_r = A_{-r}$), the radiation pattern for an even number of elements, n_e, may be expressed as

$$\frac{1}{2} E_{S,2N} = \sum_{r=1}^{N} A_r \cos (2r - 1) \frac{\psi}{2} \cdot \tag{1a}$$

and for an odd number, n_0, as

$$\frac{1}{2} E_{S,2N+1} = \sum_{r=0}^{N} A_r \cos 2r \frac{\psi}{2} , \tag{1b}$$

where $n_e = 2N$, $n_0 = 2N+1$, and where the excitation of the rth element is proportional to A_r for $r \neq 0$ and to $2A_0$ for $r = 0$. In each case $\psi/2 = \pi d/\lambda \sin \theta = \pi du/\lambda$, where d/λ is the center-to-center element spacing in terms of free space wavelength, and $u = \sin \theta$, where θ is the angle measured from the broadside direction:

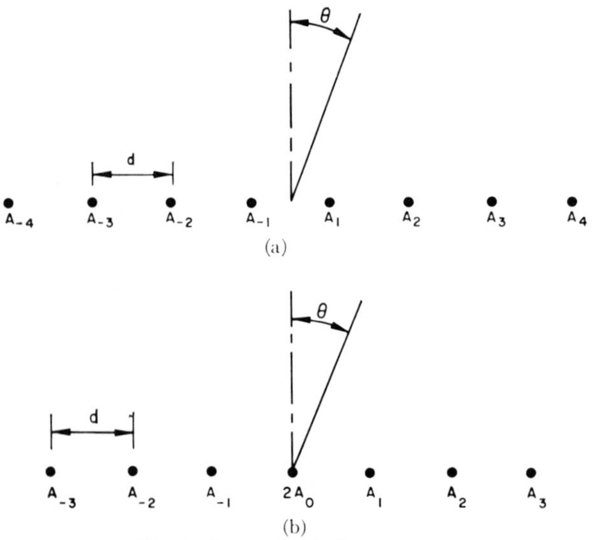

Fig. 1—Array of radiating elements.

Following Dolph, each amplitude pattern may also be represented by an appropriate Tchebycheff polynomial or by a single trigonometric term.

$$\frac{1}{2} E_{S,2N} = T_{2N-1}(x) = \cos \left[(2N - 1) \text{ arc} \cos x \right] \tag{2a}$$

$$\frac{1}{2} E_{S,2N+1} = T_{2N}(x) = \cos (2N \text{ arc} \cos x), \tag{2b}$$

where $x = x_0 \cos \psi/2$, and x_0 is defined by (3)

$$T(x) \big|_{x=x_0} = R_S, \tag{4}$$

where

$$R_S = \frac{\text{amplitude of main lobe of sum pattern}}{\text{amplitude of secondary lobes}} .$$

III. Theoretical Consideration of the Difference Pattern

The difference pattern is antisymmetric and, characteristically, for broadside operation, has a null at $\theta = 0°$ with main lobes of equal magnitude on each side of 0° together with the usual secondary maxima or sidelobes.

Again consider the configurations shown in Fig. 1, except now we will use B_r to designate the feeding coefficient of the rth element. Since $B_r = -B_{-r}$, the difference patterns may be expressed as

$$\frac{E_{D,2N}}{2} = i \sum_{r=1}^{N} B_r \sin \left[(2r - 1) \frac{\psi}{2} \right], \tag{6a}$$

and

$$\frac{E_{D,2N+1}}{2} = i \sum_{r=0}^{N} B_r \sin \left(2r \frac{\psi}{2} \right). \tag{6b}$$

With the substitution $y = \sin \psi/2$, (6a) may be written

$$\frac{E_{D,2N}}{2} = i \sum_{r=1}^{N} B_r' y^{2r-1},$$

which is a real polynomial of degree $2N-1$. In the case of an odd number of elements, a corresponding *real-polynomial* representation does not appear possible,[1] and therefore this case will not be treated from the theoretical standpoint. However, the excitation envelopes for the two cases differ insignificantly for arrays having a reasonably large number of elements, and so the approximate method outlined subsequently applies equally well to either case.

[1] Representations of the forms

$$\frac{E_{D,2N+1}}{2} = i\sqrt{1 - y^2} \sum_{r=1}^{N} B_r' y^{2r-1}$$

or

$$\frac{E_{D,2N+1}}{2} = Z^{-N} \sum_{r=1}^{N} B_r'(Z^{N+r} - Z^{N-r})$$

where $Z = e^{i\psi}$ and $y = \sin \psi/2$, are of course possible.

By definition, the characteristics of the optimum difference patterns are essentially the same as for the conventional Dolph (sum) pattern, and the proof of the optimum property follows the same line of reasoning. For simplicity, the case of $d/\lambda = \frac{1}{2}$ is treated, which results in a range of y from -1 to 1 for θ over the visible region.

Theorem:

Consider a class of odd, real polynomials of degree n,

$$F_n(y) = a_1 y + a_3 y^3 + \cdots + a_n y^n$$

which satisfy the following conditions:

1) All the roots of $F_n(y)$ are real and lie in the interval $(-1, 1)$.

2) The $F_n(y)$ are normalized so as to make the magnitude of the first left-hand and the first right-hand extrema from the origin (the two innermost extrema) equal to a given constant R (See Fig. 2).

3) All $F_n(y)$ have the same ratio R of the magnitudes of the innermost extrema to the magnitude of the largest other extremum for y in the range $(-1, 1)$.

Let the particular function $F_n{}^\circ(y)$ satisfy the above conditions and have the further property that all subsidiary extrema, *i.e.*, all extrema except the innermost pair in the interval $(-1, 1)$, are of magnitude unity. Then $F_n{}^\circ(y)$ has the smallest distance from the origin to the first nulls, $y_{\pm 1}$.

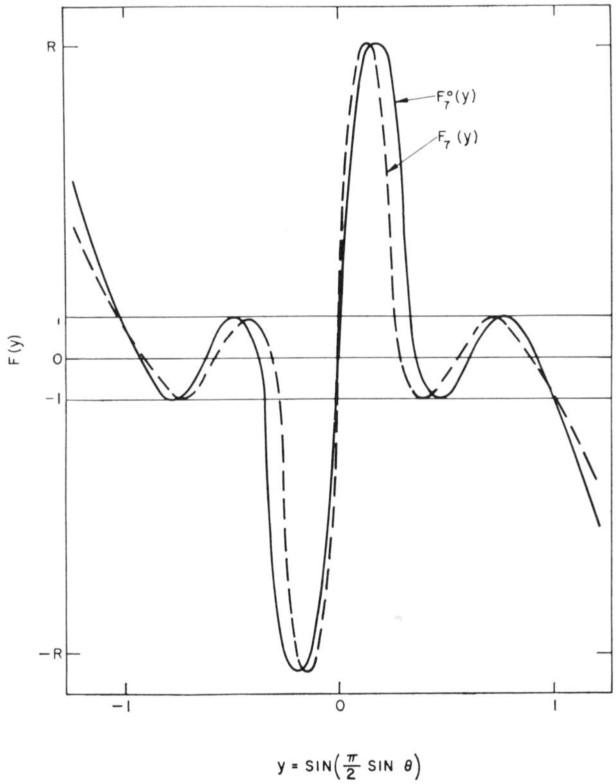

Fig. 2—Sketch of optimum and narrower-than-optimum polynomial.

The proof is equivalent to that given by Dolph [1] in his treatment of the Tchebycheff array. Let $D_n(y)$ be the difference between $F_n{}^\circ(y)$ and any other $F_n(y)$; then

$$D_n(y) = F_n{}^\circ(y) - F_n(y).$$

Hence $D_n(y)$ is a polynomial of degree not greater than n. Now assume that the above theorem is false, *i.e.*, that $|y_{\pm 1}| < |y_{\pm 1}{}^\circ|$, where $y_{\pm 1}$ and $y_{\pm 1}{}^\circ$ are first nonzero roots of $F_n(y)$ and $F_n{}^\circ(y)$, respectively. Since no subsidiary extrema of $F_n(y)$ may exceed unity in magnitude, and all subsidiary extrema of $F_n{}^\circ(y)$ must have that value, it follows that $F_n(y)$ and $F_n{}^\circ(y)$ must intersect in not less than $n+2$ points. Thus by counting the number of intersection points, $D_n(y)$ must have not less than $n+2$ zeros. Since $D_n(y)$ is a polynomial of degree not greater than n, it follows that $D_n(y) = 0$. As a result, if the function $F_n{}^\circ(y)$ exists, then it is optimum in the sense that it will have the least distance between first nulls for the given ratio R.

Corollary: One of the above class of functions, the function $F_n{}^\circ(y)$, has the (normalized) slope of greatest magnitude at the origin. The proof follows exactly as above and will not be repeated here.

Thus, it can be shown that the postulated function $F_n{}^\circ(y)$ is optimum from the standpoint of a monopulse application in that it displays *both* the lowest sidelobe ratio for a given beamwidth as well as the greatest rate of change in signal amplitude with angle about the boresight direction. This latter characteristic is of importance in obtaining good error sensitivity in the monopulse system.

Unfortunately, no known polynomial has the required characteristics; indeed the function is of the class considered by Sinclair and Cairns[2] in their treatment of the conventional optimum distribution for an even number of elements which are spaced less than one-half wavelength apart. Methods of solution involving trial and error, and analog techniques, as well as an iterative technique for improving the characteristics of a trial polynomial, are indicated in their paper. However, the operation must be carried out in entirety for each new attempted design. In the material below, a simple method for obtaining a close approximation to the optimum difference pattern from existing Dolph-Tchebycheff distribution tables is outlined. In addition, the method is applicable for obtaining difference patterns having the specialized sidelobe characteristics of other directive pattern distributions, *e.g.*, the Taylor distribution for continuous line sources [5] and the Taylor modified sin u/u distribution [4].

IV. An Approximate Method for Obtaining the Optimum Difference Pattern

Upon comparing (1) and (6) one observes, in particular, that the trigonometric factor in each term of

[2] See [3], discussion on p. 53, (18) through (26).

159

(6) is (Transmutation I) $f_1(r)$ times the derivative with respect to $\psi/2$ of the trigonometric factor of the corresponding term of (1), and (Transmutation II) $f_2(r)$ times the integral of the corresponding trigonometric term of (1). A brief consideration shows that application of Transmutation II (*i.e.*, an integrative technique), to a Dolph-Tchebycheff distribution generating function yields an aperture illumination even more highly peaked centrally. In turn, although such a distribution may be used to obtain an antisymmetrical pattern, the resulting pattern in general will not have the nulls required for an optimum antisymmetrical pattern as defined in Section III. However, a practical application of Transmutation II will be described subsequently. On the other hand Transmutation I (*i.e.*, a differentiative technique), results in a distribution less highly peaked than the generating function; the resulting pattern will be compared with the optimum pattern.

Transmutation I

The application of Transmutation I to (1) and (2) yields alternate expressions for the difference (odd) amplitude patterns which result from a Dolph-Tchebycheff pattern used as a generating function. Thus upon applying $d/d(\psi/2)$ to (1) and (2), and again letting $x = x_0 \cos\psi/2$, $y = \sin\psi/2$, we obtain

$$\frac{E_{D,2N}}{2} = -i\sum_{r=1}^{N} A_r(2r-1)\sin\left[(2r-1)\frac{\psi}{2}\right], \tag{7a}$$

$$\frac{E_{D,2N+1}}{2} = -i\sum_{r=0}^{N} A_r(2r)\sin\left(2r\frac{\psi}{2}\right); \tag{7b}$$

$$\frac{E_{D,2N}}{2} = -i\sqrt{x_0^2 - x^2}\,\frac{d}{dx}T_{2N-1}(x)$$

$$= -(2N-1)\sqrt{\frac{x_0^2 - x^2}{1 - x^2}}\sin\left[(2N-1)\arccos x\right]$$

$$\equiv P(2N-1; x), \tag{8a}$$

$$\frac{E_{D,2N+1}}{2i} = -i\sqrt{x_0^2 - x^2}\,\frac{d}{dx}T_{2N}(x)$$

$$= -2N\sqrt{\frac{x_0^2 - x^2}{1 - x^2}}\sin\left[2N\arccos x\right] \tag{8b}$$

$$\equiv P(2N; x).$$

Eqs. (8a) and (8b) are transcendental functions of x; however, as indicated previously, (7a) may be expressed as an odd polynomial in y, where $y = \sin\psi/2$. In order to obtain the corresponding odd polynomial (in y) form of (8a), write

$$\sqrt{x_0^2 - x^2} = x_0 y, \qquad x = x_0\sqrt{1 - y^2}.$$

Since the associated Tchebycheff polynomial [6] of degree $2N-2$

$$S_{2N-2}(2x) = \frac{\sin\left[(2N-1)\arccos x\right]}{\sqrt{1 - x^2}}$$

is even in terms of

$$x = x_0\sqrt{1 - y^2},$$

it is also even in terms of y. Then

$$P(2N-1; x) = -(2N-1)yS_{2N-2}(2x)$$

is an odd polynomial in y. In a similar manner it may be shown that (7b), and therefore (8b), are transcendental functions of y of the form

$$E^{(1)}_{D,2N+1} = \sqrt{1 - y^2}\sum_{r=0}^{N} a_r y^{2r-1}.$$

Although the polynomial representation is possible in the even cases, the behavior for both the even and odd cases is more clearly displayed by the transcendental representation in terms of x (8). Inspection of these equations shows that the sidelobe maxima differ in magnitude only through variation of the factor

$$\sqrt{\frac{x_0^2 - x^2}{1 - x^2}}.$$

This variation is small for the values of x_0 usually employed, whenever $|x| < 1-\epsilon$, *i.e.*, in the sidelobe regions. For example, a plot of $P(7; x)$ against y is shown in Fig. 3, from which it is evident that the magnitudes of the subsidiary extrema decrease only slightly as $|y| \to 1$.

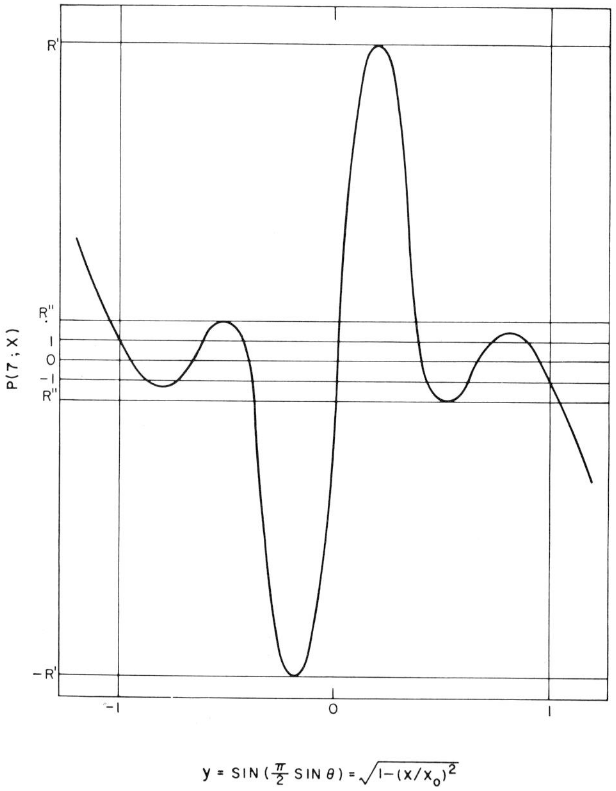

Fig. 3—Plot of the approximately optimum transcendental $P(7, x)$.

The coefficients B_r are most easily obtained by equating the right side of (6) to the right side of (7). Thus,

$$B_r = (2r - 1)A_r, \quad (9a)$$

for an even number of elements and

$$B_r = 2rA_r \quad (9b)$$

for an odd number of elements. An extensive table [7] of values for A_r has been published.

Characteristics of the near-optimum difference pattern, *e.g.*, beamwidth, sidelobe ratio, and location of the main lobes in terms of corresponding characteristics of the parent or generating Tchebycheff pattern are derived in Appendix I and plotted in Figs. 4(a) through 4(c). From these graphs, and using published results for the Tchebycheff pattern distribution function, a difference pattern having the desired sidelobe ratio may be determined.

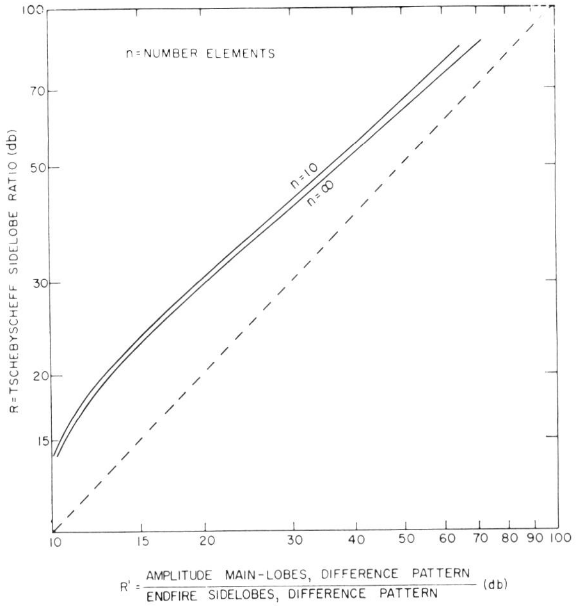

(a)

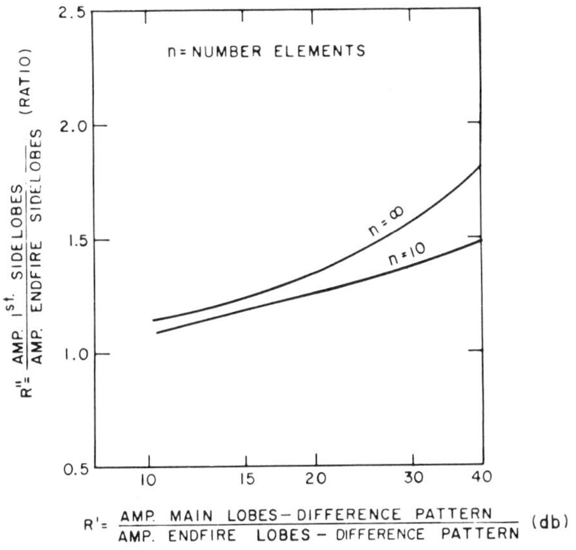

(b)

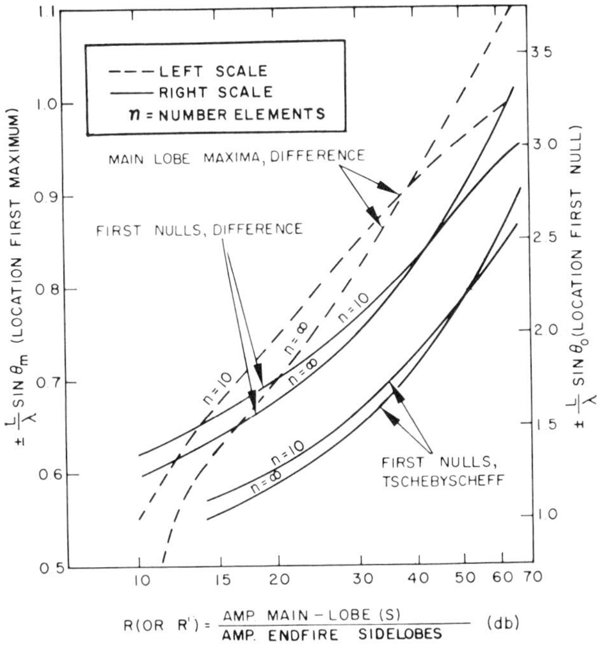

(c)

Fig. 4—(a) Sidelobe relationship between parent and transmuted Tchebycheff patterns. (b) Sidelobe variation of transmuted pattern. (c) Directivity of Tchebycheff and difference pattern.

161

The patterns obtained by this transmutation method and represented by (8) are only *nearly* optimum. In principle, one could find a distribution which would yield a pattern slightly better in the sense that for a given value of R the beamwidth would be narrower than that represented by $P(x)$. From a practical standpoint, the pattern represented by $P(x)$ is more than an adequate approximation because of the very slight variation from equal sidelobe magnitudes as subsequently shown in Appendix I and plotted in Fig. 4(b). Indeed, for an actual antenna it is suggested that it would be more fruitful to optimize the complete difference pattern (*i.e.*, the array pattern times the element pattern), as has been done by Sinclair and Cairns [3] for the conventional Tchebycheff pattern.

Example 1—The Tchebycheff Generating Function

Consider the space factor of an eight-element, one-half wavelength spaced broadside array with real feeding coefficients B_r. Require the first sidelobe to be 17.5 db below the two main lobes. Fig. 4(a) gives the sidelobe level of the generating function vs sidelobe level obtained with the use of Transmutation I. The two curves shown indicate the slight variation in design sidelobe level with number of elements. From this figure a Dolph-Tchebycheff distribution for a 27-db sidelobe level ($R_S=22.2$) sum pattern must be used as a generating function.

For the sum pattern case one finds

$$x_0 = 1.150,$$
$$A_1 = 8.25,$$
$$A_2 = 6.82,$$
$$A_3 = 4.56,$$
$$A_4 = 2.66.$$

The coefficients B_r for the difference amplitude pattern are then found by use of (9) and, after normalization to $B_4=1.00$, are

$$B_1 = 0.44,$$
$$B_2 = 1.10,$$
$$B_3 = 1.23,$$
$$B_4 = 1.00.$$

The ratio of amplitudes of the first sidelobe (lobe nearest broadside) and the lobe at endfire is shown from Fig. 4(b) to be 1.45, or 3.2 db.

Table I gives information on this example and those to follow. Fig. 5 shows the theoretical pattern for this first example with $d=\lambda/2$. An array constructed by Tang [8] was used to obtain the experimental patterns for this example and the ones to follow. This array is a standing wave array with micrometer adjustable irises for controlling the slot coupling. The slots are longitudinal and are on the center line of the waveguide broad wall. For each case eight slots were used with a slot spacing $d=0.7\lambda$. In Fig. 6 the experimental pattern is shown, as well as the theoretical pattern of Fig. 5 adjusted for the slot spacing of 0.70λ and an assumed element factor (power) $=\cos^2\theta$.

Example 2

In order to see how theoretical values compare with experimental results for a more stringent sidelobe level, the case of a transmuted −35.5-db Dolph-Tchebycheff generating function is shown in Table I.

Example 3—The Binomial Generating Function

A familiar limiting application of the Dolph-Tchebycheff method to the sum pattern is to require that the ratio of the main lobe to the sidelobe amplitude level be infinite. The result, of course, is that the feeding coefficients are proportional to the coefficients of a binomial series. Application of Transmutation I to the binomial distribution generating function leads to a difference pattern with no sidelobes for a spacing $d=\lambda/2$; the theoretical and experimental patterns for $d=0.70\lambda$ are shown in Fig. 7.

V. Extension to Other Pattern-Generating Functions

The general procedure of transmuting a known conventional pattern distribution to obtain the difference pattern distribution which will produce approximately similar sidelobe characteristics, is, of course, not limited to the Tchebycheff case. In general, the success of the procedure is based on the fact that the spacing between zeros and the shapes of the sidelobes in all conventional

TABLE I
Applications of Transmutation No. I

Example Number	Width Between Nulls Each Main Lobe for $d=\lambda/2$	(Theoretical) First Side-lobe at	(Experimental) First Side-lobe at	(Theoretical) Width of Each Main Lobe for $d=0.70\lambda$	(Experimental) Full Width of Each Main Lobe for $d=0.70\lambda$	Generating Function Used
1	25°	−17.5 db	−16 db	17.5°	18°	−27 db Dolph-Tchebycheff
2	28.5	−24 db	−21 db	20°	20°	−35.5 db Dolph-Tchebycheff
3	—	—	—	—	—	$-\infty$ db Dolph-Tchebycheff
4	25°	−15 db	−14 db	17.5°	18.5°	−23 db Taylor's Modified $\frac{\sin x}{x}$

162

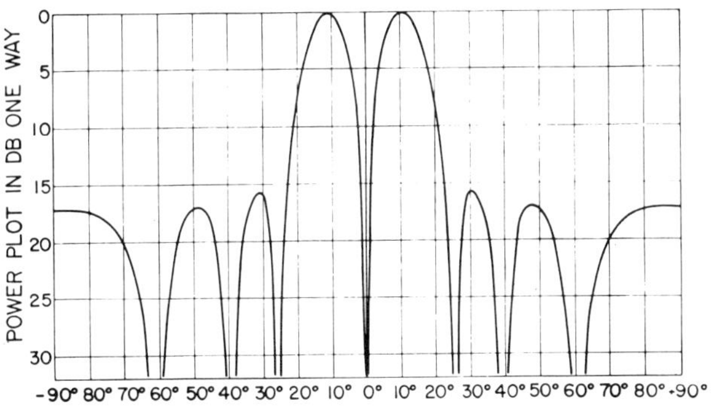

Fig. 5—Computed difference power pattern for eight-element array with spacing $d=\lambda/2$ and an isotropic element factor; differentiative technique applied to 27-db Dolph-Tchebycheff generating pattern.

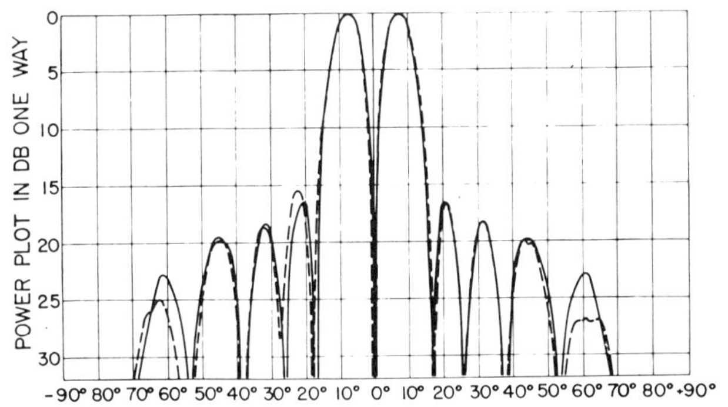

Fig. 6—Experimental and computed difference pattern for eight-element array with spacing $d=0.7\lambda$ and with an assumed element factor (power) $=\cos^2\theta$; differentiative technique applied to 27-db Dolph-Tchebycheff generating pattern.

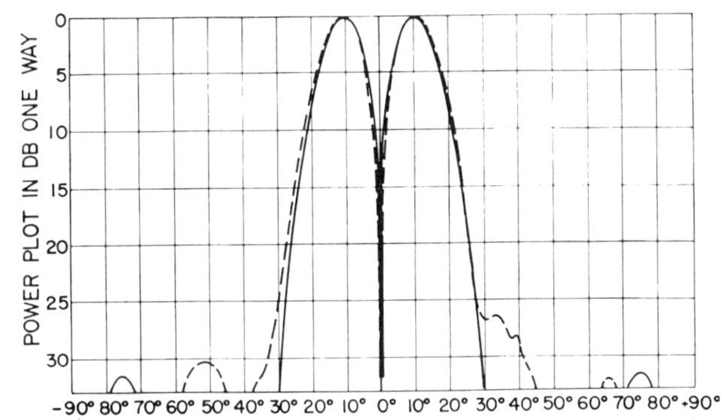

Fig. 7—Experimental and computed difference pattern for eight-element array with spacing $d=0.7\lambda$ and with assumed element factor (power) $=\cos^2\theta$; differentiative technique applied to a binomial distribution generating pattern.

directive patterns is an almost uniform function of ψ, where $\psi=kd\sin\theta$. As a result, the maximum of a given sidelobe of a pattern obtained by the differentiative technique is centered approximately at a zero of the generating or parent function, and its amplitude tends to be proportional to the magnitudes of the maxima on either side of the corresponding zero in the parent function. Except for a second-order effect resulting from the nonuniform spacing of nulls in the parent pattern, the sidelobe decay characteristic of the difference pattern is essentially that of the parent pattern. However, since differentiation of the single main lobe in the parent pattern results in a pair of main lobes, each of lower amplitude, in the difference pattern, the latter pattern always has an over-all sidelobe level higher than the corresponding parent pattern.

The general procedure works equally well for continuous line-source distributions as for arrays of discrete elements. For a continuous distribution $g(z)$ and a line source of length L, the radiation pattern is given by

$$F(u) = \int_{-L/2}^{L/2} g(z)e^{-iku z}dz, \tag{10}$$

where $u=\sin\theta$.

The corresponding difference pattern may be obtained from

$$F_D(u) = \frac{d}{du}F(u) = c\int_{-L/2}^{L/2} zg(z)e^{-iku z}dz, \tag{11}$$

thus requiring the new aperture distribution $h(z)$ where

$$h(z) = zg(z). \tag{12}$$

An example involving the Taylor modified $\sin \pi u/\pi u$ pattern [4] is given below.

Example 4—The Taylor Generating Function

In order to obtain a difference pattern with substantially lower sidelobes far from the two central beams, a Taylor modified $\sin \pi u/\pi u$ distribution may be used as a generating function. The corresponding sidelobe level and beamwidth characteristics of parent and transmuted Taylor patterns are obtained in Appendix II and plotted in Fig. 8. For an example, a Taylor generating pattern corresponding to a first sidelobe level of 23 db was used, resulting in a theoretical difference pattern first sidelobe at 15 db. Fig. 9 shows computed and experimental patterns for an array approximating the required line-source distribution with an element spacing of 0.7λ.

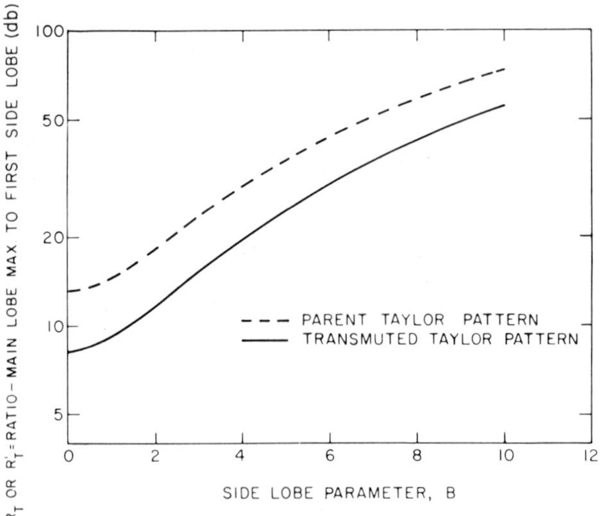

(a)

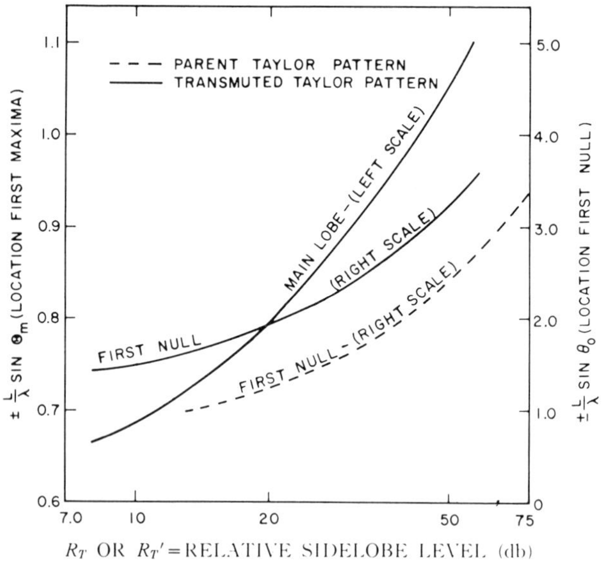

(b)

Fig. 8—(a) Parent and transmuted Taylor pattern sidelobe levels. (b) Directivity of parent and transmuted Taylor patterns.

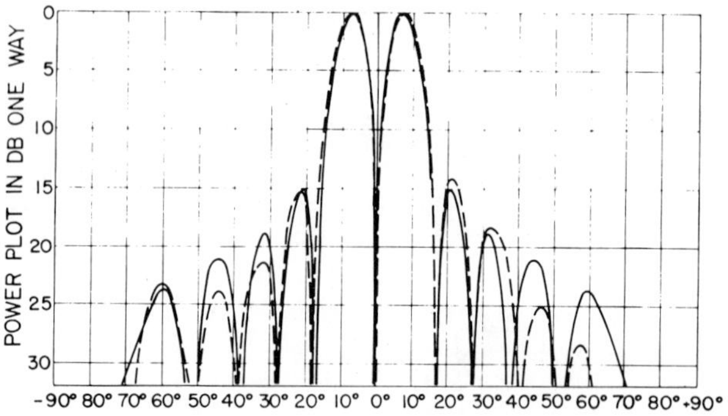

Fig. 9—Experimental and computed difference pattern for eight-element array with spacing $d=0.7\lambda$, with assumed element factor (power) $=\cos^3\theta$; differentiative technique applied to a Taylor-modified $\sin \pi u/\pi u$ generating pattern.

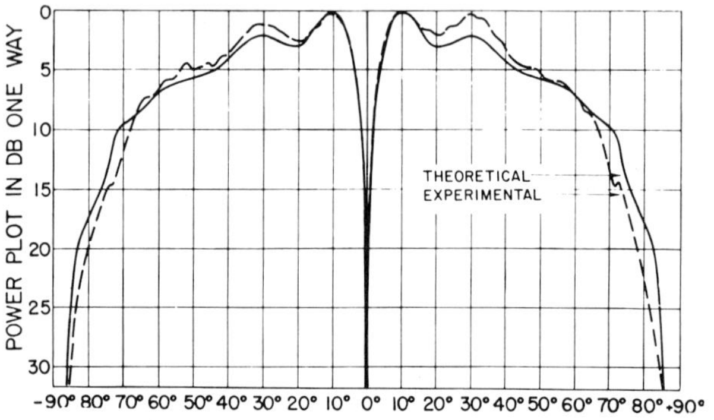

Fig. 10—Experimental and computed difference pattern for eight-element array with spacing $d=0.7\lambda$ and with assumed element (power) factor $=\cos^2\theta$; integrative technique applied to a uniform distribution generating function.

VI. The Integrative Transmutation[3]

Previously it was mentioned that the integrative method of Transmutation II results in an aperture distribution more highly peaked in magnitude than that of the generating function. Because the resulting aperture distribution is both sharply peaked in magnitude and antisymmetric in phase, there is a very sharp discontinuity in the aperture distribution at the center of the array. The effect of the described distribution is to produce an antisymmetric radiation pattern with a high level of radiation everywhere except for a sharp null at broadside. This type of pattern would be desirable for those applications where the specifications require that the radiation level of the difference pattern be high compared with the sum pattern in every direction except near broadside.

Derivation of the expression which relates the feeding coefficients for Transmutation II to the feeding coefficients of the parent pattern is strictly analogous to the procedure used in (7) through (9). Thus

$$C_r = \frac{A_r}{2r - 1} \qquad \text{for } n = n_0, \tag{13a}$$

$$C_r = \frac{A_r}{2r} \qquad \text{for } n = n_e,\ r \neq 0, \tag{13b}$$

and

$$C_0 = 0 \tag{13c}$$

where C_r = feeding coefficient for the rth element when Transmutation II is applied.

An example of this is given in Fig. 10.

[3] A. Ksienski, Hughes Aircraft Co., private communication. Dr. Ksienski has independently utilized the integrative method in the synthesis of sector beams.

VII. Conclusions

A synthesis method for antisymmetric (difference) patterns used in monopulse antennas has been presented. Linear arrays and line sources have been considered here, but the extension of the techniques to two-dimensional antennas is straightforward. The antisymmetric patterns resulting from the application of transmutation No. I to Dolph-Tchebycheff aperture distributions are nearly optimum in the Dolph-Tchebycheff sense. Application of the transmutation to other distributions used as generating functions yields antisymmetrical patterns which are analogous to the symmetrical patterns obtained from the generating functions. This was illustrated in the case of Taylor's sin $\pi u/\pi u$ pattern. A special case of interest which utilizes Transmutation II was illustrated.

A highly desirable feature of the method is that the excitation coefficients for the difference patterns may be very simply obtained from tables of aperture coefficients applicable to the parent functions. Thus any necessary compromises in the design of a monopulse antenna can be made in a more direct manner.

Appendix I

Derivation of Tchebycheff Parameters

The sidelobe level of the derived near-optimum difference pattern may be determined as follows. From (2) and (8)

$$\frac{E_{S_n}}{2} = T_{n-1}(x) = \cos\left[(n-1)\operatorname{arc\,cos} x\right] \tag{14}$$

$$\frac{E_{D_n}}{2} = -i(n-1)\sqrt{\frac{x_0^2 - x^2}{1 - x^2}}\sin\left[(n-1)\operatorname{arc\,cos} x\right] \tag{15}$$

where n stands for either $n_e = 2N$ or $n_0 = 2N+1$. The value of x_0 is given in terms of the sidelobe ratio, R, of the parent function (14),

$$x_0 = \cosh\left[\frac{\operatorname{arc\,cosh} R}{n-1}\right] \tag{16}$$

or, for large R and n,

$$x_0 \approx 1 + \frac{1}{2}\left[\frac{\ln 2R}{n-1}\right]^2. \tag{17}$$

For the maxima of the right side of (15) nearest endfire $u \approx 1$, $x \approx 0$,

$$\frac{E_D}{2i} \simeq -(n-1)x_0 \tag{18}$$

where the expressions are exact for n even. The value of (15) for the principal maxima near $u = 0$ is found by differentiating (15) and equating the result to zero. Let

$$x_0 = 1 + \alpha$$
$$x = 1 + \delta;$$

then for α and δ small (15) becomes

$$\frac{E_{D_n}}{2} \simeq -(n-1)\frac{\sqrt{\alpha - \delta}}{\sqrt{\delta}}\left[\sinh\{(n-1)\sqrt{2\delta}\}\right]. \tag{19}$$

Differentiating (19) with respect to δ and equating the result to zero gives approximately

$$\tanh \beta\Delta_{max} = \beta\Delta_{max}(1 - \Delta_{max}^2) \tag{20}$$

where

$$\beta = \ln 2R$$

$$\Delta_{max} = \sqrt{\delta_{max}/\alpha}.$$

For $R \geq 20$, it may be shown that, to a good approximation,

$$\tanh \beta\Delta_{max} \simeq 1$$

or

$$1 - \Delta_{max}^2 \approx \frac{1}{\beta\Delta_{max}}. \tag{21}$$

The roots of this expression are given by

$$\Delta_{max} = \mp 2\left(\frac{1}{3}\right)^{1/2}\cos\left[\frac{\cos^{-1}\frac{2.6}{\beta} + 2p\pi}{3}\right] \tag{22}$$

where $\beta \geq 2.6$ and where $p = 0, 1, 2$. The required root is that one in the range $0 < \Delta_{max} < 1$ which is nearest to 1. Expanding the cosine term in (22) gives for this root

$$\Delta_{max} = \cos\left[\frac{1}{3}\sin^{-1}\frac{2.6}{\beta}\right] - \frac{1}{\sqrt{3}}\sin\left[\frac{1}{3}\sin^{-1}\frac{2.6}{\beta}\right]$$

$$\simeq 1 - \frac{0.644}{\beta}. \tag{23}$$

Substitution into (19) and comparison of the result with (18) gives for the ratio, R', of the difference pattern main-beam amplitude to the endfire sidelobe amplitude

$$R' \approx \frac{1}{x_0}\frac{\sinh(\beta - 0.644)}{[\beta - 1.644]^{1/2}}. \tag{24}$$

This expression is plotted in Fig. 4(a).

In a similar manner, the ratio, R'', of the amplitude of the first sidelobe to that of the endfire sidelobe can be determined approximately as

$$R'' \simeq \frac{1}{x_0}\sqrt{1 + \frac{(\beta)^2}{20}}. \tag{25}$$

This expression is plotted in Fig. 4(b).

165

The locations of the first nulls in the difference pattern are given by

$$\sin \theta_{0_D} \simeq + \frac{\lambda}{L} \frac{n}{n-1} \sqrt{\frac{1+\left(\frac{\beta}{\pi}\right)^2}{1+\frac{1}{2}\left(\frac{\beta}{n-1}\right)^2}} \qquad (26)$$

and in the parent pattern by

$$\sin \theta_{0_P} \simeq \pm \frac{1}{\sqrt{2}} \frac{\lambda}{L} \frac{n}{n-1} \sqrt{\frac{1+2\left(\frac{\beta}{\pi}\right)^2}{1+\frac{1}{2}\left(\frac{\beta}{n-1}\right)^2}} \cdot (27)$$

166

The locations of the principal maxima for the difference pattern are given by

$$\sin \theta_{\max} \simeq \pm \frac{\lambda}{\pi L} \frac{n}{n-1} \sqrt{\frac{1.3\beta - 0.41}{1+\frac{1}{2}\left(\frac{\beta}{n-1}\right)^2}} \cdot \qquad (28)$$

These last three expressions are plotted in Fig. 4(c).

Appendix II

The Taylor Modified sin $\pi u/\pi u$ Pattern

The far-field Taylor or parent pattern is given by

$$F(u) = \frac{\sin \frac{\pi L}{\lambda}\sqrt{u^2 - B^2}}{\frac{\pi L}{\lambda}\sqrt{u^2 - B^2}} \qquad B \leq |u|$$

$$= \frac{\sinh \frac{\pi L}{\lambda}\sqrt{B^2 - u^2}}{\frac{\pi L}{\lambda}\sqrt{B^2 - u^2}} \qquad 0 \leq |u| \leq B \quad (29)$$

where L is the total aperture length, $u = \sin\theta$, and B is a parameter determining the ratio of amplitudes of main-beam to first sidelobe. As shown in Fig. 9 the sidelobe level decreases inversely with u. The aperture distribution required to produce this pattern is given by

$$g(z) = I_0\left[\pi B \sqrt{1 - \left(\frac{2z}{L}\right)^2}\right], \qquad (30)$$

where $I_0(\zeta)$ is the modified Bessel function, and z is the distance along the aperture measured from the aperture center.

The derived difference pattern is given by

$$F_D(u) = \frac{d}{du} F(u)$$

$$= \frac{u}{(u^2 - B^2)}\left[\cos \frac{\pi L}{\lambda}\sqrt{u^2 - B^2} - \frac{\sin \frac{\pi L}{\lambda}\sqrt{u^2 - B^2}}{\frac{\pi L}{\lambda}\sqrt{u^2 - B^2}}\right]$$

$$|u| \geq B$$

$$= \frac{-u}{B^2 - u^2}\left[\cosh \frac{\pi L}{\lambda}\sqrt{B^2 - u^2} - \frac{\sinh \frac{\pi L}{\lambda}\sqrt{B^2 - u^2}}{\frac{\pi L}{\lambda}\sqrt{B^2 - u^2}}\right]$$

$$0 \leq |u| \leq B. \quad (31)$$

Again, because of the more rapid decay of the sin ζ/ζ factor, as compared with that of the cosine term, the sidelobe level eventually decreases inversely with u, as u increases from B.

The aperture distribution required to produce the difference pattern, (31), is just

$$g_D(z) = zI_0\left[\pi B \sqrt{1 - \left(\frac{2z}{L}\right)^2}\right]. \qquad (32)$$

Characteristics of the Taylor pattern and the derived difference pattern are given in Fig. 8.

Acknowledgment

The authors wish to acknowledge the many hours of computation by Miss Carol Hasson, and the excellent experimental pattern work by C. H. Nonnemaker.

Bibliography

[1] C. L. Dolph, "A current distribution for broadside arrays which optimizes the relationship between beam width and sidelobe level," Proc. IRE, vol. 34, pp. 335–348; June, 1946.
[2] H. J. Riblet, "Discussion on 'A current distribution for broadside arrays which optimizes the relationship between beam width and sidelobe level,' " Proc. IRE, vol. 35, pp. 489–492; May, 1947.
[3] G. Sinclair and F. V. Cairns, "Optimum patterns for arrays of nonisotropic sources," IRE Trans. on Antennas and Propagation, vol. AP-1, pp. 50–61; February, 1952.
[4] T. T. Taylor, "One-Parameter Family of Line Sources Producing Modified (sin $\pi u/\pi u$) Patterns," Hughes Aircraft Co., Culver City, Calif., Tech. Memo. No. 324; September, 1953.
[5] T. T. Taylor, "Design of Line Sources for Narrow Beamwidth and Low Sidelobes," Hughes Aircraft Co., Culver City, Calif., Tech. Memo. No. 316; July, 1953.
[6] NBS, Appl. Math. Ser. 9, "Tables of Chebyschev Polynomials $S_n(x)$ and $C_n(x)$," U. S. Government Printing Office, Washington, D. C.; December, 1952.
[7] "Tschebyscheff Antenna Distribution, Beamwidth, and Gain Tables," Navord Rept. 4629; 1958.
[8] R. Tang, "A Slot with Variable Coupling and Its Application to a Linear Array," Hughes Aircraft Co., Culver City, Calif., RLM(M) 56-15; July, 1956.

DESIGN OF A TWELVE-HORN MONOPULSE FEED

Leon J. Ricardi
and
Leon Niro

Lincoln Laboratory[1]

Massachusetts Institute of Technology

Introduction

The general requirements of any tracking antenna system usually include the need for maximum boresight, or reference gain, and maximum angular sensitivity. In direct conflict with the former is the fact that the beam must be squinted off boresight in order to produce an error signal. In fact, all amplitude comparison error producing feed systems derive their signals from beams which are offset from boresight. If the amplitude of the signal received through each beam is compared in a sequence, the system is referred to as a sequential lobing type tracking system. If the signals received through the individual beams are compared simultaneously, then the system is referred to as a simultaneous lobing, or monopulse, type tracking system. Considering only those types of antenna systems which consist of a reflector, or arrangement of reflectors, and an illuminating feed, the four-horn monopulse feed is most commonly used. Although this type of feed has many advantages, the more logical design requires a compromise between angular and range sensitivity which results in an undesirable amount of spillover associated with the difference pattern and a reduction in the error sensitivity. The use of a twelve-horn feed arrangement permits a more satisfactory compromise and results in a considerable reduction in the amount of spillover. In this paper, there will be presented a well-known method of analysis as a means of comparing the theoretical advantages of the twelve-horn feed over the four-horn feed. A practical application of this theory with regard to the design of a Cassegrainian feed system will be presented together with an alternate arrangement which provides simultaneously sequential lobing type tracking signals and a communication channel.

Range, Error and Angular Sensitivity

The parameters most commonly used in describing the performance of a tracking system are range, error and angular sensitivity. Range sensitivity, as the words logically suggest, is the sensitiveness of the radar system to a change in range to the target being tracked. Generally speaking, it is proportional to the magnitude of the signal received from the target; hence, it is directly proportional to the gain of the antenna in the direction of the target. In the case of sequential lobing, or conical scan type tracking antenna systems, it is directly proportional to the maximum gain of the antenna reduced by the cross-over level. In a monopulse antenna system, it is proportional to the sum channel gain.

The error sensitivity is, by definition, the rate of change of the error signal as a function of the angular displacement of the target from the boresight axis of the antenna when the latter is pointing toward the target. With respect to a monopulse antenna system, it is the derivative of the receiver difference voltage with respect to the bearing angle when the latter is zero. Finally, the angular sensitivity is the sensitiveness of the antenna system as a function of its angular position with respect to the target. It is dependent upon the gain of the transmitter in the direction of the target when illuminating it and upon the error sensitivity upon reception of the return signal. In order to determine the error sensitivity, one can assume that the radiation pattern of the secondary beam produced by the reference channel can be expressed as follows:

$$E\,(\theta) = \sqrt{G}\;\frac{\sin N\theta}{N\theta} \qquad (1)$$

where G is the maximum gain of the beam and θ is the direction angle referred to the boresight axis, as indicated in Figure 1. The secondary beams which produce the error signals are considered to be squinted off boresight, an amount equal to θ_o and can be expressed as follows:

$$E_1\,(\theta) = \sqrt{G}\;\frac{\sin N(\theta + \theta_o)}{N(\theta + \theta_o)} \qquad (2)$$

$$E_2\,(\theta) = \sqrt{G}\;\frac{\sin N(\theta - \theta_o)}{N(\theta - \theta_o)} \qquad (3)$$

In a monopulse system, the error signal, $E_\Delta(\theta)$, is equal to the difference between E_1 and E_2. The error sensitivity is equal to the slope of $E_\Delta(\theta)$ at $\theta = 0$. Taking the difference between Equations 2 and 3, differentiating with respect to θ, and evaluating at $\theta = 0$, there results

$$\frac{\text{Error Sensitivity}}{2N\sqrt{G}} = \frac{\sin N\theta_o - N\theta_o \cos N\theta_o}{N^2\,\theta_o^{\,2}} \qquad (4)$$

which is clearly dependent upon the gain of the individual beams and the squint angle, θ_o. The derivation of Equation 4, presented here, assumes that the Gain, G, of the individual beams is constant; whereas in actual practice, it will decrease as the value of θ_o is increased. For all practical purposes, this assumption will not introduce an appreciable error in the results obtained by Equation 4, or which are plotted in Figure 2.

In summary, the angular and range sensitivity are given by

$$\text{Angular Sensitivity} = \sqrt{G_{trans.}} \times \text{Error Sensitivity} \qquad (5)$$

$$\text{Range Sensitivity} = [\sqrt{G_{trans.}}]^2 \qquad (6)$$

where $G_{trans.}$ is equal to the maximum antenna gain obtained in the sum, or reference channel, and the error sensitivity, for a monopulse antenna, can be given by Equation 4. The reader will notice that Equations 5 and 6 can be applied to any type of antenna tracking system and will express the expected performance in terms of these well known figures of merit.

The Four-Horn Monopulse Feed

The most common type of monopulse antenna system consists of a feed horn array, as illustrated in Figure 3, illuminating a paraboloid. The feed horns are excited through a comparator network which is schematically represented in Figure 4. At each circled junction, there is a magic tee, a folded tee, or a similar type hybrid junction which has the following properties:

1. Signals incident upon the sum arm are divided equally and delivered in phase to the side arm output terminals.

2. Signals incident upon the difference arm are divided equally and delivered 180° out of phase at the same side arm output terminals.

3. Signals incident upon the sum arm do not appear at the difference arm and conversely signals incident upon the difference arm do not appear at the sum terminal.

The diagrams shown at the terminals in the figure represent the horn excitation realized when signals are impressed upon these terminals. For example, signals incident upon the sum channel will excite all four horns in phase and produce an on-axis beam and a primary illumination pattern commensurate with the total horn aperture. Signals incident upon the azimuth error terminal will excite all horns equally, but with horns 1 and 4 180° out of phase with horns 2 and 3. In this manner, one may consider horns 1 and 4 combining to produce a secondary beam squinted from the boresight axis, θ_o in one direction and horns 2 and 3 producing a similar beam squinted θ_o in the opposite direction off the boresight axis. The RF comparator circuit merely takes the difference in the signals received by these two beams and presents it at the azimuth error terminal. This error signal is usually characterized as having an amplitude proportional to the angle between the boresight axis and the target, and a phase of either 0°, or 180°, dependent upon whether the beam produced by the horns 1 and 4, or horns 2 and 3, points more closely toward the target. The present analysis will consider each horn to have a square aperture so as to permit the use of dual, or circular, polarization (i.e., A = B in Fig. 2).

The squint angle, θ_o, can be calculated by first determining the distance, d, between the focus of the paraboloid and the effective center of radiation. The effective center of radiation of a square horn aperture is at its geometric center. Therefore,

$$\theta_o = D_v \frac{d}{f} \qquad (7)$$

where f = focal length of the paraboloid and D_v is the beam deviation factor which can be determined experimentally, or obtained from the literature.[2] In most practical applications, D_v is approximately equal to 0.95. The relative gain of the sum channel, or on-axis beam of the paraboloid, is dependent upon the illumination taper provided by the feed horn which would consist of all four horns energized equally, and in phase. Assuming normal waveguide excitation of their respective apertures and vertically polarized signals, the primary patterns of the feed horns are given by:

$$E_\Sigma(\gamma) = \frac{\sin(kB\sin\gamma)}{Bk\sin\gamma}\cos\gamma \qquad (8)$$

$$E_\Sigma(\xi) = \frac{\cos^2[k\frac{A}{2}\sin\xi]}{1-[\frac{2A}{\lambda}\sin\xi]^2} \qquad (9)$$

$$E_\Delta(\gamma) = 2j\frac{[1-\cos(kB\sin\gamma)]}{k\sin\gamma}\cos\gamma \qquad (10)$$

$$E_\Delta(\xi) = 2j\frac{A\sin(kA\sin\xi)}{\pi[1-4(\frac{A}{\lambda})^2\sin^2\xi]} \qquad (11)$$

where $k = \frac{2\pi}{\lambda}$.

Calculating the edge illumination as a function of horn aperture, A, reference to Figure 5 will enable one to determine the relative gain of the secondary beam as a function of horn aperture.[3] Equation 8 will be used to calculate this gain because prior experience has shown that the optimum feed horn aperture size will be greater than that which yields the maximum gain. Since the directivity in the H-plane, as given by Equation 9, results in a higher edge illumination for the same aperture size compared to that given by Equation 8, the final optimization will produce a more satisfactory illumination in the H-plane. Assuming a 66 wavelength diameter paraboloid having an f/D ratio equal to 0.6, the angle subtended by the reflector, at its focus, is 90°; therefore, the value of $\gamma = 45°$ is substituted into Equation 8 and the edge intensity determined as a function of A. The relative voltage gain (in db) is determined from Figure 5 and the result plotted in Figure 6. The corresponding range sensitivity is determined by merely plotting the square of the relative voltage gain.

In order to determine the error sensitivity as a function of the dimension A, it is necessary to derive a relationship between θ_o, A, and evaluate N as given in Equation 1. Estimating the half-power beamwidth to be equal to 1°, then $N\theta$ is approximately equal to 80 and N = 160. The relationship between θ_o and A is given by

$$\theta_o = 0.9 \frac{A}{2f} 57.3° \qquad (12)$$

$$\theta_o = 0.65 \frac{A}{\lambda} \qquad (13)$$

Substituting the value of N = 160 and Equation 13 into Equation 4, one determines the error sensitivity as a function of feed horn aperture which is then plotted in Figure 6. Using the values thus far obtained, the angular sensitivity is calculated in accordance with Equation 5 and plotted in Figure 6. Normalizing the curve of angular sensitivity, its intersection with the curve of range sensitivity determines that feed horn aperture size which will yield optimum overall system sensitivity. As indicated in Figure 6, this should be approximately 0.59λ square. Substituting this value of A into Equations 8, 9, 10 and 11, the resulting primary patterns are calculated and shown in Figure 7.

In the previous calculation, the effect of space attenuation upon the aperture illumination has not been taken into account. However, with a paraboloid whose f/D ratio = 0.6, the modification of the feed horn taper, as it would appear in the aperture of the paraboloid, is relatively small and will not alter the results appreciably. The gain and range sensitivity shown in Figure 6 were determined as a function of the E-plane aperture; however, it will be noted that the compromise horn aperture results in only a negligible reduction in range sensitivity and is probably not a true reflection of the expected results since the reflector aperture will be illuminated in a more optimum manner in the H-plane. Notice that the compromise values of range sensitivity and angular sensitivity are not too different from their optimum values.

<u>The Twelve-Horn Monopulse Feed</u>

Referring to the primary patterns shown in Figure 7, it will be noted that the difference pattern produces a considerable amount of spillover which represents a reduction in the signal received in the error channels. Since the feed horn aperture which produces one of the error beams is approximately one half the feed horn aperture producing the sum pattern, one can estimate that the edge intensity has increased to approximately 3 db. Referring to Figure 5, we see that this produces a loss in gain of approximately 1-3/4 db and, hence, the value of G in Equation 4 should be reduced by this amount. The twelve-horn monopulse system, shown in Figure 8, produces a sum pattern which is identical to those shown in Figure 7 and a difference pattern which has much less spillover as shown in Figure 9 (remember that the edge of the reflector coincides with the direction angle $\gamma = \xi = 45°$). In order to understand this more completely, let us proceed to analyze this feed system in the same manner as was done with the four-horn monopulse feed. The diagram shown next to each terminal in Figure 8 indicates the excitation of the feed horn if signals were impressed upon that particular terminal.[4] When producing the sum pattern, the feed horn array is excited in exactly the same manner as a four-horn monopulse; therefore, the gain and range sensitivity as given in Figure 6 apply in the same manner. The azimuth and elevation error patterns are produced by an array of eight horns, four generating the beam squinted to one side of the boresight and four generating the beam squinted to the other side. The total aperture of these four horns is identical to that realized by the array of horns that produce the sum pattern. It follows, therefore, that the edge illumination is very nearly the same when signals are impressed upon either the sum or the error channels. The center of phase, however, is displaced from the boresight axis by a full horn width, instead of the half horn width as realized in the case of the four horn monopulse. The corresponding value of θ_o, as given by Equation 13, must be modified as follows:

$$\theta_o = 1.3 \frac{A}{\lambda} \text{ in degrees} \qquad (14)$$

The resulting error sensitivity, range sensitivity, and angular sensitivity are plotted in Figure 10. The feed horn aperture yielding the optimum overall system sensitivity is 0.56λ. Notice that the angular sensitivity curve has not been normalized, indicating an increase in error signal for a given target displacement from the boresight axis, as compared with that obtained with the four-horn feed analyzed above. This increase in error signal is approximately 2.6 db which, when added to the increase in gain (refer to Equation 4) obtained because the degree of spillover has been reduced, should theoretically increase the error signal four, or even more, db. This increase in error signal is compared to the signal received from a four-horn monopulse design, as described previously, and with a given target displacement from the boresight axis.

It is certainly true that the foregoing analysis is not rigorous and is by no means complete. In particular, the optimization was carried out for E-plane primary patterns; repeating these calculations for H-plane primary patterns would yield slightly different results making necessary still further compromises. However, this method of determining system performance is based upon accepted logic[5,6,7] and does demonstrate the superiority of the twelve-horn over the four-horn feed.

Experimental Data

The reader should note that in the example presented above, an f/D ratio equal to 0.6 resulted in an optimum horn aperture approximately 0.56λ square. Reducing the f/D ratio would necessitate reducing the horn aperture which could result in dimensions which would not permit the propagation of energy unless it were ridge, or dielectric, loaded. Conversely, increasing the f/D ratio makes it possible to increase the horn aperture but at the expense of increasing the mechanical complexity and usually results in an undesirable configuration. The use of a Cassegrainian feed system makes it possible for the designer to select, arbitrarily, the f/D ratio of the paraboloid and the feed horn apertures by choosing the appropriate magnification factor associated with the Cassegrainian feed system. In short, the image of the primary feed produced at the focus of the paraboloid, by the hyperboloid, is reduced in size in direct proportion to the magnification factor, M, of the Cassegrainian system. This factor is given by

$$M = \frac{e + 1}{e - 1} \tag{15}$$

where e is the eccentricity of the hyperboloid. Details of the method of calculating the range sensitivity and angular sensitivity have been presented in the literature[6] and are not appropriate to our present discussion.

Some time ago these authors were faced with the problem of designing a single antenna feed capable of providing a communications channel, monopulse tracking error signals, and sequential lobing error signals. The feed would be used to illuminate a 60-foot diameter paraboloid at a frequency of approximately 8 kmc. The communications receiver was to employ a MASER type front end and, therefore, it was desirable to minimize all waveguide losses. Since the Cassegrainian feed system places the feed point at the vertex of the paraboloid and eliminates the necessity to install the MASER at the focus of the paraboloid, or to include a length of waveguide between the vertex and the focus of the paraboloid which could contribute a considerable amount of loss, it was decided to use this type feed system. An optimization of feed horn size, similar to that described previously, was carried out and a resulting feed horn aperture determined. However, in horn arrays where the horn apertures are two or more wavelengths square the side lobes in the primary radiation pattern are unusually high. For this reason the desired pyramidal horns were divided into sectorial horns each having an aperture approximately 3.4 wavelengths by 0.85λ. Thus, each of the individual twelve horns were made up of four sectorial horns which had an integral power divider and a single input for a total of twelve inputs and 48 horns (refer to Figure 12). Typical H-plane radiation patterns of the sum and error channels are shown in Figure 11. It will be remembered that one of the major advantages of the twelve-horn monopulse feed was the reduction in spillover energy in the primary radiation patterns of the error channel. The radiation patterns shown in Figure 11 certainly demonstrate the reduction in spillover; however, it has not been possible to obtain data determining the increase in error signal as predicted by the foregoing analysis because of the lack of time and availability of equipment. This will be investigated in the very near future.

Alternate Applications

During the development of the feed described above, the need for a combination sequential lobing and communications feed took precedence over the monopulse feed. Specifically, it was desired to have an antenna which provided a communications channel having the maximum possible on-axis gain and a sequential lobing channel, which could be operated simultaneously on a non-interfering basis. Referring to Figure 8, it will be noticed that if equal power is delivered to the auxiliary azimuth sum channel and the azimuth error channel, there will result an asymmetrical excitation across horns 1 through 8. Similarly, if equal power is impressed upon the auxiliary elevation sum and elevation error terminals, an asymmetrical distribution will be produced across the apertures of horns 9, 10, 11, 12 and 2, 6, 3 and 7. Further investigation shows that the resulting primary patterns are satisfactory in that the illumination pattern is relatively smooth and the edge illumination is approximately -6 db. Due to the asymmetrical excitation of the horn apertures, the center of phase is no longer located at the center of the horn array and, therefore, this type of illumination will produce a squinted secondary beam. Varying the phase between the signals impressed upon the auxiliary sum and the error channels will vary the asymmetry in the field distribution across the horn apertures and change the location of the effective center of radiation. In this manner it is possible to produce a set of sequential lobing type beams which, as will be noted by reference to Figure 8, are produced by energizing terminals which are isolated from the sum channel by at least 30 db.

A further investigation into the character of the primary patterns, with this type excitation, yielded the result that exciting the outer horns in sequence, together with the center four horns operating in the difference mode, would yield satisfactory illumination patterns and also provide the required isolation from the sum channel. The schematic representation of the necessary circuitry is shown in Figure 13 and is pictured in Figure 12b. The crossover level obtained with the sequential lobing feed arrangement such as this is not considerably different from that obtained by merely placing four horns about a center feed horn and exciting the four horns separately; however, the primary illumination pattern produces less spillover tending toward increasing the gain and, in effect, raising the crossover level (refer to Figure 11).

In the design of a system like this, it is desirable to be able to predict the resulting beam squint as a function of the horn aperture size. One method of determining this consists of resolving the field across the horn apertures into distributions having even and odd symmetry. The secondary radiation pattern for each symmetrical distribution can be calculated in a straightforward manner. Linear addition of the resulting secondary patterns should yield a reasonably accurate picture of the squinted beam produced by that particular feed excitation.

Conclusions

Based upon the methods of analysis presented here, it has been shown that the twelve-horn monopulse feed system can theoretically increase the signal, received in the error channels, approximately 4 db over that which would be received by an optimum four-horn monopulse. This is due primarily to a reduction in the amount of spillover associated with the difference pattern and to the increase in error sensitivity produced by the error generating beams. At the same time, there is a tendency toward increasing the range sensitivity since the compromise horn aperture is more nearly that which produces maximum range sensitivity. It has also been shown that changing the comparator circuit associated with the twelve-horn monopulse feed will result in a communications and sequential lobing system which can operate simultaneously on a non-interfering basis.

Acknowledgments

The possibility of using twelve horns in a monopulse feed system was first suggested to one of the authors during a conversation with members of the Wheeler Laboratories staff. The final comparator circuitry shown in Figure 8 was derived by Dr. Shepard Holt who, throughout the course of development, has been very helpful in analyzing and discussing the results obtained. Several members of the Lincoln Laboratory staff and the Sperry Gyroscope Corporation contributed advice, encouragement and help during the course of the program.

References

1 Operated with support from the U. S. Army, Navy and Air Force.

2 S. Silver, "Microwave Antenna Theory and Design", Radiation Laboratory Series, Vol. 12, P. 488.

3 The curves shown in this figure were derived by one of the authors using conventional techniques for a circular aperture having the illumination taper shown.

4 This circuit was originally suggested to the authors by Dr. Shepard Holt.

5 M. K. Kales, "Optimum Design Criterion for Simultaneous Lobing Antennas", NRL-R-3451 dated 20 April 1949.

6 Robert W. Martin & Leon Schwartzman, "A Monopulse Cassegrainian Antenna", 1960 IRE National Convention.

7 D. R. Rhodes, "Introduction to Monopulse", McGraw Hill, 1959.

171

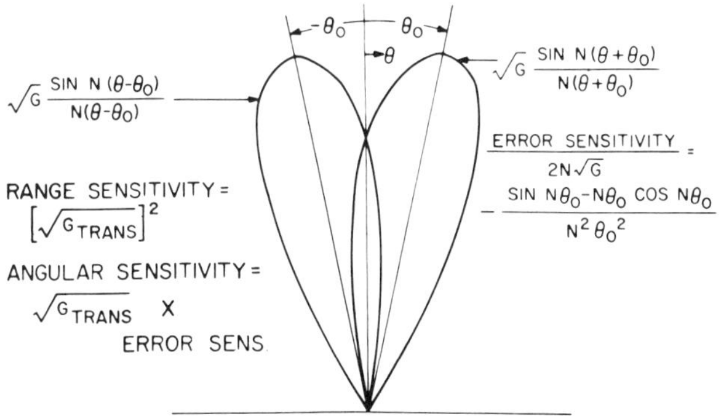

FIG. 1 TRACKING BEAM CONFIGURATION

172

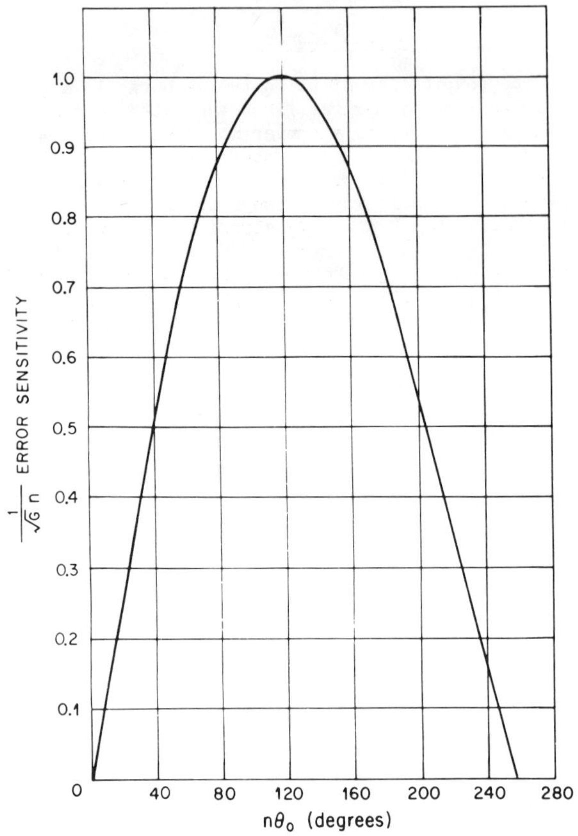

FIG. 2 ERROR SENSITIVITY

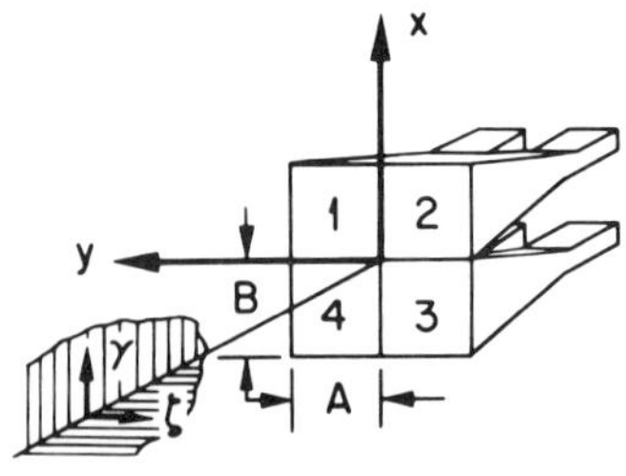

FIG. 3 FOUR - HORN FEED

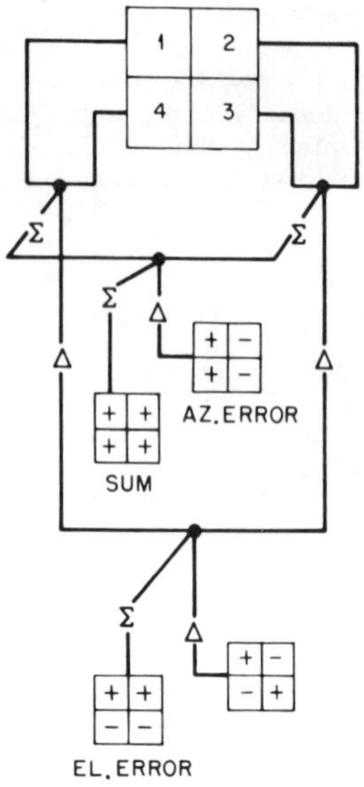

FIG. 4 COMPARATOR CIRCUIT

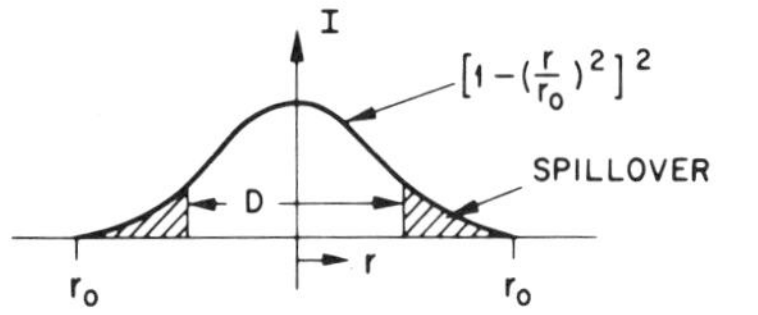

ILLUMINATION TAPER OVER A CIRCULAR APERTURE

FIG. 5A

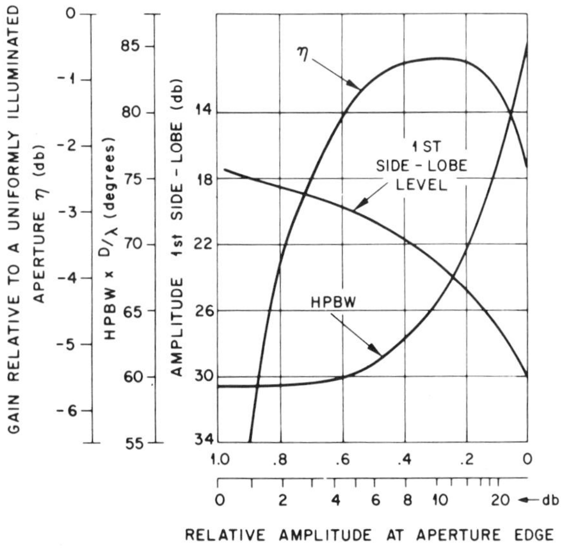

FIG. 5B CHARACTERISTICS OF CIRCULAR APERTURES.

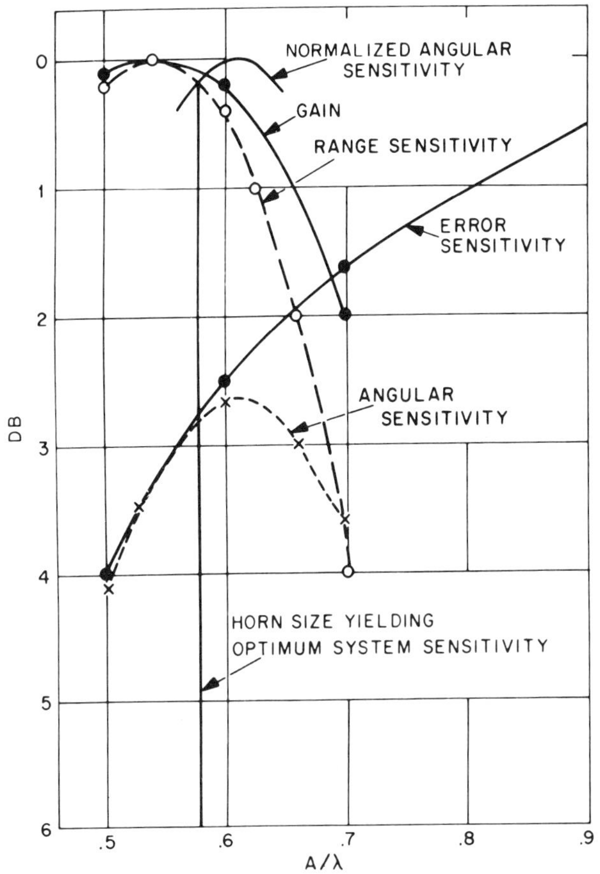

FIG. 6 SYSTEM CHARACTERISTICS vs. HORN APERTURE

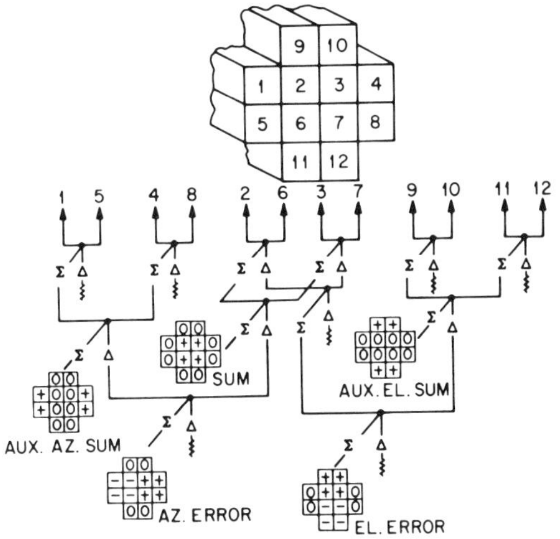

FIG. 8 TWELVE - HORN MONOPULSE FEED

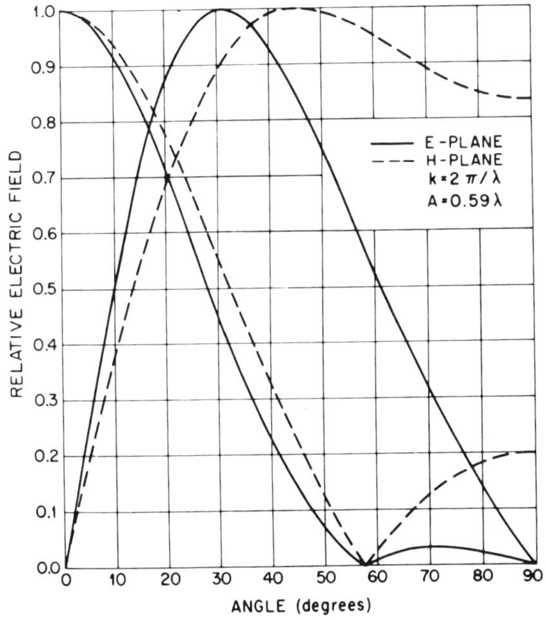

FIG. 7 PRIMARY PATTERNS OF HORN FEED

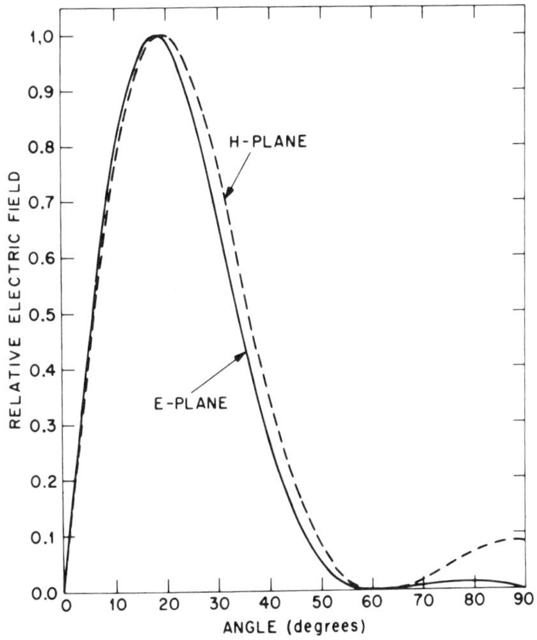

FIG. 9 DIFFERENCE PATTERN, TWELVE-HORN MONOPULSE

173

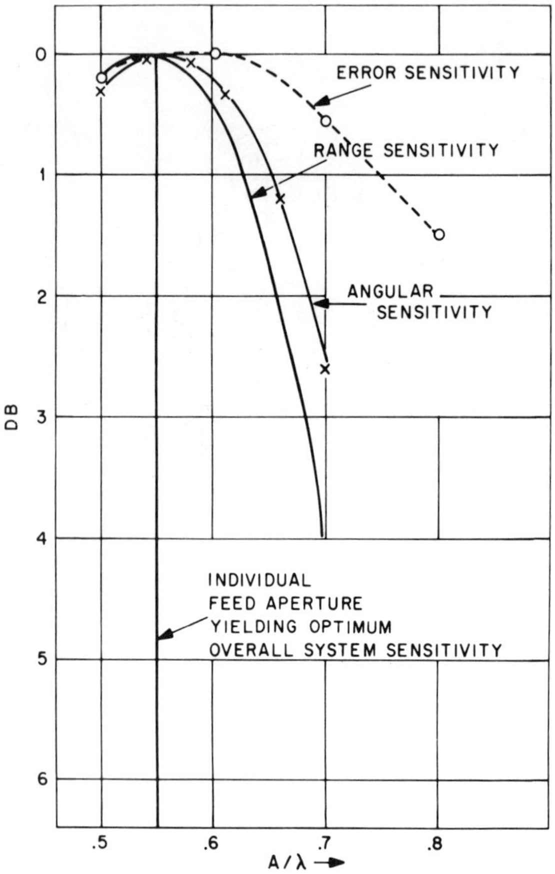

FIG. 10 SYSTEM CHARACTERISTICS, TWELVE-HORN MONOPULSE

(a)

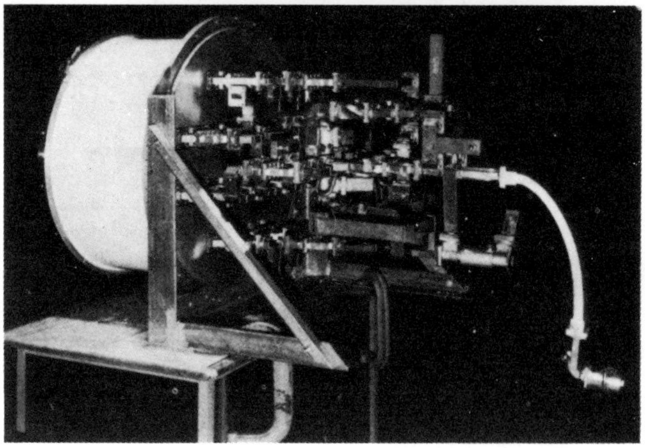

(b)

Fig. 12. (a) Twelve horn feed. (b) Twelve horn feed.

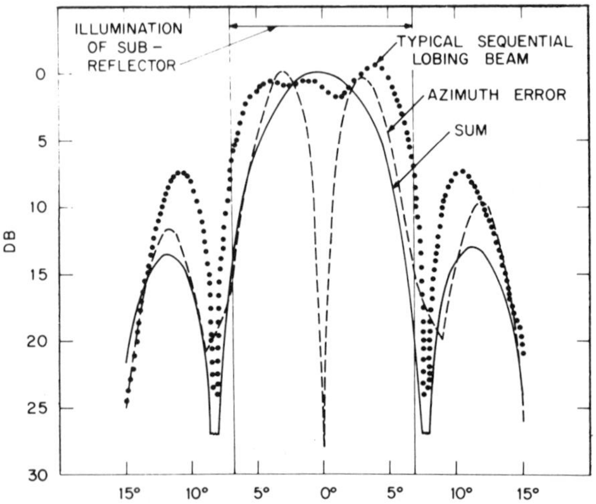

FIG. 11. 12 HORN FEED H-PLANE PRIMARY PATTERNS

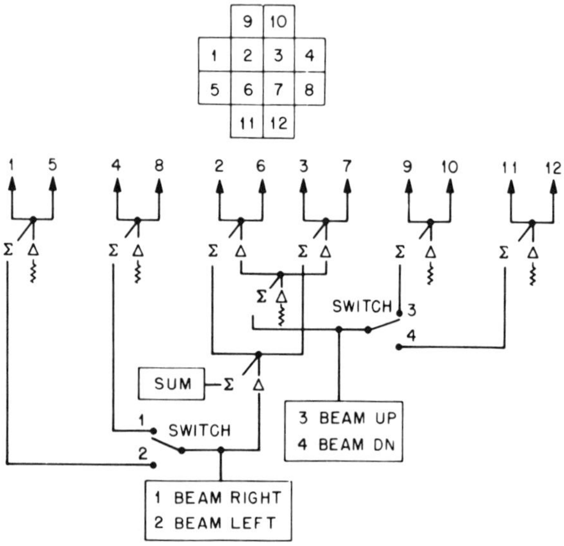

FIG. 13 AUXILLIARY SEQUENTIAL LOBING CIRCUIT.

12.

A MONOPULSE ANTENNA HAVING INDEPENDENT OPTIMIZATION OF THE SUM AND DIFFERENCE MODES

Peter W. Hannan and Patricia A. Loth
Wheeler Laboratories
Smithtown, N. Y.

Summary

A monopulse antenna for a tracking radar has been developed which overcomes the usual need for compromise between sum and difference mode performance. The reflector of this antenna is illuminated by an unusual feed which enables independent optimization of all three modes: sum, azimuth difference, and elevation difference.

The feed achieves suitably tapered illumination in all the modes by selective excitation of four stacked horns in one plane, and by utilization of three waveguide modes in each horn in the other plane. Measurements of the antenna confirm an improvement in both the azimuth and elevation difference modes of over 3 db increase in peak gain, and over 10 db further sidelobe suppression, while retaining the optimum sum-mode performance.

Introduction

In recent years, the possibility of independent optimization of the sum and difference modes of a monopulse antenna feed has raised considerable interest. One particular technique for obtaining this was presented at the 1959 IRE National Convention (Ref. 1). To be described in this paper is a different technique, whereby a practical feed for a monopulse antenna achieves independent optimization of the sum and both difference modes. In addition, the measured performance of an operational monopulse antenna which incorporates this feed is presented.

The Need For Independent Control

An example of the situation existing in the ordinary monopulse antenna is shown in Fig. 1. The antenna is an amplitude-comparison type comprising a four-horn cluster which illuminates a main aperture containing a reflecting dish. In the sum mode, a monopulse comparator excites all four horns in the same polarity, creating an illumination of the dish similar to that obtained with a simple feed. To get the best overall monopulse performance, it is customary to choose a feed size such that the sum mode has nearly maximum antenna gain. This corresponds to a sum illumination that has a well-tapered distribution across the dish, as indicated. A typical case is the optimum design studied by M. L. Kales of the Naval Research Laboratory, and reported elsewhere (Ref. 2).

In each of the difference modes, the comparator excites one pair of horns in a polarity opposite to that of the other pair. The resulting illuminations each have two main regions of opposite polarity, giving rise to the two-lobed antenna radiation patterns that are required for angle tracking. Unfortunately, when the size of the feed is chosen as noted above, both of the difference illuminations are about twice as wide as the reflecting dish, as indicated. Thus the peaks of the difference illuminations occur near the edge of the dish, and about half of the power is wasted in spillover radiation. As a consequence of this, the difference patterns of the antenna exhibit low gain and high sidelobes.

Fig. 2 illustrates the basic concept needed to solve this fundamental problem of monopulse antennas. By approximately doubling an appropriate feed dimension for each difference mode, the difference illuminations are essentially confined to the dish, and optimum performance in both difference modes is obtainable. Optimum sum-mode performance requires approximately the original dimensions. Thus the feed should be different for each of the three modes, in the manner indicated in Fig. 2. It is evident that a means for controlling the feed independently in the three modes is necessary to achieve optimum performance in all three modes.

175

Operation of Multihorn-Multimode Feed

A photograph of the comparator-feed assembly developed as part of a monopulse antenna is shown in Fig. 3. At the rear is a network of waveguide hybrid junctions from which extend the three ports for the sum, azimuth difference, and elevation difference modes. At the front are four rectangular horns stacked in a single direction. Appearing in the open end of each horn is a correcting lens that is employed in this particular case because the antenna happens to require a feed aperture many wavelengths in size.

Fig. 4 illustrates more clearly the significant aspects of this feed system. There are a total of eight hybrid junctions; the rearmost four selectively excite inputs to the four horns in a manner dependent on the port to which a signal is applied. Each of the other four hybrid junctions is closely associated with one of the four horns, and furnishes even or odd excitations to its horn. Within each horn is a generator of natural horn modes similar to natural waveguide modes. When even excitation is applied, the one-mode and the three-mode are generated; these appear at the open end of the horn with relative amplitude and phase such that the combination is as indicated in Fig. 4. When odd excitation is applied, the two-mode is generated, and appears as indicated.

Fig. 5 shows the resulting field distribution across the feed aperture for the sum, azimuth difference, and elevation difference modes. Indicated by the dotted lines are the equivalent areas of the feed which are excited in the three modes of operation. The correspondence between these areas and the desired ones given in Fig. 2 may be noted.

Performance of the Multihorn-Multimode Feed

The measured radiation patterns of the feed system are given by the heavy curves in Fig. 6. These patterns correspond to the illumination of the reflecting dish. On the left are shown half of the E-plane sum pattern and half of the azimuth difference pattern; on the right are shown halves of the H-plane sum and elevation difference patterns. The angle scale is normalized to the average overall dimension of the multihorn-multimode feed, in wavelengths. Also indicated is the extent of the reflecting dish aperture; all the patterns are normalized to their peak values, so the edge-taper values are easily seen. It is apparent that well-tapered illumination is provided in all three modes.

For comparison, the difference patterns are shown by dotted curves for an ordinary four-horn cluster whose sum patterns are essentially the same width as those for the multihorn-multimode feed. The improved characteristics obtained with independent control are evident, especially in the H-plane (elevation plane) where the sidelobes are low. The latter result occurs, in spite of the large feed aperture, as a consequence of the smooth field distribution provided by the multimode technique across the H-plane of the feed aperture.

Performance of Complete Antenna

A photograph of the monopulse antenna which includes the multihorn-multimode feed appears in Fig. 7. This antenna is a Cassegrain double-reflector type, permitting the comparator-feed assembly to be conveniently located in the rear. A minimum-blocking design (Ref. 3) is employed; hence the feed is nearly as large as the sub dish, which is small compared with the main dish.

Fig. 8 shows the measured radiation patterns of the complete antenna. The presentation is the same as that of Fig. 6 except that the angle scale is normalized to the diameter of the main reflecting dish, in wavelengths, and the amplitudes of all the patterns are normalized to the peak of the sum pattern. The antenna "efficiency" at this sum peak is 53%, neglecting about 0.3 db excess attenuation associated with the experimental waveguide assembly. ("Efficiency" is used here as the ratio of measured gain to the gain which would be obtained if the same main aperture were uniformly illuminated, with no spillover or other losses.) As expected, the peak gain of both difference patterns is within 3 db of the sum peak gain, and this is obtained while retaining high sum gain.

For comparison, the difference patterns for an ordinary monopulse antenna with a four-horn cluster are shown by the dotted curves; an improvement of more than 3 db in difference peak gain is obtained with the independently controlled feed. (The actual improvements are 4.3 and 4.7 db, as indicated, but these are probably not typical because the size of the four-horn cluster happened to be set exactly for maximum sum gain rather than the more usual, slightly larger size which would yield a small increase of difference gain).

The voltage slopes of the center of the difference patterns should increase nearly as much as the peak voltages. In this case the ratios of the slopes are over 1.4 (actually the ratios are 1.52 and 1.57).

Perhaps of most interest is the improvement of sidelobe suppression in the difference patterns. With the independently controlled feed, the first difference sidelobes are more than 23 db below the corresponding difference peaks; with the ordinary feed the suppression is only about 13 db. This 10 db improvement of first-sidelobe suppression is supplemented by a significant improvement in suppression of the wide-angle difference sidelobes, because of the greatly reduced spillover from the feed.

Conclusion

Independent optimization in all three modes of operation of an amplitude-monopulse antenna of the reflecting-dish type has been achieved. The multihorn-multimode feed has proven to be an effective and practical technique for this purpose. Measurements of the complete antenna have demonstrated a significant improvement in the radiation patterns of both difference modes compared with the ordinary monopulse antenna, while retaining good performance in the sum mode. Since the difference modes are mainly responsible for the angle-tracking function of a monopulse radar, an increase in the angular accuracy is one of the benefits which is available.

Acknowledgments

The development reported in this paper was performed for the Bell Telephone Laboratories in connection with a prime contract for the Army Ordnance Corps. Direction of this work by BTL has been furnished by P. L. Hammann, R. L. Mattingly, and T. W. Madigan. At WL, contributions to this development have been made by J. E. Becker, R. E. Puttre, I. Koffman, and W. A. Elliott.

References

(1) J. P. Shelton, "Improved Feed Design for Amplitude Monopulse Radar Antenna", IRE National Convention Record, Part I Antennas and Propagation, pp. 93-102; March 1959.

(2) D. R. Rhodes, "Introduction to Monopulse", McGraw-Hill, pp. 98-101; 1959.

(3) P. W. Hannan, "Microwave Antennas Derived from the Cassegrain Telescope", IRE Transactions on Antennas and Propagation, vol. AP-9, pp.140-153 ; March 1961.

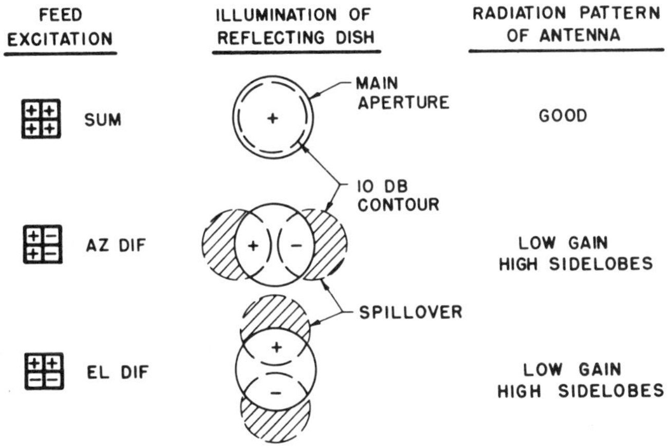

Fig. 1. Ordinary feeding of monopulse antenna.

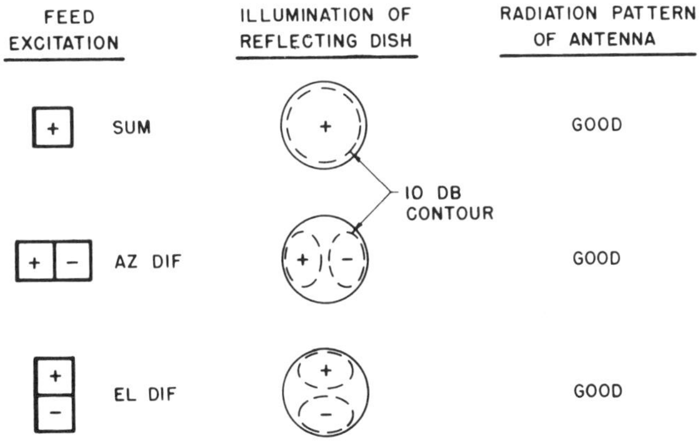

Fig. 2. Optimum feeding of monopulse antenna.

Fig. 3. Multihorn-multimode feed assembly.

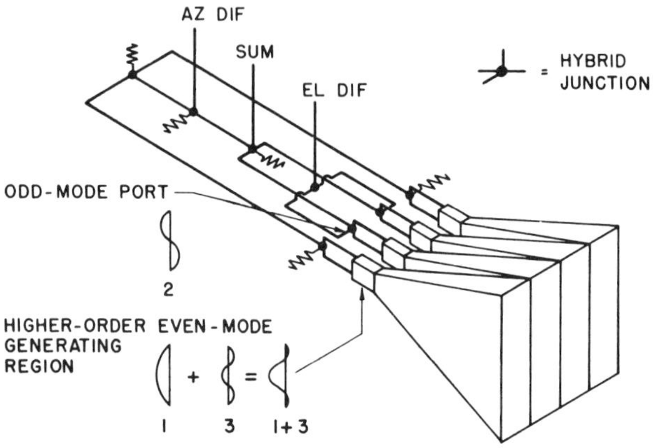

Fig. 4. Circuit of multihorn-multimode feed assembly.

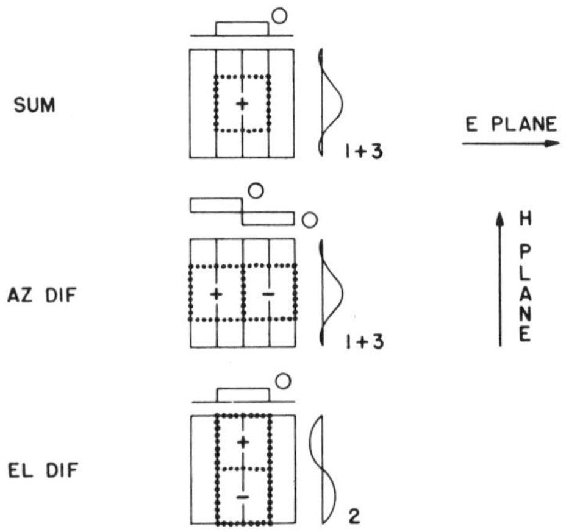

Fig. 5. Multihorn-multimode feed excitation.

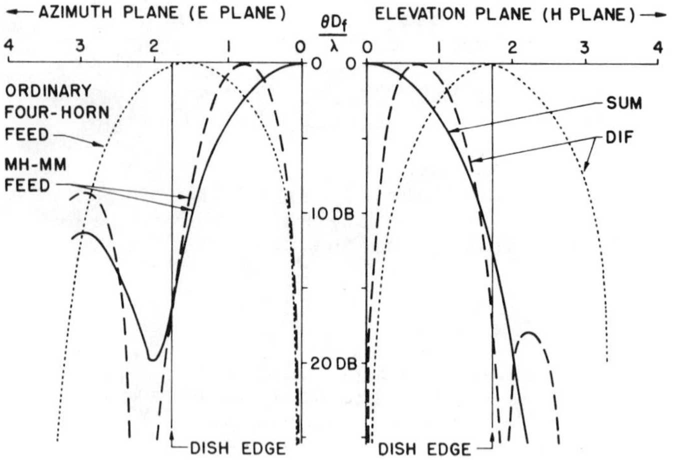

Fig. 6. Patterns of multihorn-multimode feed.

Fig. 7. Antenna with multihorn-multimode feed.

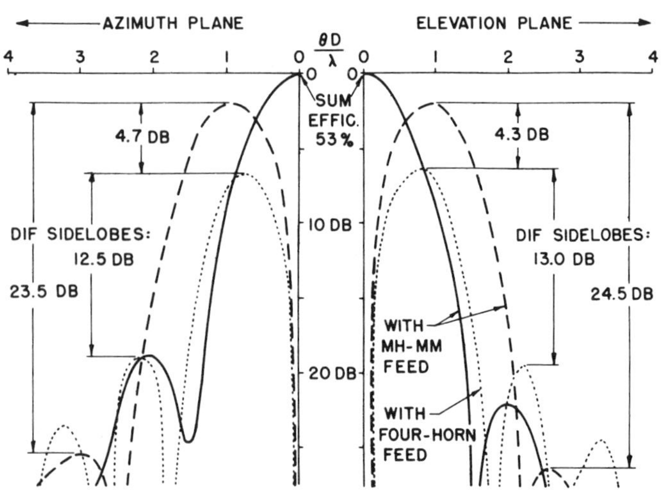

Fig. 8. Patterns of antenna with multihorn-multimode feed.

Maximum Gain in Monopulse Difference Mode*

It is well known that maximum theoretical gain of a simple antenna of the large aperture type is obtained if the aperture is uniformly illuminated with no spillover.[1] For a monopulse antenna, it has been shown that maximum theoretical slope gain in the difference mode is obtained with odd linear illumination having no spillover.[2]

It is generally accepted that slope gain is the most significant measure of performance in the difference mode, because the slope at the center of the difference pattern is a factor in the ability of a monopulse antenna to accurately track a target. However, the gain at the peaks of the difference pattern is also significant, since it may affect the ability of a monopulse antenna to achieve angle lock-on of a distant target. Furthermore, peak gain is a term which is universally employed in the antenna field, and is a standard measurement in the testing of an antenna. In this note the aperture illumination which provides maximum theoretical peak gain in the difference mode will be determined.

Consider first the field which a distant source would create on the aperture of an antenna during reception. If the source is located off the antenna axis, the distribution of this field may be separated into even and odd components. Since only the odd component couples to the difference mode of the antenna, only this component is of interest here. The shape of this odd field distribution is a simple sine curve, as indicated in Fig. 1.

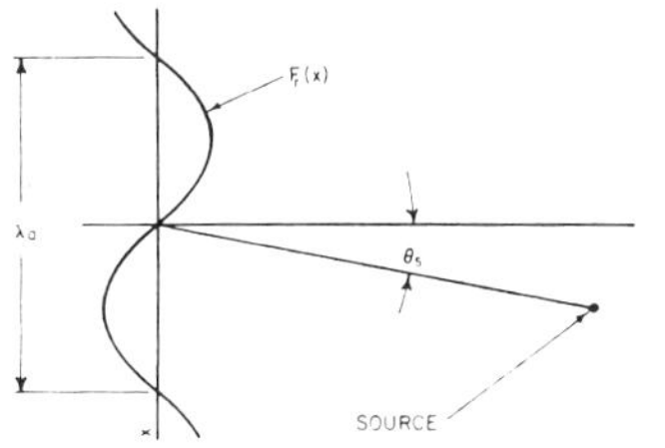

Fig. 1—Odd component of field received from a distant off-axis source.

It is described by the following relationships:

$$F_r(x) = c_1 \sin kx \tag{1}$$

$$k = \frac{2\pi}{\lambda_a} = \frac{2\pi \sin \theta_s}{\lambda} \tag{2}$$

where $F_r(x)$ is the amplitude of the received field distribution, c_1 is a constant, x is the aperture coordinate, λ_a is the wavelength of the sine distribution across the aperture, θ_s is the angle of the source, and λ is the free-space wavelength.

What shape should the aperture illumination of this antenna (during transmission) have in order to obtain a maximum signal (during reception) from the above field distribution? By analogy with the "matched filter criterion"[3] the illumination should have exactly the same shape as the field distribution. Thus an aperture illumination having a sine shape should yield maximum gain the difference mode. The wavelength of the sine illumination should equal the wavelength of the field distribution, which, in turn, is directly related to the angle of the source. The following relationship describes this matching aperture illumination:

$$F_t(x) = c_2 \frac{\sin kx}{\left(\int_{-L/2}^{L/2} \sin^2 kx dx\right)^{1/2}} \tag{3}$$

where $F_t(x)$ is the amplitude of the aperture illumination during transmission, C_2 is a constant, and L is the extent of a one-dimensional aperture. The denominator term conserves power across the aperture.

When the antenna aperture is finite in extent, there is one source angle which permits the greatest signal to be obtained. This angle is the angle of the peak of the difference pattern. Correspondingly, there is one wavelength of the sine illumination which yields the maximum peak gain with a given finite aperture. This may be determined in the following manner.

The voltage received by an antenna is given by

$$E = c_3 \int_{-L/2}^{L/2} F_t(x) F_r(x) dx \tag{4}$$

where E is the received voltage and c_3 is a constant. This is easily demonstrable by means of superposition and reciprocity, or by analogy to filter theory. Now combining (1), (3), and (4), the received voltage is

$$E = c_4 \int_{-L/2}^{L/2} \frac{\sin^2 kx}{\left(\int_{-L/2}^{L/2} \sin^2 kx dx\right)^{1/2}} dx. \tag{5}$$

Evaluating this integral gives the following simple relation for received voltage as a function of k and L:

$$E = \frac{c_4}{2}\left(L - \frac{\sin kL}{k}\right)^{1/2}. \tag{6}$$

In order to obtain the relation between k and L which gives the maximum received voltage, (6) is differentiated with respect to k and the derivative is set equal to zero. This gives

$$\tan kL = kL. \tag{7}$$

Solving this transcendental equation gives

$$\frac{kL}{2} = 0.715\pi$$

$$= \frac{\pi L}{\lambda_a} = 128.7°. \tag{8}$$

Thus the wavelength of the sine illumination for maximum peak gain in the difference mode has been determined.

The resulting illumination is shown, normalized to its peak, in Fig. 2. This illumination is a truncated sine curve having an edge taper of 2.15 db, and happens to closely resemble that which exists in many monopulse antennas of the type in which a parabolic reflector is illuminated by a monopulse feed. However, in those cases the gain in the difference mode is far below its maximum possible value, because of the large amount of spillover. Naturally, there is no spillover for this theoretical case.

The antenna pattern may be calculated for this illumination by means of the usual techniques,[4] or may be inferred from the truncated-sine illumination. The pattern, normalized to its peak, is shown in Fig. 3.

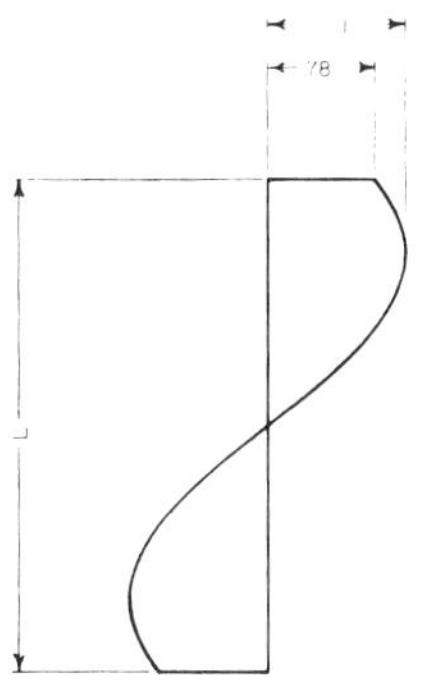

Fig. 2—Aperture illumination for maximum gain at the difference peak.

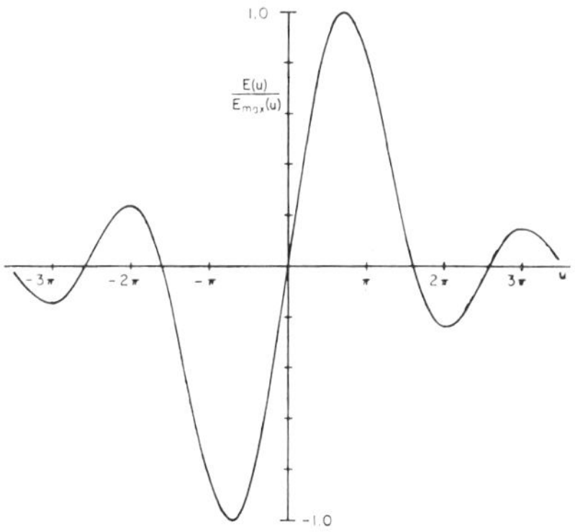

Fig. 3—Difference pattern having maximum peak gain.

Its value, normalized to the unit-amplitude illumination of Fig. 2, is given by the following formula:

$$E(u) = \frac{1}{2}\frac{\sin(u - u_m)}{u - u_m} - \frac{1}{2}\frac{\sin(u + u_m)}{u + u_m}$$

where

$$u = \frac{\pi L \sin \theta}{\lambda} \tag{9}$$

and

$$u_m = \frac{\pi L \sin \theta_m}{\lambda} = 0.715\pi$$

* Received by the PGAP, December 19, 1960.

[1] S. Silver, "Microwave Antenna Theory and Design," M.I.T. Rad. Lab. Ser., McGraw-Hill Book Co. Inc., New York, N. Y., vol. 12, pp. 177–178; 1949.

[2] G. M. Kirkpatrick, General Electric Co., ASTIA Doc. No. AD 18458, pp. 13–15; August 1952. Also see E. J. Powers, ASTIA Doc. No. AD 231217, pp. 8–15, 66–72; July, 1959.

[3] J. H. Van Vleck and D. Middleton, "A theoretical comparison of visual, aural, and meter reception of pulsed signals in the presence of noise," *J. Appl. Phys.*, vol. 17, pp. 943–944, 964; November, 1946.

[4] J. F. Ramsay, "Fourier transforms in aerial theory—Part II, Fourier sine transforms," *Marconi Rev.*, vol. 10, p. 17; January–March, 1947.

where $E(u)$ is the pattern amplitude, θ is the pattern angle, and θ_m is the angle of the pattern peak. This pattern is simply the difference of two ordinary maximum-gain patterns, squinted to the difference peak angles. It is interesting that the squint angle is such that the first sidelobe peak of one of these "lobe" patterns coincides with the main peak of the other; this condition would be expected to correspond to maximum peak gain in the difference pattern if two such lobe patterns could be independently created with a single antenna structure.

The value for difference peak gain may be expressed by comparing it with ordinary peak gain. This ratio is determined as follows:

$$\frac{g}{g_0} = \frac{E^2(u_m)}{\frac{1}{L}\int_{-L/2}^{L/2} \sin^2\left(0.715\pi \frac{2x}{L}\right) dx} \quad (10)$$

where g is the maximum difference peak gain, and g_0 is the ordinary maximum peak gain obtained with uniform illumination of the same aperture, both cases, of course, assuming no spillover. The denominator represents the power of the unit-amplitude illumination of Fig. 2. Evaluating this expression yields

$$\frac{g}{g_0} = \sin^2 0.715\pi = 0.609. \quad (11)$$

Thus the maximum possible gain in the difference mode of a monopulse antenna is 2.15 db below the maximum possible gain in the sum mode of the same antenna. It is interesting that this number exactly equals the 2.15-db edge taper mentioned in connection with Fig. 2. It may also be noted that this gain ratio is reasonably close to the 3-db ratio which would be expected if the two previously-mentioned "lobe" components of the difference pattern were separated by a squint angle great enough so that only one component contributed significantly at the difference peak.

In conclusion, it is appropriate to mention that a knowledge of the illumination which provides maximum theoretical peak gain in the difference mode of a monopulse antenna can be helpful in the design of such an antenna, even if one chooses to avoid this illumination because of the high sidelobes it yields. Furthermore, the value of the theoretical peak gain can provide a useful standard for rating the efficiency of a monopulse antenna. Of course, the results presented here apply exactly only to a one-dimensional aperture or to a rectangular aperture with separable distributions.

It is perhaps also worth noting that the "matching" method employed to determine the optimum illumination involved a simple physical process rather than a complex mathematical one. The method is applicable to other similar problems; for example, the illumination for ordinary maximum peak gain and for maximum difference slope gain, mentioned at the beginning of this communication, may be solved by inspection with this method.

Peter W. Hannan
Wheeler Labs.
Smithtown, N. Y.

APPENDIX I

SOME FURTHER NOTES ON THE SPATIAL INFORMATION AVAILABLE FROM MONOPULSE RADAR

Dean B. Anderson, Senior Technical Specialist
and
Donald R. Wells, Research Engineer

May 15, 1961

REVISED DECEMBER 3, 1962

Submitted to: 7th Annual Radar Symposium
University of Michigan
Ann Arbor, Michigan

Autonetics A DIVISION OF NORTH AMERICAN AVIATION, INC., DOWNEY, CALIFORNIA

ABSTRACT

Radar information concerning a target echo signature is statistical and structural in nature. Target position data are usually the only spatial structural information determined by a conventional angle detection radar. The authors' intent is the delineation of a monopulse radar which extracts additional spatial information, namely, extent, shape, and rotation from the target echo signature.

Monopulse angle detection radar derives the even, odd, and quadrupolar components of the antenna diffraction pattern. The echo signal component which is correlated between the even and odd channels leads to angular position data. This paper considers the echo signal component which is uncorrelated between the even and either the odd or quadrupolar ports. A product detector in quadrature with the monopulse angle detector responds to the uncorrelated echo signal component which is interpreted as a measure of target extent. Further, the influence of target extent and distribution on the antenna diffraction patterns is considered. The analogies of this instrumentation with the optical and radio stellar interferometer are cited.

CONTENTS

ILLUSTRATIONS

I. INTRODUCTION

Radar information of a target echo signature is statistical and structural in nature. The statistical information, not considered further herein, is related to the reliability of detection and the accuracy with which parameters can be estimated. Structural information, of which only spatial will be considered further, relates to the degree of ambiguity and the resolution of different targets in position. Conventional angle detection radars determine only position data from the available spatial structural information.

Some other spatial aspects of the target echo signature, which are available from monopulse radar, are interpreted as (1) a measure of the target extent, (2) the target shape as determined by its anisotropic scattering, and (3) the target rotation as determined by modulation of the extent signal. The mechanism by which a monopulse antenna responds to a distributed target echo and the adjunct instrumentation necessary for a monopulse radar to extract these additional spatial features will be delineated.

Target radial extent and radial velocity are obviously revealed in the time-angular frequency domain. Target rotation about itself and extent transverse to the target radius vector are manifested in the aperture-spatial frequency domain. Target tumbling about an axis normal to the radius vector and target spinning about the radial appear as modulation of the odd and quadrupolar monopulse signal channels respectively.

Scanning an antenna diffraction pattern across a target which subtends an angle appreciably larger than the beamwidth produces the familiar width increase of the observed response. A technique which yields a measure of extent even when the target subtends a small fraction of the antenna beamwidth will be described for a monopulse radar wherein position, extent, and shape data are available simultaneously.

The monopulse concept resembles interferometry, which has been used to measure source throughout the electromagnetic spectrum for a long time.[1-5] The Michelson stellar interferometer[6] was the first instrument developed capable of star diameter measurement. The interferometer of either the optical or radio astronomer records the received intensity at the confluence from separate apertures as interference fringes. The visibility of these fringes defines the degree of coherence[7] of the field which is the amplitude of the extended source diffraction pattern.

There are some distinctions between monopulse and interferometry techniques. Optical and celestial radio sources are generally incoherent, and their detection is proportional to the field intensity. An interferometer is composed of two or more widely separated apertures which have numerous interference lobes within the envelope of the single aperture diffraction pattern. Thus, resolution is improved, while ambiguity is impaired. The interferogram record shows interference fringe maximum and minimum sequentially as the source passes through the lobes. On the other hand, the radar signal is coherent, and the detection of the reflected echo is proportional to the field amplitude. The monopulse antenna is composed of two apertures arranged to cause a one-fringe maximum within the single aperture diffraction pattern because unambiguous resolution is paramount. A hybrid junction is provided at the monopulse antenna transmission line confluence where separate ports simultaneously yield the interference fringe maximum and minimum.

---

1 L. L. McCready, J. L. Pawsey, and R. Payne-Scott, "Solar Radiation at Radio Frequencies and its Relation to Sunspots", Proc. Roy. Soc (London) A Vol. 190, pp 357-375, August 1947

2 E. Wolf, "A Macroscopic Theory of Interference and Diffraction of Light from Finite Sources II Field with a Spectral Range of Arbitrary Width," Proc. Roy. Soc. (London) A Vol. 230, pp 246-265

3 J. L. Pawsey and R. N. Bracewell, "Radio Astronomy," Oxford University Press, Chapter II, 1955

4 M. Born and E. Wolf, "Principles of Optics," Pergamon Press, Chapters VII and X, 1959

5 A. R. Stokes, "A Numerical Fourier Analysis Method for the Correction of Widths and Shapes of Lines on X-Ray Power Photographs," Proc. Phys. Soc., Vol. 61, p 382

6 A. A. Michelson and F. G. Pease, "Measurement of the Diameter of γ Orionis with the Interferometer," Astro Phys., Journ., Vol. 53 p 249, 1921

7 F. Zernike, "The Concept of Degree of Coherence and its Application to Optical Problems," Physica, Vol No. 8, p 785-795, August 1938

Antenna diffraction patterns are measured in the Fraunhofer region with a point source. The general effects of an arbitrary source distribution on the observed antenna diffraction pattern have been considered by the radio astronomer.[8-9] The observed antenna response to a distributed source may be evaluated with the convolution integral or transformed into Fourier spatial frequency components and evaluated by a spectral product. Thus, the observed response is a smoothed reproduction of the true source distribution because the antenna aperture functions in a manner analogous to a low-pass filter. Complete restoration of the true source distribution is impossible because source spatial frequencies outside the antenna spectral sensitivity function are irretrievably lost. Only the Fourier spatial frequencies within the antenna aperture spectral sensitivity cutoff will be available to deduce the bounds of the source extent.

This paper will consider the manner by which a monopulse antenna responds to the target distribution designated in terms of the symmetry within even, odd, and quadrupolar functions. An heuristic discussion of the response to these distribution components will suggest methods of obtaining a measure of target extent, shape, and rotation. The results of a monopulse radar elevation extent experiment are illustrated.

---

8 S. Matt and J. D. Kraus, "The Effect of Source Distribution on Antenna Diffraction," Proc. IRE, Vol 43, pp 821-825, July 1955

9 R. N. Bracewell and J. A. Roberts, "Aerial Smoothing in Radio Astronomy," Aust. Jour. Phys., Vol 7, pp 615-640, December 1954

II. THE MONOPULSE CONCEPT

There are many implementations of monopulse angle detection radar which are now formulated into a consistent unified theory.[10] A short discussion of the apposite aspects of the antenna angle sensing and the angle detection functions is prudent before the target extent influence is considered.

The terms "amplitude" and "phase," which relate to the physical aperture arrangements, are classifications used to describe a monopulse antenna angle sensing function. Two apertures with a common phase center and squinted diffraction patterns are termed amplitude, while two apertures with separated phase centers and collimated diffraction patterns are termed phase sensing. Usually, the signal from both antennas is combined in a hybrid junction labyrinth where the even, elevation odd, azimuth odd, and quadrupolar components are extractedfrom the aperture illumination. Data from the labyrinth in the even-odd form are preferred because transmission through active components and amplifiers does not impair the boresight stability.

After appropriate amplification, the echo signal is detected in an angle detector. There are three classes of angle detection characterized by the input signal form, namely, (1) the ratio of odd-to-even components, (2) the logarithm of the amplitude ratio, and (3) the sine of the phase difference. The echo signals in each channel are normalized to eliminate any detected target angle dependence on radar system and target parameters.

Further consideration herein of monopulse radar will concentrate on the even-odd form of antenna sensing and the odd-to-even ratio angle detection form for a single coordinate because of the implementation simplicity, its general acceptance, and an ease of analysis. Conversion to other forms of antenna sensing and angle detection is easily accomplished by utilizing already well-established techniques.

The Fourier transform properties of symmetrical functions are briefly reviewed here to clarify the influence of a distributed source on the monopulse concept. Consider the aperture distribution function f(x) which is related to the antenna diffraction pattern g(u) by the Fourier transform

---

[10] D. R. Rhodes, "Introduction to Monopulse," McGraw-Hill Book Co., 1959

$$g(u) = \int f(x)\, e^{jux}\, dx \tag{1}$$

and its inverse transform

$$f(x) = \frac{1}{2\pi} \int g(u)\, e^{-jux}\, du \tag{2}$$

where $u = \frac{\pi D}{\lambda} \sin\theta$ and $x = \frac{2z}{D}$.

These functions g (u) and f (x) are separated into real-imaginary and even-odd components as

$$f(x) = f_{er}(x) + j f_{ei}(x) + f_{or}(x) + j f_{oi}(x) \tag{3}$$

which transforms into

$$g(u) = g_{er}(u) + j g_{ei}(u) + j g_{or}(u) - g_{oi}(u) \tag{4}$$

Note that only the odd function spaces transform in time quadrature.

The monopulse hybrid junction labyrinth extract the even Σ, odd Δ, and quadrupolar Q components of the aperture distribution function as

$$\Sigma \qquad f_e(x, y) = \frac{1}{4}\left[f(x, y) + f(x, -y) + f(-x, y) + f(-x, -y)\right] \tag{5}$$

$$\Delta_{el} \qquad f_o(x, y) = \frac{1}{4}\left[f(x, y) + f(-x, y) - f(x, -y) - f(-x, -y)\right] \tag{6}$$

$$\Delta_{az} \qquad f_o(x, y) = \frac{1}{4}\left[f(x, y) + f(x, -y) - f(-x, y) - f(-x, -y)\right] \tag{7}$$

$$Q \qquad f_q(x, y) = \frac{1}{4}\left[f(x, y) + f(-x, -y) - f(-x, y) - f(x, -y)\right] \tag{8}$$

In a conventional monopulse radar, the echo signal components Σ, Δ_{el} and Δ_{az} are converted to a lower frequency for amplification and normalized by an instantaneous AGC actuated by Σ while the Q component is absorbed in a load. The normalized Δ components are processed by a product detector where the reference signal phase is provided by the Σ channel which yields an angle output Re F (u). The angle detected output is written in terms of the aperture and diffraction functions as

$$F(u) = \frac{\Delta(u)}{\Sigma(u)} = \frac{-g_{oi}(u) + j\, g_{or}(u)}{g_{er}(u) + j\, g_{ei}(u)} \rightleftharpoons \frac{f_{or}(x) + j\, f_{oi}(x)}{f_{er}(x) + j\, f_{ei}(x)}$$

$$= \frac{\left[g_{ei}(u)g_{or}(u) - g_{er}(u)g_{oi}(u)\right] + j\left[g_{oi}(u)g_{ei}(u) + g_{or}(u)g_{er}(u)\right]}{g_{er}^2(u) + g_{ei}^2(u)} \tag{9}$$

and in polar form is written as

$$= \left[\frac{g_{or}^2(u) + g_{oi}^2(u)}{g_{er}^2(u) + g_{ei}^2(u)}\right]^{1/2} \angle \tan^{-1} \frac{g_{oi}(u)g_{ei}(u) + g_{or}(u)g_{er}(u)}{g_{ei}(u)g_{or}(u) - g_{er}(u)g_{oi}(u)} \tag{10}$$

The angle detector will resolve one component determined by the reference phase. The physical aperture arrangement and the hybrid junction type determine the value of a differential phase shifter necessary to yield the real part of the angle detection function. Monopulse antennas are usually designed to focus at infinity so that $g_{ei}(u) = 0$; thus, equations (9 - 10) reduce to

$$F(u) = \frac{-g_{oi}(u) + jg_{or}(u)}{|g_{er}(u)|} = \frac{\left[g_{or}^2(u) + g_{oi}^2(u)\right]^{1/2}}{|g_{er}(u)|} \angle \tan^{-1} - \frac{g_{or}(u)}{g_{oi}(u)} \tag{11}$$

The antenna aperture physical arrangement is usually designed to maximize either the amplitude or phase sensing while sensitivity to the other form is eliminated. Thus, for an amplitude arrangement $g_{oi}(u) = 0$, and the angle detection function becomes

$$\text{Re}\, F_a(u) = \frac{-g_{oi}(u)}{|g_{er}(u)|} \tag{12}$$

and for the phase arrangement $g_{oi}(u) = 0$, so

$$\text{Re } F_p(u) = \frac{j\, g_{or}(u)}{|g_{er}(u)|} \tag{13}$$

The only difference between the mechanization of the angle detector for even-odd data from either antenna sensing form is the reference phase shift value.

The individual element aperture distribution and the corresponding diffraction patterns for amplitude and phase sensing physical arrangements are shown in Fig. 1 and 2 respectively. Each function has been resolved into the even and odd components in both the aperture and diffraction spaces. The aperture distributions selected for illustration have readily recognized Fourier transforms; these are also the distributions which result in the maximum angular sensitivity[11-12] for monopulse antenna arrangements confined to the same width. The odd-even ratio angle detection functions for both arrangements are also illustrated. Note that odd-even angle detection function for either antenna sensing has a trivial difference over the principle lobe, while in the side-lobe region the periodicity differs. Therefore, no further distinction between amplitude and phase sensing will be cited. The angle detection functions for the simple transform pairs used are $\text{Re } F_a(u) = (\frac{1}{u}) - \cot u$ and $\text{Re } F_p(u) = \tan \frac{u}{2} = \frac{1}{\sin u} - \cot u$.

With this rudimentary introduction to the monopulse concept, the effect of target angular extent on the radar-tracking function and the means to obtain a target extent measure follows directly. The block diagram of one form of conventional two-axis monopulse radar is illustrated in Fig. 3 together with the additional functional blocks necessary to implement the method used to experimentally obtain a target extent measure.

---

[11] G. M. Kirkpatrick, "Aperture Illuminations for Radar Angle-of-Arrival Measurements," IRE Trans. of PGANE, Vol PGAE-9, pp 20-27, Sept 1953

[12] E. J. Powers, Jr. "Analysis and Synthesis of a General Class of Difference Patterns," MIT Research Lab of Electronics TR #8, July 30, 1959

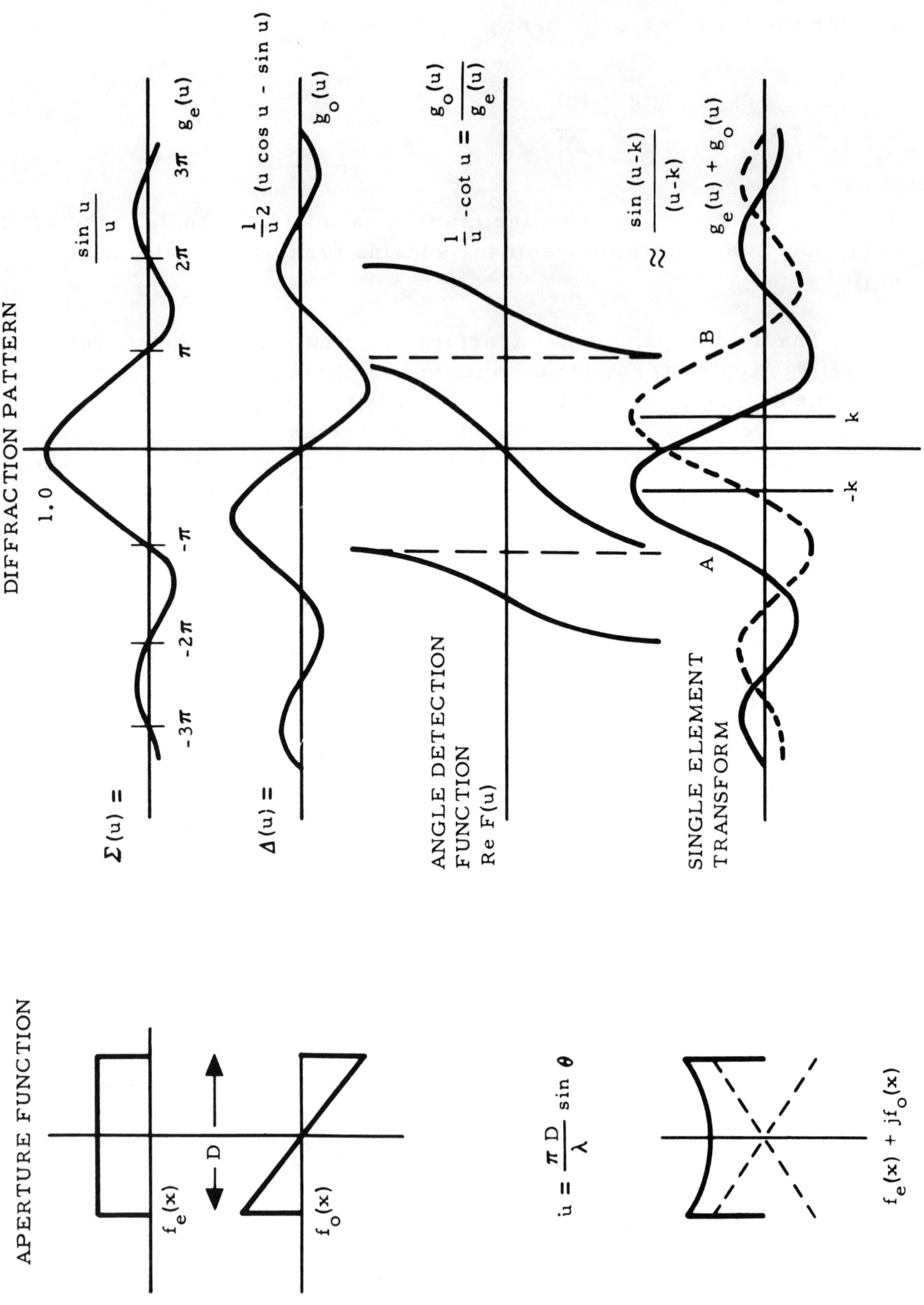

Fig. 1. Transforms in Amplitude Sensing Monopulse Antennas

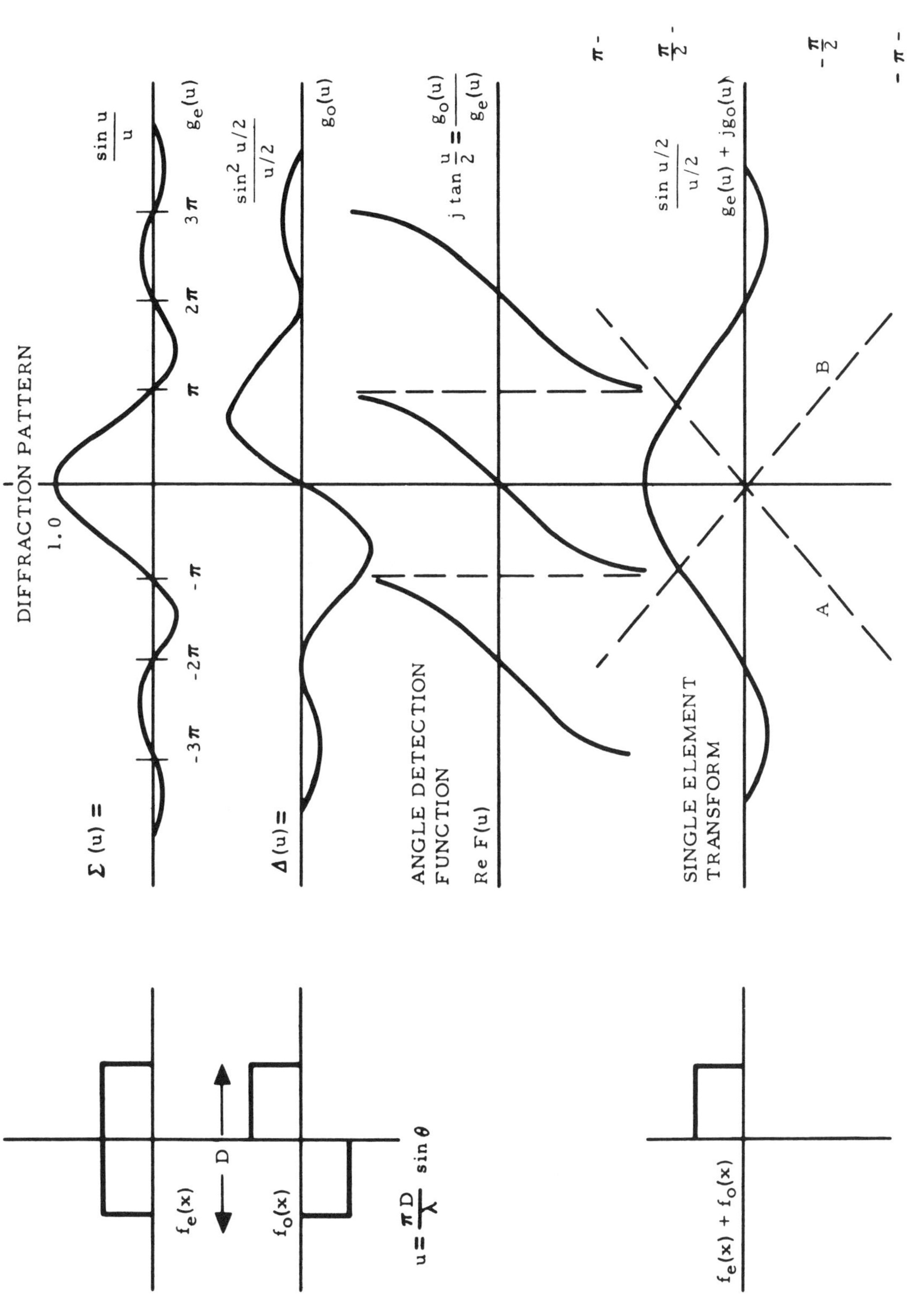

Fig. 2. Transforms in Phase Sensing Monopulse Antennas

This convolution integral describes the smoothing process. For a point source of zero extent at $u = 0$, the observed distribution reduces to the Fraunhofer diffraction pattern $G(u) = g(u)$.

The monopulse antenna diffraction patterns $g_e(u) = \frac{\sin u}{u}$ and $g_o(u) = \frac{\sin^2 \frac{u}{2}}{\frac{u}{2}}$ will be used to evaluate (14) because their integrals Si (x) and Cin (x) have been tabulated.[13] Consider a unit rectangular source distribution even about $u = 0$ with total extent $w = \frac{\pi DL}{2\lambda R}$ where the other parameters are: total antenna aperture width D, total target width L, target range R, and the radiated wavelength λ so that

$$h_e(u) = 1, \quad -\frac{w}{2} < u < \frac{w}{2} \tag{15}$$

$$h_e(u) = 0, \quad \text{elsewhere}$$

Also consider a unit rectangular source distribution odd about $u = 0$ with total extent $w = \frac{\pi DL}{2\lambda R}$ so that

$$h_o(u) = -1, \quad -\frac{w}{2} < u < 0$$

$$h_o(u) = +1, \quad 0 < u < \frac{w}{2} \tag{16}$$

$$h_o(u) = 0, \quad \text{elsewhere}$$

Then, the observed response of an even antenna diffraction pattern to an even source distribution $G_{ee}(u_c)$, (14) becomes

$$G_{ee}(u_c) = \frac{1}{w} \int_{-\frac{w}{2}}^{\frac{w}{2}} \frac{\sin(u - u_c)}{(u - u_c)} du$$

$$= \frac{1}{w} \left[\mathrm{Si}\left(\frac{w}{2} - u_c\right) + \mathrm{Si}\left(\frac{w}{2} + u_c\right) \right] \tag{17}$$

[13] J. D. Kraus, "Antennas," McGraw-Hill Book Co., 1950, pp 535-539

Equation (17) has been evaluated for several values of w and is plotted in Fig. 4, which shows that the antenna beamwidth is not a sensitive indicator of extent except when it is large compared with the beamwidth. Now consider the observed response of an odd antenna diffraction pattern to an even source distribution: $G_{oe}(u_c)$ (14) becomes

$$G_{oe}(u_c) = \frac{1}{w}\int_{-\frac{w}{2}}^{\frac{w}{2}} \frac{1 - \cos(u - u_c)}{(u - u_c)}\, du$$

$$= \frac{1}{w}\left[\mathrm{Cin}\left(\frac{w}{2} - u_c\right) - \mathrm{Cin}\left(\frac{w}{2} + u_c\right)\right] \tag{18}$$

Equation (18) has been evaluated and is plotted in Fig. 5 for the same values of w which shows a zero response for any extent when the antenna boresight is aligned to the target axis of symmetry and shows the typical position data off boresight. In the same way, the observed response of an even antenna diffraction pattern to an odd source distribution $G_{eo}(u_c)$ (14) becomes

$$G_{eo}(u_c) = \frac{1}{w}\int_{0}^{\frac{w}{2}} \frac{\sin(u - u_c)}{(u - u_c)}\, du - \frac{1}{w}\int_{-\frac{w}{2}}^{0} \frac{\sin(u - u_c)}{(u - u_c)}\, du$$

$$= \frac{1}{w}\left[\mathrm{Si}\left(\frac{w}{2} - u_c\right) - \mathrm{Si}\left(\frac{w}{2} + u_c\right) + 2\,\mathrm{Si}(u_c)\right] \tag{19}$$

Equation (19) has been evaluated and is plotted in Fig. 6 for three values of w, which shows a zero response for any extent when the antenna boresight is aligned to the target center. Off boresight, the response is a sensitive indicator of extent, but the antenna is not on the position of track.

Finally, consider the observed response of an odd antenna to an odd source distribution $g_{oo}(u_c)$, so (14) becomes

III. THE NATURE OF SPATIAL INFORMATION

To recognize a target, the reflected field over an angular sector must be distinguished from the surrounding background by a prominence, while the word "extent" implies that a single prominence has width between at least vaguely defined bounds, otherwise the target source distribution is completely arbitrary. The Fourier transform of the target source distribution is the target reflected diffraction pattern which is the spatial Fourier components of the source. All the spatial spectral components are necessary to fully describe the source distribution.

As the antenna scans across the target, the observed response is a smoothed version of the true source distribution. In other words, the antenna aperture functions as a low pass filter where the cutoff frequency is $\frac{D}{\lambda}$. Spatial frequencies (cycles per radian) above the cutoff are irretrievably lost, while spatial frequencies below the cutoff are the only spectral components available to describe the target boundary. When a large part of the source spatial spectrum is lost by the inquiring antenna because the beamwidth subtends an angle appreciably larger than the target, the true source distribution is smoothed to the point where the observed distribution approaches in fact the antenna diffraction pattern. An excellent treatment of the antenna smoothing phenomenon appears in Bracewell's literature.

The target angular extent and shape effect on the monopulse antenna is introduced by considering some simple even and odd source distributions for which the integrals are easily evaluated. Then, the significance of the significance of the various combinations of the even and odd source distributions are interpreted.

Consider an antenna diffraction pattern $g(u - u_c)$ and a source distribution $h(u)$ where the antenna response $G(u_c)$ is observed when the antenna centerline is displaced by an angle u_c from the target center. The observed response distribution is given by

$$G(u_c) = \frac{\int g(u - u_c)\, h(u)\, du}{\int |h(u)|\, du}$$

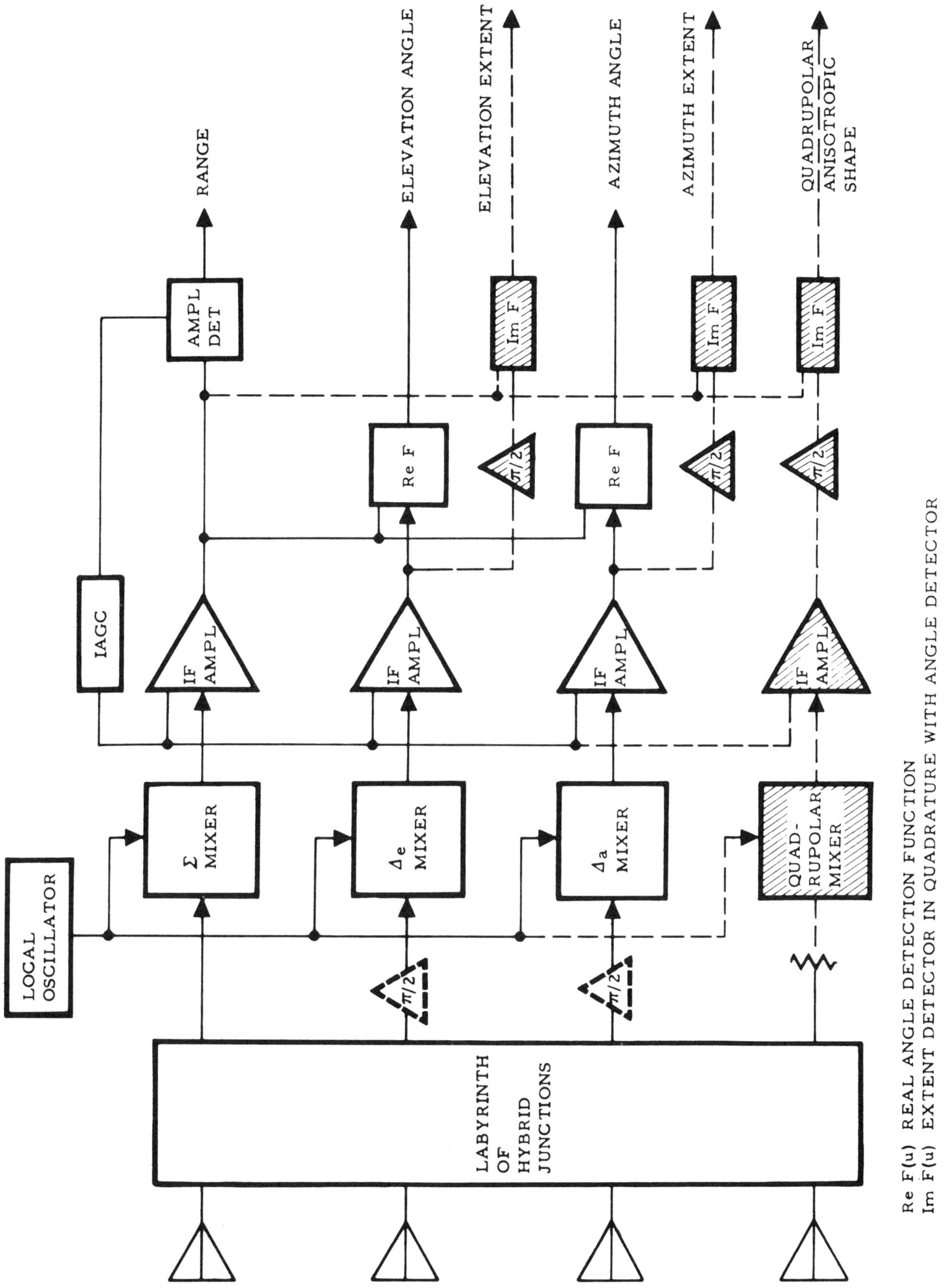

Fig. 3. A Monopulse Radar Block Diagram

$$G_{oo}(u_c) = \frac{1}{w}\int_0^{\frac{w}{2}} \frac{1-\cos(u-u_c)}{(u-u_c)}du - \frac{1}{w}\int_{-\frac{w}{2}}^{0} \frac{1-\cos(u-u_c)}{(u-u_c)}du$$

$$= \frac{1}{w}\left[\mathrm{Cin}\left(\frac{w}{2}+u_c\right) + \mathrm{Cin}\left(\frac{w}{2}-u_c\right) - 2\,\mathrm{Cin}(u_c)\right] \qquad (20)$$

Equation (20) has been evaluated and plotted in Fig. 7 for the same three values of w in Fig. 6. For this combination, the observed response is a sensitive function of extent even for small targets. Note that the extent data maximum exists while the radar is tracking.

In nature, various combinations of $h_e(u)$ and $h_o(u)$ will arise. The definition of even and odd source distributions is referenced to the antenna boresight radius vector. When the source distribution in the Fraunhofer region is even, the target position indicated is $u_c = 0$ and the angle detection function is shown in Fig. 8 for various values of w. If the target range decreases, the even source distribution becomes complex in the Fresnel region.

Figure 8 shows that monopulse radar will track properly an extended source with reduced angular sensitivity.

When the source distribution in the Fraunhofer region is odd and of the same total width, the angle detection function is shown in Fig. 9 which shows that a $h_o(u)$ source distribution will cause the proper tracking function to fail relative to the boresight. Observe that tracking will continue, but relative to a $g_e(u)$ null. This phenomena, called the Heiltigtag[14] effect, has been known since the beginning of wireless direction finding.

When both the even and odd source distributions of (15) and (16) are combined in phase, the resulting target distribution is interpreted as a target of half-width and displaced by $\frac{w}{4}$. The response observed in the angle detector is $\mathrm{Re}\,F(u_c = 0) = \tan\frac{w}{8}$, and the response in the detector in quadrature with the angle detector is $\mathrm{Im}\,F(u_c = 0) = 0$. The

---

[14] R. Keen "Wireless Direction Finding" Iliffe & Sons, 4th Edition, 1947

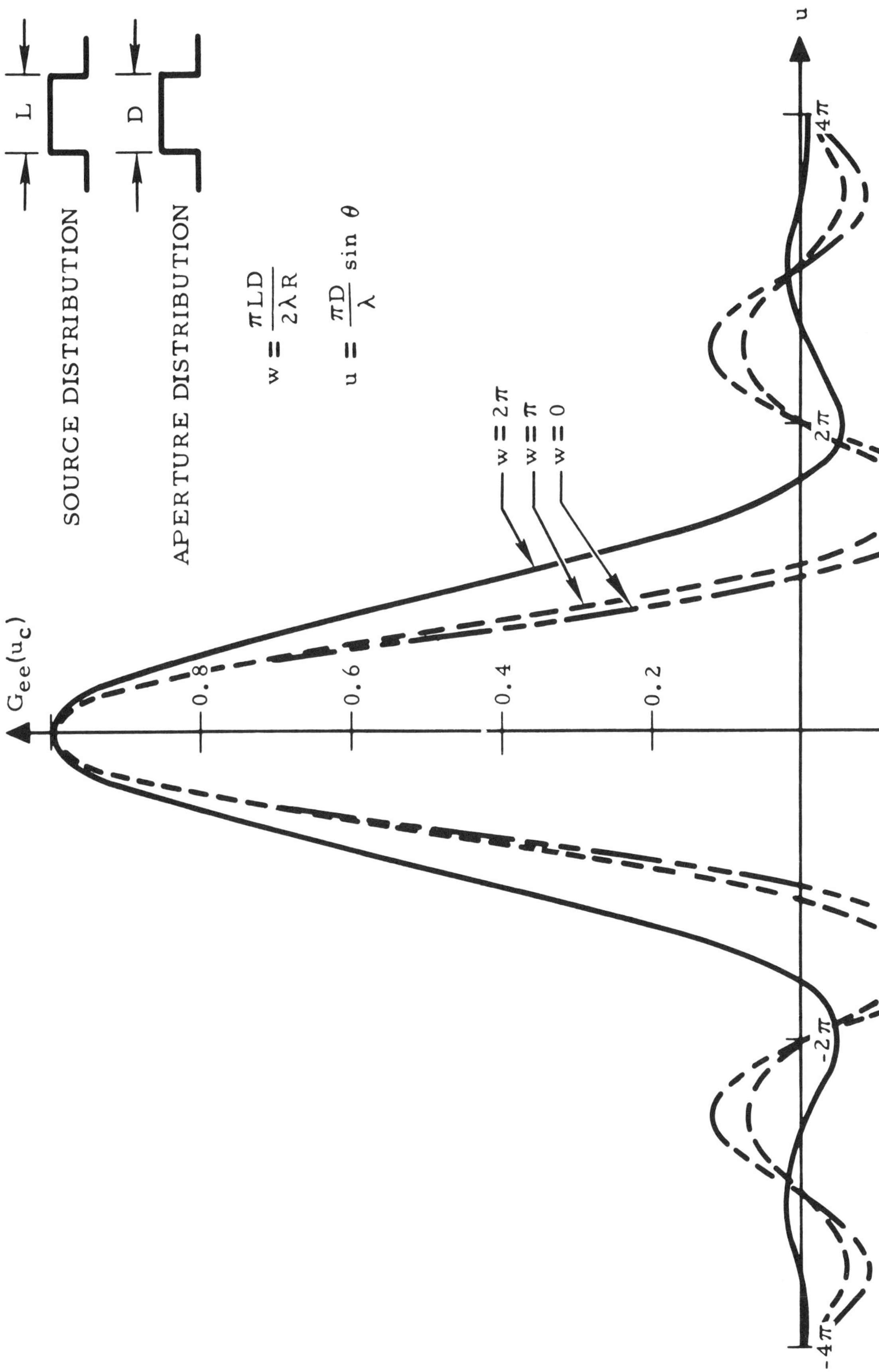

Fig. 4. The Response of an Even Antenna Diffraction Pattern to an Even Source Distribution

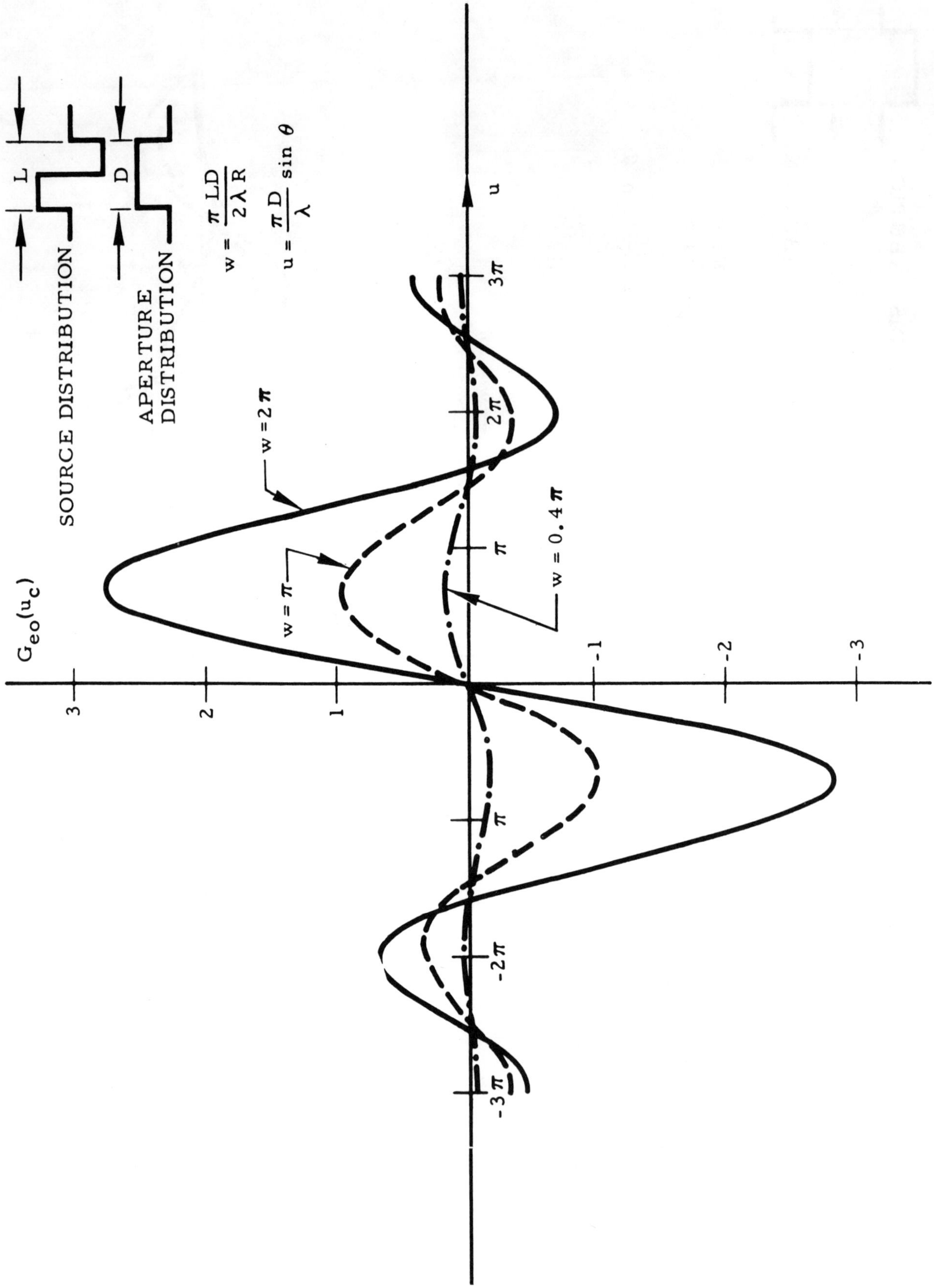

Fig. 5. The Response of an Odd Antenna Diffraction Pattern to an Even Source Distribution

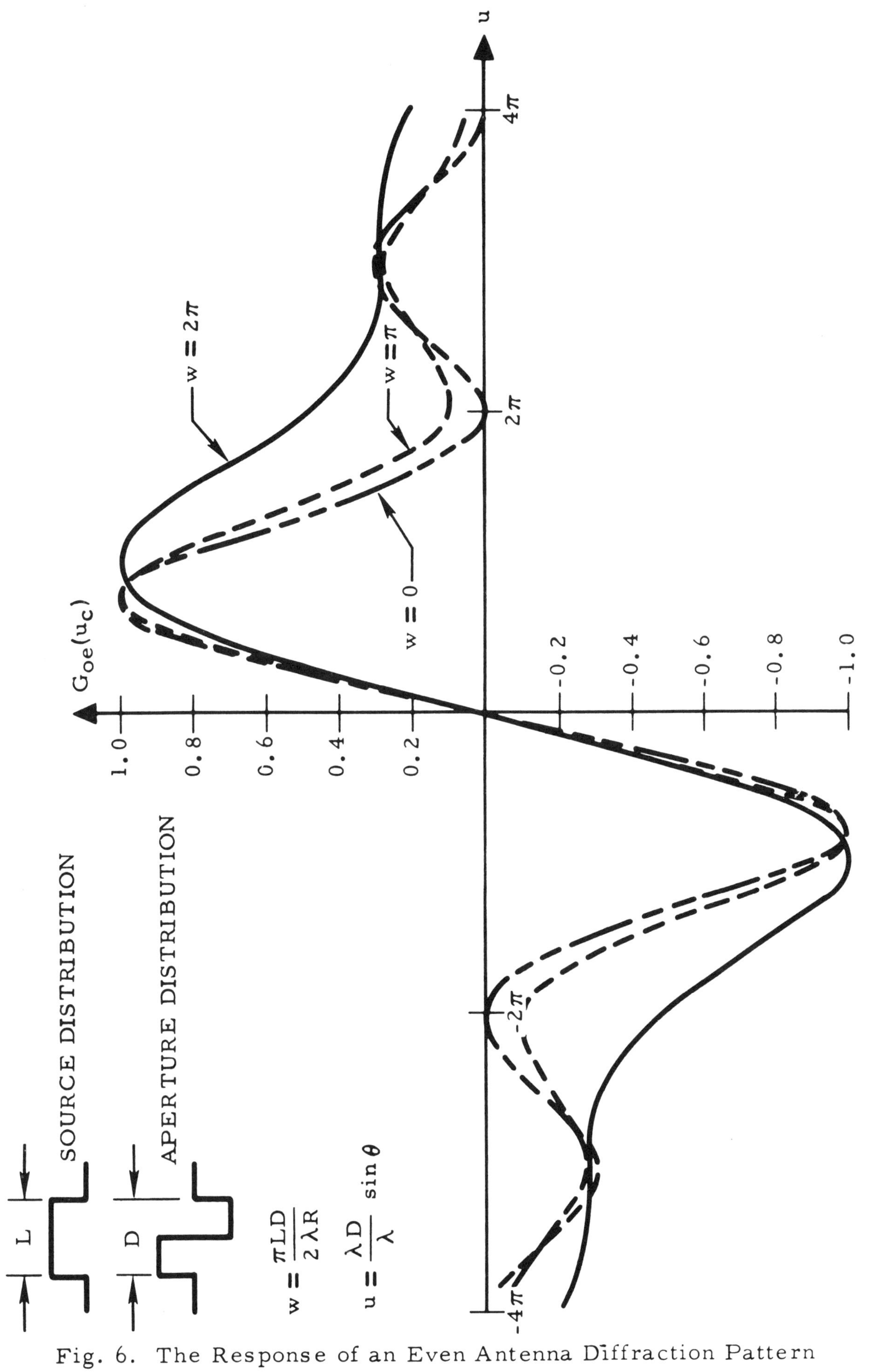

Fig. 6. The Response of an Even Antenna Diffraction Pattern to an Odd Source Distribution

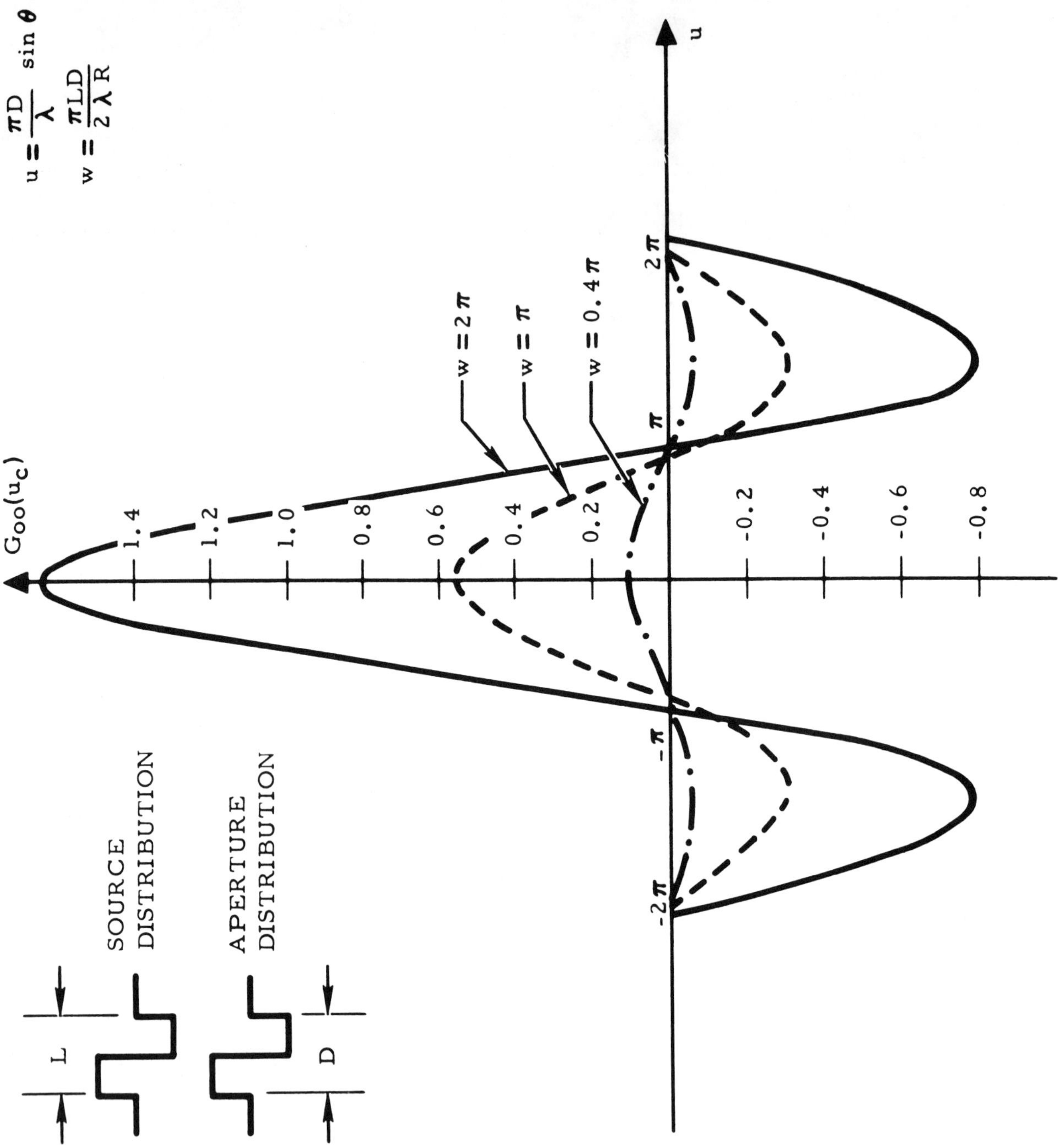

Fig. 7. The Response of an Odd Antenna Diffraction Pattern to an Odd Source Distribution

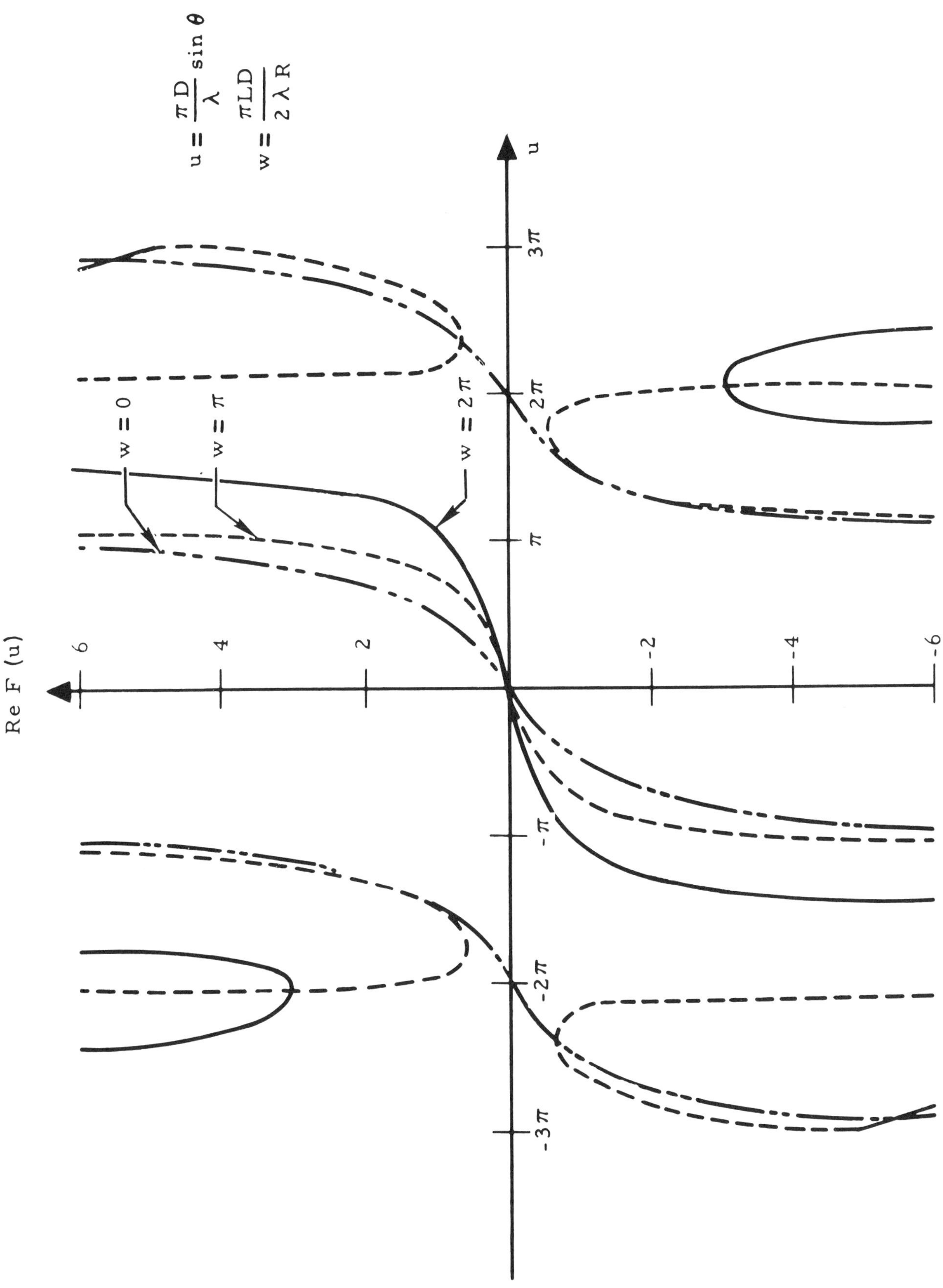

Fig. 8. The Angle Detection Function Response to an Even Source Distribution

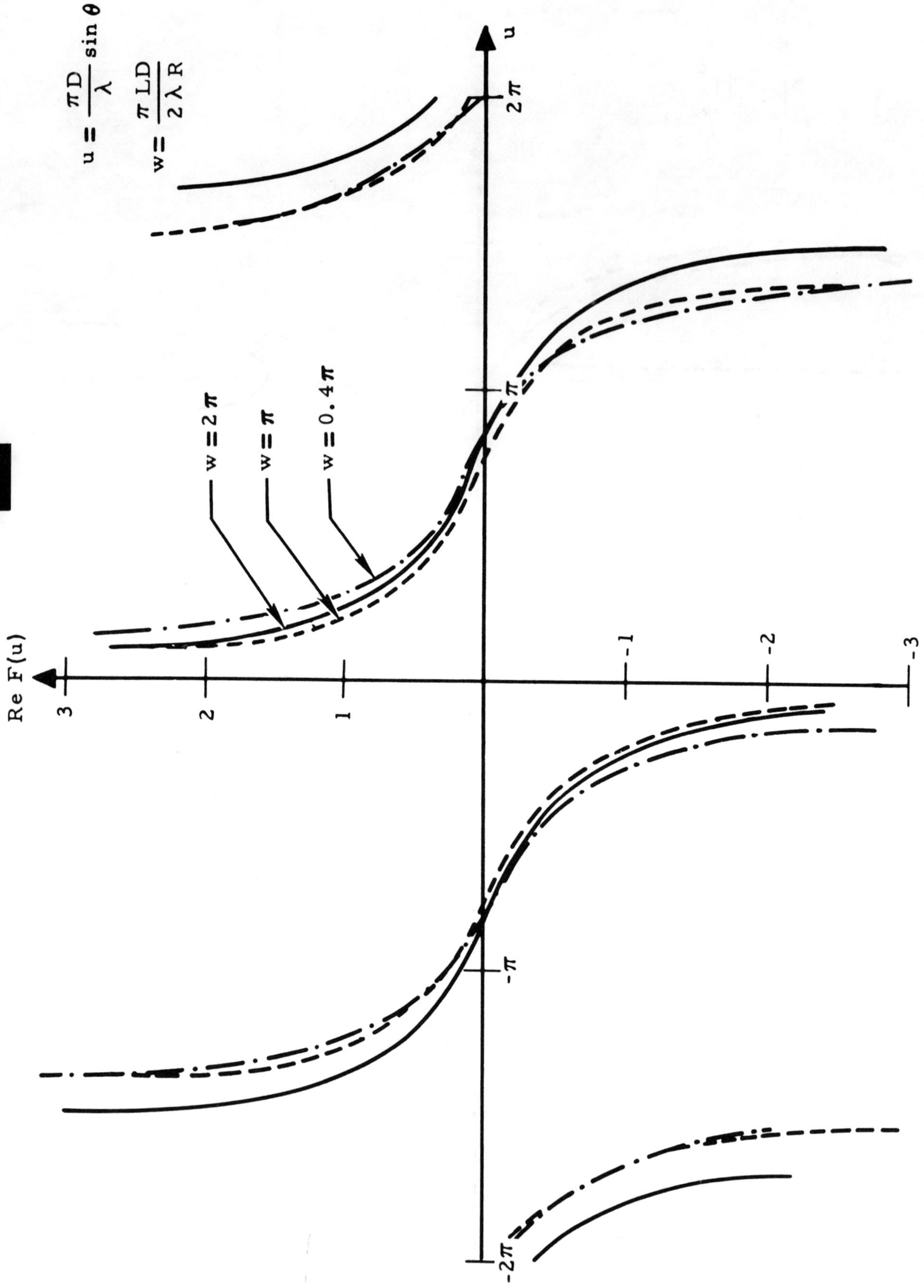

Fig. 9. The Angle Detection Function Response to an Odd Source Distribution

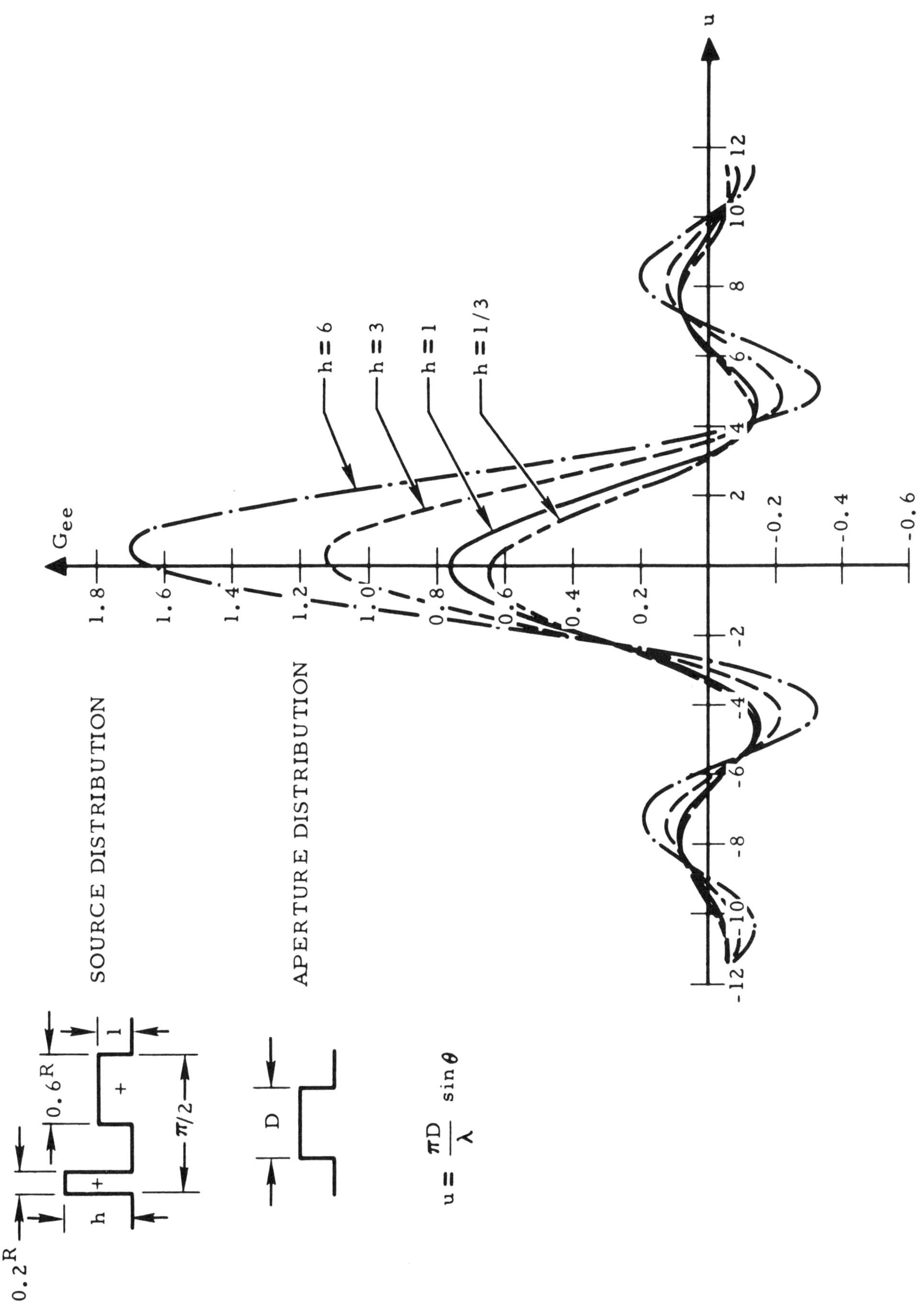

Fig. 10. The Response of an Even Antenna Diffraction Pattern to an Asymmetrical Source Distribution of Two Closely Spaced Targets In Phase

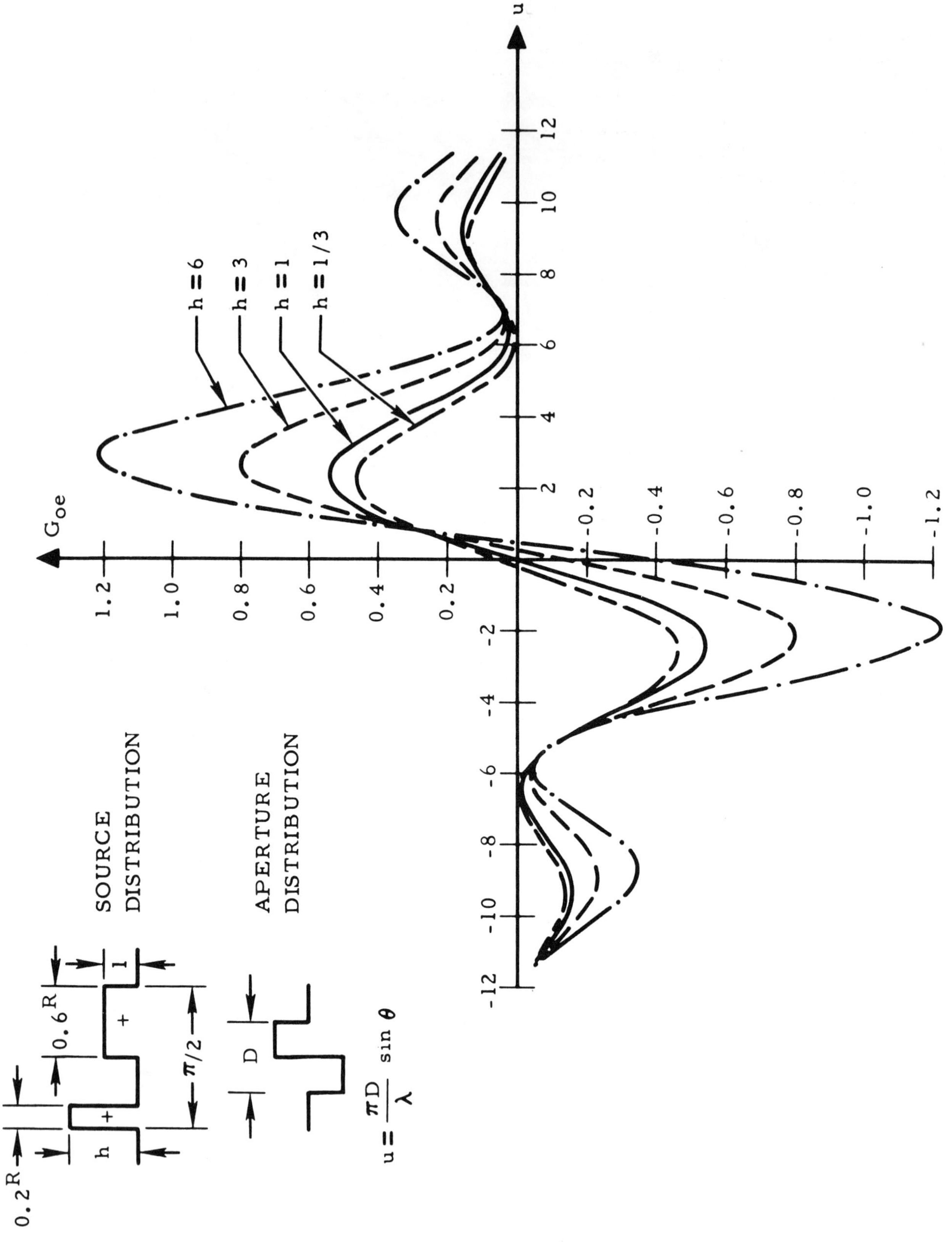

Fig. 11. The Response of an Odd Antenna Diffraction Pattern to an Asymmetrical Source Distribution of Two Closely Spaced Targets in Phase

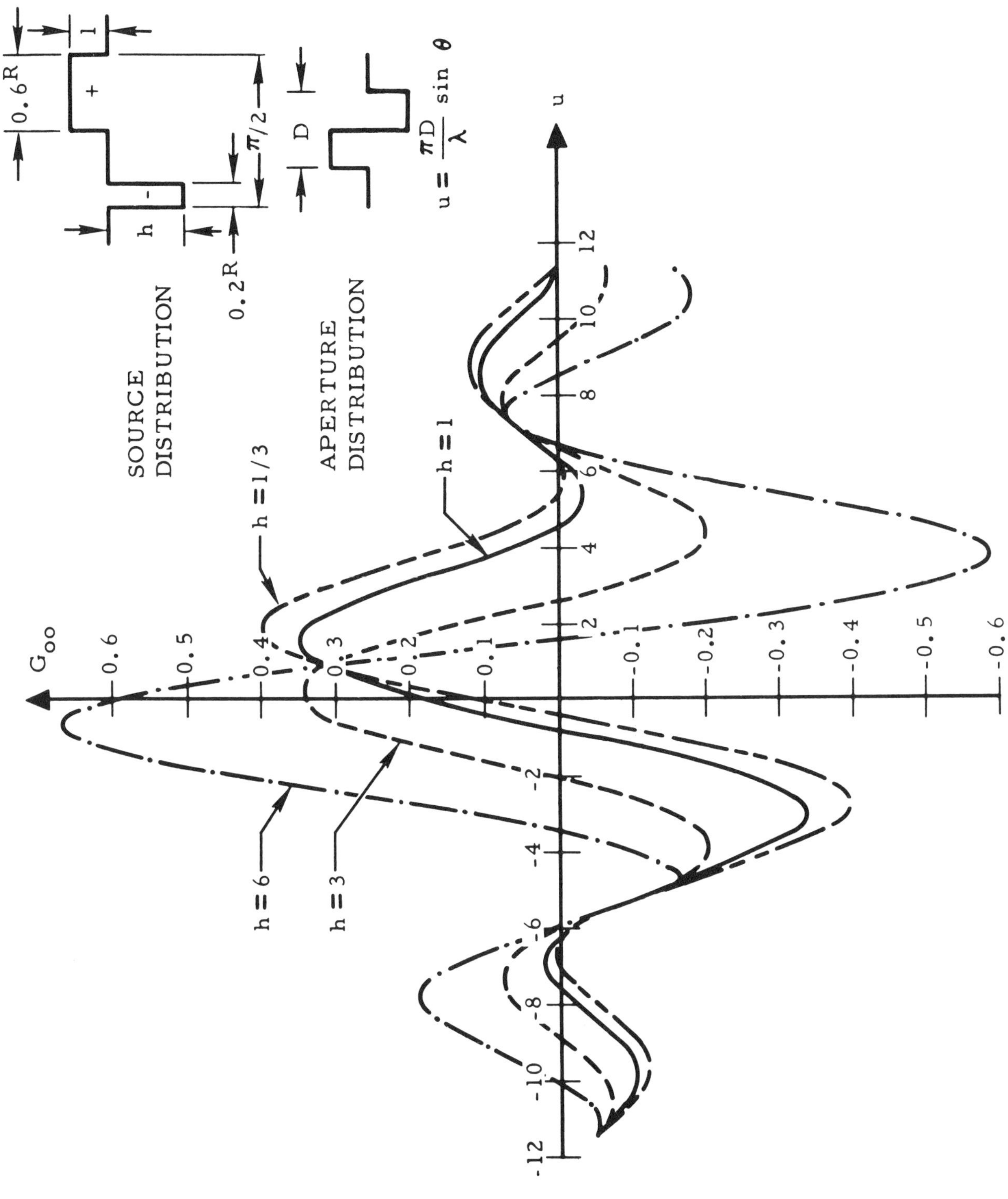

Fig. 12. The Response of an Odd Antenna Diffraction Pattern to an Asymmetrical Source Distribution of Two Closely Spaced Targets Out of Phase

even-real and odd-real source distributions correlate with target motion across the antenna lobes. The real-odd component changes sign when the target crosses the boresight. The width of these real components is proportional to the target position away from boresight.

When both the even and odd source distributions are combined in quadrature, the resulting distribution is complex with a crude progressive phase shift across the target. The odd distribution causes a target reflected diffraction pattern null to be alined with the inquiring antenna boresight. The even distribution in quadrature with the odd causes the null position to fill in by shifting its angular position. The response observed in the angle detector is $\mathrm{Re}\, F(u_c = 0) = 0$, while the response observed in the detector in quadrature with angle detector is given from the Van Cittert-Zernike theorem as

$$\mathrm{Im}\, F(u_c = 0) = \frac{1 - \dfrac{\sin w}{w}}{1 + \dfrac{\sin w}{w}} \simeq \frac{2\, \mathrm{Cin}(w)}{\mathrm{Si}(w)} \qquad (21)$$

where $w = \dfrac{\pi DL}{2\lambda R}$. The value of (21) as a function of w in terms of null depth is plotted in Fig. 14 for both linear and circular apertures. The quantity $\dfrac{\sin w}{w}$ is the measure of the degree of coherence according to the Van Cittert-Zernike Theorem in optics which is the visibility of the interference fringes. The degree of coherence between points is the absolute value of the normalized Fourier transform of the intensity function of the source. The monopulse antenna boresight null is partially filled in by the spatial incoherent echo components which is the imaginary odd source distribution. Note that in the region $w > 2.5$, the null depth measure of extent is ambiguous. The distribution $h_{oi}(u)$ is caused to fluctuate by target rotation, target internal motion, radar platform motion, and a heterogenous propagation path and so is uncorrelated with the $h_{er}(u)$ distribution. The product or phase detector instrumentation of $\mathrm{Im}\, F(u)$ has a bipolar output, the sign of which is insignificant and, therefore, the extent signal output may be squared.

The monopulse system response has been evaluated from (14) for the simplest of rectangular even and odd source distributions. It is interesting to consider a more complicated source distribution which is easily evaluated, such as two rectangular targets of differing reflection coefficients confined to a constant angular extent. Figures 10, 11, and 12 show the effects on $G_{ee}(u_c)$, $G_{oe}(u_c)$ and $G_{oo}(u_c)$ respectively for an asymmetrical source distribution of two closely spaced targets as indicated in the key. Note that the antenna does not resolve the individual targets; nevertheless, the monopulse system indicates the composite center weighted by the target comparative intensities. For this two-target example, the extent signal $G_{oo}(u)$ is very sensitive to the comparative target amplitudes as shown in Fig. 12 and 13, but for a particular antenna position near the composite weighted center, the extent detector output Im F (u) is independent of the two target comparative amplitudes.

The minimum target extent measure is controlled by the $\Sigma(u)$ signal-to-noise ratio, the amplitude and phase unbalance in the antenna transmission line confluence, and undesired couplings in the physical apertures and the hybrid junction labyrinth. The monopulse antenna boresight null depth as a function of amplitude and phase unbalance is shown in Fig. 15.

Target rotational motion may be either tumbling or spinning. Tumbling denotes that the rotational axis is normal to the radius vector, while spinning denotes that the target spins about the radius vector. Target rotation is one cause for $h_{oi}(u) = 0$. From a knowledge of λ, extent L, and the modulation frequency of L, the target rotational rate can be estimated. Target spinning will cause modulation of the azimuth and elevation extent and quadrupolar shape signal components, but the modulation phases will differ between the components. These modulation frequencies are due to $\frac{\lambda}{2}$ movement of the target extremities. Target shape, which has a large ratio of major to minor dimensions, also causes a modulation envelope to appear because of the variations in projected extent. The envelope phase in the azimuth and elevation channels is in quadrature, while the envelope in the quadrupolar channel is a double frequency. The quadrupolar port is sensitive to the spatial arrangement as shown $\begin{smallmatrix}+ & -\\ - & +\end{smallmatrix}$ or to the likeness of the hyperbolic paraboloid.

The field dependence of $G_{ee}(u)$, $G_{eo}(u)$, $G_{oe}(u)$, $G_{oo}(u)$, and the quadrupolar $G_{qq}(u)$ on the target range through the Fraunhofer into the Fresnel regions is illustrated in Fig. 16 for circular apertures. The response of an even diffraction pattern to an even source distribution has the familiar $\frac{1}{R}$ dependence in the Fraunhofer region. The response of an odd diffraction pattern to an odd source distribution has the less familiar $\frac{1}{R^2}$ dependence. The response of a quadrupolar source distribution has a $\frac{1}{R^4}$ dependence. Note the difference in periodicity of each response in the Fresnel region. The field dependence places a severe restriction on the useful target range for extent measurements except for target extents defined by the diffraction pattern principle lobe width.

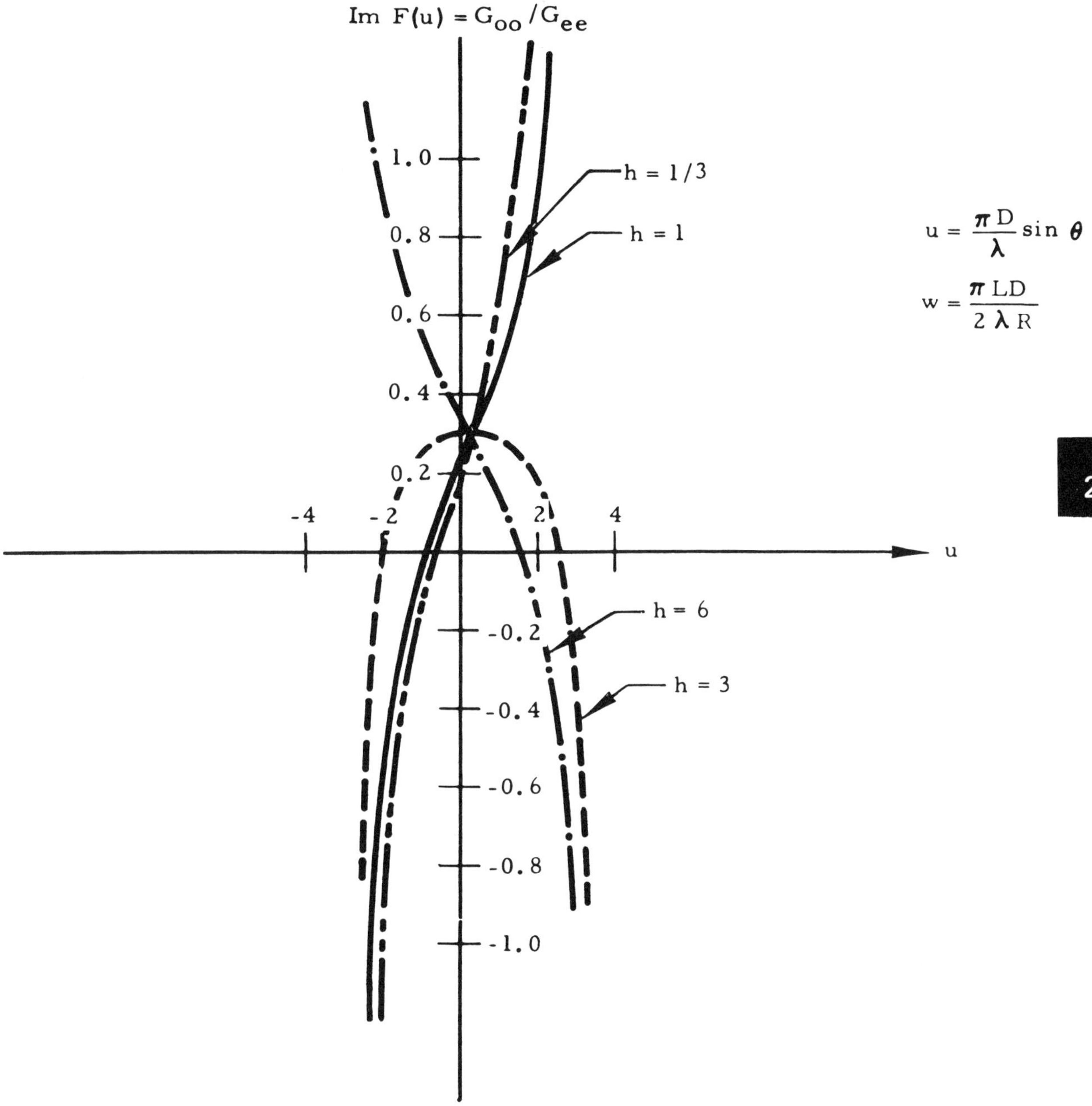

Fig. 13. The Angle Detection Function Response to an Asymmetrical
Source Distribution of Two Closely Spaced Targets In Phase

$$u = \frac{\pi D}{\lambda} \sin \theta$$

$$w = \frac{\pi LD}{2\lambda R}$$

Im $F(u) = G_{oo}/G_{ee}$

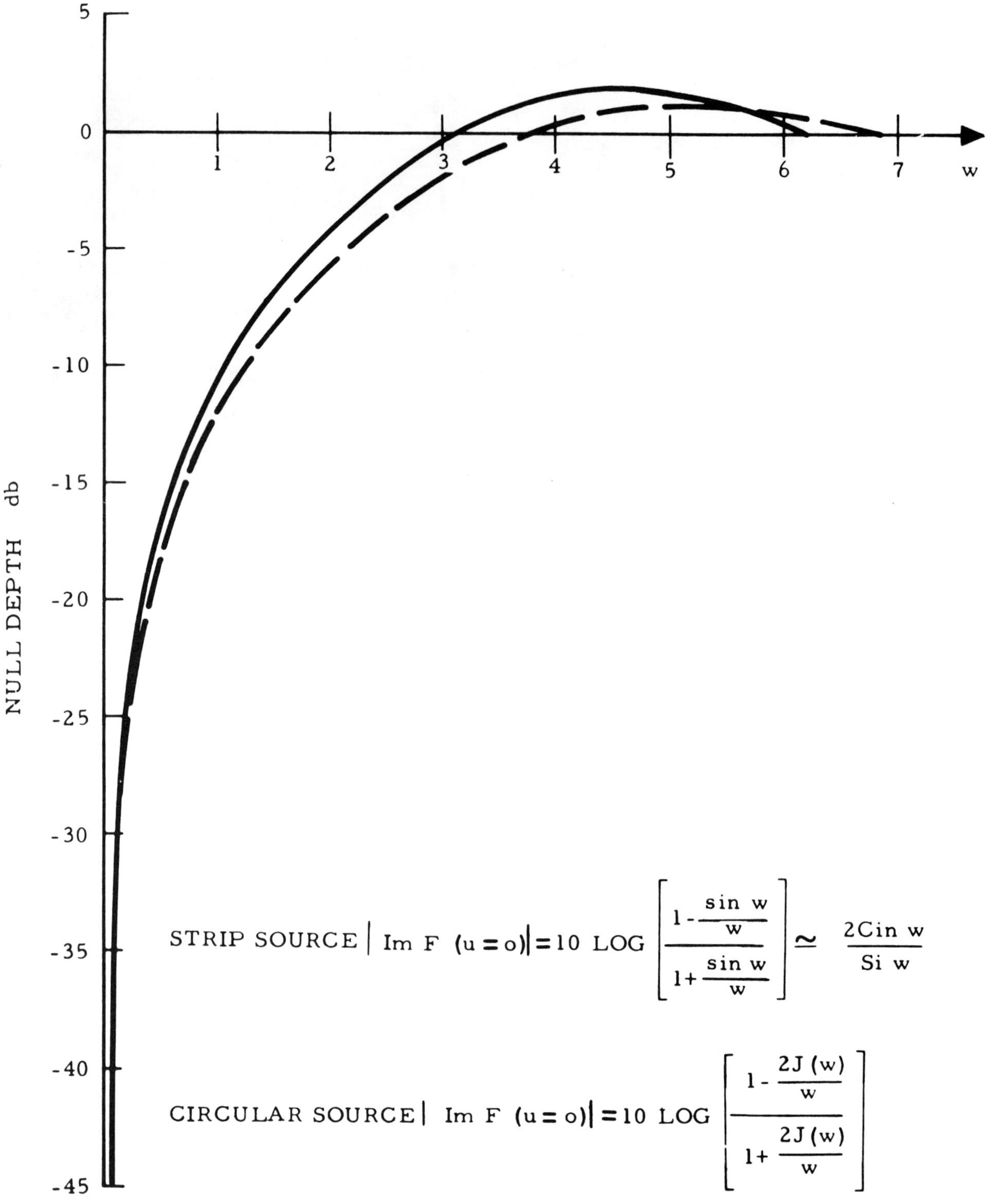

STRIP SOURCE $\left| \mathrm{Im}\ F\ (u = o) \right| = 10\ \mathrm{LOG} \left[\dfrac{1 - \dfrac{\sin w}{w}}{1 + \dfrac{\sin w}{w}} \right] \simeq \dfrac{2\mathrm{Cin}\ w}{\mathrm{Si}\ w}$

CIRCULAR SOURCE $\left| \mathrm{Im}\ F\ (u = o) \right| = 10\ \mathrm{LOG} \left[\dfrac{1 - \dfrac{2J\,(w)}{w}}{1 + \dfrac{2J\,(w)}{w}} \right]$

$$w = \frac{\pi LD}{2\lambda R}$$

Fig. 14. Monopulse Boresight Null Depth Dependence on Source Extent

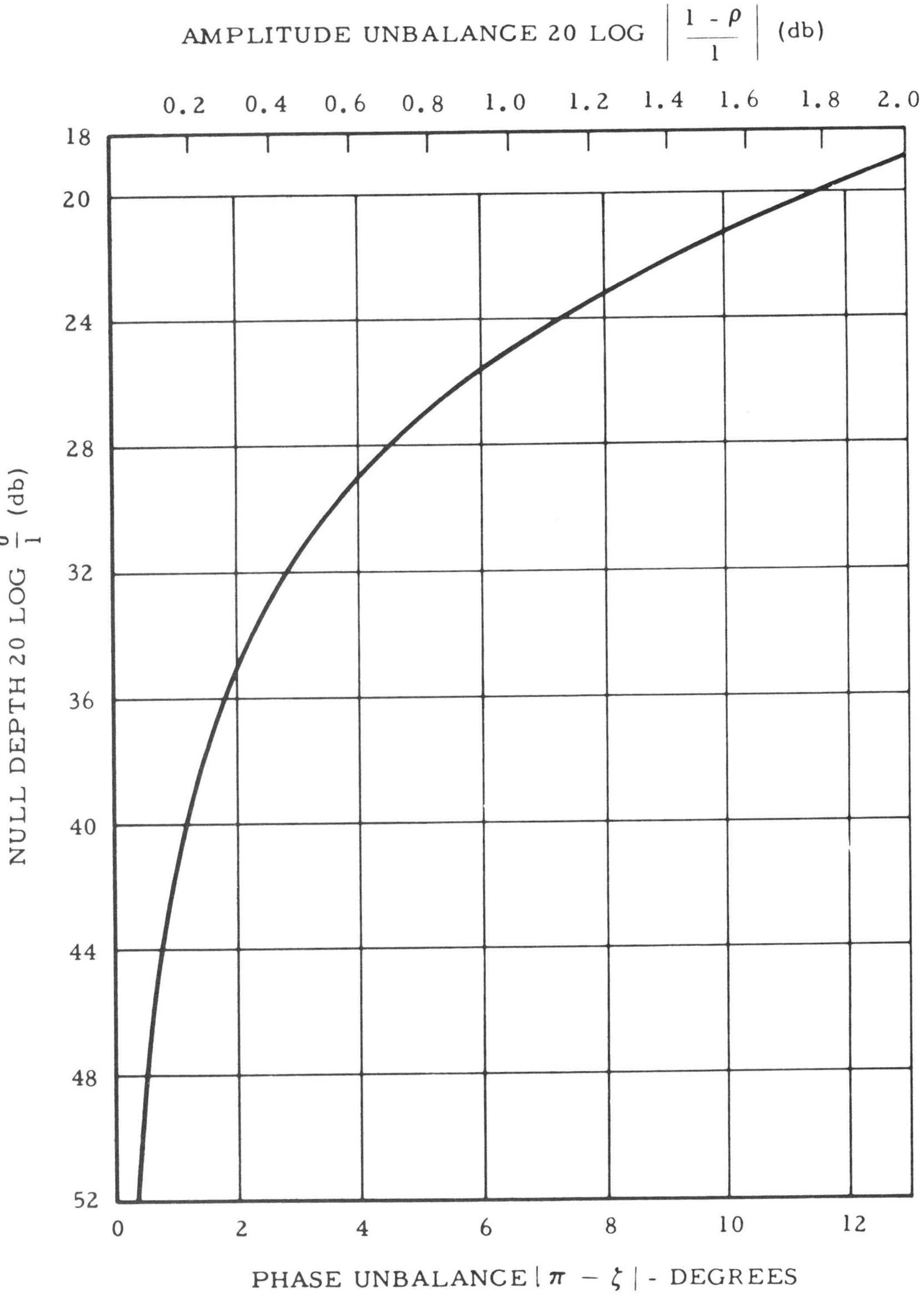

PLOT OF $\frac{1-\rho}{1+\rho} = \delta$ AND $\delta = \tan \frac{\zeta}{2}$

Fig. 15. The Monopulse Null Depth Dependence on the Amplitude and Phase Unbalance in the Monopulse Hybrid Junction Labyrinth

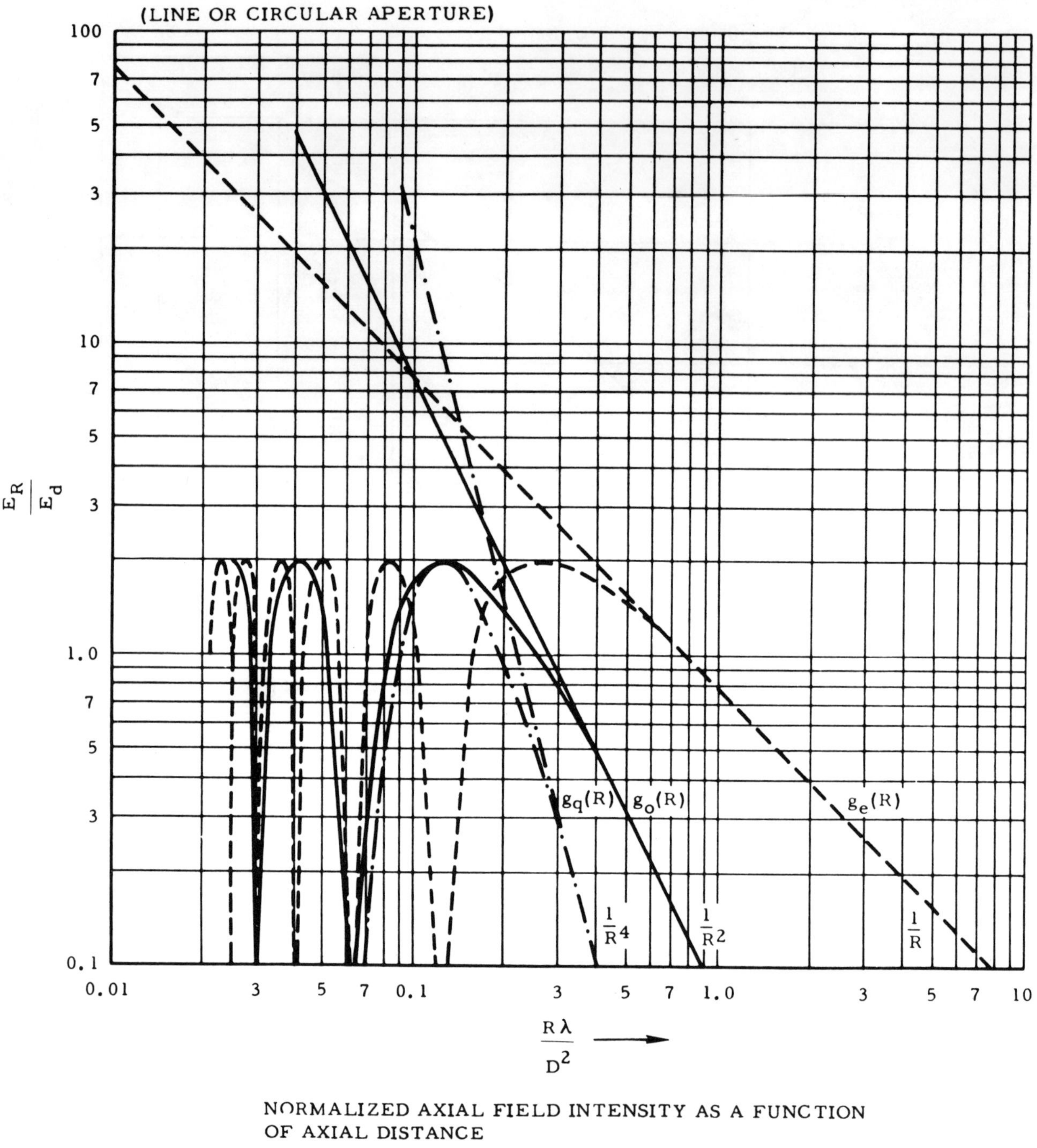

Fig. 16. Field Dependence as a Function of Range in the Transition Zone Between Fraunhofer and Fresnel Regions for Even, Odd, and Quadrupolar Aperture Functions

IV. EXPERIMENTAL RESULTS

The analytical results have been experimentally verified by modification in the elevation channel of a crude research model monopulse radar as suggested by the implemention in Fig. 3. The results are illustrated in a monopulse B-scan radar display of Fig. 17. The display subtends a range of 50 mi and 90 degrees azimuth and covers the sea, mountainous terrain, and cultural targets. The range detector output, elevation angle detector bipolar signals, and the elevation extent detector bipolar signals appear in separate crude photographs. A coherent ringing synthetic echo is included in the photographs. Note that one synthetic echo is coincident in range with natural clutter and that the lack of spatial correlation causes a prominent display in the detector in quadrature with the angle detector. The azimuth scan sweeps across the sector at grazing incidents, so elevated targets show as increased light intensity. Some natural extended targets, both black and white, are displayed in Fig. 17. The suppression of the synthetic echo point source and the amplifier recovery from intensity echoes is far from perfect. The physical mechanism is admittedly crass, but nevertheless validates the concept of extent measure by a monopulse radar.

A MONOPULSE B-SCAN RADAR MAP
AZIMUTH SCAN 90°
RANGE 50 MILES

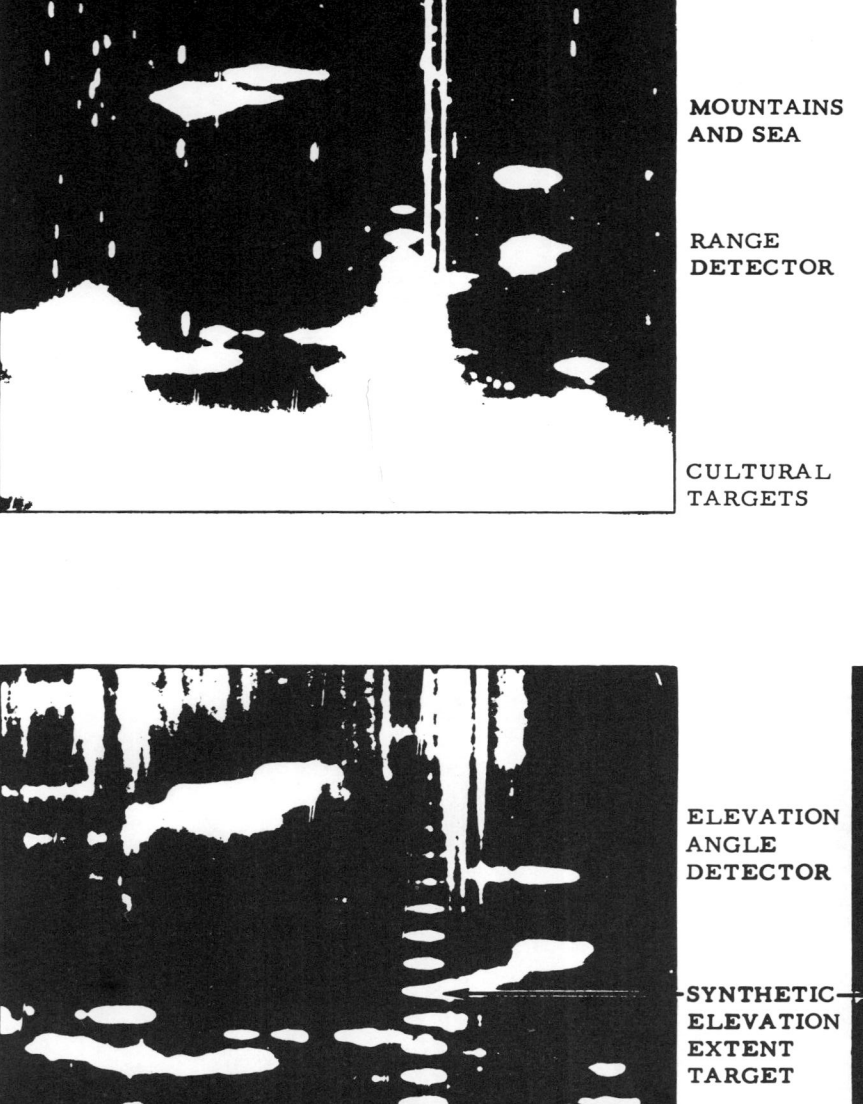

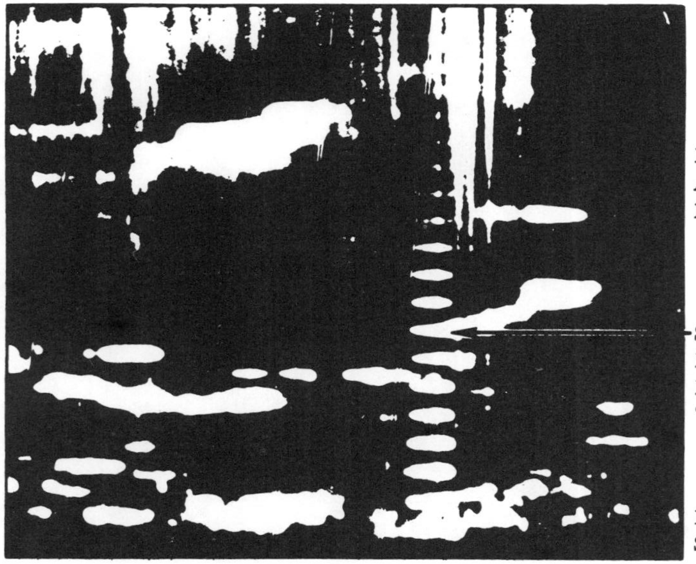

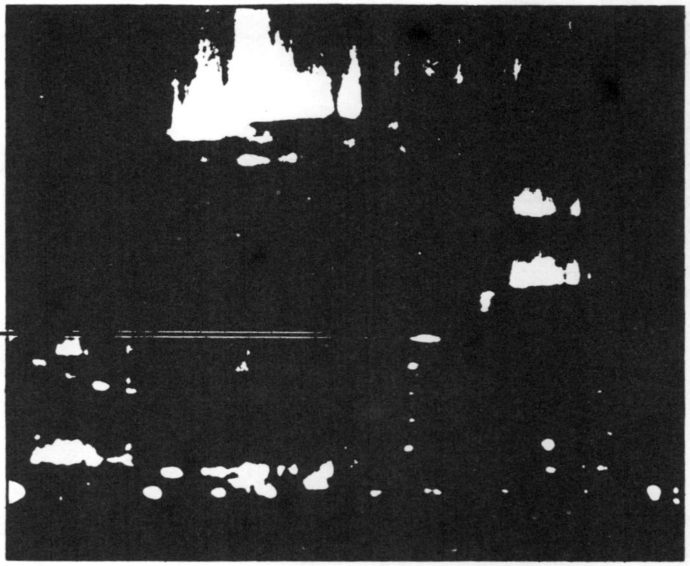

V. DISCUSSION

A conventional monopulse angle detection radar responds to the even-real and odd-real target source distributions to derive target angular position data. If a product or phase detector is incorporated into the radar in quadrature with the angle detector, it responds to the target extent.

The target extent data arise from an odd-imaginary target source distribution. This component of the target diffraction pattern has a null directed toward the inquiring radar antenna.

If the radar angle detector has any sensitivity to the odd-imaginary target component, it is recognized as target noise, glint, or scintillation which will contribute to tracking errors depending upon the magnitude and statistics. On the other hand, the same component in the extent and shape detectors yields useful information, but is somewhat impaired by the target distribution.

Modulation of the extent and shape signal components is interpreted as target tumbling or spinning. The measurement of the target non-unity aspect ratio and the major axis orientation is possible utilizing information available in the quadrupolar port. Thermal noise and monopulse components limit the maximum range, and ambiguous results in the Fresnel region restrict the minimum range for the implementation suggested.

The intent of this paper will have been fulfilled if it stimulates a more comprehensive application of the spatial structural aspects of the target echo signature data available in monopulse radar antenna. It is suggested that the techniques used by the prior art in interferometry can be translated into monopulse radar.

15.

Thermal-Noise Errors in Simultaneous-Lobing and Conical-Scan Angle-Tracking Systems*

JEAN A. DEVELET, JR.†, MEMBER, IRE

Summary—Relationships for rms angular errors are developed for certain common active space probe and satellite angle-tracking systems. The only source of error considered is the thermal and shot noise of the receiver, bandlimited by the tracking servo noise bandwidth. If additional smoothing after angular readout is performed, only the special case of many samples averaged over a time long compared to the reciprocal servo noise bandwidth is considered.

These thermal-noise errors are by no means the usual practical accuracy limitations of an angle-tracking system. They do, however, set bounds on minimal signal strength allowable for the desired tracking accuracy.

The received signals were assumed to be sinusoidal of constant peak amplitude with the information, if any, contained in phase or frequency modulation. This is the most common signal in space probe or satellite tracking.

I. Introduction

THIS PAPER presents for system designers an analysis of receiver noise errors in two common angle tracking systems, simultaneous lobing and conical scan. The work herein is oriented toward space probe and satellite tracking in which an active source is being tracked. The source is assumed to have only phase or frequency modulation imposed. The peak amplitude of the signal is assumed constant. In passive tracking, as in fire control and search radars, the relative importance of effects such as glint and scintillation usually outweighs the fundamental limitation of thermal noise.[1,2]

The typical approximations of square IF pass bands and white Gaussian noise are used throughout. In addition, the antenna beam patterns are assumed Gaussian in the regions of interest. Two types of detection which represent the extremes usually encountered are square-law (correlation in the simultaneous-lobing case) and coherent. The former is simpler in that a reference need not be constructed, while the latter can, under certain conditions, yield the best form of detection possible as regards maximum sensitivity. Square-law or correlation detection has another advantage for large index phase or frequency modulations where phase-lock or coherent detection may become inefficient due to lack of fixed spectral lines. In general, the performance of most detection schemes will require signal powers bounded by the square-law and coherent cases.

* Received by the PGSET, January 25, 1961; revised manuscript received, February 12, 1961.

† Space Technology Labs., Los Angeles, Calif.

[1] R. H. Delano, "A theory of target glint or angular scintillation in radar tracking," Proc. IRE, vol. 41, pp. 1778–1784; December, 1953.

[2] R. H. Delano, "The effect of AGC on radar tracking noise," Proc. IRE, vol. 44, pp. 801–810; June, 1956.

II. Principal Results

This section gathers together the important relationships developed throughout the text. This compact summary will allow use of the results of this paper without wading through a myriad of detail.

One assumption is made throughout to make the mathematics tenable and should be emphasized. The AGC (automatic gain control) in all systems is assumed to perform perfectly, regardless of signal level. If this is not so, a more detailed inspection of the particular system may be required. The relationships herein, however, can certainly aid initial system synthesis.

219

All results are given in terms of the tracking-loop noise bandwidth defined as follows:

$$B_N = \int_0^\infty \left| \frac{\theta_{bs}(j\omega)}{\theta_t(j\omega)} \right|^2 df \text{ cps}$$

where:

B_N = one-sided (video) servo noise bandwidth (cps)

$\frac{\theta_{bs}(j\omega)}{\theta_t(j\omega)}$ = transfer function of the closed loop servomechanism

θ_{bs} = boresight angle referred to some arbitrary zero reference

θ_t = target angle referred to the same reference as θ_{bs}.

This formulation allows general results without the added complication of considering the servomechanism details.

A. Simultaneous Lobing

See Figs. 1–8 (page 44).

1) *Phase Comparison*

a) *Coherent detection:*

$$\theta^2_{rms} = \left[\frac{\lambda}{\pi D \prod_{k=0}^{N} J_0(\phi_k)} \right]^2 \frac{\Phi B_N}{S} \text{ radians}^2. \quad (23)$$

b) *Correlation detection:*

$$\theta^2_{rms} = \left(\frac{\lambda}{\pi D}\right)^2 \left(1 + \frac{\Phi BW_{IF}}{S}\right)\left(\frac{\Phi B_N}{S}\right) \text{ radians}^2. \quad (24)$$

2) *Amplitude Comparison:*

a) *Coherent detection:*

$$\theta^2_{rms} = \left\{ \frac{5\psi^2}{2\, DB_c\beta^2 \ln 2 \ln 10 \left[\prod_{k=0}^{N} J_0(\phi_k) \right]^2} \right\} \left(\frac{\Phi B_N}{S}\right) \text{ radians}^2. \quad (25)$$

b) *Correlation detection*:

$$\theta^2_{rms} = \left(\frac{5\psi^2}{2\ DB_c\beta^2 \ln 2 \ln 10}\right)\left(1 + \frac{\Phi BW_{IF}}{S}\right)\left(\frac{\Phi B_N}{S}\right) \text{radians}^2. \quad (26)$$

3) *Symbol Definitions*:

θ^2_{rms} = mean-square error in (radians)^2 due to receiver noise in either azimuth *or* elevation (since the channels are uncorrelated, the total error is twice θ^2_{rms})

λ = wavelength of the received signal in units consistent with D

D = feed horn separation [see Fig. 2(b)]

Φ = $KT_0(F - 1)$

where

$KT_0 = 10^{-20.4}$ watts/cps

F = noise figure of the preamplifier system

B_N = servo noise bandwidth (cps)

S = total received signal power on boresight in units consistent with Φ

ϕ_k = the kth modulation index of the information phase or frequency modulation in radians [see (1) and (12)]

BW_{IF} = a single channel intermediate frequency amplifier noise bandwidth (cps)

ψ = *sum* pattern 3-db beamwidth (radians)

DB_c = crossover point expressed in decibels down from the maximum point of the *individual* beam pattern

β = ratio of sum pattern 3-db beamwidth to individual pattern 3-db beamwidth. This quantity is a function of DB_c and is plotted in Fig. 4.

B. Conical Scan

See Figs. 9–15.

1) *Coherent Detection*:

$$\theta^2_{rms} = \left\{\frac{5\alpha^2}{DB_c \ln 2 \ln 10\left[\prod_{k=0}^{N} J_0(\phi_k)\right]^2}\right\}\left(\frac{\Phi B_N}{S}\right) \text{radians}^2. \quad (38)$$

2) *Square-Law Detection*:

$$\theta^2_{rms} = \left(\frac{5\alpha^2}{DB_c \ln 2 \ln 10}\right)\left(1 + \frac{\Phi BW_{IF}}{2S}\right)\left(\frac{\Phi B_N}{S}\right) \text{radians}^2. \quad (39)$$

3) *Symbol Definitions*: All symbols are identical to the simultaneous-lobing case. However, in conical scan there is only one beam of 3-db-width α radians. DB_c is the number of decibels below maximum the beam crosses the boresight axis.

C. Effects of Additional Smoothing

If the raw angular readouts are further smoothed, the errors given by (23)–(26), (38) and (39) are reduced. For the special case of small target motion during the smoothing interval, two useful situations have been evaluated.

First, if many samples separated in time by much less than $1/B_N$ seconds are taken for a period of time $\tau \gg 1/B_N$ seconds, the following improvement results:

$$\theta^2_{srms} = \frac{\theta^2_{rms}}{2\ B_N\tau} \quad (40)$$

where:

θ^2_{srms} = smoothed mean-square angular error

θ^2_{rms} = raw mean-square angular error given by (23)-(26), (38) and (39)

τ = smoothing time in seconds (equals number of samples × time between samples).

Secondly, if N samples are taken, separated in time by much greater than $1/B_N$ seconds, the following improvement results:

$$\theta^2_{srms} = \frac{\theta^2_{rms}}{N}. \quad (41)$$

A more detailed treatment of additional smoothing is beyond the scope of this paper. In general the correlation function associated with θ is required for exact solution. This in turn requires a knowledge of the details of the specific servo transfer function.

III. Analysis

A. Simultaneous Lobing

1) *The Tracking Loop*: The block diagram of Fig. 1 will be the assumed model for the angle-tracking loop. Only one of the two identical orthogonal channels will be considered in the analysis. Since the noise in the azimuth and elevation channels is uncorrelated, the total mean-square error will be twice the azimuth or elevation mean-square error.

Referring to Fig. 1, it can be seen a closed servo loop exists which tends to drive the error $\theta_t - \theta_{bs}$ to zero. The receiver amplifies the signals from the antenna hybrid [see Section III-A, 2)] in identical sum and difference channels. A product detector then produces an error signal which drives a servomechanism designed for the proper dynamic tracking. The details of the servomechanism need not concern us here as will be seen later. However, one does need to know the servo noise bandwidth at the signal level of concern.

An AGC (automatic gain control) circuit normalizes signal levels to sum channel signal strength and maintains equal gain in both channels. Its purpose is to remove the servo system gain variations with signal strength. It will be assumed for this paper that the AGC performs this normalizing function perfectly to minimum usable signal levels.

2) *The Antenna*: The antenna gain constant relates the voltage output in the difference channel to the angular deviation of the target off boresight. There are two types of simultaneous-lobing systems which are of general

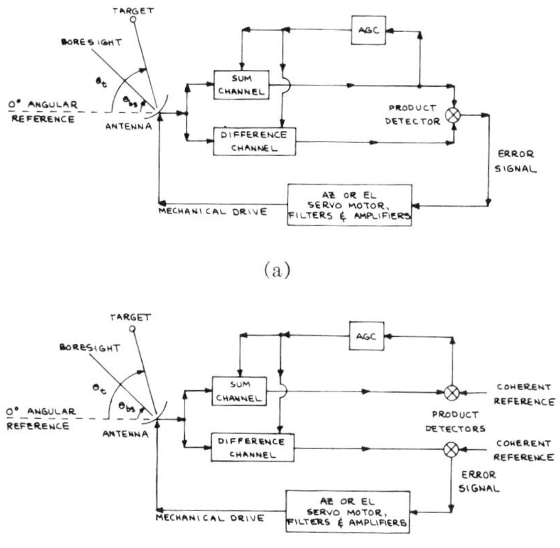

Fig. 1—Simplified block diagram of one channel of the tracking loop for simultaneous lobing. (a) Correlation detection. (b) Coherent detection.

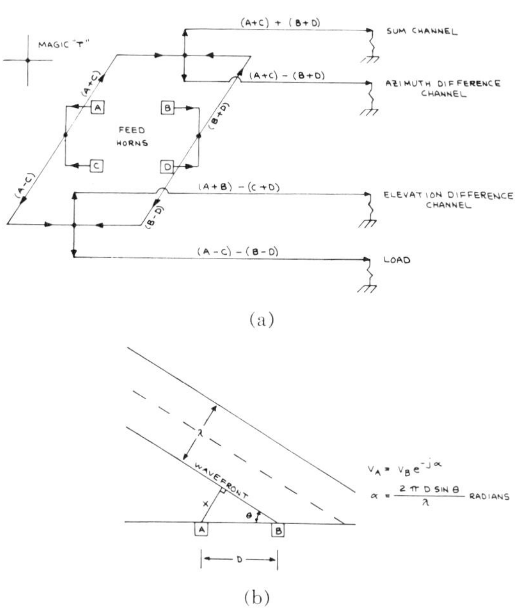

Fig. 2—Simultaneous-lobing antenna signal processing and phase-comparison wavefront geometry. (a) Received signal processing. (b) Wavefront geometry—azimuth or elevation channel.

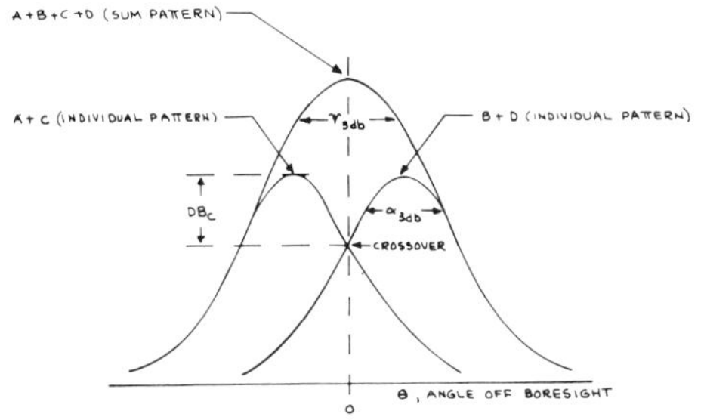

Fig. 3—Typical amplitude-comparison beam pattern.

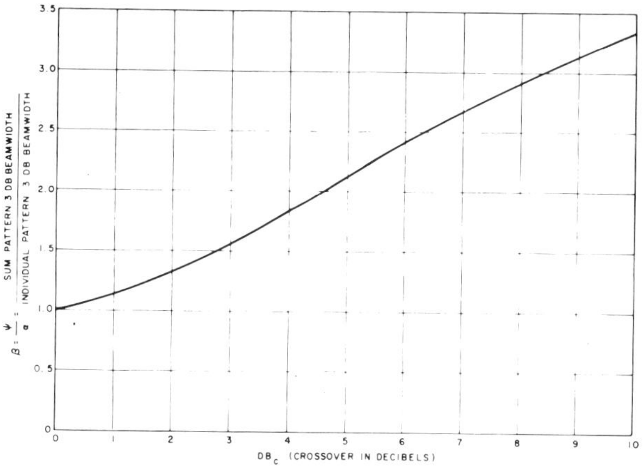

Fig. 4—Ratio of sum pattern beamwidth to individual pattern beamwidth for an amplitude-comparison simultaneous-lobing antenna system.

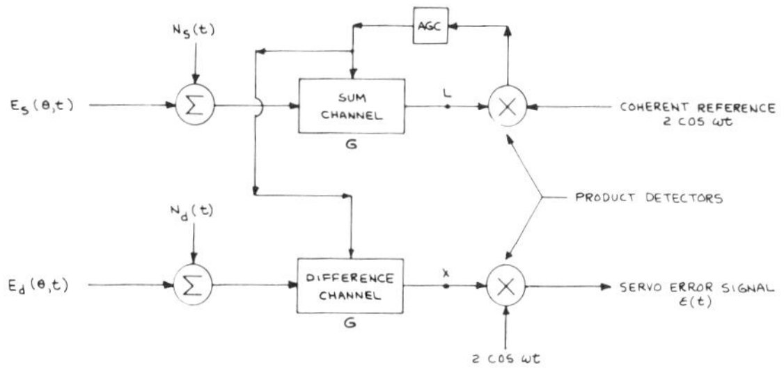

Fig. 5—Coherent-detection system, simultaneous-lobing tracker.

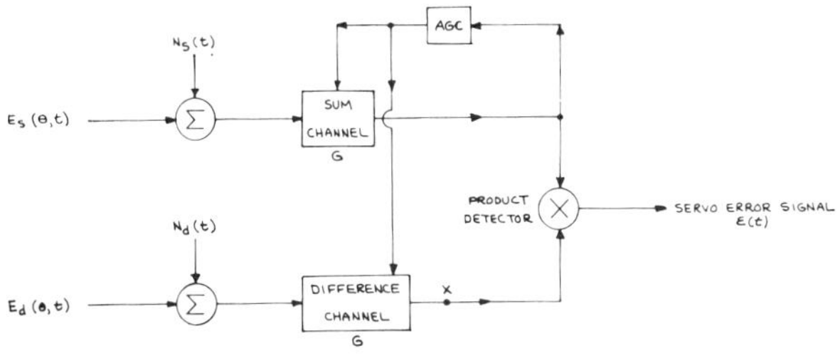

Fig. 6—Correlation-detection system, simultaneous-lobing tracker.

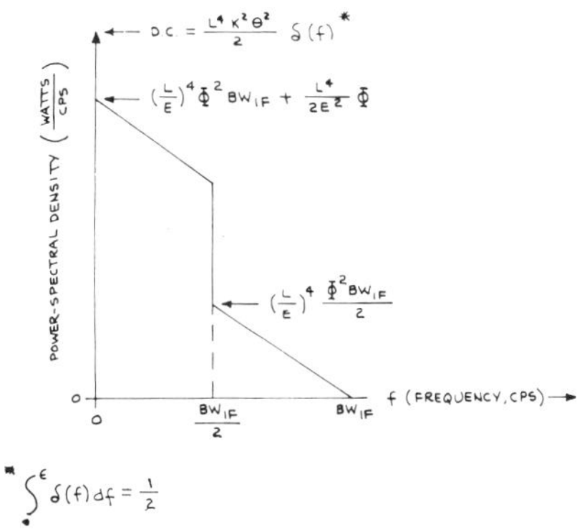

Fig. 7—Power-spectral density of the error voltage for correlation detection.

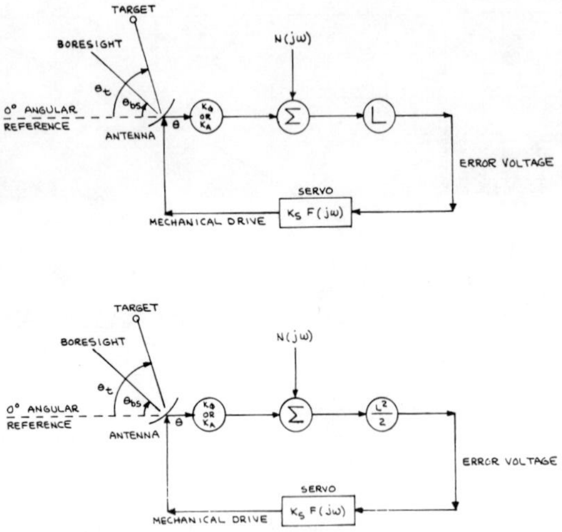

Fig. 8—Model tracking loops for (a) coherent and (b) correlation detection.

222

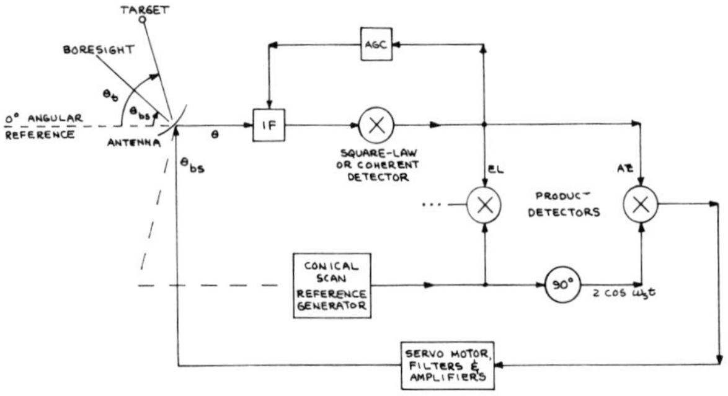

Fig. 9—Conical-scan angle-tracking system.

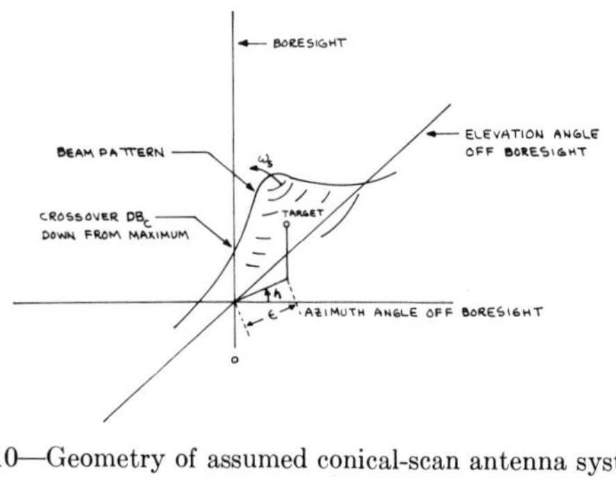

Fig. 10—Geometry of assumed conical-scan antenna system.

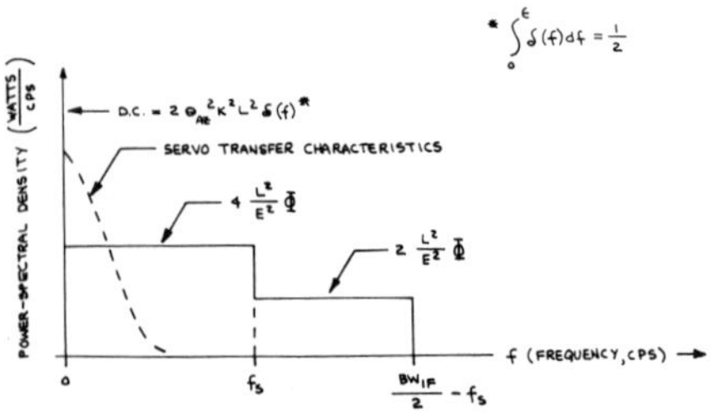

Fig. 11—Pertinent power-spectral density of error signal for coherent detection.

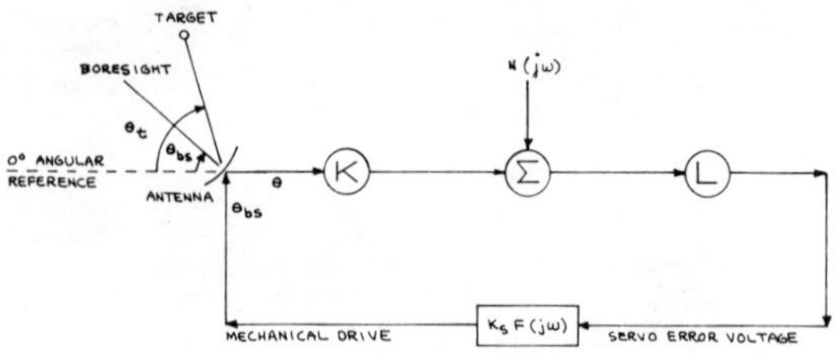

Fig. 12—Model conical-scan tracking loop, coherent detection.

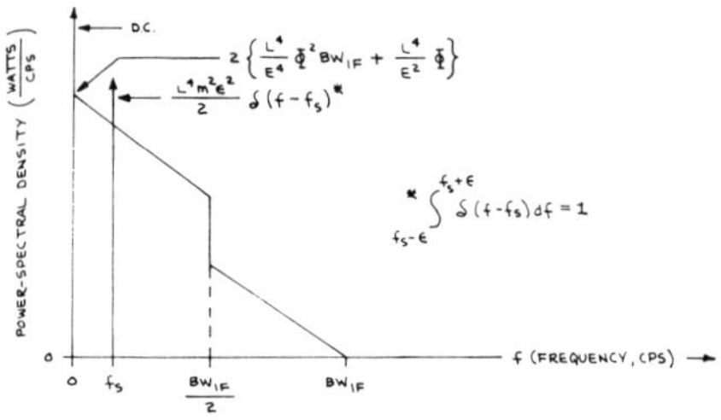

Fig. 13—Power-spectral density of the square-law detector output.

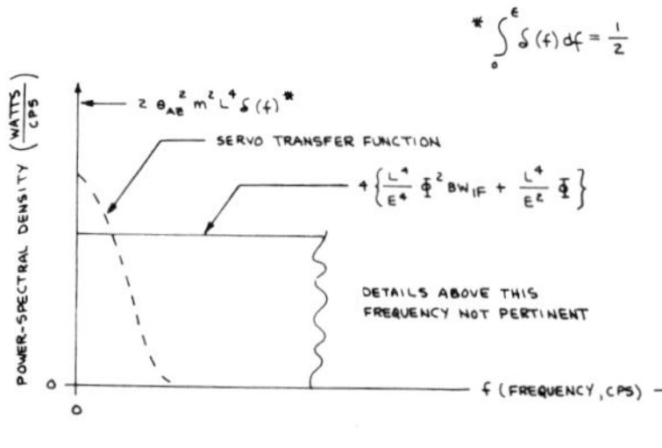

Fig. 14—Power-spectral density of the error signal for conical-scan square-law detection.

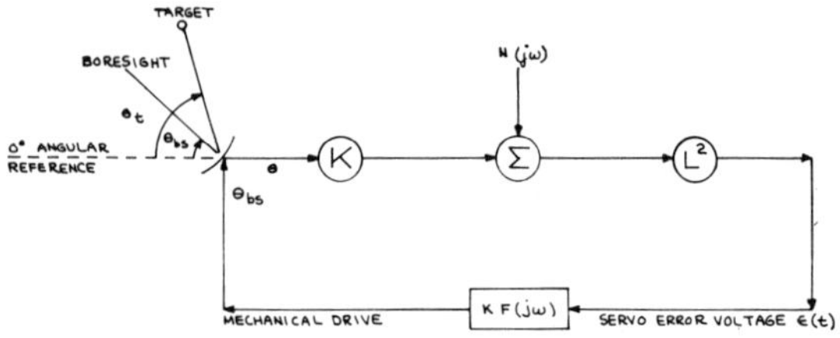

Fig. 15—Model conical-scan tracking loop, square-law detection.

interest. One is a phase-comparison system in which the phase difference between the voltage in two adjacent antenna horns generates an error signal. The second is an amplitude-comparison system in which the amplitude difference between two adjacent horn voltages generates an error signal.

a) *Phase comparison*: Fig. 2(a) depicts a simultaneous-lobing antenna feed-horn system. Fig. 2(b) shows the geometric relationships which determine the phase difference voltages in a pair of horns.

To a very good approximation, the voltage received in each horn is identical in magnitude. The reason is that in a phase-comparison system, the horns are spaced relatively far apart, and each has its own reflector producing identical parallel symmetric beams. Thus, the sum channel voltage becomes:

$$E_s(\theta, t) = Re\,\{Ee^{j[\omega t + \phi(t)]}\} \tag{1}$$

where:

Re = real part of [ ]

E = peak amplitude of the sum channel received signal (volts)

$\phi(t)$ = angle modulation containing telemetry or other information.

Future development will drop the Re [ ] notation and the ωt in the exponent, these quantities being understood.

The voltage in a difference channel can be obtained by reference to Fig. 2 and a simple phasor diagram.

$$E_d(\theta, t) \cong \frac{E}{2}\alpha(\theta)e^{j[\phi(t)+(\pi/2)]} \tag{2}$$

where:

$$\alpha = \frac{2\pi D \sin\theta}{\lambda} \text{ radians}$$

D = distance between feed horns

θ = wavefront angle off boresight

λ = wavelength.

Note that (2) is valid only for small α.

b) *Amplitude comparison*: In the amplitude comparison system, the horns are relatively close, usually in the center of a parabolic reflector. This results in equal phase existing in all horns, but due to the sharp focus of the parabola, different amplitudes will exist depending on the target angle off boresight. Fig. 3 shows a typical amplitude comparison beam pattern.

The calculations involving the channel outputs are more involved for this system and will be merely stated. Assuming that the beam shapes of Fig. 3 are Gaussian and identical, it may be shown that:

$$E_s(\theta, t) = \frac{E}{2}\exp\left(\frac{(DB_c)\ln 10}{20}\right) \cdot \left\{\exp\left(\frac{-2\ln 2}{\alpha^2}\left[\theta - \frac{\alpha}{2}\sqrt{\frac{\ln 10}{10\ln 2}(DB_c)}\right]^2\right) + \exp\left(\frac{-2\ln 2}{\alpha^2}\left[\theta + \frac{\alpha}{2}\sqrt{\frac{\ln 10}{10\ln 2}(DB_c)}\right]^2\right)\right\}\cdot e^{j\phi(t)} \tag{3}$$

$$E_d(\theta, t) \cong E\left(\frac{2\,DB_c \ln 2 \ln 10}{5}\right)^{1/2}\left(\frac{\theta}{\alpha}\right)e^{j\phi(t)} \tag{4}$$

where:

$E_s(\theta)$ = sum channel voltage as a function of angle off boresight (volts)

$E_d(\theta)$ = difference channel voltage either azimuth or elevation vs angle off boresight (volts)

θ = angle off boresight (radians)

α = 3-db beamwidth of an individual lobe (radians)

DB_c = number of db down crossover is from the peak of an individual lobe

E = voltage in the sum channel (2 × received power)$^{1/2}$ (volts)

$\phi(t)$ = angle modulation containing telemetry or other information.

Eq. (4) is valid for small angles off boresight.

The 3-db beamwidth ψ of the sum pattern is usually of interest. It is a function of α and DB_c and is obtained by solution of a transcendental equation. Fig. 4 graphs ψ/α vs DB_c for the usual range of values encountered. For DB_c greater than approximately 3 db, a dip occurs in the sum pattern on boresight. This may not be desirable in certain systems.

3) *Detection*: Having calculated the error voltages produced by the antenna, the next step is to determine the output error voltage applied to the antenna servo as a function of the antenna error signal and the additive receiver noise. There are two cases of major interest: coherent detection and correlation detection.

a) *Coherent detection*: The model assumed for the detection system is shown in Fig. 5. It will be assumed that the AGC in the sum channel maintains the amplitude of the output voltage due to E at a constant value L. Thus, the gain of each channel becomes

$$G = \frac{L}{E}. \tag{5}$$

The noise introduced in the sum and difference channels is assumed to originate as thermal and shot noise in the first stages of the IF strips. In addition, it is clear that $\overline{N_s(t)\,N_d(t)} = 0$ and $\overline{N_s(t)} = \overline{N_d(t)} = 0$. The one-sided power-spectral density of $N_s(t)$ and $N_d(t)$ will be assumed identical and equal to:

$$\Phi = KT_0(F - 1)\,\frac{\text{watts}}{\text{cps}} \tag{6}$$

223

where:

$K = 1.38 \times 10^{-23}$ joules/°K
$T_0 = 290°K$
F = noise figure of the amplifier measured at T_0.

The IF amplifiers being identical have a noise bandwidth BW_{IF}.

The voltage at point x on Fig. 5 now becomes:

Amplitude Comparison

$$E_x(\theta, t) \cong L\left(\frac{2\,DB_c \ln 2 \ln 10}{5}\right)^{1/2}\left(\frac{\theta}{\alpha}\right) \cdot \cos[\omega t + \phi(t)] + N_d(t)\left(\frac{L}{E}\right). \quad (7)$$

Phase Comparison

$$E_x(\theta, t) = L\left(\frac{\pi D \sin\theta}{\lambda}\right) \cdot \cos\left[\omega t + \phi(t) + \frac{\pi}{2}\right] + N_d(t)\left(\frac{L}{E}\right)$$

which for small θ

$$E_x(\theta, t) = L\left(\frac{\pi D\theta}{\lambda}\right) \cdot \cos\left[\omega t + \phi(t) + \frac{\pi}{2}\right] + N_d(t)\left(\frac{L}{E}\right). \quad (8)$$

Notice the 90° phase shift in (8) which does not appear in (7). This may be compensated for by a requisite shift in the reference. $N_d(t)$ can be represented by the usual in phase and quadrature notation as follows:

$$N_d(t) = X(t)\cos\omega t + Y(t)\sin\omega t \quad (9)$$

where:

$$\overline{X(t)^2} = \overline{Y(t)^2} = \Phi BW_{IF}.$$

Note that the one-sided power-spectral density of $X(t)$ or $Y(t)$ is 2Φ watts/cps.

Thus, it can be seen the output of the product detector for each situation becomes:

Amplitude Comparison

$$\epsilon(t) = L\left(\frac{2\,DB_c \ln 2 \ln 10}{5}\right)^{1/2}\left(\frac{\theta}{\alpha}\right)\overline{\cos\phi(t)} + X(t)\left(\frac{L}{E}\right). \quad (10)$$

Phase Comparison

$$\epsilon(t) = \frac{L\pi D\theta}{\lambda}\overline{\cos\phi(t)} + X(t)\left(\frac{L}{E}\right). \quad (11)$$

Cos $\phi(t)$ in (10) and (11) is averaged, because it is assumed that the servo bandwidth is much less than the modulating frequencies contained in $\phi(t)$. This is essentially saying that the only component which contributes to the error voltage is the carrier of the received signal, or more generally the spectral line to which the receiver is locked.

For the special and important case where $\phi(t)$ can be represented as the sum of sinusoids, we have:

$$\phi(t) = \sum_{k=0}^{N}\phi_k \sin(\omega_k t + \gamma_k). \quad (12)$$

If the ω_k in (12) are *not* commensurable (a good approximation for most modulations), expansion and use of identities yield:[3]

$$\overline{\cos[\phi(t)]} = \prod_{k=0}^{N} J_0(\phi_k). \quad (13)$$

Eqs. 10 and 11 allow the model of the tracking loop to be established in more quantitative form as will be shown in the following section.

b) *Correlation detection*: The model assumed for this detection system is shown in Fig. 6. Notice that no coherent reference need be derived in this instance.

The assumptions as to the AGC function will be identical to those made in the previous section. Therefore

$$G = \frac{L}{E}. \quad (14)$$

The identical assumption as to the character of the noise will also remain as before.

Restating prior results the voltage at point x is:

Amplitude Comparison

$$E_x(\theta, t) = L\left[\left(\frac{2\,DB_c \ln 2 \ln 10}{5}\right)^{1/2}\left(\frac{\theta}{\alpha}\right)\right] \cdot \cos[\omega t + \phi(t)] + N_d(t)\left(\frac{L}{E}\right). \quad (7)$$

Phase Comparison

$$E_x(\theta, t) = L\left(\frac{\pi D\theta}{\lambda}\right)\cos\left[\omega t + \phi(t) + \frac{\pi}{2}\right] + N_d(t)\left(\frac{L}{E}\right). \quad (8)$$

In the sum channel for small angles off boresight, we have:

$$E_s(\theta, t) = L\cos[\omega t + \phi(t)] + N_s(t)\left(\frac{L}{E}\right). \quad (15)$$

Note that here a 90° phase shift must be accomplished for phase comparison as before; however, over the wide band of the receiver signal. Performing this phase shift, and letting the bracketed portion of (7) and (8) equal $k_A\theta$ and $k_\phi\theta$, respectively, in order to prevent exhaustive rewriting, we obtain for the LF error voltage:

Amplitude Comparison

$$\epsilon(t) \cong \frac{L^2 k_A \theta}{2} + \frac{L^2}{E} N_d(t)\cos[\omega t + \phi(t)] + \left(\frac{L}{E}\right)^2 N_d(t) N_s(t). \quad (16)$$

[3] Internatl. Telephone and Telegraph Corp., "Reference Data for Radio Engineers," Stratford Press, Inc., New York, N. Y., 4th ed., p. 1065; 1957.

Phase Comparison

$$\epsilon(t) \cong \frac{L^2 k_\phi \theta}{2} + \frac{L^2}{E} N_d(t) \cos[\omega t + \phi(t)] + \left(\frac{L}{E}\right)^2 N_d(t) N_s(t). \tag{17}$$

From (16) and (17), it is a routine matter to find the correlation function $\overline{\epsilon(t)\,\epsilon(t+\tau)}$ and thereby the power spectral density of the error output. This power spectral density is shown in Fig. 7.

An assumption made in deriving the spectra was that the correlation time of $\phi(t)$ was much larger than the correlation time of the perturbing IF noise.

4) *RMS Noise Error*: The background generated until now will allow us to construct a more quantitative tracking loop than that of Fig. 1. Fig. 8 depicts the final tracking loops which were derived with the aid of (10), (11), (16) and (17).

The closed-loop response of the models in Fig. 8 to $N(j\omega)$ is easily derived and is stated below for each case:

Coherent Detection

$$\theta(j\omega) = \frac{N(j\omega)}{k}\left(\frac{kLK_sF(j\omega)}{1 + kLK_sF(j\omega)}\right). \tag{18}$$

Correlation Detection

$$\theta(j\omega) = \frac{N(j\omega)}{k}\left[\frac{k\frac{L^2}{2}K_sF(j\omega)}{1 + k\frac{L^2}{2}K_sF(j\omega)}\right]. \tag{19}$$

It can be seen that the term in parenthesis is merely the closed-loop transfer function $\theta_{bs}(j\omega)/\theta_t(j\omega)$ of the tracking system. Henceforth, this quantity will be referred to as $Z_t(j\omega)$. Its detailed structure need not be known for this analysis.

It is well known[4] that the mean-square error due to thermal noise may be found by the following relationship:

$$\theta_{rms}^2 = \int_0^\infty \lim T \to \infty \left(\frac{N(j\omega)N^*(j\omega)}{k^2 T}\right) |Z_t(j\omega)|^2 \, df. \tag{20}$$

Since $G(f)$ the

$$\lim T \to \infty \left(\frac{N(j\omega)N^*(j\omega)}{T}\right)$$

is merely the one-sided power spectral density of the receiver noise, and further since $Z_t(j\omega)$ has a cutoff frequency much less than BW_{IF} in the usual case, we may substitute the zero frequency spectral density $G(o)$ for $G(f)$. Thus, (20) becomes:

$$\theta_{rms}^2 = \frac{G(o)}{k^2}\int_0^\infty |Z_t(j\omega)|^2 \, df. \tag{21}$$

[4] J. L. Lawson and G. E. Uhlenbeck, "Threshold Signals," RDL Ser., McGraw-Hill Book Co., Inc., New York, N. Y., vol. 24, ch. 3; 1950.

However the integral in (21) is immediately recognized as the video (one-sided) noise bandwidth of the closed-loop tracking function

$$Z_t(j\omega) = \frac{\theta_{bs}(j\omega)}{\theta_t(j\omega)}.$$

Consequently:

$$\theta_{rms}^2 = \frac{G(o)B_N}{k^2} \tag{22}$$

where B_N = one-sided servo noise bandwidth. All that remains now is to substitute k and $G(o)$ in (22) for the various cases of interest. The results are the following:

a) *Phase comparison*:

Coherent Detection

$$\theta_{rms}^2 = \left[\frac{\lambda}{\pi D \cos\phi(t)}\right]^2 \frac{\Phi B_N}{S} \text{ radians}^2$$

$$= \left[\frac{\lambda}{\pi D \prod_{k=0}^{N} J_0(\phi_k)}\right]^2 \frac{\Phi B_N}{S} \text{ radians}^2. \tag{23}$$

Correlation Detection

$$\theta_{rms}^2 = \left(\frac{\lambda}{\pi D}\right)^2 \left(1 + \frac{\Phi BW_{IF}}{S}\right)\left(\frac{\Phi B_N}{S}\right) \text{ radians}^2. \tag{24}$$

b) *Amplitude comparison*:

Coherent Detection

$$\theta_{rms}^2 = \left\{\frac{5\psi^2}{2\,DB_e\beta^2 \ln 2 \ln 10\left[\prod_{k=0}^{N} J_0(\phi_k)\right]^2}\right\}\left(\frac{\Phi B_N}{S}\right) \text{ radians}^2. \tag{25}$$

Correlation Detection

$$\theta_{rms}^2 = \left(\frac{5\psi^2}{2\,DB_e\beta^2 \ln 2 \ln 10}\right)\left(1 + \frac{\Phi BW_{IF}}{S}\right)\left(\frac{\Phi B_N}{S}\right) \text{ radians}^2. \tag{26}$$

B. Conical Scan

1) *The Tracking Loop*: Fig. 9 depicts a typical conical-scan tracking loop. The two types of detection which bracket the performance of most conical-scan systems are coherent and square-law. The operation of the conical-scan angle-tracking system is similar in some ways to simultaneous lobing except for the method of deriving an error signal from the antenna. The antenna beam is revolved conically in space producing amplitude modulation on the received signal. The phase of this modulation and its amplitude are functions of the target offset from boresight and can be used to correct the antenna position.

2) *Derivation of the Antenna Gain Constant*: The antenna gain constant will be computed with similar assumptions as in the simultaneous lobing situation. That is, the beam

is Gaussian and has a crossover at DB_c. Fig. 10 shows the geometry of the situation.

The conical-scan radian frequency is ω_s. For a target ϵ radians off boresight and making an angle η with respect to the reference azimuth axis, it can be shown that the received signal becomes:

$$E_r(t) \cong E[1 + m\epsilon \cos(\omega_s t - \eta)] \cos[\omega t + \phi(t)] \qquad (27)$$

where:

E = peak amplitude of the signal voltage received with target on boresight (note this voltage is DB_c down from maximum)
m = slope of the antenna pattern at crossover in volts/volt-radian
ϵ = radians off boresight
ω_s = radian scan frequency rad/sec
η = angle target makes with reference azimuth axis (radians)
ω = carrier radian frequency rad/sec
$\phi(t)$ = information modulation radians.

The approximation in (27) occurs when m is assumed a constant. This is a good approximation for the small values of ϵ which occur in practice.

The quantity m may be derived from the functional relation for beam shape which is stated below.

$$E(\theta) = E \exp\left[\left(\frac{DB_c \ln 10}{20}\right) - \frac{2 \ln 2}{\alpha^2} \cdot \left(\theta - \frac{\alpha}{2}\sqrt{\frac{DB_c \ln 10}{10 \ln 2}}\right)^2\right] \qquad (28)$$

where:

E = signal voltage received with target on boresight (note this voltage is DB_c down from maximum)
α = 3-db beamwidth radians
θ = angle off boresight
DB_c = beam crossover in db.

Since $m = E'(\theta)/E$ evaluated at $\theta = 0$, we find from (28):

$$m = \frac{1}{\alpha}\left(\frac{2\ DB_c \ln 2 \ln 10}{5}\right)^{1/2}. \qquad (29)$$

The final signal voltage amplified by the IF amplifiers can now be established as:

$$E_r(t) \cong E\left[1 + \frac{\epsilon}{\alpha}\left(\frac{2\ DB_c \ln 2 \ln 10}{5}\right)^{1/2} \cos(\omega_s t - \eta)\right] \cdot \cos[\omega t + \phi(t)]. \qquad (30)$$

3) *Detection*: In the IF amplifiers, the receiver noise of power-spectral density Φ adds to the received signal given by (30). This resultant is then detected by either coherent or square-law techniques. As in simultaneous lobing, it will be assumed the AGC maintains constant signal voltage amplitude by modifying the predetection gain.

a) *Coherent detection*: In coherent detection it is assumed that a local reference ($2 \cos \omega t$) is generated by phase-lock techniques. The signal plus noise is then multiplied by this local reference to yield the detected output. If this is done and one assumes the same form for the noise as previously, the following output error voltage obtains:

$$\epsilon(t) = L\left[1 + \frac{\epsilon}{\alpha}\left(\frac{2\ DB_c \ln 2 \ln 10}{5}\right)^{1/2} \cos(\omega_s t - \eta)\right] \cdot \cos \phi(t) + \left(\frac{L}{E}\right) X(t) \qquad (31)$$

where:

L = AGC controlled limit level
$X(t)$ = noise voltage of power-spectral density 2Φ watts/cps.

Note that (31) contains a dc term $L\ \overline{\cos \phi(t)}$, a term of frequency equal to the conical scan, and a noise term $(L/E)X(t)$. The term at the conical-scan frequency will be further coherently detected at the scan frequency in separate orthogonal azimuth and elevation error channels. The references for this operation are obtained from the conical-scan reference generator (see Fig. 9).

If it is assumed the frequencies contained in $\cos \phi(t)$ are much higher than ω_s, the average value of $\cos \phi(t) = \overline{\cos \phi(t)}$ is all that concerns the narrow servo-bandwidth which follows detection. Thus the final error voltage becomes for the azimuth channel:

$$\epsilon(t) = \epsilon \cos \eta \frac{L\ \overline{\cos \phi(t)}}{\alpha}\left(\frac{2\ DB_c \ln 2 \ln 10}{5}\right)^{1/2} + \overbrace{\left(\frac{L}{E}\right) X(t) 2 \cos \omega_s t}^{LN(t)}. \qquad (32)$$

Note from Fig. 10 that $\epsilon \cos \eta$ is equal to the azimuth error of the antenna. Let this quantity be called θ_{az}. In addition, let k be equal to the following:

$$k = \frac{\overline{\cos \phi(t)}}{\alpha}\left(\frac{2\ DB_c \ln 2 \ln 10}{5}\right)^{1/2}, \qquad (33)$$

as before, if $\phi(t)$ is the sum of many sinusoids which are *not* commensurable. That is:

$$\phi(t) = \sum_{k=0}^{N} \phi_k \sin(\omega_k t + \gamma_k);$$

then it may be shown by expansion[3]

$$\overline{\cos \phi(t)} = \prod_{k=0}^{N} J_0(\phi_k).$$

The pertinent power-spectral density of the error signal given by (32) is shown in Fig. 11.

As in the simultaneous-lobing situation, (32) and (33) allow an accurate model to be established for the coherent-detection conical-scan tracking loop. Fig. 12 depicts this model loop.

b) *Square-law detection*: A square-law detector operates on the signal plus noise in the IF strip with the following law:

$$e_\phi(t) = [v(t)]^2. \tag{34}$$

Substituting for $v(t)$ the received signal plus noise, we have:

$$e_\phi(t) = \left\{L\left[1 + \frac{\epsilon}{\alpha}\left(\frac{2\ DB_c \ln 2 \ln 10}{5}\right)^{1/2} \cos(\omega_s t - \eta)\right] \cdot \cos[\omega t + \phi(t)] + \left(\frac{L}{E}\right)V_N\right\}^2 \tag{35}$$

where:

$V_N = X(t) \cos \omega t + Y(t) \sin \omega t$

L = limit level voltage amplitude.

If certain assumptions are made which are valid in most practical situations, the power-spectral density of $e_\phi(t)$ given by (35) is readily computed. Pertinent LF components are depicted in Fig. 13. The assumptions utilized in computing this power-spectral density are the following:

1) $\omega_c >\!\!>\!\!> \omega_s$ and $\omega_s <\!\!<\!\!< BW_{IF}$.

2) $\frac{\epsilon}{\alpha}\left(\frac{2\ DB_c \ln 2 \ln 10}{5}\right)^{1/2} <\!\!<\!\!< 1.$

3) $[\phi(t + \tau) - \phi(t)] \approx 0$ for those values of τ for which $R(\tau) = \overline{X(t + \tau)\, X(t)} = \overline{Y(t + \tau)\, Y(t)}$ has appreciable value.

4) Twice the conical-scan frequency is rejected by the filtering.

The third assumption is the most questionable, but for channels of reasonable bandwidth for good fidelity of telemetry reception, it is valid.

Since $\omega_s <\!\!<\!\!< BW_{IF}$ in the following, it is not worth taking into account the small reduction in noise density about f_s as compared to zero frequency.

The square-law detector is followed by a coherent detector utilizing the conical-scan reference generator (see Fig. 9). The resulting voltage becomes for the azimuth channel:

$$\epsilon(t) = 2 \cos \omega_s t \cdot e_\phi(t)$$

$$= \frac{L^2}{\alpha} \cdot \left(\frac{2\ DB_c \ln 2 \ln 10}{5}\right)^{1/2} \epsilon \cos \eta + L^2 N(t). \tag{36}$$

The LF power-spectral density of this voltage is easily determined from Fig. 13 and is shown in Fig. 14. Note, as before, that $\epsilon \cos \eta = \theta_{az}$.

As in previous situations, (36) and Fig. 14 may be combined to form a model tracking loop. Fig. 15 depicts such a loop for square-law detection.

The constant k in this instance becomes:

$$k = \frac{1}{\alpha}\left(\frac{2\ DB_c \ln 2 \ln 10}{5}\right)^{1/2}.$$

4) *RMS Noise Error*: Since the model tracking loops developed here are identical in form to those developed for simultaneous lobing, the formulation for rms angular error will merely be restated below:

$$\theta^2_{rms} = \frac{G(o)B_N}{k^2} \tag{22}$$

where, as before:

$G(o)$ = zero-frequency power-spectral density of the noise.

B_N = one-sided servo noise bandwidth.

k as defined in Sections III-B, 3)-a) and III-B, 3)-b).

Substitution of the power-spectral densities and k's developed in Sections III-B, 3)-a) and III-B, 3)-b) in (22), we finally obtain the mean-square error relationships for conical scan:

Coherent Detection

$$\theta^2_{az} = \left\{\frac{5\alpha^2}{DB_c \ln 2 \ln 10\left[\prod_{k=0}^{N} J_0(\phi_k)\right]^2}\right\} \frac{\Phi B_N}{S} \text{ radians}^2. \tag{38}$$

Square-Law Detection

$$\theta^2_{az} = \left(\frac{5\alpha^2}{DB_c \ln 2 \ln 10}\right)\left(1 + \frac{\Phi BW_{IF}}{2S}\right)\frac{\Phi B_N}{S} \text{ radians}^2. \tag{39}$$

C. *Additional Smoothing*

Quite often, the angular readouts from an angle-tracking system are further smoothed. In general, if additional smoothing is accomplished it is necessary to know the correlation function of the angle noise to find an exact solution.

For the special case of straight averaging as the smoothing technique, two extremes will be considered.

First, consider the case in which many samples are taken separated in time by much less than the reciprocal noise bandwidth of the servo-loop. In addition, the number of samples taken are assumed to stretch over a time long compared to the reciprocal noise bandwidth of the servo-loop. In this instance, the resultant mean-square angular error can be shown to be:[5]

$$\theta^2_{saz} = \frac{\theta^2_{az}}{2B_N\tau} \tag{40}$$

[5] J. A. Develet, Jr., "Noise on AGS Subcarriers," Space Technology Labs., Inc., Los Angeles, Calif., unpublished STL Internal. Memo. 7250.4-2; October 25, 1960.

where:

θ^2_{saz} = smoothed resultant angle error (radian^2)

θ^2_{az} = raw mean-square angle error as given by previous developments (radian^2)

B_N = one-sided servo noise bandwidth (cps)

τ = smoothing time (seconds).

The second situation of interest is the case where N independent samples of angular data are taken and averaged. To insure independence, the time separation should be much greater than the reciprocal servo noise bandwidth. In this instance the smoothed mean-square angle noise becomes:

$$\theta^2_{saz} = \frac{\theta^2_{az}}{N} \text{ radians}^2. \tag{41}$$

Eqs. (40) and (41) assume the target does not move appreciably over the additional smoothing interval.

More complex additional smoothing than the above special cases is felt beyond the scope of this paper.

IV. Acknowledgment

The author wishes to acknowledge valuable suggestions and reference material offered by S. Evans and D. D. Pidhayny of Aerospace Corporation, and G. Kasai of Space Technology Laboratories.

J. A. Develet, Jr.

Jean A. Develet, Jr. (S '51—A '55—M '59) was born in St. Albans, N. Y., on February 5, 1933. He received the B.E.E. and M.E.E. degrees in 1953 and 1954, respectively, from the Polytechnic Institute of Brooklyn, N. Y.

While a graduate student at Polytechnic he was a junior research Fellow engaged in pulse-echo fault location studies for the Signal Corps. He entered the U. S. Army in 1954 and served with the Corps of Engineers at Fort Belvoir, Va., as Chief of the Utilities Section, The Engineer Test Unit. There, he was responsible for the planning and reporting of field tests on such diverse items as liquid oxygen generators for guided missile support, electrical power generators of 3-kw to 5000-kw capacity, fire-fighting equipment, and water purification plants for field armies.

In 1956 he joined the technical staff of Ramo-Wooldridge Corporation, Los Angeles, Calif., where he was employed in multimegawatt modulator construction, video receiver design, high-power millimicrosecond pulse circuitry, and traveling-wave-tube and microwave measurements, in addition to power supply design for traveling-wave tubes. He became a member of the technical staff of Space Technology Laboratories, Los Angeles, in 1959, working in system analysis pertaining to space communications, tracking, and telemetry; he also worked on the ground station evaluation studies for the Pioneer V space probe experiment, remote tracking station time synchronization, and atmospheric and ionosperic refraction, as well as communication system synthesis and modulation techniques.

Mr Develet is a member of Sigma Xi, RESA, Tau Beta Pi, and Eta Kappa Nu.

Optimum Feeds for All Three Modes of a Monopulse Antenna I: Theory*

PETER W. HANNAN†, SENIOR MEMBER, IRE

Summary—**In a monopulse antenna for use in a tracking radar, the requirements for the sum and difference modes are not the same. For the amplitude-comparison type of monopulse having an antenna whose main aperture is illuminated by a feed, these independent requirements can be met by a feed designed to fulfill two conditions. One is that excitation of the feed aperture in the difference modes be effectively about twice as wide as in the sum mode. The other is that the shapes of all the feed excitations be free of any avoidable irregularities.**

The amount of improvement available when these conditions are met depends on the optimization point desired for each mode, as well as the design of the reference antenna. In a typical case, the gains and slopes in the difference modes should increase by several db. Furthermore, the near sidelobes and the spillover radiation in the difference modes should decrease by an order of magnitude. A small increase of gain in the sum mode is also available, and finally, the criticalness of positioning the feed may be appreciably reduced.

I. Introduction

IN THE FIELD of tracking radars, the need for antennas of increased angular accuracy has led from sequential lobing and conical scan to simultaneous lobing or monopulse. As the requirements become even more demanding, it becomes necessary to incorporate advances in the design of monopulse antennas to more fully realize their potentialities. One such advance is possible in the amplitude-monopulse type of antenna consisting of a reflector or lens illuminated by a feed.

For an ordinary single-mode antenna, the optimum design of a simple feed for maximum antenna gain is well known [1], [2]. In the case of a monopulse antenna, there are usually a sum and two difference modes, and it has not been customary to design a feed for simultaneous optimum performance in all three of these modes. The particular compromise that is made may depend on the system requirements, and the relative importance of the various modes. One choice often made involves a "crossover level" which maximizes a certain product of sum gain and difference slope [3], [4]; in this case, neither of these quantities achieves its optimum value. In addition, some of the other antenna characteristics, such as sidelobe levels, spillover radiations, and criticalness to misalignment, are considerably poorer than they need be with an optimum design in every mode.

It is the purpose of this paper to discuss in detail the limitations involved in the usual compromise design, and to describe the degree of improvement available if each mode can be optimized independently. In a companion paper, means to achieve this independent optimization in all three modes will be presented.

* Received by the PGAP, November 22, 1960; revised manuscript received, March 27, 1961. The material in this paper is the result of antenna development work for the Bell Telephone Laboratories on Army Ordnance Corps projects. Preparation of this paper has been supported mainly by a subcontract with BTL in connection with prime contract No. DA-30-069-ORD-1955.

† Wheeler Labs., Inc., Smithtown, N. Y.

II. Description of Monopulse Antenna

It is assumed that the reader has some familiarity with monopulse [5], [6]; however, it is desirable to describe the basic monopulse antenna system which is considered in this paper. This is shown in Fig. 1, together with the terminology to be used. The essential components are a comparator, a feed, and a focusing element.

There are three regions of the antenna system which are of special interest. When signals are provided to the feed by the comparator, there exists a field across the aperture of the feed which is herein termed the *feed excitation*. This, in turn, creates a field across the aperture of the focusing element which is called the *main aperture illumination*. The resulting field at a large distance from the antenna may be described by its distribution as a function of angle; this is the *antenna pattern*.

A. Modes of Operation

In describing the operation of a monopulse antenna, there are two different concepts which are commonly employed. One concept considers the signals which may exist within the comparator that would excite different portions of the feed aperture and create antenna patterns consisting of simple lobes displaced from the antenna axis. This may be called the *lobe* concept; it stems from sequential-lobing and conical-scanning antennas. The lobe concept is helpful in understanding the properties of a monopulse antenna, and for the ordinary monopulse antenna it is often employed as the principal means of analysis. However, it does not provide a convenient basis for the optimization discussed in this paper. In addition, the lobe signals represent an intermediate stage in the system, and do not have as direct an operational significance as may be desired.

The concept which is employed in this paper considers the signals existing in the three channels at the input of the comparator. Since these three channels cor-

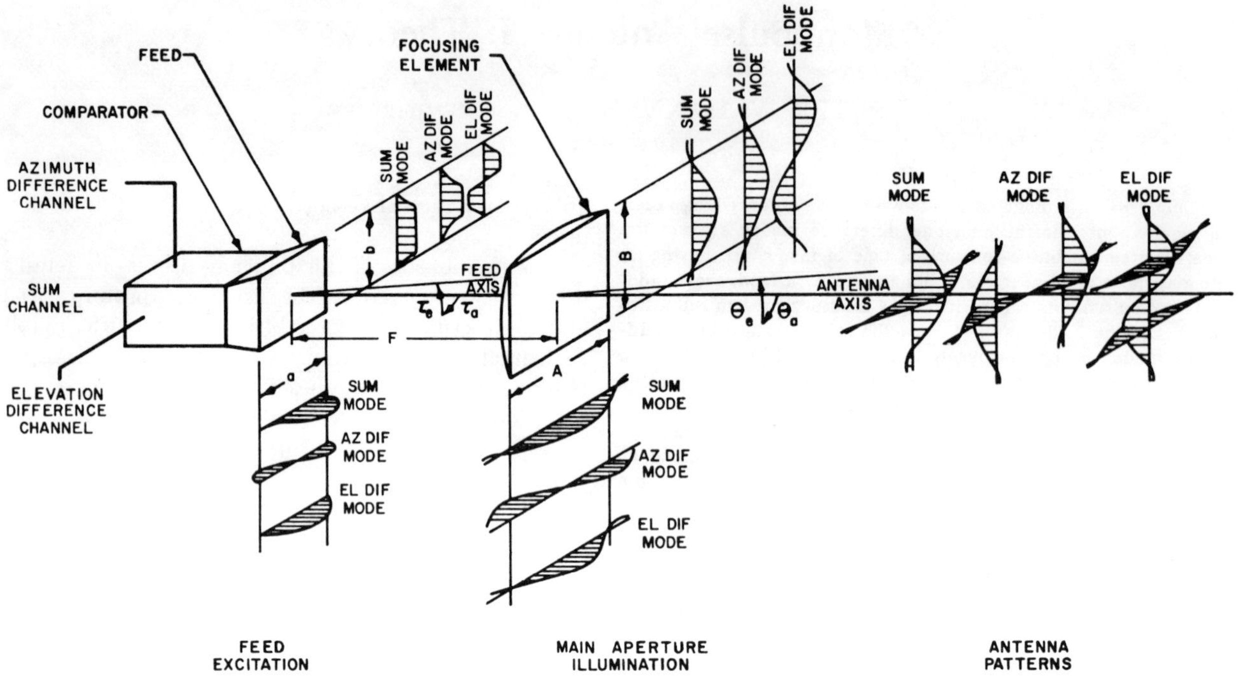

Fig. 1—Monopulse antenna system.

respond to three distinct modes of operation, this viewpoint may be referred to as the *mode* concept. These modes are called the *sum mode*, the *azimuth difference mode*, and the *elevation difference mode*, in accordance with the type of action occurring within the comparator during reception.

When coupled to the transmitter of a radar system, the sum mode provides illumination of the distant target; when coupled to a receiver it provides range information and a reference signal. The azimuth and elevation difference modes are coupled to receivers whose signals, when combined with the reference sum signal, provide azimuth and elevation angle information, respectively. Although only the sum mode actually exists in transmission, it is common practice to consider all three modes in transmission for ease of analysis; by reciprocity the antenna patterns are the same whether obtained in transmission or reception. Shown in Fig. 1 are a sample set of field distributions during transmission for the sum, azimuth, and elevation modes, in each of the three regions of interest.

B. Assumptions and Limitations

A number of assumptions are made in this paper which simplify the calculations and presentation. As shown in Fig. 1, the feed and main apertures are rectangular, and each field distribution is assumed to be separable into the product of a horizontal and a vertical distribution. In the sum mode, both the horizontal and the vertical distributions are of the even type. In each of the difference modes, the distribution in one direction is of the odd type while in the other direction it is of the even type. The antenna is, of course, assumed to be free of any dissipation losses. It is also assumed that both the feed aperture and the main aperture are large compared with a wavelength, and that the feed is much smaller than the main aperture. As a result of all these assumptions, both the main illumination and the antenna patterns are relatively easy to calculate, and can be presented in a single dimensionless form. Many practical antennas do not correspond exactly to the assumptions made, and in these cases the numerical results obtained herein are applicable only as approximations; however, the basic concepts and trends are generally valid.

The analysis presented in this paper is specifically developed for the *amplitude-comparison* type of monopulse, as typified by the antenna shown in Fig. 1. A similar approach could be applied to the phase-comparison type [7]. However, in a phase-comparison antenna of the kind involving a feed and a focusing element, improvements analogous to those suggested in this paper are difficult to implement, and are beyond the scope of this presentation.

III. Need for Different Widths of Feed Excitation in Sum and Difference Modes

One of the requirements for optimizing a monopulse feed in all three modes is that the width of feed excitation be different in the sum and difference modes. This can be shown in several ways, each furnishing a different insight into the basic processes involved in obtaining good antenna performance. In each case, the sum and one difference mode will be investigated; the other difference mode would have the same requirement as the one considered.

A. Focal Field

The first viewpoint is shown in Fig. 2, which illustrates the region around a feed during reception. When a plane wave is incident on the focusing element, it is focused to a compact area [8] which allows a relatively small feed to capture most of the available power. It can be seen intuitively (as well as proven rigorously) that for maximum signal, the size of the feed aperture should be about equal to the size of the main lobe of the focused wave, when the power in the sidelobes of the focused wave is intentionally wasted in the interests of a practical design. For the sum mode, indicated in Fig. 2(a), this situation corresponds to the standard design of a simple single-mode antenna for maximum gain.

In the difference mode, the antenna pattern has two main peaks of opposite polarity which are displaced equal amounts off the antenna axis. When a wave is incident on the antenna from the direction of one peak, it is focused to an area off the center of the feed, as indicated in Fig. 2(b); a similar area on the opposite side would correspond to a wave incident from the direction of the other peak. To obtain the proper signal from each wave, the feed aperture may be divided into two halves, connected to the symmetrical arms of a hybrid junction. It can be seen that for maximum signal at each peak, each half aperture should be about equal to the size of the main lobe of one focused wave, just as it is in the case of the sum mode. Thus the over-all feed-aperture dimension in the odd plane of the difference mode should be about twice as great as that for maximum sum gain, and in the even plane it should be about equal to that for maximum sum gain. It is also interesting to note that, upon analysis of the output from the complete circuit including the hybrid junction, the maximum difference gain should be about 3 db below the maximum sum gain.

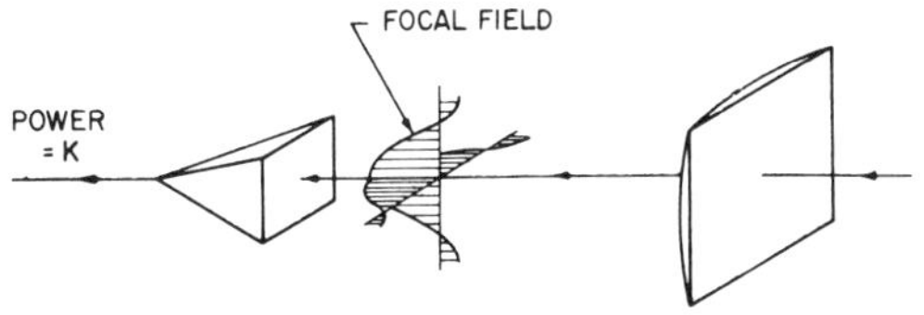

(a) Sum mode.

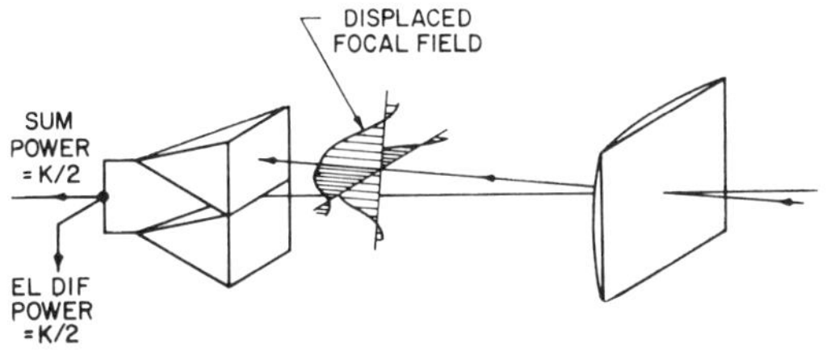

(b) Difference mode.

Fig. 2—Optimum feed sizes; focal-field viewpoint.

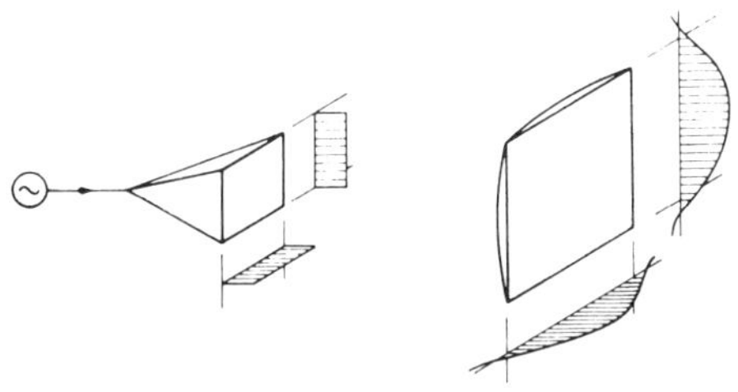

(a) Sum mode.

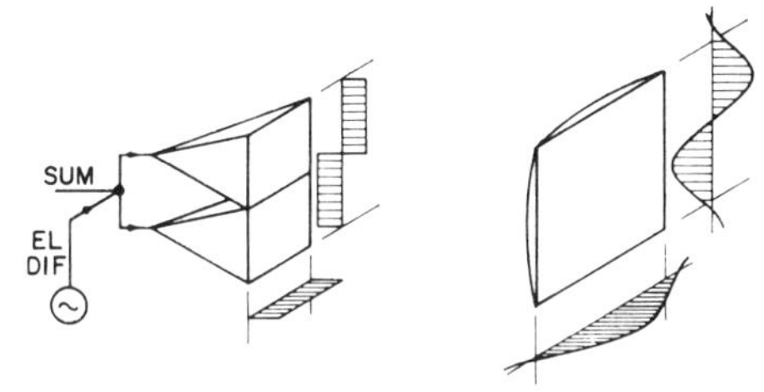

(b) Difference mode.

Fig. 3—Optimum feed sizes; illumination viewpoint.

B. Main Aperture Illumination

Another viewpoint considers the illumination of the main aperture by the feed during transmission; this is illustrated in Fig. 3 for a hypothetical feed having "constant" excitation in both planes. In order to obtain maximum gain in the sum mode, the feed size should be such that the illumination is tapered down at the edge of the main aperture by about 10 db, as shown in Fig. 3(a). This is the well-known [1], [2] optimum taper which balances excessive spillover loss against excessive waste of the main aperture, to achieve maximum gain in a simple, single-mode antenna. In some designs, the illumination is tapered further down so that the discontinuity at the edge of the main aperture is reduced: this lowers the level of the sidelobes in the antenna pattern, at the expense of a small reduction of gain.

For the case of the difference mode, the considerations of maximum gain and low sidelobes lead to a similar conclusion: namely, the illumination of the main aperture should be appreciably tapered down at the edge. In addition, some of the special problems of the difference mode, such as criticalness to feed tilt and edge asymmetries, place a premium on low illumination at the edge of the main aperture. For simplicity, it may be assumed that the difference illumination should be tapered down by about the same amount as the sum illumination, say 10 db, as shown in Fig. 3(b). However, in the odd plane of the difference mode, the feed is only about half as directive as in the sum mode; therefore the feed aperture dimension in this odd plane should be about twice as great as it is for the sum mode. In the even plane, the feed aperture dimensions for the two modes should be about equal. This conclusion is, of course, the same as that reached previously.

It is instructive to consider the situation in which the feed size is optimized for the sum mode, but cannot be increased for the difference mode. Under these circumstances the difference illumination reaches a maximum close to the edge of the main aperture, as shown in Fig. 4(a), and the large discontinuity at this location creates high sidelobes in the corresponding antenna pattern. At least half of the power in the difference mode goes into spillover, so that there is about 3-db loss in the difference signal compared with the optimum condition; thus, the difference peak gain would be about 6 db below the sum gain. The high illumination of the edge of the main aperture makes the difference mode sensitive to antenna misalignment and edge asymmetries, and the large amount of spillover is likely to contribute to additional sidelobes. These "spillover sidelobes" may be especially offensive in permitting spurious signals to enter the difference channel; examples of this are coherent signals such as those caused by reflection from the ground and reflection from asymmetrical antenna surfaces, or incoherent signals such as thermal radiation from the ground.

If the feed size were optimized for a difference mode without decreasing it for the sum mode, the sum illumination would be too narrow. As indicated in Fig. 4(b), optimization in one difference mode causes about half the main aperture to be wasted in the sum mode; thus a reductirn of about 3 db in sum gain would result. Attempting to optimize the feed size in both difference modes would create additional losses.

While it is true that a feed size might be utilized which strikes a compromise between optimum sum-mode and optimum difference-mode performance, the defects mentioned above would still be present to a large degree. Thus, it is evident that the ordinary feed design having the same effective size in the sum and difference modes imposes a limitation which degrades the antenna performance in a number of ways, many of which are likely to have a significant effect on system performance.

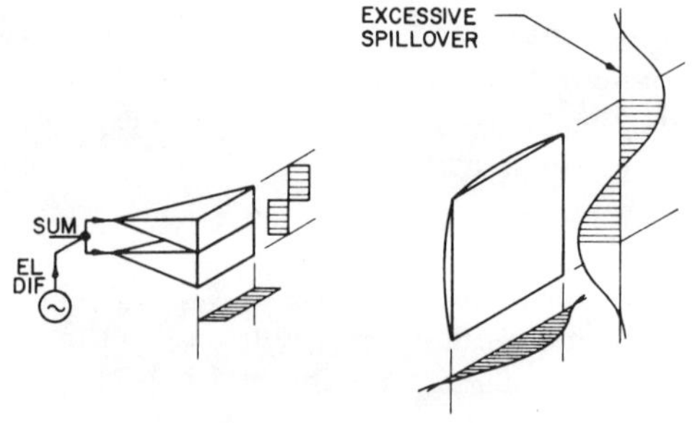

(a) Difference mode when feed is optimized for sum mode.

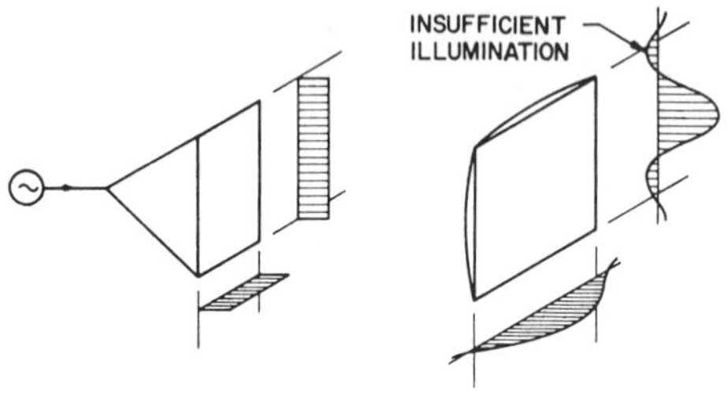

(b) Sum mode when feed is optimized for one difference mode.

Fig. 4—Non-optimum feed sizes; illumination viewpoint.

C. Antenna Pattern Characteristics

Up to this point, the viewpoints chosen have merely illustrated the need for different feed-aperture sizes in the sum and difference modes, and have not yielded quantitative estimates of antenna performance. The final viewpoint to be presented is one whereby numerical values are obtained for a number of significant characteristics of the antenna pattern. These values are plotted in Figs. 5–8 (pp. 448–449) as a function of normalized feed aperture size, for the case of "constant" shapes of excitation of the feed aperture. As will be discussed in Section IV, these are not the best shapes for the feed excitation. However, they are relatively simple to calculate, and provide a typical case for study of the effect of feed aperture size on the antenna patterns; in addition, the useful concept of effective aperture is ordinarily related to a "constant" shape.

For the exact definitions of the various antenna pattern characteristics, the reader is referred to Section VII. The curves themselves are drawn as accurately as possible, because they are likely to be helpful during the design of an antenna. Each curve originates from an explicit formula; these formulas appear in the report [9] from which this paper stems. A few of the formulas also appear in [10] and [11].

Sum Mode: In Fig. 5, a set of curves are shown for several characteristics of the antenna pattern in the sum mode. These characteristics include gain ratio (efficiency), sidelobe ratios, and beamwidth ratio. A curve for spillover power ratio is also presented; since the spillover cannot be included in a general calculation of sidelobes, some indication of the "spillover sidelobe" effect is desirable. Also shown for reference is a dashed curve of edge taper voltage ratio for the illumination of the main aperture.

It should be understood that all of these quantities are determined on the basis of a one-dimensional geometry. However as mentioned in Section II-B, the field strength of the antenna radiation is separable into the product of a horizontal and a vertical distribution, in the usual manner for a large rectangular aperture. As

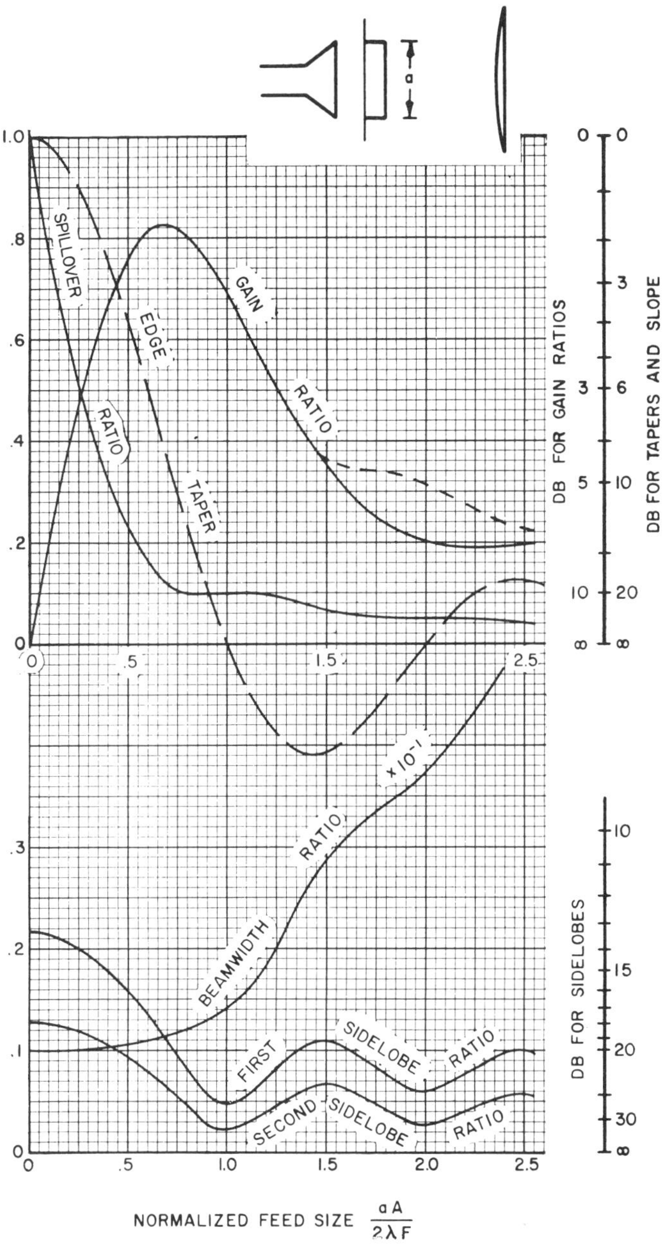

Fig. 5—Sum-pattern characteristics vs feed size.

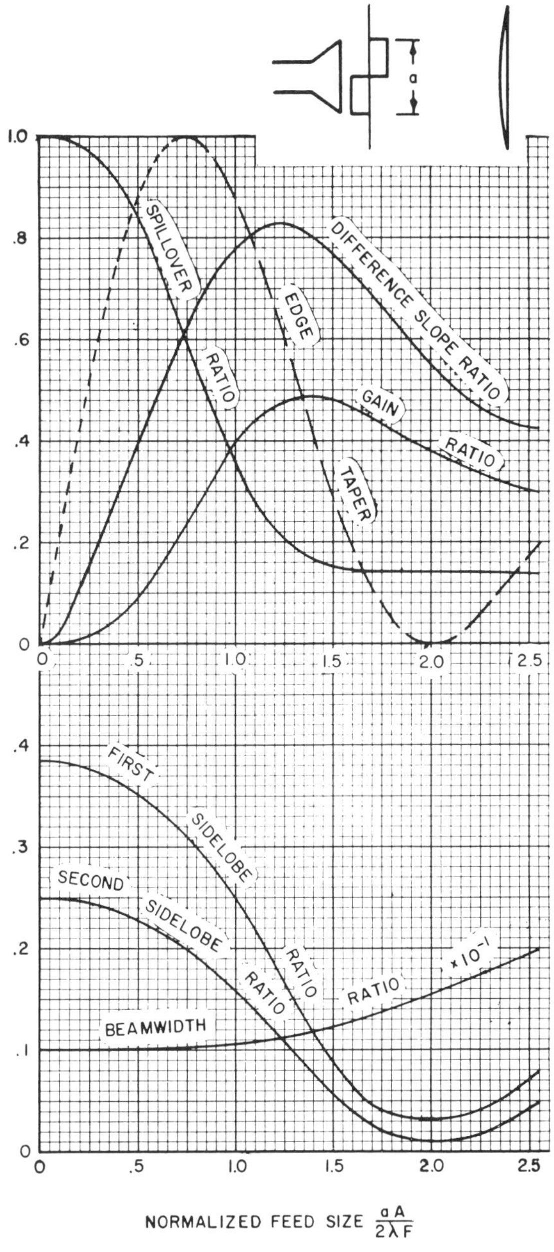

Fig. 6—Difference-pattern characteristics vs feed size.

a result, the two-dimensional sum-mode value for gain ratio may be obtained by multiplying together the appropriate one-dimensional sum-mode values for each of the two planes involved. In the case of spillover, this multiplication rule applies instead to the fraction of power not spilled over, *i.e.*, one minus the spillover ratio. The other pattern characteristics given in Fig. 5 are also valid for the two-dimensional case, when restricted to the appropriate cardinal plane.

It may be seen that the curve of gain ratio reaches a maximum at a normalized feed size of 0.68, while the curves for the relative levels of both the first and second sidelobes reach a minimum at a normalized feed size of 1.00. If one were to define an optimum feed size for the sum mode in terms of the best combination of these two properties, then an intermediate size, say about 0.80, would be chosen. Assuming that the antenna requirements are the same for both planes of the sum mode, the same result would apply to both dimensions of the feed aperture.

Difference Mode: In Fig. 6, all but one of the curves show properties for the difference mode which are similar to those properties given in Fig. 5 for the sum mode. As before, these are given for a one-dimensional geometry; in this case, only the odd plane of the difference mode is considered. The two-dimensional difference-mode values for gain ratio (difference peak efficiency) and one-minus-spillover-ratio are obtainable by mul-

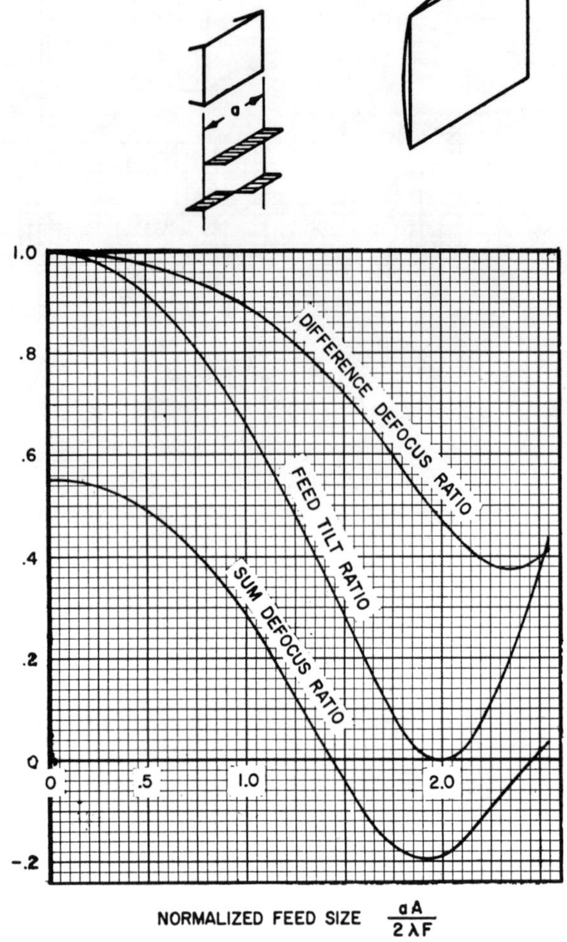

Fig. 7—Effects of misalignment on patterns vs feed size.

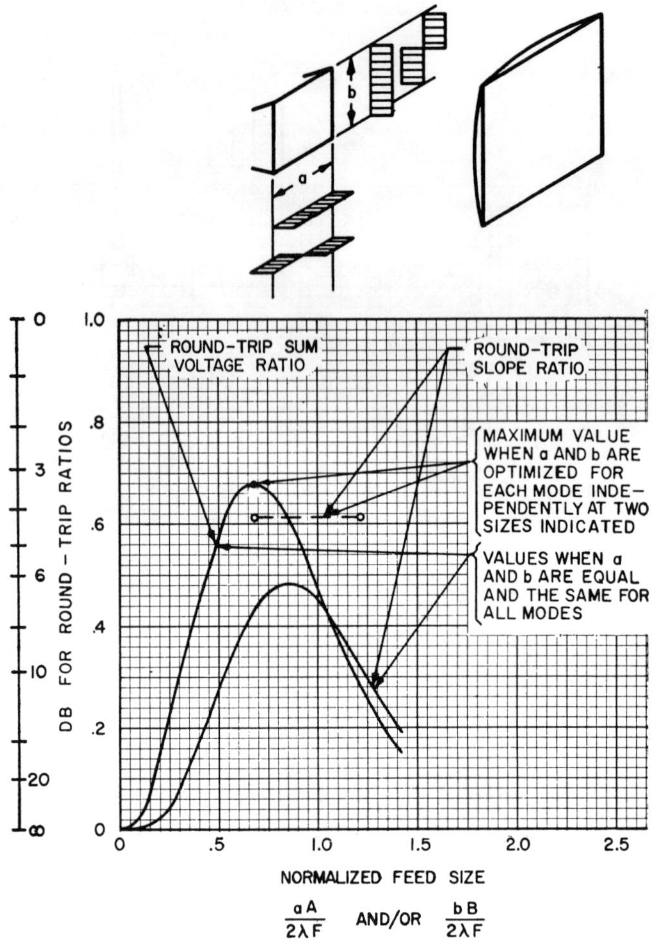

Fig. 8—Round-trip pattern characteristics vs feed size.

tiplying the one-dimensional difference-mode value in the odd plane (Fig. 6) by the appropriate one-dimensional sum-mode value in the even plane (Fig. 5). The beamwidth ratio, sidelobe ratios, and edge taper are valid for the two-dimensional case when restricted to the odd plane of the difference mode.

The curve of gain ratio reaches a maximum at a normalized feed aperture size in the difference plane of about 1.37, which is approximately twice the size for maximum sum gain. The first and second sidelobe ratios are minimum at a normalized feed size of 2.00, so an optimum size in the difference plane, considering both gain and sidelobe level, might be at about 1.60; these feed sizes also are twice those obtained for the sum mode. In the other plane, the feed aperture size would be optimized on the basis of the sum-mode curves, and would be about equal to the size obtained previously for the sum mode.

As mentioned in Section III-B, spillover may be an important contribution to sidelobes in the difference mode. It may be seen from the curve of spillover ratio that at a normalized feed size of 1.60 the spillover power is less than 1/3 of its value at a feed size of 0.80, and has been reduced to an approximate, practical minimum. Actually, as will be discussed in Section IV, a more ideal shape for the feed excitation would permit a further reduction in spillover.

There is one additional characteristic of the antenna pattern in the difference mode which is shown in Fig. 6; this is the difference slope, or the slope of the difference pattern on the axis of the antenna. This characteristic is of fundamental importance in a monopulse antenna because it determines the angle sensitivity, or the ability to accurately track the angle of a distance source under conditions in which errors may be caused by spurious signals of either an incoherent nature (receiver noise, antenna noise, etc.) or a coherent nature (plumbing asymmetries, antenna asymmetries, ground reflections, etc.). The difference slope is presented in Fig. 6 as a voltage ratio, normalized with respect to the maximum possible difference slope (see Section VII). As before, it is given for a one-dimensional geometry; the two-dimensional case may be determined for the odd plane of the difference mode by multiplying the difference slope ratio by the square root of the sum-mode gain ratio in the other plane. It may be observed from the curve that the difference slope reaches a maximum for a normalized feed aperture size of about 1.23 in the odd plane. This is somewhat smaller than the size for maximum gain at the difference peak, but is still considerably larger than that for maximum sum gain.

Misalignment Factors: In Fig. 7 some quantities are shown which are concerned with the degradation of the monopulse characteristics caused by certain misalignments or misadjustments within the antenna. While these do not have the basic significance of the characteristics discussed so far, they are, nevertheless, of great interest to the antenna designer. One of these quantities is called the feed-tilt ratio, and indicates the amount by which the central null of the difference pattern is filled in when the feed is tilted away from the direction of the center of the main aperture. The difference-signal component thus created is in quadrature with the normal signal and does not shift the angular position of the central minimum of the difference patterns; however, as has been described elsewhere [12], it can combine with a phase difference between the channels of the plumbing or IF circuits to cause a boresight error. The feed-tilt ratio is presented here as the rate of change of spurious difference voltage with feed-tilt angle for a small amount of tilt, normalized with respect to the maximum possible value (see Section VII). Under the assumptions of Section II-B, this ratio is applicable to both a one-dimensional and a two-dimensional geometry without any reference to the other plane. In the two-dimensional case, the tilt angle to be considered is the component (see Fig. 1) in the odd plane of the particular difference mode of interest.

It is apparent from the curve of feed-tilt ratio that if the feed aperture dimension in the difference plane is small, the feed-tilt ratio is large and the feed may have to be carefully aligned to obtain good performance. However, if the normalized feed size increases to a value of 2.00, the feed-tilt ratio is zero and the alignment becomes noncritical. This is a consequence of the zero in amplitude of illumination at the edge of the main aperture; under this condition a small sideways shift of the illumination causes no significant change. (The second-order shape of the curve at this point corresponds to the second-order shape of the illumination curve, and the latter is an unusual property of the "constant" shape of difference excitation assumed at the feed aperture. A more customary shape would be predominantly first-order and would go through a reversal of sign at this point, but the noncritical property would still exist.) In many ordinary monopulse antennas, it is necessary to provide a feed-tilt adjustment because tolerances cannot be held tight enough to rely on proper alignment of the feed. If the feed aperture were enlarged toward the noncritical point, it might be possible to eliminate the feed-tilt adjustment. On the other hand, one should remember that certain errors in the comparator may also be compensated by a feed-tilt adjustment; if this is eliminated, another type of compensation may be required.

The other quantities shown in Fig. 7 describe an effect of defocusing the antenna, or of moving the feed toward or away from the main aperture. The particular effect of interest is the change of phase between the on-axis sum and difference signals. In a pure amplitude-comparison monopulse antenna, an incoming wave creates sum and difference signals which are in-phase or out-of-phase with each other at the output of an idealized comparator. If the antenna is then defocused, this phase relationship is modified, and the monopulse system becomes a mixture of amplitude-comparison and phase-comparison. As described elsewhere [13], it is desirable to keep this phase change, called the hybrid-monopulse angle, within limits. In Fig. 7, two curves are shown which are labeled difference defocus ratio and sum defocus ratio, respectively. These ratios yield the rate of a phase change with axial feed displacement for small departures from the focal point, normalized with respect to the maximum possible rate of change in the difference mode (see Section VII). The difference between the two phase changes is the hybrid-monopulse angle. Under the assumptions of Section II-B, this is applicable to both a one-dimensional and a two-dimensional geometry, without any reference to the other plane. It may be seen that by choosing certain feed aperture dimensions in the sum and difference modes, the values for the difference and sum defocus ratios can be made equal, and therefore the hybrid-monopulse angle can be made noncritical to small amounts of defocusing. The ratio of feed sizes is, however, appreciably greater than two; for instance, a sum size of 0.68 would require a difference size of about 2.09.

The properties shown in Fig. 7 may be considered together with all of those of Fig. 6 in determining an optimum size for the feed aperture in the difference mode. The size chosen will, of course, depend on the relative importance of the various properties in any particular application. In a typical case, a consideration of all the properties might give the same result as that obtained with only the first six properties in Fig. 6. Thus, the normalized feed size in the odd plane would have an optimum value of about 1.60 for the difference mode, roughly twice as great as the optimum value for the sum mode.

Round-Trip Performance: Earlier in this paper, mention was made of a compromise feed size which would provide the best performance for the ordinary situation in which the sizes cannot be different in the sum and difference modes. It is instructive to examine such a design, in order to compare it with the more ideal one in which the size for each mode may be optimized independently. For the sake of simplicity and correlation with other investigations of this subject, the optimization will be confined to two particular quantities; one is the on-axis round-trip sum voltage (sum voltage squared) and the other is the on-axis round-trip slope (sum voltage times difference slope). Fig. 8 shows curves of these two quantities in normalized form, as a function of normalized feed aperture size, for a two-dimensional case in which the sizes are the same for the sum and both difference modes, and the hypothetical "constant" excitation is assumed for both planes.

A feed size midway between the peaks of these two curves may be chosen as representing the best compromise, and it appears that only a small degradation is obtained [3]. However, in the case of the round-trip slope curve, one factor is the one-way one-dimensional difference slope, and the latter is far from its true optimum value.

The values which the two round-trip quantities would attain when the feed size is optimized independently in the sum and difference modes are indicated in Fig. 8, by

means of the black dot and the dashed horizontal line. For the round-trip slope a significant increase is thereby obtained; the actual amount is approximately 2.5 db. There would also be, of course, a marked improvement in other properties discussed previously, such as sidelobes and spillover in the difference modes. Furthermore, these latter benefits could be considerably greater if the feed size were optimized with all of these quantities in mind.

IV. Need for Efficient Shapes of Feed Excitation in Sum and Difference Modes

The preceding section has shown the desirability of allowing the feed aperture *size* to be different for the sum and difference modes. There is another aspect of the feed excitation which has an important effect on antenna performance: namely, the *shape* of the excitation. This will be illustrated by the same three viewpoints employed previously.

A. Focal Field

In Fig. 9, the field of an incoming wave in the focal region near the feed is shown. The shape of the major lobe of this field is smooth and tapers in a simple manner to zero on each side. It is natural to expect that for high gain, the shape as well as the size of the feed excitation should be similar to that of the focused field. (As a matter of interest, the "matched filter" criterion [14] could probably be employed to prove that the greatest signal is received when the shapes are identical.) By way of example, two simple hypothetical shapes of feed excitation may be considered. In Fig. 9(a), cosine and sine shapes are shown for the sum and difference modes, respectively. Since these closely approximate the shape of the major lobe of the focal field, the gain should be close to the maximum obtainable by complete utilization of this major lobe. In Fig. 9(b), double-cosine shapes are shown; here the gain should be significantly reduced because of the dissimilarity in the region of the "holes" between the double cosines. It

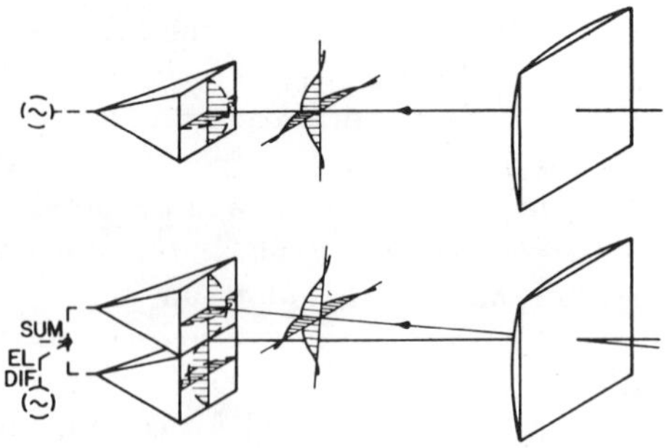

(a) An efficient set of shapes.

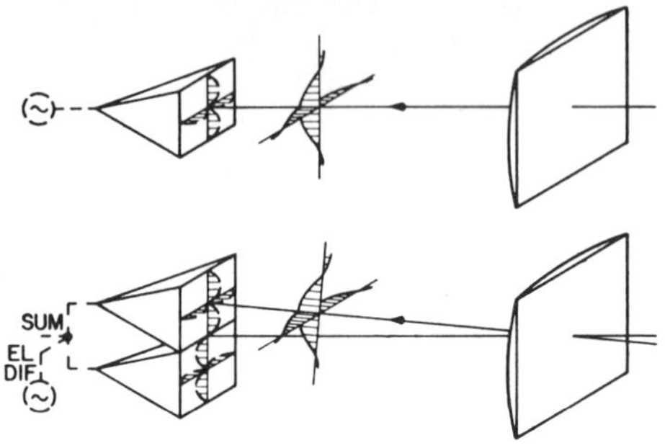

(b) An inefficient set of shapes.

Fig. 9—Comparison of feed-excitation shapes; focal-field viewpoint.

might also be added that this situation is characterized by an inherent reflection of the incoming wave by the feed, and the reflected power is capable of causing additional deficiencies in the performance of the antenna [15].

B. Main Aperture Illumination

Fig. 10 presents the viewpoint which evaluates the illumination of the main aperture, and the same examples of feed-excitation shapes are considered. With the cosine and sine shapes illustrated in Fig. 10(a), the radiation pattern of the feed has low sidelobes, and the power wasted in spillover beyond the edges of the main aperture is minimized. With the double-cosine shapes illustrated in Fig. 10(b), the sidelobes of the feed pattern are high, and it becomes necessary to waste an appreciable amount of power in spillover. Thus it is evident that the former shapes are more efficient than the latter; in addition, the excess spillover existing in the latter case may aggravate certain problems, as mentioned in Section III-B.

C. Antenna Pattern Characteristics

The final viewpoint, given in Fig. 11, presents quantitative values for a few particular properties of the antenna pattern. These properties are gain ratio in the sum mode, slope ratio in the difference mode, and the spillover ratios for both modes, for those particular feed sizes which yield maximum gains or slopes, as indicated. The values are given for a one-dimensional case; as before, two-dimensional results are obtainable by multiplying together the proper one-dimensional values.

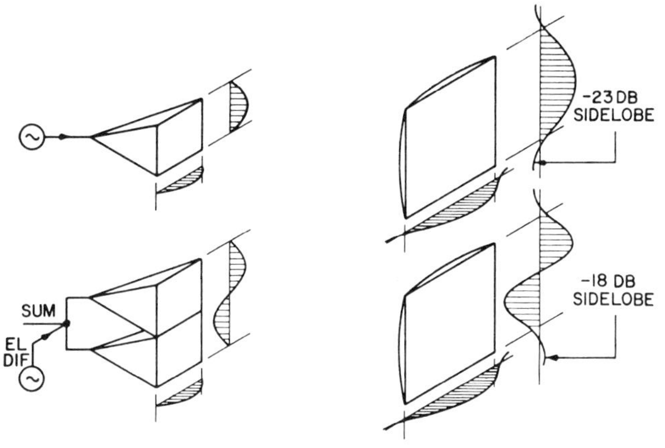

(a) An efficient set of shapes.

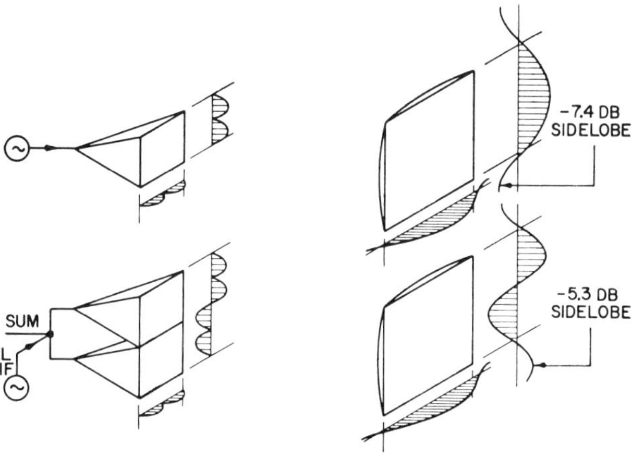

(b) An inefficient set of shapes.

Fig. 10.—Comparison of feed-excitation shapes; illumination viewpoint.

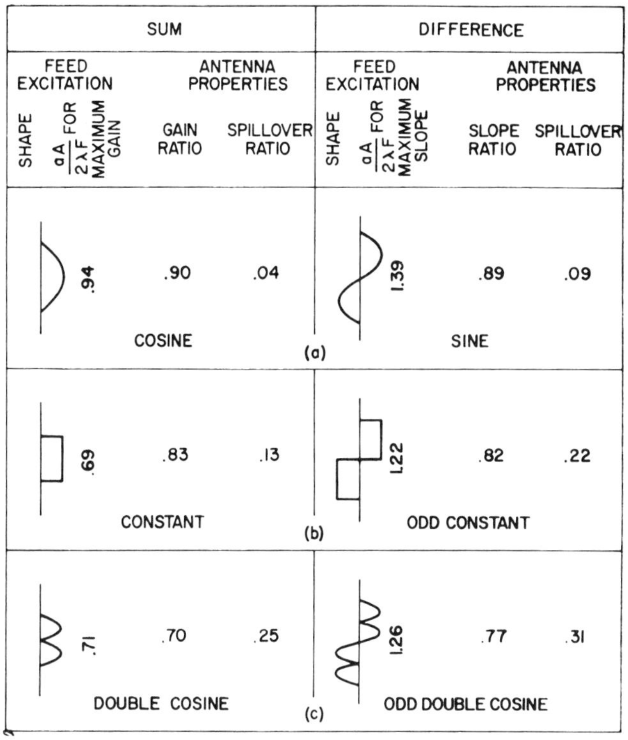

	Sum: Feed excitation — Shape	Sum: Feed excitation — aA for 2λF maximum gain	Sum: Antenna properties — Gain ratio	Sum: Antenna properties — Spillover ratio	Difference: Feed excitation — Shape	Difference: Feed excitation — aA for 2λF maximum slope	Difference: Antenna properties — Slope ratio	Difference: Antenna properties — Spillover ratio
(a)	Cosine	.94	.90	.04	Sine	1.39	.89	.09
(b)	Constant	.69	.83	.13	Odd constant	1.22	.82	.22
(c)	Double cosine	.71	.70	.25	Odd double cosine	1.26	.77	.31

Fig. 11—Antenna-pattern characteristics for various shapes of feed excitation.

Three examples are chosen to illustrate the effect of shape of feed excitation; these are the single and double cosine and sine types mentioned just previously, plus the constant types. It is apparent that the single sine and cosine shapes (a) are the best and the double-cosine shapes (c) are the worst, while the constant shapes (b) are intermediate between these two. The differences among the various results are large enough to have a significant effect on antenna performance. For example, in going from the double shapes (c) to the single shapes (a), in only the one dimension, there is an increase of 1.1 db in sum gain (see also [2]) and 1.3 db in difference slope, and a reduction by a factor of 0.16 in sum spillover and 0.28 in difference spillover. These spillover reductions would be even more striking if a larger feed size had been chosen.

V. An Ideal Feed Excitation

In Fig. 12, an "ideal" feed excitation is shown in two dimensions, for all three modes. This is not to be regarded as a unique ideal, but rather as a typical one representing a kind of composite of the previously-discussed concepts. For example, the difference excitations are about twice as wide in the odd planes as are the effective sum excitations, and the shapes of the excitations are free of any avoidable irregularities. If a feed were designed to have all these features, the resulting antenna performance should be very close to optimum in all three modes of the monopulse antenna.

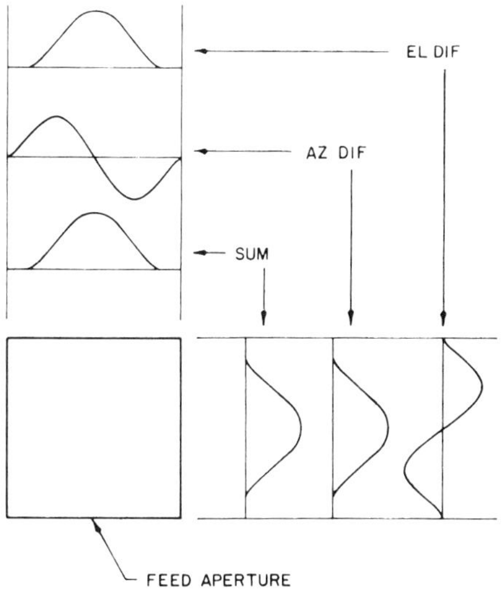

Fig. 12—An ideal feed excitation.

VI. Discussion

It is instructive to estimate the improved performance obtainable with a feed such as that shown in Fig.

12. Suppose that an ordinary feed had an effective normalized size of 0.80 in both planes and all three modes, and that with the new feed this size becomes 1.60 in the odd plane of the difference modes. Suppose further that the ordinary feed had shapes of excitation which were double cosine by constant for the even types and sine by odd constant for the odd. The curves and tables in this paper may now be employed to estimate the improvement, in terms of the complete two-dimensional quantities.

In the sum mode the gain should increase by about 1.4 db and the spillover ratio should (it is estimated) be reduced from about 0.3 to less than 0.03. These changes are a result of the improved excitation shape. In the difference modes the peak gains should be increased by an average of about 3.5 db and the slopes by an average of about 1.9 db, while the spillover ratio should (it is estimated) be reduced on the average from about 0.5 to less than 0.05. These changes are mainly a result of the doubled feed size in the odd planes.

238

The first sidelobe ratios in the difference modes should go from about 10.5 db to about 23.7 db, because of the doubled feed sizes. This 13-db improvement may be complemented by an improved suppression of the far sidelobes of about 10 db in all the modes, as a consequence of the spillover reduction. Finally, the criticalness of positioning the feed should be reduced to about $\frac{1}{4}$ in tilt and about $\frac{1}{2}$ in focusing, because of the doubled feed sizes.

The analysis in this paper has been confined to a rectangular or square main aperture. Although no calculations have been made for a circular aperture, it is believed that somewhat more improvement in difference gain, slope, and spillover, and somewhat less improvement in the near difference sidelobes, would be obtained. It is concluded, therefore, that for any amplitude-monopulse antenna involving a large feed and a focusing element, a feed excitation such as that shown in Fig. 12 should provide significantly better performance than can be obtained with an ordinary monopulse feed.

VII. Definitions

Gain Ratio:

For the sum mode, gain ratio is the on-axis antenna gain relative to the maximum possible gain. The latter, reference gain, is that gain which would be obtained if the main aperture were uniformly illuminated with no spillover. (An alternate term for gain ratio which is often used for microwave antennas is "efficiency".) Normally, the on-axis sum gain is also the peak gain; for the range of feed sizes where this is not true, the peak gain is also shown by means of a dashed addition to the curve in Fig. 5.

For the difference mode, gain ratio is the antenna gain at the peak of the difference pattern relative to the maximum possible sum gain. (The above definition of difference gain ratio employs the maximum possible sum gain as the reference quantity because this allows one to compare difference gain with sum gain directly on the graphs. It would also be possible to define difference gain ratio employing the maximum possible difference gain [16] as the reference quantity; in this case the difference gain ratio would be greater than that given on the curve in Fig. 6 by the factor 1/0.609, or 2.15 db.)

Spillover Ratio:

For each mode, spillover ratio is the ratio of power radiated by the feed into the region outside the edges of the main aperture, to total power radiated by the feed. (Spillover is likely to contribute to sidelobes in the antenna radiation pattern.)

Edge Taper:

For each mode, edge taper is the ratio of voltage at the edge of the main aperture to the maximum voltage on the main aperture, when the main aperture is illuminated by the feed. (In the case of the difference mode, a small feed may radiate a pattern whose peak is outside the edge of the main aperture; in this case the value of the edge voltage relative to the peak radiated voltage is indicated by the dotted portion of the curve in Fig. 6.)

Sidelobe Ratio:

For each mode, the sidelobe ratio is the ratio of voltage at the peak of the designated sidelobe of the antenna radiation pattern to the peak voltage of the main lobe of the same antenna pattern. The sidelobe ratios are calculated directly from the main aperture illumination without including any effect of spillover.

Beamwidth Ratio:

For the sum mode, the beamwidth ratio is the angle between half-power points of the antenna pattern with a specified feed size, relative to the angle with a very small feed (*i.e.*, uniform illumination of the main aperture). In the latter, reference case, the angle is 0.887 λ/A radians.

For the difference mode, the beamwidth ratio is the angle between the outer half-power points of the antenna pattern with a specified feed size, relative to the angle with a very small feed (*i.e.*, odd, linear illumination of the main aperture). In the latter, reference case, the angle is 2.03 λ/A radians.

Difference Slope:

Slope of the center of the antenna pattern in the difference mode, assuming no asymmetries. In other words, rate of change of the difference voltage with antenna angle, where the angle is small and the difference voltage contains no quadrature components.

Relative Difference Slope:

Difference slope relative to maximum possible sum voltage.

Difference Slope Ratio:

Ratio of the difference slope to the maximum possible difference slope. The latter could be obtained if the main aperture were illuminated by an odd, linear voltage pattern [17], [18] with no spillover; in this reference case the relative difference slope (see previous definition) would be $\pi A/\sqrt{3}\,\lambda$ voltage ratio per radian.

Feed-Tilt Ratio:

The rate of change of the on-axis difference signal with feed-tilt angle, where the amount of feed tilt is small, relative to the rate of change which would be obtained with any very small feed (*i.e.*, odd, linear illumination of the main aperture). In the latter, reference case, the rate of change of difference signal with feed-tilt angle, if taken relative to the difference slope, would be $6\lambda F/\pi A^2$.

Difference Defocus Ratio:

The rate of change of phase of the near-axis difference signal with axial displacement of the feed in wavelengths, where the phase change is caused only by the phase curvature across the main aperture resulting from a small amount of displacement of the feed from the focal point, relative to the rate of change which would be obtained with a very small feed (*i.e.*, odd, linear illumination of the main aperture). In the latter, reference case, the rate of change of difference phase with axial feed displacement in wavelengths is $3\pi A^2/20F^2$ radians.

Sum Defocus Ratio:

The rate of change of phase of the on-axis sum signal with axial displacement of the feed in wavelengths, where the phase change is caused only by the phase curvature across the main aperture resulting from a small amount of displacement of the feed from the focal point, relative to the rate of change which would be obtained with a very small feed operating in the difference mode (*i.e.*, odd, linear illumination of the main aperture). (The above definition employs the small-feed rate of change in the difference mode as the reference quantity because this simplifies the calculation of the hybrid monopulse angle. It would also be possible to define the sum defocus ratio employing the small-feed rate of change in the sum mode as the reference quantity; in this case the sum defocus ratio would be greater than that given on the graph by the factor 9/5, as indicated.

Round-Trip Sum Voltage Ratio:

The round-trip sum voltage ratio is the square of the on-axis sum-pattern voltage, relative to the maximum possible value. The latter, reference value, would be obtained if the main aperture were uniformly illuminated with no spillover. For the two-dimensional square case given in Fig. 8, the round-trip sum voltage ratio is equal to the square of the one-dimensional sum gain ratio given in Fig. 5.

Round-Trip Slope Ratio:

The round-trip slope ratio is the product of on-axis sum-pattern voltage and on-axis difference slope, relative to the maximum possible value. The latter, reference value, could be obtained if the main aperture had uniform illumination with no spillover in the sum mode and in the even plane of the difference mode, and had odd, linear illumination with no spillover in the odd plane of the difference mode. For the two-dimensional square case given in Fig. 8, the round-trip slope ratio is equal to the 3/2 power of the one-dimensional sum gain ratio given in Fig. 5 times the difference slope ratio given in Fig. 8.

VIII. Acknowledgment

R. L. Mattingly has been principally responsible for the direction of this effort by Bell Telephone Laboratories, and the author would like to express his appreciation for the encouragement he has given it.

At Wheeler Laboratories, this work has been directed by H. A. Wheeler, D. Dettinger, N. A. Spencer, and P. A. Loth. The author would like to acknowledge the cooperation of the Wheeler staff in the preparation of this paper, especially H. A. Wheeler, H. W. Redlien, and D. S. Lerner for their helpful comments and suggestions, S. I. Warshaw and B. J. Karafin for their computational effort, and C. J. Stona for his work on the illustrations.

IX. Bibliography

[1] C. C. Cutler, "Parabolic-antenna design for microwaves," Proc. IRE, vol. 35, p. 1286; November, 1947.
[2] J. W. Crompton, "On the optimum illumination taper for the objective of a microwave aerial," *Proc. IEE*, vol. 101, Pt. 3, (also see ASTIA Doc. No. AD 12571); November, 1954.
[3] M. L. Kales, "Optimum Design Criterion for Simultaneous-Lobing Antennas," Naval Res. Lab., unpublished Rept. R-3451, ASTIA Doc. No. ATI 61934, pp. 18–24; April, 1949.
[4] D. R. Rhodes, "Introduction to Monopulse," McGraw-Hill Book Co., Inc., New York, N. Y., pp. 98–101; 1959.
[5] R. M. Page, "Monopulse Radar," 1955 IRE National Convention Record, pt. 8, pp. 132–134.
[6] J. F. P. Martin, Bell Telephone Labs., unpublished Memo. No. MM-47-2730-7; January, 1947.
[7] H. W. Redlien, "A Unified Viewpoint for Amplitude and Phase Comparison Monopulse Tracking Radars," Wheeler Labs., unpublished Rept. No. 845 to Bell Telephone Labs.; March 1959.
[8] P. A. Matthews and A. L. Cullen, "A study of the field distribution at an axial focus of a square microwave lens," *Proc. IEE*, vol. 103, Pt. C, pp. 449–456; July, 1956.
[9] P. W. Hannan, "Principles of Feed Design for Optimum Performance in All Three Modes of an Amplitude-Monopulse Radar Antenna," Wheeler Labs., unpublished Rept. No. 884; August, 1960.
[10] J. F. Ramsay, "Fourier transforms in aerial theory—part II, Fourier sine transforms," *Marconi Rev.*, vol. 10, pp. 17–22; January-March, 1947.
[11] B. Berkowitz, "Antennas fed by horns," Proc. IRE, vol. 41, pp. 1761–1765; December, 1953.
[12] H. W. Redlien, "Theory of Monopulse Operation—Introduction," Wheeler Labs., unpublished Rept. No. 434B to Bell Telephone Labs., (also ASTIA Doc. No. AD 305996); January, 1953.
[13] H. W. Redlien, "Theory of Monopulse Operation—Effects of a Defocused Antenna," Wheeler Labs., unpublished Rept. No. 495A to Bell Telephone Labs., (also ASTIA Doc. No. Ad 59288); August, 1953.
[14] J. H. Van Vleck and D. Middleton, "A theoretical comparison of visual, aural, and meter reception of pulsed signals in the presence of noise," *J. Appl. Phys.*, vol. 17, pp. 943–944; November, 1946.
[15] P. W. Hannan, "Reflections in microwave antennas and their harmful effects," 1954 IRE National Convention Record, pt. 1, pp. 39–45.
[16] P. W. Hannan, "Maximum gain in monopulse difference mode," IRE Trans. on Antennas and Propagation, vol. AP-9, pp. 314–315; May, 1961.
[17] G. M. Kirkpatrick, "Aperture illuminations for radar angle-of-arrival measurements," IRE Trans. on Aeronautical and Navigational Electronics, vol. AE-9, pp. 20–27; September, 1953.
[18] G. M. Kirkpatrick, General Electric Co., unpublished report, ASTIA Doc. No. AD 18458, pp. 13–15; August, 1952.

Optimum Feeds for All Three Modes of a Monopulse Antenna II: Practice*

PETER W. HANNAN†, senior member, ire

Summary—In an amplitude-monopulse antenna whose main aperture is illuminated by a feed, it is desired to control the feed excitation independently in the sum, azimuth difference, and elevation difference modes. It is possible to describe a feed system, comprising an infinite array of radiators and hybrid junctions, which demonstrates the principle of complete independent control of the three modes.

There are several practical feeds which approach a hypothetical ideal one in varying degrees. One representative type is the "twelve-horn feed;" although it has some disadvantages, it is versatile and quite useful. However the most attractive type for many applications appears to be one having a combination of multiple-horn excitation in one plane and multimode-waveguide excitation in the other. With a simple form of the "multihorn-multimode" feed, the ideal is substantially realized, and a major improvement in monopulse performance can be obtained.

* Received by the PGAP, November 22, 1960; revised manuscript received, March 27, 1961. The material in this paper is the result of antenna development work for the Bell Telephone Laboratories on Army Ordnance Corps projects. Preparation of this paper has been supported mainly by a subcontract with BTL in connection with prime contract No. DA-30-069-ORD-1955.
† Wheeler Labs., Smithtown, N. Y.

240

I. Introduction

In a previous paper [1] a study was made of the improved performance available in an amplitude-monopulse radar antenna comprising a feed and a focusing element. This improvement arises from the possibility of independently optimizing the feed in each of the three modes of operation, rather than compromising between them. There was presented a hypothetical "ideal" feed excitation which would essentially optimize all three modes, but no information was given regarding the attainment of such a feed. It is the purpose of this paper to discuss the means by which this theoretical feed may be reduced to practice.

Some early monopulse antennas have already obtained improved performance by reducing the amount of compromise between the modes. This has often occurred as a fortunate by-product of a feed design intended mainly to achieve some other benefit, such as shielding from nearby objects, a simple comparator structure, or a short focal length. As a result, the optimizing process has not usually been complete. Antennas known to the writer which belong in this category have been developed by Bell Telephone Laboratories [2], North American Aviation Company, Hughes Aircraft Corporation [3], and Radio Corporation of America [4].

About seven years ago, Wheeler Laboratories made an investigation of this problem, in connection with the design of a monopulse tracking antenna [5], [6], for Bell Telephone Laboratories. A principle was conceived and a number of design techniques were worked out for feeds which would give optimum performance in all three modes. The principle was subsequently applied in a study by Wheeler Laboratories, in which the basic merits of the amplitude and phase types of comparison were analyzed and compared [7].

An investigation of feeds for optimum monopulse performance has occupied the attention of several other groups. In the first published material known to the writer [8], the principle was recognized and experiments made with a particular technique which approaches optimum performance in more than one mode. A number of interesting techniques have been studied at Naval Research Laboratory and General Electric Company. A feed [9] for use in an operational antenna has been developed by Lincoln Laboratory which applies the principle to all three modes of operation.

Several additional monopulse tracking antennas have been designed by Wheeler Laboratories for Bell Telephone Laboratories. Some of these antennas [10] incorporate feeds which have achieved substantially optimum performance in all three modes of operation. While it is not proposed to describe the details of their design and measured performance in this paper, it is possible to present the basic methods which are involved. Various techniques will be discussed, and a comparison of their performance will be calculated.

II. Principle of Independent Control of Excitation in the Sum and Difference Modes

It is interesting to inquire whether an excitation of the ideal type is theoretically realizable. Ordinary methods of exciting a feed yield distributions which are limited to having the same width in all modes, and having shapes which are good for one mode but not for another. Upon consideration of the problem, it becomes evident that *independent control* of the excitation in the sum and difference modes is required to fulfill the objectives.

Fig. 1 presents a scheme which, in principle, is capable of achieving any degree of independent control which may be desired. As shown, the feed aperture is divided into many small sections, and each section is excited independently in the sum, azimuth difference, and elevation difference modes. The amplitude of each excitation is determined by the amount of coupling existing in the appropriate coupling element of the appropriate distribution network, while the necessary even or odd character of the excitation is provided by the series of hybrid junctions comprising the comparison network.[1]

It should be mentioned that any undesired irregularities or ripples in the excitation caused by a nonuniform field of an individual radiator can be effectively smoothed by making the spacings small compared to a wavelength. Of course the problems of impedance of the radiators, as well as coupling between radiators, would be very considerable. However, in principle, these problems could be solved.

Since the above scheme enables an ideal feed excitation to be approximated as closely as is desired, it may be stated that such an excitation is *theoretically realizable*.

Actually, the degree of independent control provided by the scheme of Fig. 1 is more extensive than that required to achieve the "ideal" excitation previously [1] indicated. In that case, the excitation was shown as being the product of the horizontal and vertical distributions, and the sum distribution was assumed to be utilized for the even plane of the difference mode. The scheme of Fig. 1, while capable of providing both these features, is not limited to doing so.

III. Practical Methods for Achieving Independent Control of Excitation

The scheme shown in Fig. 1 is theoretically able to provide independent control of excitation in the three

[1] This scheme demonstrates independent control not only for a feed in a focusing type of antenna, but also for an array type of antenna [7]; in the latter case independent control of the monopulse modes is also desirable. An alternate scheme for independent control of a practical array has been suggested by Mattingly [11].

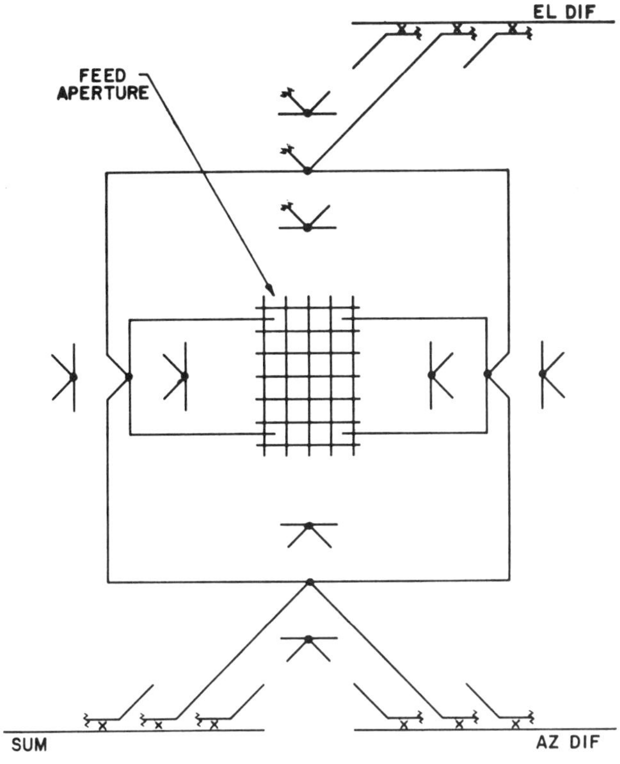

Fig. 1—A scheme for demonstrating complete independent control of the sum and both difference modes.

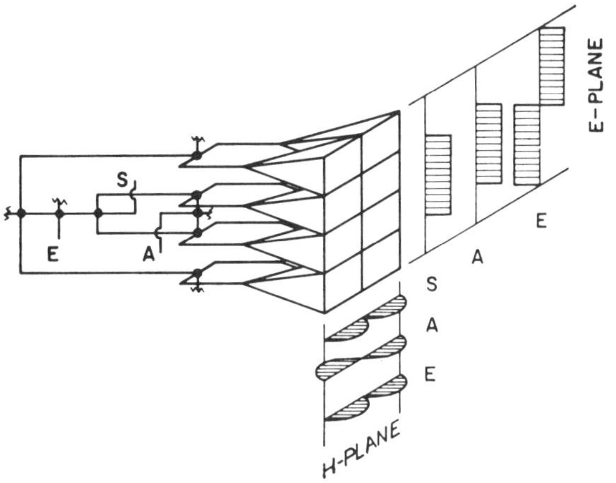

(a) Eight-horn feed for control in E plane.

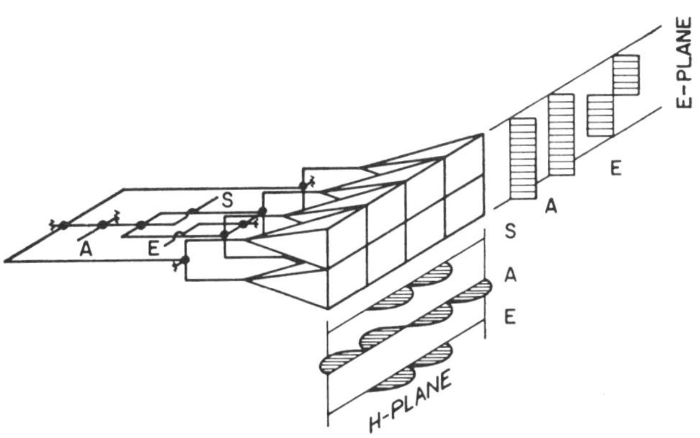

(b) Eight-horn feed for control in H plane.

Fig. 2—Single-plane multihorn feeds.

modes of operation. However, the large number of closely-spaced apertures and the multitude of circuits imply a low probability for achieving a satisfactory working design with a reasonable amount of effort. It is natural, therefore, to investigate some simpler schemes which may be more practical to develop, even though their feed excitations are not completely ideal.

There are two techniques which are considered here: one is termed the multihorn method and the other is called the multimode method. As an introduction, these techniques are first presented for the case in which independent control is the objective in the sum mode and *only one* difference mode, *i.e.*, control in only one plane. For simplicity, it will be assumed that the azimuth and elevation planes correspond to the H and E planes, respectively.

A. Single-Plane Multihorn Feeds

In Fig. 2, the multihorn method is illustrated, for either E-plane or H-plane control of 8 horns. Also shown is a network of hybrid junctions which would properly excite these horns. It is evident that in the plane of control, the desired two-to-one width of excitation is available. The shapes, however, are not ideal, particularly in the H plane. (This defect is also present in an ordinary four-horn cluster having no independent control.) Naturally, the two-to-one width cannot be obtained in the plane that lacks independent control, so that this feed would be far from optimum in a system requiring all three monopulse modes.

B. Single-Plane Multimode Feeds

Fig. 3 illustrates the multimode method, for either E-plane or H-plane control of two horns by means of three waveguide modes. (The terms "multimode method" and "waveguide modes" refer to the natural modes of propagation in a waveguide or horn; this should not be confused with the three so-called modes of operation, which are the sum, azimuth difference, and elevation difference modes.) Also shown in the figure are the two waveguide-mode combinations which in this case were chosen for the even excitation. As before, the desired two-to-one width of excitation is available in the plane of control. In addition, the shape of the sum mode in the plane of control is rather good. Some of the other shapes, however, are not ideal, especially those existing in the case of E-plane control. Again, of course, the two-to-one width cannot be obtained in the plane that lacks independent control, so that this feed would also be far from optimum in a system requiring all three modes of monopulse operation.

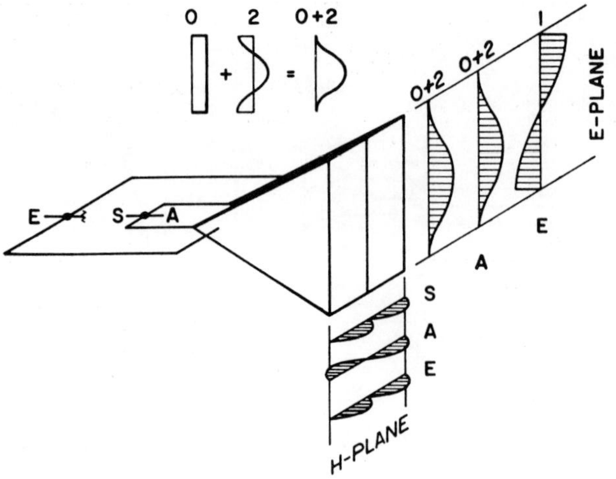

(a) Triple-mode feed for control in E plane.

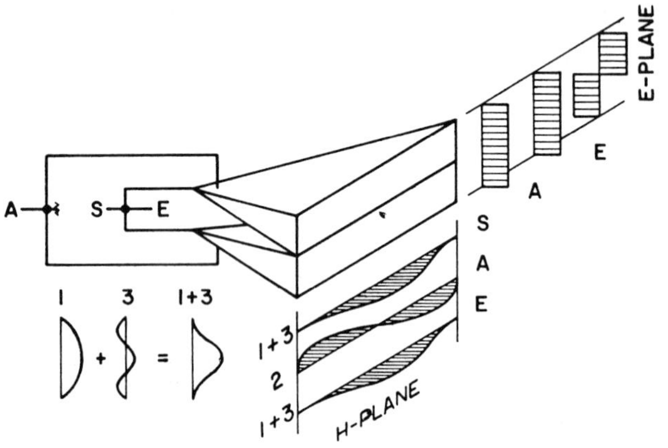

(b) Triple-mode feed for control in H plane.

Fig. 3—Single-plane multimode feeds.

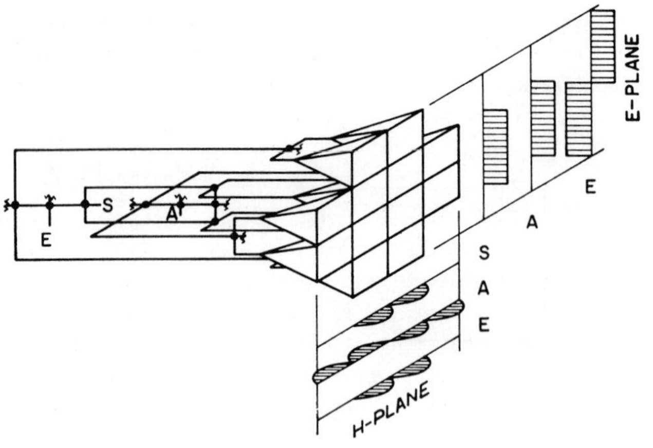

(a) Twelve-horn feed.

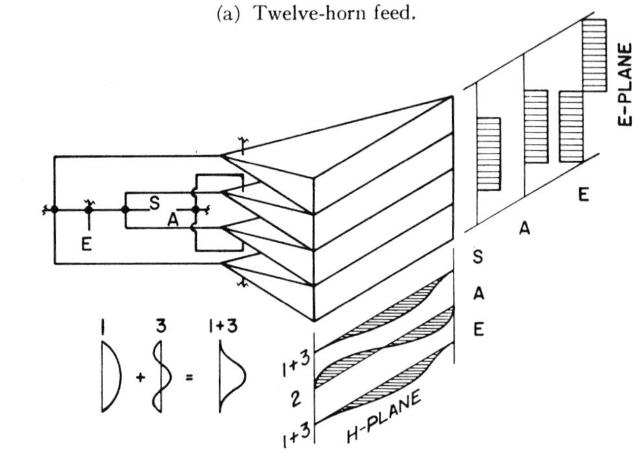

(b) Four-horn triple-mode feed.

Fig. 4—Dual-plane feeds.

C. Dual-Plane Multihorn Feed

Now consider the application of these methods towards obtaining independent control in the sum mode and *both* difference modes. Fig. 4(a) shows a multihorn method for the case of twelve horns, together with an appropriate network of hybrid junctions. In this case the desired two-to-one widths of excitation are available in both planes, but again, the shapes of excitation fall short of the ideal. It is interesting to note that this multihorn configuration is basically similar to the scheme of Fig. 1; however, the limited number of apertures and lack of amplitude control prevent complete optimization from being realized. On the other hand, a very substantial improvement in performance would be gained in comparison to the ordinary four-horn cluster; this will be illustrated quantitatively in a later section of this paper.

It should be realized that there are a number of other multihorn configurations available. For instance, the partitions in the outer eight horns of the twelve-horn feed may be removed to yield an eight-horn feed. This would give somewhat better performance in the elevation difference mode and would eliminate four hybrid junctions. However, these advantages would have to be weighed against the more complicated excitation shapes that would exist in such a feed. Another configuration would involve the addition of partitions or more horns to the twelve-horn feed. This would permit a closer approach to the scheme of Fig. 1, thereby allowing more complete optimization; however, the physical structure would become more complex.

D. Dual-Plane Multimode Feed

Having combined the single-plane multihorn methods of Fig. 2 into the dual-plane multihorn configuration of Fig. 4(a), it is natural to ask whether the multimode methods of Fig. 3 can be similarly combined. On investigation of this question, one concludes that, while there is no apparent theoretical limitation on such a process, it is quite unattractive because of the practical difficulties involved in properly controlling the many waveguide modes required. In addition, the elevation difference mode would still have a rather poor shape, unless even more waveguide modes were employed. However, it is of interest to note that a hypothetical scheme employing an unlimited number of waveguide modes would provide an approach to complete independent control leading to the same theoretical conclusions as did the scheme of Fig. 1.

E. Dual-Plane Multihorn-Multimode Feed

There are still further combinations available among the methods presented in Figs. 2 and 3; these combinations involve a multihorn method in one plane and a multimode method in the other. Fig. 4(b) presents such a case, in which the better plane of each method has been retained. The result is a feed excitation in which not only are the desired two-to-one widths available in both planes, but the shapes are effectively ideal in the H plane and depart only a moderate amount from the ideal in the E plane.

The feed as shown comprises four identical horns, each utilizing three waveguide modes. As might be expected, there are a number of alternate multihorn-multimode configurations which might be considered. For instance, the outer two horns might be reduced in width and excited with only the 1-mode; this would reduce the physical size of the feed at the expense of a somewhat more complicated set of excitation shapes. Another example involves the use of more than the four horns shown, thereby permitting greater control of the excitation in the E plane. This would allow a closer approximation to the ideal shapes, at the expense of a more complex excitation network.

Returning to the simple configuration shown in Fig. 4(b), it is interesting to mention the factors relating to its usefulness in a real antenna system. As regards the practicability of this scheme, it may be noticed that the configuration of the hybrid network is less complex than that of Fig. 4(a), and is confined essentially to a single plane. The generation of the proper waveguide modes is certainly a problem of design; however, these modes lend themselves to techniques which are quite feasible, and can yield results very close to the ideal. Thus the "multihorn-multimode" method appears to be most attractive as a practical scheme for realizing almost completely the benefits obtainable with independent control of the sum and both difference modes.

IV. Comparison of Performance for Several Practical Feeds

FEED EXCITATION WITH $\left(\frac{aA}{2\lambda F}\text{ OR }\frac{E_3}{E_1}\right)$ AND $\left(\frac{bB}{2\lambda F}\right)$ FOR MAXIMUM SUM GAIN	SUM GAIN RATIO	SUM SPILLOVER RATIO	AZ DIF SLOPE RATIO	AZ DIF SPILLOVER RATIO	EL DIF SLOPE RATIO	EL DIF SPILLOVER RATIO
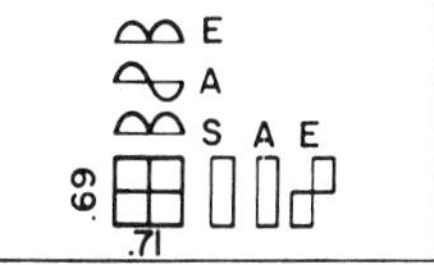	.58	.34	.52	.72	.48	.76
(a) FOUR-HORN						
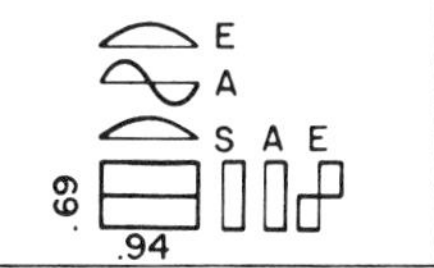	.75	.16	.68	.50	.55	.69
(b) TWO-HORN DUAL-MODE						
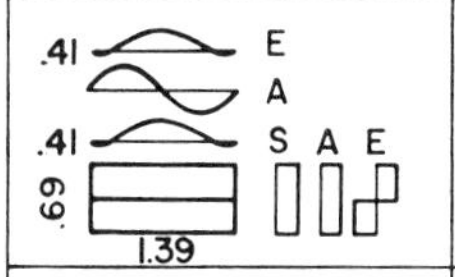	.75	.17	.81	.20	.55	.69
(c) TWO-HORN TRIPLE-MODE						
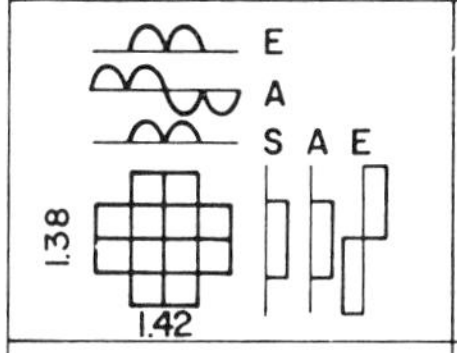	.58	.34	.71	.37	.67	.38
(d) TWELVE-HORN						
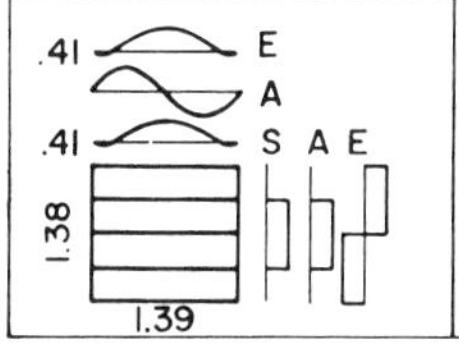	.75	.17	.81	.20	.75	.22
(e) FOUR-HORN TRIPLE-MODE						

Fig. 5—Antenna-pattern characteristics for various practical feeds.

It is instructive to compare a few practical feeds on a quantitative basis. Fig. 5 presents such a comparison, based on calculated values for three properties: sum gain ratio, difference slope ratio, and spillover ratio. For simplicity, the size and the mode ratio of each feed is chosen to yield maximum sum gain; as discussed in the previous paper [1], this is a somewhat smaller size than that which might be chosen as an optimum when several properties are considered. As a result, the listed gains and spillovers are somewhat greater than they might be in a more typical case. Furthermore, the slopes turn out to be appreciably different from those which would be obtained with a larger feed. The assumptions made regarding a rectangular aperture, a large feed in wavelengths, etc.,[2] are the same as were made in the previous paper, and all the definitions and symbols are the same. The values listed are the complete two-dimensional quantities, as obtained by multiplying together the proper one-dimensional values. As before, it is assumed that the azimuth and elevation planes correspond to the H and E planes, respectively.[3]

[2] Since practical antenna defects (aperture blocking, reflections, cross-polarization, "space-attenuation," dissipation, tolerances, etc.) are neglected, the listed gains and slopes are likely to be higher than the values encountered in actual practice. On the other hand, the assumed large feed in wavelengths may have somewhat more spillover loss, and hence less gain, than one about a wavelength in size, particularly in those cases involving poor shapes of excitation.

[3] In the case of feeds giving lower spillover in the H plane than in the E plane, the harmful effects of the ground might be minimized by reversing this assumption.

243

A. Four-Horn Feed

The ordinary four-horn cluster is shown in Fig. 5(a). This has, of course, the defects associated with an excessively narrow difference excitation and the poor double-cosine shape for the H-plane even excitation. The main virtue of this feed appears to be its conceptual simplicity [5], [6], including its susceptibility to the historical lobing method of analysis. It also has a relatively small aperture size.

B. Two-Horn Dual-Mode Feed

In Fig. 5(b), another common feed is shown; in this type the E-plane partition, present in the four-horn cluster, has been removed. This allows two waveguide modes to propagate, and the H plane is characterized by a single cosine for the even excitation and a sine for the odd excitation. The single cosine is not only an efficient shape, but its effective width is narrower than that of a double cosine of the same over-all dimensions. As a consequence, the over-all feed-aperture size in the H plane can be increased, thereby allowing the azimuth-mode excitation to widen toward the desired value. However this width falls short of the ideal two-to-one difference-to-sum condition, and no further improvement can be made with only the two modes which are available. In addition, of course, the elevation-mode excitation remains too narrow. This type of feed [12] has found most use [3] in applications which place a premium on a simple mechanical structure and relatively short focal length.

C. Two-Horn Triple-Mode Feed

In Fig. 5(c) the feed shown in Fig. 3(b) is analyzed. Here the H-plane dimension of the feed has been set to maximize the difference slope, and then the mode ratio of the even excitation (E_3/E_1) has been set so that the sum gain is also maximized. While the performance in the sum and azimuth difference modes is excellent, that in the elevation difference mode, of course, remains poor. This type of feed is most helpful in applications [4] which require a very short focal length; it is believed that RCA was the first to incorporate the triple-mode technique in a monopulse antenna.

It is evident that the feeds of both Part B and Part C suffer from lack of independent control in the elevation plane. It should be noted that in certain antenna applications there may also be a problem caused by the different antenna patterns which would occur in the azimuth and elevation planes. These defects are virtually eliminated by the incorporation of independent control in both planes. The two practical feeds which achieve this are to be shown in Part D and Part E.

D. Twelve-Horn Feed

In Fig. 5(d) the feed shown in Fig. 4(a) is analyzed. Here the dimensions of every horn are chosen to be identical, for simplicity and ease of calculation. This happens to yield the two-to-one width condition, so with the size set to maximize sum gain, the difference slope is fairly close to the maximum value obtainable with this feed. As expected, the performance is good in all three modes, but is still somewhat less than optimum because of the inefficient shape of the feed excitation in the H plane. In many applications, this defect, together with the complexity of the circuitry, would tend to limit the attractiveness of the twelve-horn feed. On the other hand, it may be the only possible solution in some cases. One such case might be a diversity-polarization antenna using dual-polarized horns; perhaps another case is a low-frequency antenna in which a dipole feed array must replace the horn array. There are other, special uses for independent control in monopulse antennas where the twelve-horn method may be the most practical solution.[4] It is believed that the twelve-horn method was first employed by Lincoln Laboratory in a feed [9] developed for a versatile monopulse antenna.

E. Four-Horn Triple-Mode Feed

In Fig. 5(e) the feed shown in Fig. 4(b) is analyzed. Here the H-plane dimensions and mode ratios are set as they are in Part C, and the E-plane dimensions as in Part D. It is evident that this feed yields the best performance in all three modes when compared with any of the others. Furthermore, the performance is rather close to that obtainable with the "ideal" feed [1].

The numerical results given in Fig. 5 illustrate the advantages to be gained with a practical form of independent control. In going from the ordinary four-horn cluster to the four-horn triple-mode feed, for instance, the sum gain has increased by 1.1 db and the difference slopes have increased by 3.9 db, while the total sum spillover has decreased to about $\frac{1}{2}$ and the total difference spillovers to about $\frac{1}{4}$ of their original values. The reduction of difference spillover is particularly significant, since originally it was so great. Although not listed in Fig. 5, various other properties of the difference modes, such as gain, sidelobe level, and criticalness to misalignment are, as discussed in the previous paper [1], also greatly improved.

[4] For example, certain antenna systems obtain sequential lobing by combining the sum and difference signals of a monopulse antenna. Where independent control is available, the lobing performance may be greatly improved. With a twelve-horn feed and comparator, this feature is relatively straightforward to provide, as follows. Those ports of the comparator which would furnish even excitation of the outer horns in Fig. 4(a) are not terminated but are properly coupled to the difference ports, and these latter ports are combined alternately in-phase and out-of-phase with the sum ports in the proper manner. The result is a sequential lobing system having high crossover level and gain together with low sidelobes and spillover, compared with a lobing system using a four-horn feed. Lincoln Laboratory has worked out [9] some alternate schemes which instead provide independent channels for sequential lobing and communication.

V. Discussion

The analyses in these papers have assumed that the feed aperture is large compared with a wavelength. In many antennas this simplifying condition does not exist, and it is appropriate to ask whether the conclusions obtained herein would be significantly altered. It happens that when the feed aperture becomes small compared with a wavelength, one particular theoretical approach indicates that the main illumination would appear to resemble the optimum, regardless of the widths and shapes of the feed excitation. In such a case, therefore, one might be led to expect that a "simple" feed, not incorporating independent control, would yield good monopulse performance. While it is possible that some of the benefits of independent control could be obtained in this manner, the electrically small size of such a feed would introduce difficulties and defects which would considerably degrade the antenna performance. In the ordinary single-beam antenna, these problems are well-known; they include narrow bandwidth of impedance match, high-power limitations, and the likelihood of excessive cross polarization. For a monopulse antenna, these difficulties are usually aggravated, and additional feed problems, such as defocusing errors and coupling between radiators, are encountered. Thus it is believed that an electrically small feed should be avoided.

Another possibility exists for achieving the benefits of independent control without the additional circuits or waveguide modes described in this paper. Certain devices, when placed just in front of an ordinary monopulse feed, have the property of affecting the radiation of the feed differently for the various modes of operation. For example, a thin metal fin in the central H plane of a feed affects only the difference mode having odd symmetry about the fin. Such a fin was employed about seven years ago in a monopulse antenna [5], [6] designed by Wheeler Laboratories; it narrowed an excessively wide radiation from a four-horn feed in one difference mode, as well as improving the impedance match in this mode. However, it was found that the fin could not be designed to provide an effect equivalent to the ideal two-to-one feed size without introducing severe defects in various other properties of the nonopulse antenna. More recently, a similar limitation on the benefits of a fin was found in the case of a two-horn triple-mode feed. Although there may be some exceptions, it is probably fair to consider these experiences typical. Therefore it is believed that "parasitic" feed devices, while quite useful for certain limited purposes, are not a reliable means for designing an optimum monopulse feed.

Returning to the question of feed size, there remains to be considered the more common case, in which the feed aperture is neither large nor small compared with a wavelength. Under these conditions, the ordinary feed design may incorporate to some degree those desirable radiation properties obtainable with independent control. Nevertheless, when an independently controlled feed is designed, an improvement approaching that indicated in previous portions of this paper is likely to be obtained. About five years ago, for example, measurements were performed at Wheeler Laboratories with a monopulse antenna which initially utilized an ordinary four-horn cluster as the feed. The over-all aperture dimensions of this feed were about $3\lambda/2$, which was electrically just large enough to yield acceptable performance in a particular application, and the edge taper of the main illumination was about 11 db in the sum mode for maximum gain with a circular main aperture. Upon substitution of a feed having the same effective aperture size but a more efficient excitation shape, the antenna gain in the sum mode increased by about 1.0 db. When a further substitution was made in which one dimension of the feed aperture was increased to approach the desired two-to-one condition, the gain in the corresponding difference mode increased by more than 3.0 db and the sidelobe level changed from 12 db to 18 db below the difference peak. Thus the expectation that a wavelength-size feed aperture would encounter almost the full benefit from independent control has been experimentally confirmed for a particular case.

In some more recent antenna designs by Wheeler Laboratories, the multihorn-multimode technique has been applied to obtain simultaneous optimum performance in the sum and both difference modes. One case [10] involves a feed aperture very large compared with a wavelength, while another has over-all aperture dimensions of about 2 wavelengths. Both cases have yielded the benefits expected from independent control, and have proven satisfactory from a practical standpoint. It is concluded that the incorporation of a practical form of independent control in the feed of a monopulse antenna provides a worthwhile degree of improvement. In addition, the low spillover and noncritical nature of an antenna of this type suggests that independent control may permit some unusual antenna applications in which monopulse would ordinarily not be feasible.

VI. Acknowledgment

The author would like to thank R. L. Mattingly, who has been principally responsible for the direction of this effort by BTL, for the encouragement he has given it.

At Wheeler Laboratories, this work has been directed by H. A. Wheeler, D. Dettinger, N. A. Spencer, and P. A. Loth. The author would like to acknowledge the cooperation of the Wheeler staff in the preparation of this paper, especially H. A. Wheeler, H. W. Redlien, and D. S. Lerner for their helpful comments and suggestions, S. I. Warshaw and B. J. Karafin for their computational effort, and C. J. Stona for his work on the illustrations.

VII. Bibliography

[1] P. W. Hannan, "Optimum feeds for all three modes of a monopulse antenna I: Theory," IRE Trans. on Antennas and Propagation, vol. AP-9, pp. 444–454; September, 1961.
[2] H. W. Redlien, Wheeler Labs., unpublished Rept. No. 557, ASTIA No. AD 116472; October, 1952.
[3] W. A. Snyder, Hughes Aircraft Co., unpublished paper, *Abstracts of 5th Annual Symp. on USAF Res. and Dev.*, held at University of Illinois, ASTIA No. 90397; October, 1955.
[4] "Final Report on Instrumentation Radar AN/FPS-16 (XN-2)," Radio Corporation of America, unpublished Rept., ASTIA No. AD 250500, pp. 4-123, 4-125.
[5] P. A. Loth, Wheeler Labs., unpublished Rept. No. 686; September, 1959.
[6] P. W. Hannan, Wheeler Labs., unpublished paper, *Record of the 6th Ann. Radar Symp.*, held at University of Michigan, ASTIA No. AD 318072, pp. 165–194; June, 1960.
[7] H. W. Redlien, "A Unified Viewpoint for Amplitude and Phase Comparison Monopulse Tracking Radars," Wheeler Labs., unpublished Rept. No. 845 to Bell Telephone Labs.; March, 1959.
[8] J. P. Shelton, "Improved feed design for amplitude monopulse radar antenna," 1959 IRE National Convention Record, pt. 1, pp. 93–102.
[9] L. J. Ricardi and L. Niro, "Design of a twelve-horn monopulse feed," 1961 IRE International Convention Record, pt. 1, pp. 93–102.
[10] P. W. Hannan and P. A. Loth, "A monopulse antenna having independent optimization of the sum and difference modes," 1961 IRE International Convention Record, pt. 1, pp. 57–60.
[11] R. L. Mattingly, Bell Telephone Labs., unpublished Memo. No. 27495-42; August, 1959.
[12] P. G. Smith and C. E. Brockner, "Waveguide Hybrid Network for Monopulse Comparator," U. S. Patent No. 2,759,154; November, 1954- August, 1956.
[13] H. Jasik, "Antenna Engineering Handbook," McGraw-Hill Book Co., Inc., New York, N. Y., pp. 25–29, 25–30; 1961.

The Future of Pulse Radar for Missile and Space Range Instrumentation*

DAVID K. BARTON†, SENIOR MEMBER, IRE

Summary—An account of instrumentation radar development is given, and advantages and disadvantages of radar as compared to other instruments are discussed. Capabilities of present monopulse radars are described, based upon actual test data from the AN/FPS-16. This radar has a range of 200 miles on echo targets of one-square-meter cross section and can track to an accuracy of 0.1 mil in angle and 5 yards in range. The next generation in instrumentation radar is represented by the AN/FPQ-6, now under development, which will extend accurate tracking to ranges in excess of 500 miles on echo targets and will track existing beacons beyond the moon. An important capability not yet exploited in pulsed-instrumentation radars is the coherent pulsed-Doppler velocity-measurement channel which will equal the accuracy of microwave CW systems in radial-velocity data. Provisions for adding this fourth tracking channel to both AN/FPS-16 and AN/FPQ-6 are being made, and suitable beacons are being designed. An important advantage of pulsed-Doppler radar is the ability to share a single coherent beacon in multiple-station operation, providing highly accurate, three-coordinate velocity and position data without special interstation communication links over ground paths.

Beyond the immediate developments of the AN/FPQ-6, there are three major areas of improvement which will greatly extend radar performance. Solid-state maser preamplifiers will increase sensitivity of microwave radars by a factor of nearly one hundred within one or two years. Microwave antennas are already under construction in the 100- and 300-ft diameter class. Multiple-tube transmitters will make available ten to one hundred times the presently used average power, and signal-processing techniques are available to code the transmissions for accurate measurement of range and velocity. As a result of these developments, the improvement in radar performance should proceed at a more rapid pace even than that of the past ten years.

The unique ability of radar in acquiring and tracking uncooperative objects has been appreciated for some years, and examples of actual tracks are presented to show some of the interesting data which can be extracted from satellite echo signals.

* Received by the PGMIL, June 2, 1961.

† RCA Missile and Surface Radar Dept., Moorestown, N. J. Development and production of the AN/FPS-16 was carried out under cognizance of the Navy Dept. on Contracts NOas-55-869c and NOas-56-1026d. The AN/FPQ-6 and AN/TPQ-18 are currently being developed under cognizance of the Navy Bureau of Weapons, Contract NOw-61-0428.

I. Background and Status of Instrumentation Radar

THE future needs of space-range instrumentation will be filled by evolution of present missile-instrumentation systems and by introduction of new equipment and techniques made possible by the many scientific programs in the areas of radar, communications and astronomy. Pulsed-tracking radar instruments can be expected to play a major role in future systems, just as they have in the past. Before discussing the lines along which instrumentation radar evolution is taking place, it is appropriate to review briefly the history of present radar systems, and establish the status and capabilities of the equipment now in use.

Early U. S. Systems

248 The development of high-altitude research rockets and guided missiles in the United States brought together instrumentation methods from many fields. From the aircraft test field came internal instrumentation, radio telemetry and recording devices. Antiaircraft artillery evaluation has made extensive use of theodolites, and these were brought to bear on missiles in the earliest stages of development. Various fixed-camera methods were derived from other ordnance projects as well as from the field of astronomy, which also contributed the long focal-length telescope. Electronic instrumentation came from the fields of navigation, direction finding, fire control, close-support bombing, and from specialized methods such as the Doppler radio techniques used by the Germans in the V-2 program. Since the program got under way toward the end of World War II, it was logical that military equipment, surplus and captured, would play a significant role, and this was certainly true in the case of radar equipment.

The SCR-584 Radar

One radar in particular, the SCR-584, proved particularly valuable to the test ranges and has maintained its usefulness to this day. The primary reasons for this were its availability (some fifteen-hundred sets had been built by the end of the war), reliability (a tremendous production and field engineering effort had followed the original MIT Radiation Laboratory development work), and flexibility. Designed for control of 90-mm AA guns and similar weapons, the SCR-584 was equipped with alternate forms of data output, an oversized modulator and power supply, and ample space for additions and modifications that proved necessary. Furthermore, development of instantaneous electronic-plotting boards (operating in rectangular coordinates) had already been carried out for use in close-support bombing and in mortar location. Transponder beacons were also available from the close-support bombing program, as was a standard radar modification for long-range tracking of beacons. Thus, the SCR-584 was ready for the earliest V-2 and Wac Corporal firings at the White Sands Proving Ground and was in use at other rocket ranges. Along with the radar, of course, there were used the famous German Askania theodolite, Bowen-Knapp fixed cameras, and Doppler radio system.

Modifications and Refinements

The flexibility of the original SCR-584 has already been mentioned. Dozens of important field modifications were made in order to incorporate design advances to this radar at various times in the range instrumentation programs. These were concerned primarily with extending the range of tracking and range rates, synchronizing adjacent radars for reliability of operation without interference, recording of position and signal-strength data and providing control links to the missile through modulation of radar pulse rate or pulse code. Ultimately these and other improvements were incorporated through factory modification in the AN/MPQ-12 radar, which first became available in 1949, and in the improved AN/MPQ-18, which followed the MPQ-12 in design and has been operational since late 1954. These radars have been used primarily at the White Sands Missile Range, but similar modifications have been applied to the SCR-584 for use at the Pacific Missile Range, NOTS (China Lake), the Atlantic Missile Range, Edwards Air Force Base, and the Gulf Test Range. Other nomenclature applied to modified SCR-584 radars includes AN/MPG-2 (referring to an X-band set using the MC607 kit on a standard SCR-584), tracking radar Mod II (similar to the MPQ-18), and AN/MPS-25 (a C-band version used at the Pacific Missile Range). In each case the basic SCR-584 pedestal, modulator and power supply, and 30-cycle conical scan is employed, while most other major components have been replaced or extensively modified. In some cases, the cost of modification has run to two or three times the original cost of the radar.

Parallel to evolution of improved radars has been a great effort in system design, involving chain radar operation, real-time computing, plotting and data transmission, precision data recorders, communications systems, transponder beacons, and control and telemetry equipment The effect of this effort, which will not be described in detail here, has been to provide further uses for radar data and to keep a continuing pressure towards development of radars with greater precision and range, so that significant data can be supplied to the instrumentation system as a whole.

The AN/FPS-16 Instrumentation Radar

By 1953, military development programs had made possible the design of highly precise tracking radars using the "monopulse" method [1], with adequate range and accuracy to serve as primary missile-range instrumentation. Development of such radars was carried out on a triservice program under cognizance of the Navy Bureau of Aeronautics. The first experimental model, designated AN/FPS-16 (XN-1), was completed and tested by June, 1956, and installed early in 1957 at Patrick Air Force Base for use on various missile and satellite programs. This model (see Fig. 1) was housed in a trailer, with the antenna on a solid base nearby. A second model, designated AN/FPS-16 (XN-2), was built as a fixed-station radar and served as a prototype for subsequent production of 46 sets for use in all parts of the world. The XN-2 equipment (Fig. 2) was installed at Grand Bahama Island in June, 1957, and has served since that date in support of all test programs at the Atlantic Missile Range. The distribution of production models of the AN/FPS-16 is shown in Fig. 3. Major characteristics of the radar already described in the literature [2] are repeated in the Appendix for reference.

Fig. 1—Instrumentation radar AN/FPS-16 (XN-1).

Fig. 2—Radar set AN/FPS-16 (XN-2) at Grand Bahama Island.

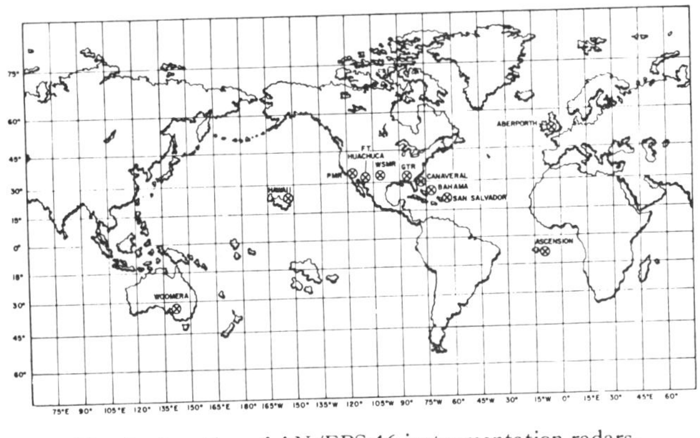

Fig. 3—Location of AN/FPS-16 instrumentation radars.

AN/FPS-16 Modifications

As with the venerable SCR-584, the new AN/FPS-16 has been modified successively to increase its range, accuracy and adaptability to new requirements. Some of the major modifications will be listed here to indicate the current status of different versions of the radar.

1) *AN/FPS-16 (XN-3).* A fairly extensive modification program was initiated by the Signal Corps soon after the AN/FPS-16 program began. The changes involved the use of a 3-Mw tunable klystron to replace the original magnetron transmitter, addition of a circularly-polarized mode of operation to supplement linear polarization in beacon tracking, replacement of the boresight camera with a closed-loop TV system, and inclusion of a data corrector to overcome mechanical limitations of the antenna pedestal and extend the bandwidth of the data system to 20 cps. The XN-3 radar remains in operation at Moorestown, N. J., as a base for further modifications, and three field radars, designated AN/FPS-16 A-X, are operating at the White Sands Missile Range. The circular polarizer and data correction provisions have also been applied to several PMR and AMR radars.

2) *Range System Modifications.* Field modifications to provide continuous tracking to 500 miles have been installed in several radars, and such trackers are standard equipment on the later production sets. In addition, a digital range tracker with a 5000-mile range and provision for automatic resolution of 500-mile ambiguities has been developed and installed in five radars. The digital range tracker makes possible the acquisition of beacon-equipped satellites and missiles as soon as they rise above the horizon, and provides for continuous tracking as long as the target stays within line-of-sight range. Tracking can proceed without interruption beyond 5000 miles, on strong enough signals, with ambiguities at intervals of 8192 miles.

3) *16-ft Reflector.* Using the original pedestal, the reflector size has been increased from 12 to 16 ft on five radars, and the 16-ft reflector is standard on late production units.

4) *AN/MPS-25.* A new trailerized model of the radar, using the same basic circuits as the AN/FPS-16, has been built, and seven units are in use.

5) *Acquisition Console.* Added acquisition displays and controls have been placed in eleven sets, primarily to assure rapid acquisition on programs such as Mercury.

6) *Improved Receivers.* Improved crystal mixers giving a 7-db system noise figure and parametric amplifiers giving 4–5 db have been developed for use with the various models of the AN/FPS-16 radar.

Although the possibilities of further piecemeal modification of the AN/FPS-16 are far from exhausted, a program for a major step forward in an integrated design was undertaken late in 1960. This program calls for combining the major modifications developed for the AN/FPS-16 with a newly-developed antenna and pedestal, to produce both fixed-station equipment (AN/FPQ-6) and transportable units (AN/TPQ-18). These developments will be discussed more fully in a later portion of this paper.

Advantages and Disadvantages of the Pulsed Radar Method

Pulsed tracking radars, starting with the SCR-584 and proceeding through AN/FPS-16 to AN/FPQ-6, have been adopted as standard instrumentation on all the world's missile ranges. The reasons for this choice, along with some of the negative factors to be overcome, can be discussed in a general way here. The principal advantages of the radar method are as follows:

1) Complete data are obtained at a single station.

2) Data are available in electrical form.

3) Reduction of data to rectangular coordinates (or other three-coordinate system) is simple.

4) As a result of 2) and 3), very rapid data processing is possible with small-analog computers, and real-time reduction of precise data is feasible with large-scale digital machines.

5) Operation is possible in all weather conditions.

6) Operation is possible either by reflection from a noncooperating target, or by use of a transponder carried by the target. In the latter case, range is usually limited only by line-of-sight.

7) Interference and multipath propagation troubles are minimized by the narrow angular beam and the narrow range tracking gate.

8) The measured coordinates always intersect at right angles, minimizing geometrical dilution of precision as found in triangulation and base-line systems.

9) Ionospheric effects are negligible if frequencies above 3000 Mc are used.

10) Measurements do not depend upon availability of continuous past history of the trajectory, as with CW Doppler systems.

11) Control and telemetry channels are available when a transponder is used.

12) Reliable equipment is available in production quantities with a high degree of standardization.

13) The radar can be operated and maintained by as few as two men, and maintenance personnel are available from both commercial and military organizations. Manpower required for data transmission and processing is also reduced to a minimum.

Chief limitations of radar methods as currently used are as follows:

1) The radar may be a large, complex and rather expensive instrument.
2) Mechanical error in the antenna pedestal and noise in the tracking servo loop must be contended with.
3) Power consumption of the radar is relatively high.
4) Velocity is generally obtained from differentiated position measurements, with resulting errors depending on smoothing.
5) Tropospheric refraction introduces error in the radar measurement.
6) Tropospheric attenuation is present at the higher radar frequencies.
7) No information is normally obtained on target aspect (attitude).
8) Tracking at low altitudes is affected by ground-clutter reflection in the antenna beam and side-lobes, and line-of-sight limitations apply.

In many cases, these limitations may be overcome by improved siting, operational procedures, or refinements in design, as will be discussed later. Many of the above limitations also apply with equal or greater force to instrumentation systems other than radar.

Performance of Monopulse Radars

The post-war development of monopulse tracking radars has led to a number of highly-refined types of equipment, used in both tactical and instrumentation systems. The AN/FPS-16 is in many ways typical of all such radars, and since it is an unclassified equipment, as well as the only such radar designed specifically for range instrumentation, its performance will be discussed here as indicating the current capability of field radar equipment. In a later section, some of the experimental advances leading to much longer-range radar of comparable accuracy will be covered.

In the AN/FPS-16 development, significant advances were made toward easing the following limitations of the radar method:

1) *Mechanical and Servo Error.* Completely new pedestal design, based on analysis of errors in radar pedestals previously developed, brought total mechanical error below the 0.1 mil level. Careful electrical design reduced servo and target scintillation errors to negligible factors, while the rms extent of target glint for echo tracking is held to about one-fourth the extent of the target itself. Test data substantiating this is given in [2] and [3].

2) *Velocity Data.* The spectrum of position errors was controlled to permit the use of lower smoothing time in derivation of velocity. This was made possible by achievement of servo bandwidths up to 5 cps in angle and 10 cps in range.

3) *Tropospheric Refraction.* Correction formulas were derived to minimize this error.

4) *Tropospheric Attenuation.* A medium frequency (5400 to 5900 Mc) was found to reduce effects of attenuation and masking in clouds, rain, and moist atmosphere, as compared with higher frequencies. At the same time, beamwidths in this band can be made narrow enough, with reasonable antenna size, to reduce error at low-elevation angles, as compared to *S*-band operation.

5) *Low-Altitude Tracking.* Sidelobes were reduced to the point where accurate tracking was possible within 3° of the ground. Below this angle, use of special screens or radar fences may further extend the tracking range.

The net result of these developments was to make possible radars for instrumentation which are capable of a full order of magnitude improvement in instrumental accuracy over that obtained with modified SCR-584's. Range at which precise tracking is possible on similar targets is improved almost 10:1 over the original SCR-584, and 3:1 over the most advanced modifications of this radar used in range instrumentation. Angular precision of 0.03 mil and absolute accuracy of 0.1 mil are available after correction for propagation effects. Range precision of 1 to 2 yards rms and absolute accuracy of 5 yards rms are available after correction of propagation effects. Table I gives the results in terms of space positions.

Analysis of Errors in Monopulse Radar

Within the past few years a great deal of work, both theoretical and practical, has gone into the analysis of radar tracking errors. References [2], [3] and [4] present the results of such analysis on the AN/FPS-16 radar, and the following discussion summarizes these results and illustrates the type of data obtained in verification of the theoretical findings.

It has been common practice to represent radar-angle errors as the sum of three components, as shown in Fig. 4. As radar development proceeded and tracking range was extended, it became necessary to break the errors down into a large number of small components, analyze how each one varied with operating conditions, and devise tests to verify each such relationship. This has been accomplished for the AN/FPS-16, with the error breakdown in angle shown in Table II.

TABLE I
Errors in Monopulse Radar Position Measurement

Range from Radar (n.m.)	Precision (ft rms)	Accuracy (ft rms)
5–10	5	15
10–20	7	20
20–40	10	30
40–100	15	50
100–200	30	100
200–400	60	200
400–1000	150	400
1000–2000	300	1000
2000–4000	600	2000
4000–10,000	1500	4000

Note: Based on coordinate having largest error.

251

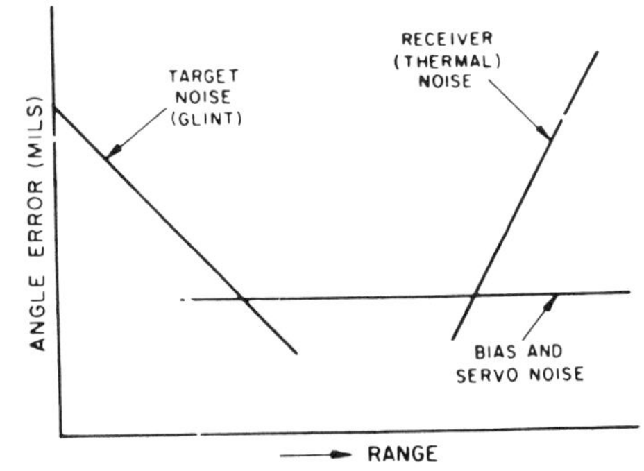

Fig. 4—Theoretical radar errors vs range.

TABLE II
Inventory of Angle Error Components

	Bias	Noise
Radar-dependent tracking errors (deviation of antenna from target)	Boresight axis collimation Axis shift with: RF and IF tuning Receiver phase shift Target amplitude Temperature Wind force Antenna unbalance Servo unbalance	Receiver thermal noise Multipath (elevation only) Wind gusts Servo electrical noise Servo mechanical noise
Radar-dependent translation errors (errors in converting antenna position to angular coordinates)	Levelling of pedestal North alignment Static flexure of pedestal and antenna Orthogonality of axes Solar heating	Dynamic deflection of pedestal and antenna Bearing wobble Data gear nonlinearity and backlash Data take-off nonlinearity and granularity
Target-dependent tracking errors	Dynamic lag	Glint Dynamic lag variation Scintillation Beacon modulation
Propagation errors	Average refraction of troposphere Average refraction of ionosphere	Irregularities in tropospheric refraction Irregularities in ionospheric refraction
Apparent or instrumentation errors (for optical reference)	Telescope or reference instrument stability Film emulsion and base stability Optical parallax	Telescope, camera or reference instrument vibration Film-transport jitter Reading error Granularity error Variation in optical parallax

Detailed measurements of the "fixed" components (those whose rms values are the same for most operating conditions) were made and verified by over-all system measurements under controlled conditions. The results were as follows:

Component	Bias (mils)	Noise (mils)
Mechanical	0.04 }	
Electrical	0.04 }	0.025
Data system	0.01	0.035
Total (rms) "fixed" error	0.06	0.04

In order to evaluate those components whose values vary with operating conditions, a series of equations was derived to describe the dependence of error upon various radar, target and environmental factors. A summary of these equations for radar angle measurement is given in Table III. Experimental data were taken under conditions simulating missile tracking, as well as during aircraft tracking. For simulation of missile- and space-tracking conditions and beacon operations, a series of tracks were run on small reflective targets carried by free balloons. Fig. 5 shows the results of a typical tracking run made with the AN/FPS-16 on a 6-in metal sphere carried aloft inside a free balloon. In this figure, azimuth tracking noise is plotted against range (the elevation error plot is similar). The boresight film was analyzed to determine the standard deviation of the error, which is defined as the difference between the radar-pointing direction and the optical line-of-sight to the center of the balloon. Each of the points plotted on the graph represents a sample analyzed over a period of 30 seconds at intervals during the run. The circles and triangles represent two different servo bandwidths.

The lines which have been drawn in on the graph represent theoretical limits or explanations of the results. The left portion of the plot is well-fitted by a straight line of negative slope representing inverse proportionality to range. The explanation of this lies in the fact that the metal sphere is not rigidly supported inside the balloon, but can wander around a bit. Since the radar sees the sphere and the camera sees the balloon, there is an error. The line drawn at the left of Fig. 5 corresponds to a linear error of $1\frac{1}{2}$ in at the balloon.

The line rising to the right has a slope representing proportionality to the square of the range. This is the theoretical lower limit of error due to thermal noise in accordance with the formula appearing in Table III. The points are in fairly good agreement with this line.

Running horizontally at about 0.01 mil is a line which represents the lower limit set by the error in recording and reading the data. In some of the runs, it has been found that the points in the center region did, in fact, closely approach this line. In Fig. 5, this is not the case. This run has been purposely selected because it illustrates another interesting and important effect, namely, random errors due to propagation anomalies or atmospheric blobs of varying index of refraction.

Muchmore and Wheelon [16] have derived the propagation-noise formula given in Table III. If the following typical values are taken:

$$\overline{\Delta N} = 0.5$$
$$L_0 = 50 \text{ ft}$$
$$L = \text{total range to target,}$$

TABLE III

SUMMARY OF TRACKING ERROR EQUATIONS

Thermal Noise (for $S/N > 1$)	$\sigma_t = \dfrac{\theta}{\sqrt{\dfrac{2S}{N}\dfrac{f_r}{\beta_n}}}$ radians	(1)
Multipath	$\sigma_n = \dfrac{\theta_p}{\sqrt{8A_s}}$ radians	(2)
Target Glint	$\sigma_s \cong \dfrac{1}{4}\dfrac{L_t}{R}$ radians	(3)
Servo Lag	$\sum = \dfrac{\omega_t}{K_v} + \dfrac{\dot{\omega}_t}{K_a}$ radians	(4)
Propagation noise	$\sigma_p \cong 2\Delta N \times 10^{-6}\sqrt{L/L_0}$ radians	(5)

Symbols:

θ = radar beamwidth (rad)
S/N = signal-to-noise power ratio
f_r = repetition rate (pps)
β_n = servo bandwidth (cps)
ρ = ground reflectivity (voltage ratio)
A_s = sidelobe attenuation (power ratio)
L_t = span of target } (any consistent units)
R = range of target }
ω_t = target angle velocity (rad/sec)
$\dot{\omega}_t$ = target angle acceleration (rad/sec²)
K_v = velocity error constant (1/sec)
K_a = acceleration error constant $\cong 2.5\beta_n^2$ (1/sec²)
ΔN = rms variation in tropospheric refractivity (N-units, or $[n-1]\times 10^6$)
L = length of tropospheric anomalies } (any consistent units)
L_0 = scale length of anomalies }

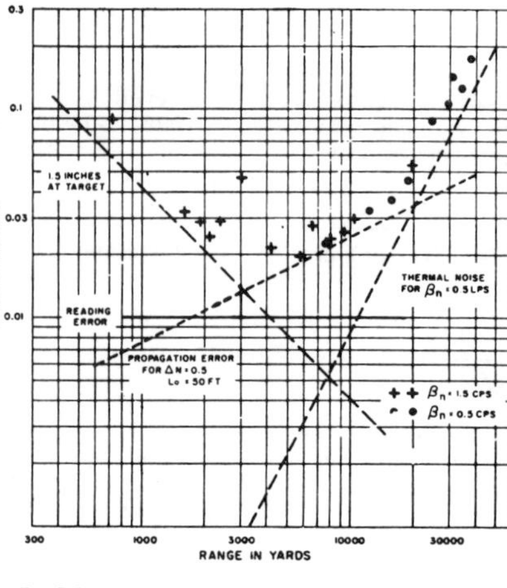

Fig. 5—Measured angle tracking noise vs range.

the straight line which appears in the center portion of Fig. 5 is obtained; this line has a positive slope indicating proportionality to the square root of range. This sets a lower limit of precision in the center region of the graph.

In this explanation is correct, it is believed that this is the first time the effect of atmospheric propagation anomalies has been observed with a single radar. This suggests the possible use of a tracking radar for observation and measurement of this effect.

Fig. 5 extends out only to a range of about 25 miles. This limitation is due, not to the radar, but to inability of the camera to see the balloon at longer range. The radar has tracked these balloons (or rather the 6-in metal spheres inside) out to 70 nautical miles. The maximum range on a target of one-square-meter radar cross section is 150 miles for the unmodified radars, and 200 miles for those currently in production.

Plots obtained from aircraft tracking runs are similar to the one shown for a metal sphere, except that the glint or wander of the electrical center of the target is greater (about $\frac{1}{4}$ of the target dimension in each coordinate), and the thermal noise line moves out further to the right (that is, toward longer range), because of the larger echoing area.

A large amount of additional experimental data is available in [2] and [3].

Range of Echo Tracking

The classical radar equation can be used to calculate the ratio of the received signal power to the radar-receiver noise, and from this ratio the ability of the radar to detect, lock-on and track may be estimated. In its generally used form, the radar equation gives the SNR in the IF amplifier (after the first stage or two) as follows:

$$S/N = \frac{P_t G_0^2 \lambda^2 \sigma}{(4\pi)^3 R^4 K T_0 B \overline{NF}_0 L}, \qquad (6)$$

where

- S/N is the IF SNR during reception of the pulse,
- P_t is peak transmitted power in watts,
- G_0 is antenna gain along the axis of the beam,
- λ is wavelength,
- σ is effective radar cross section of the target,
- R is range to the target,
- K is Boltzmann's constant,
- T_0 is ambient temperature at the receiver in degrees Kelvin,
- B is receiver bandwidth in cps,
- $\overline{NF}_0$ is operating receiver noise figure, and
- L is the total system loss, arising from atmospheric attenuation, plumbing losses, transmitter power outside the receiver bandwidth, and degradation of previously listed radar parameters from their assumed values.

Note that λ, σ and R must use the same units of length, while G_0, $\overline{NF}_0$ and L are dimensionless ratios. The applicable AN/FPS-16 radar parameters in the wide-pulse mode are

$$P_t = 1.0 \text{ Mw}$$

$$G_0 = 44.5 \text{ db}$$

$$\lambda = 5.3 \text{ cm (at midband)}$$

$$B = 1.6 \text{ Mc}$$

$$\overline{NF}_0 = 11 \text{ db}$$

$$L \cong 4 \text{ db (known loss in plumbing and duplexer)}.$$

To calculate values of S/N to be expected at a given range R (in nautical miles) on a given target σ (in square meters), a number of conversion factors must be applied along with the constants $(4\pi)^3$, $K = 1.37 \times 10^{-23}$ watts/degree/cps and $T_0 = 291°$ K. By coincidence, the combined value of all conversion and other constants multiplies out to unity (actually 1.08) when λ is expressed in cm, σ in m^2, R in nautical miles, and B in cps. The resulting simplified form of the radar equation can be written

$$S/N = \frac{P_t G_0^2 \lambda^2 \sigma}{R^4 B \overline{NF}_0 L}. \qquad (7)$$

Substituting the AN/FPS-16 radar parameters, as given above, we obtain

$$S/N = \frac{\sigma}{R^4} \times 4.4 \times 10^8 \qquad (8)$$

or, in logarithmic form,

$$S/N = 10 \log \sigma - 40 \log R + 86.5 \text{ db} \qquad (9)$$

(σ must be in m^2, R in nautical miles). Fig. 6 is a plot of (9) showing the calculated SNR for various target cross sections as a function of range. The SNR attainable using the 0.25-μsec pulse and 8-Mc bandwidth, is almost exactly 7 db lower than values given in Fig. 6. Using the 0.5 μsec pulse and 8 Mc bandwidth, the SNR is also down 7 db from Fig. 6, but the longer pulse provides an improvement in visibility on the display. If the 250 kw transmitter is used, the SNR's will be reduced 6 db from the values obtained with the 1.0 Mw transmitter under similar conditions.

Beacon Tracking

For beacon tracking, two separate equations describe the ability of the radar to interrogate the beacon and its ability to detect and track the beacon response. The radar signal available at the beacon receiver is given by

$$S_b = \frac{P_t G_r G_b \lambda^2}{(4\pi)^2 R^2 L_1}, \qquad (10)$$

253

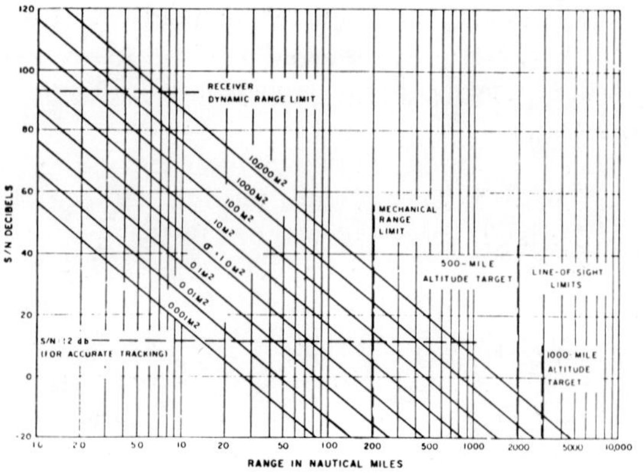

Fig. 6—Calculated values of SNR vs range for echo track for AN/FPS-16.

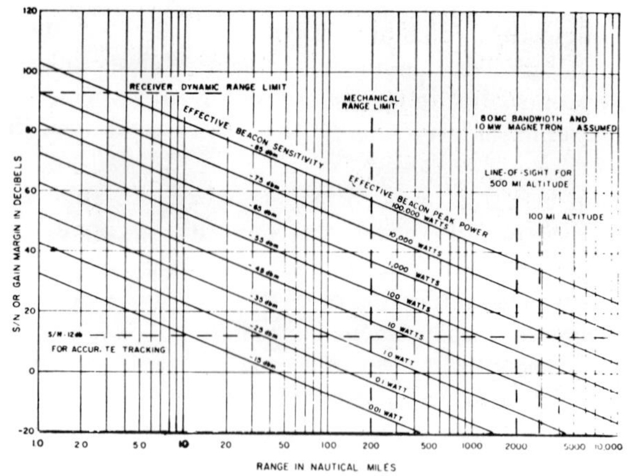

Fig. 7—Theoretical beacon tracking range AN/FPS-16.

where

G_r is the radar antenna gain,

G_b is the beacon antenna gain,

L_1 is one-way system loss for interrogation, and other symbols are as defined for (6).

If P_t is replaced by P_b, the beacon transmitter power, and L_1 by L_2, the one-way loss in response, the equation will give beacon signal available at the radar receiver. Then, since noise power referred to the radar receiver input is

$$N = KT_0B\overline{NF}_0, \qquad (11)$$

the resulting SNR at the radar is

$$S/N = \frac{P_bG_rG_b\lambda^2}{(4\pi)^2R^2L_2KT_0B\overline{NF}_0}\cdot \qquad (12)$$

Fig. 7 is a plot of gain margin for interrogation (defined as the ratio of available S_b to $S_{\min}$, the minimum power required to trigger the beacon) and SNR at the radar, for AN/FPS-16 radar parameters and various effective values of $S_{\min}$ and P_b. The curves may be used directly for omnidirectional, unity-gain beacon antennas. For antennas with gain G_b, the effective power may be found by the product P_bG_b, while effective sensitivity will be $S_{\min}/G_b$.

Of the radar parameters used in these calculations, only λ and B can be measured directly with high accuracy. Transmitter power can be measured to within a fraction of a decibel using calorimetric procedures, but this is not done as a regular maintenance test, and at any given time the power may be 1 db off in either direction. Similar considerations apply to $\overline{NF}_0$, which is influenced by crystal aging and damage, and G_0, which is originally calculated from beam patterns.

II. Current Development Programs

The first major step beyond the AN/FPS-16 class of instrumentation radars is the AN/FPQ-6 (along with the transportable AN/TPQ-18). The major characteristics of this radar, compared to those of the AN/FPS-16, are given in the Appendix. The performance is basically the same in tracking accuracy (with some improvement in precision), but the range is extended by a factor of four over the AN/FPS-16. An artist's sketch of the new radar is given in Fig. 8. Other than the new antenna and pedestal, it can be seen that the new radar represents a combination of the 3-Mw klystron transmitter, the digital-ranging system (extended to 32,000 miles), complete acquisition features, and updated designs for receivers, data handling and peripheral equipment. The AN/FPQ-6 and AN/TPQ-18 class of radars provides both an extension of missile range coverage and a major step into space instrumentation. Existing beacons can be tracked to the moon with these radars, and beacons now under development will provide wide safety margins in the application [5]. The added performance of the AN/FPQ-6 is largely due to the new and larger antenna. A Cassegrainian design has been used (Fig. 9) to combine high-aperture efficiency, low sidelobes, short waveguide runs and minimum mechanical difficulty. The mechanical resonance is to be above 15 cps to preserve the servo-bandwidth capability of the AN/FPS-16. At the same time, the design should make possible very low antenna noise pickup from the surrounding ground, looking forward to the time when low-noise receiver preamplifiers become available. Delivery of these radars to the initial sites is scheduled for March, 1962, with full operation four months later.

Pulse-Doppler System

An important capability not yet exploited in operational pulse-radar range instrumentation systems is the refined measurement of radial velocity through the Doppler effect. Search radars have used coherent MTI since the end of World War II, and numerous tracking and navigation systems have since been designed to select targets and perform measurements on the basis of Doppler shifts in the spectrum of pulsed signals. The

Fig. 8—Artists sketch of AN/FPQ-6 instrumentation radar.

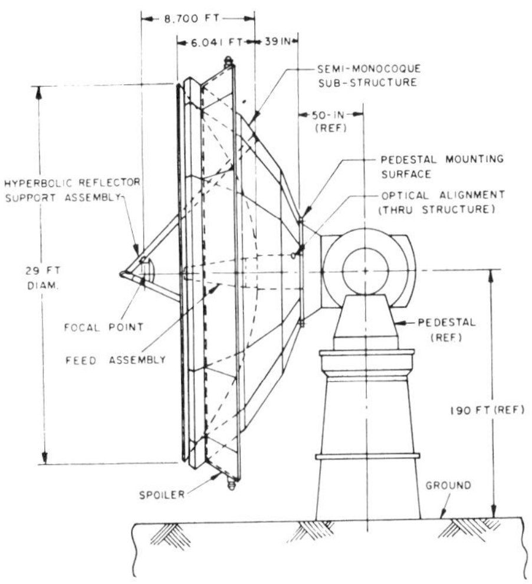

Fig. 9—Cassegrain antenna for AN/FPQ-6.

importance of applying this technique to range instrumentation stems from four major factors:

1) Accurate three-coordinate velocity data is needed for impact prediction, orbit determination, probe guidance, etc.

2) Accuracy of data based on conventional range and angle data is limited by the extreme ranges required.

3) Solutions based upon range-only (trilateration) operation are effective (see discussion in a later section) but are limited by the precision of range measurement.

4) Solutions based upon combined range and Doppler data from three-station nets will meet present and projected requirements for velocity data.

Alternate approaches to solving this problem have been proposed using CW techniques, but the pulse-Doppler system possesses unique advantages:

1) It will operate on noncooperative targets within the echo-range limitations of the pulsed radar.

2) It permits a single transponder beacon to be shared by three or more stations on a time-multiplexed basis, with each station performing a continuous track in all coordinates.

3) The ground equipment consists almost entirely of existing radar sets.

4) The ground radar provides high-system gain and permits the system to expand and extend its range continuously as radar equipment is added and improved.

5) Communications, data links and support equipment are those presently used with the radar.

6) No critical measurement links over ground paths are used in three-station radar trilateration, due to the independent echo or beacon response to each station.

The pulse-Doppler approach is, of course, subject to the same fundamental limitations on accuracy as apply to CW systems: uncertainty in the velocity of light, atmospheric refraction, stability of the basic reference oscillator, acceleration and higher derivative errors, and need for good baseline surveys. In addition, the pulsed system at conventional instrumentation radar-repetition rates is very ambiguous in its Doppler readings. At 640 pps, for instance, a *C*-band radar will have Doppler ambiguities at intervals of about 50 ft/sec, corresponding to the spacing between successive lines in the signal spectrum. Thus, it is necessary to measure independently the target radial velocity with sufficient accuracy to define the velocity intervals being observed. For the case cited, a measurement of 8 ft/sec rms will give 99.7 per cent assurance of unambiguous results, and this is easily within the capability of differentiated range data. 255

The current development program in coherent pulse-Doppler instrumentation has, as a primary objective, the design of circuits for use with the AN/FPS-16 or AN/FPQ-6 to measure Doppler velocity to the order of 0.1 ft/sec. Secondary objectives include the exploitation of velocity resolution for overcoming thermal noise (predetector integration), for reduction of ground clutter at launch, and for selection of the desired target component as stages are separated in multistage vehicles. Some additional benefits which will be derived incidentally from the coherent system modification are the assurance of exact transmitter and receiver frequency control (through use of ultra-stable oscillator and frequency synthesizer), adequate stability for wide-pulse, narrow-band operation on both echo and beacon tracks, and adaptability of both transmitter and receiver to pulse-compression schemes such as "chirp."

The initial modification for velocity measurement will involve only the addition of a stable exciter and a Doppler channel to the klystron transmitter already available in some of the AN/FPS-16's (see Fig. 10). The existing receivers, angle-tracking circuits and ranging circuits will be undisturbed, except for substitution of a STALO (stable local oscillator) signal from the exciter in place of the existing local oscillator. Ambiguity resolution for the Doppler channel will be provided by processing of range-difference data, and the Doppler circuit will go into operation only after a target has been acquired and tracked by the ranging system. Subsequent modification (Fig. 11) will provide narrow-band filters in all tracking channels to exploit the secondary objectives mentioned above. This Doppler de-

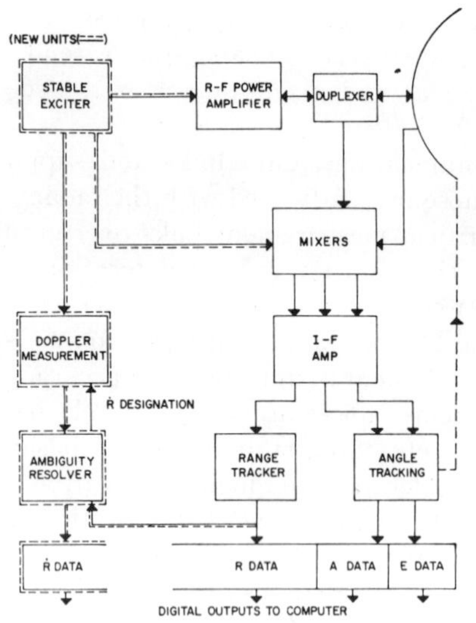

Fig. 10—AN/FPS-16 velocity measurement modification.

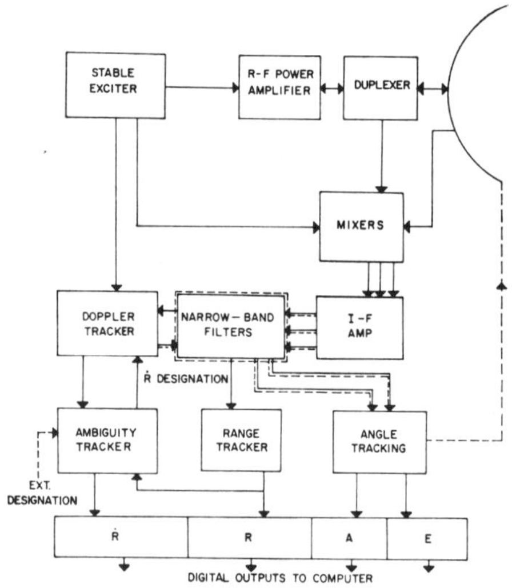

Fig. 11—AN/FPS-16 velocity tracking modification.

velopment program is now proceeding, and the stable exciter will be included as standard equipment with the klystron transmitters of the AN/FPQ-6. Coherent beacons are also under development.

Other Current Modification Programs

Going beyond the AN/FPQ-6 and the Doppler-measurement programs, there are several areas of pulsed-radar-system development which are being followed and which can be expected to result in later modifications to radars after installation in the field. In the antenna area, improved monopulse feeds are under development which will provide complete polarization diversity in reception and flexible control of transmitted polarization. Use of the Cassegrainian antenna permits the extra waveguide components to be added and provides a place for multiple-receiver preamplifiers near the feed assembly. Design of larger antennas and suitable pedestals is being carried out under several radar and communication programs and then may be applied directly where economic considerations permit. For the larger antennas, the combination of equipment for two or more radar bands on a single pedestal and reflector becomes economical, and multiband feed development is going forward.

In the transmitter area, the most significant developments are concerned with building up greater average power and pulse energy. This is done principally through the use of parallel klystron amplifiers with moderate peak-power capabilities per tube, but with cathodes designed for long-pulse operation. Where the range resolution of the present short-pulse systems must be preserved, pulse-compression circuits may be used to combine the pulse-energy advantage of the long pulse with the high-range resolution inherent in a wide spectral bandwidth. A similar advantage might be gained through coherent (predetector) integration of short pulses at high-repetition rates, but this approach is limited by the presence of ground clutter or other short-range targets, which tend to mask the desired echo from extended range in a multiple-time-around tracker.

Developments in the receiver area are chiefly concerned with improved sensitivity through use of parametric amplifiers and masers. Super-cooled paramps are under development which offer significant noise reduction as compared to the present paramp design. Ultimately, the microwave maser will eliminate receiver noise as a significant contribution to system noise, leaving sky noise, antenna noise and waveguide loss as the limiting factors in sensitivity. These developments will be discussed in more detail later.

Also under development are data-processing devices to match the capabilities of the new radars and to permit flexible-net operation in missile- and space-range applications. The details of such equipment are beyond the scope of this paper, but the general system requirements and expected performance will be covered in a later section.

III. Development of Radar for Space Instrumentation

Basic Problems and Limitations

Radar may be used in solution of three major problems in space instrumentation. First, it is applicable to surveillance and tracking of all objects in orbit around the earth, using skin return or whatever cooperative devices may be available on the satellites. Second, use of radar or radar-type ground equipment is mandatory in tracking lunar and deep space probes at ranges from 250,000 miles to the limits of the solar system, using cooperative transponders. Lastly, the radar astronomers employ radar equipment to measure directly the surface characteristics of the moon and planets. This discussion will be limited to use of tracking radar for measurement of position and velocity of satellites and probes and for obtaining other data on these targets through reception of telemetering signals or analysis of signal-strength records.

The limitations on radar performance in space applications arise from two sources: the extreme range over which the signals must travel and the necessity of operating from beneath the atmospheric blanket which covers the earth. The range limitation, coupled with limited transponder power or reflection area of practical space vehicles, forces the space radar to extremes in transmitter power, antenna gain and receiver sensitivity which go far beyond past designs in radar equipment. The atmosphere acts directly to limit the accuracy of measurement, the sensitivity of the system (through "sky noise" contributions), and the physical size and configuration of antennas which must withstand the assaults of local weather. In addition to these two types of problems, there are the force of gravity which places mechanical limits on antenna design; the presence of the earth, which obstructs line of sight and radiates noise toward the antenna; high-voltage breakdown and personnel-safety limitations on generated and radiated power; man-made, galactic, and solar noise; and many practical equipment design problems. Brief summaries of all these limitations will be given.

Range and Radar Parameters

As a background for discussion, assume that the radar is to track by skin echo a small satellite ($\approx$1.0 m^2 cross section) in a synchronous orbit around the earth. A radar which achieves a single-pulse SNR of unity at 30,000 n.m. is suitable for this task. The basic radar equation may be rewritten as

$$R_0 = \left[\frac{P_t\tau G^2\lambda^2\sigma}{\overline{NF}_0 L}\right]^{1/4} \text{n.m.,} \tag{13}$$

where R_0 is the range at which unity IF SNR is achieved and the other terms are as used in (7), with pulse width $\tau = 1/B$.

From this equation, the resulting requirement on radar parameters for $R_0 = 30{,}000$ miles on 1.0 m^2 may be expressed as

$$P_t\tau G^2\lambda^2/\overline{NF}_0 L = 8 \times 10^{17} \tag{14}$$

or, if the antenna is limited to a diameter D in feet,

$$P_t\tau D^4/\lambda^2\overline{NF}_0 L = 10^{10}. \tag{15}$$

If a minimum repetition rate f_r for satisfactory acquisition and measurement is established, the product $P_t\tau f_r = Pav$ specifies the average power required of the transmitter. Values of f_r between 10 and 100 would be typical, depending upon the exact requirements, and there might be some trade-off between higher repetition rate and reduced single-pulse S/N to perform a given function.

Another measure of required radar parameters could be obtained by assuming transponder operation to a 500 million mile range, or approximately to Saturn's orbit. The beacon-to-radar path will normally limit range in such a case, due to power output limitations in the beacon. The beacon equation (for the response link) is

$$S/N = \frac{P_b\tau G G_b\lambda^2}{R^2\overline{NF}_0 L} \times 5 \times 10^7 \tag{16}$$

for the mixed system of units used above. At a range of 5×10^8 miles, for unity SNR at the radar, the beacon equation gives

$$P_b\tau G G_b\lambda^2/\overline{NF}_0 L = 5 \times 10^9 \tag{17}$$

or, for a beacon antenna of restricted diameter D_b in feet,

$$P_b\tau G D_b^2/\overline{NF}_0 L = 5.5 \times 10^5 \text{ (independent of } \lambda\text{).} \tag{18}$$

Consider now a beacon pulse energy $P_b\tau$ of 0.01 w-sec, and an effective antenna diameter of three feet. The radar requirement becomes

$$G/\overline{NF}_0 L = 6 \times 10^6 \text{ (independent of } P_t \text{ and } \lambda\text{).} \tag{19}$$

In certain applications where tracks must be continued for protracted periods, a radar reflector provides better performance than an active beacon (it is certainly superior in long life and reliability). The equation for maximum cross section of a corner reflector is

$$\sigma_{\max} = \frac{4}{3}\pi\frac{a^4}{\lambda^2}, \tag{20}$$

where a is the side length (in the same units as σ and λ^2). A Luneberg lens may have a cross section

$$\sigma_{\max} = \frac{\pi^3 d^4}{4\lambda^2}, \tag{21}$$

where d is the diameter. If the value of $\sigma_{\max}$, for a limited to one meter, is used in calculation of parameters for 30,000 n.m. echo tracking, the equation becomes

$$P_t\tau G^2/\overline{NF}_0 L = 3.7\times10^{13} \text{ (independent of frequency)} \tag{22}$$

(for a Luneberg lens, $d = 3$ ft and the factor is 7.4×10^{13}).

Projected State-of-the-Art in Radar

The peak and average power capabilities of single transmitter tubes, past and present, are shown in Fig. 12. Extrapolating from this into the near future, and taking into account the factors of equipment size and cost and possible use of parallel tubes, the figures of Table IV were arrived at as reasonable limits for the next generation of radar equipment in several frequency bands. It becomes increasingly difficult to use higher peak and average powers at available radar sites, and even the levels shown may be difficult to achieve and use in some cases.

257

Advances in antenna development have proceeded rapidly over the years, and radio telescopes in the 250- to 1000-ft class have been built or started. It appears possible to build antennas with directive gains approaching 80 db, at which point the atmospheric anomalies begin to limit gain. The economic aspects of tracking antennas impose stricter limits in most cases, however, and the following comparison will be based on the general features of the "Haystack Hill" class of antennas [6], rather than on any "ultimate" limitations. Table V shows the assumed diameters, efficiencies and gains achievable with antennas which might be used on this type of tracking pedestal. The beamwidth is also included to emphasize two points: first, that the higher-frequency radars must receive accurate designation from search radar or computation based on prior tracking data and second, that the angular accuracy will be much better for the higher-frequency systems.

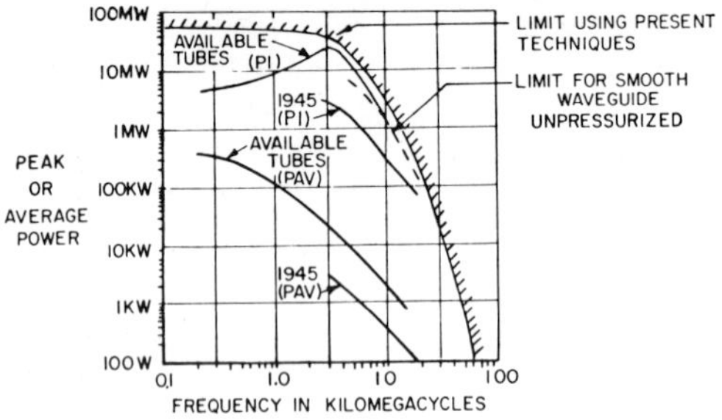

Fig. 12—Power capabilities of transmitting tubes.

TABLE IV

ESTIMATED RADAR POWER

Radar band	UHF	L	S	C	X	K
Wavelength (cm)	75	25	12	6	3	1.5
Peak power (Mw)	10	25	25	10	3	0.5
Average power (kw)	600	300	100	50	20	3
Pulse energy (w-s)	20,000	12,000	7500	1000	300	50
Pulse width (μsec)	2000	500	300	100	100	100

TABLE V

ESTIMATED ANTENNA PARAMETERS

Radar Band	UHF	L	S	C	X	K
Wavelength (cm)	75	25	12	6	3	1.5
Ant. diameter (ft)	200	150	120	120	120	100
Efficiency (percent)	50	50	50	50	50	40
Gain (db)	45	52	56.5	62.5	68.5	71.5
Bandwidth (degree)	0.9	0.4	0.24	0.12	0.06	0.04

TABLE VI

ESTIMATED INPUT NOISE (AT 10° ELEVATION)

Radar band	UHF	L	S	C	X	K
Wavelength (cm)	75	25	12	6	3	1.5
Sky temp. (°K)	50	10	15	18	20	100
Receiver temp. (°K)	10	10	10	10	10	10
Input temp. (°K)	60	20	25	28	30	110
NF_0 (db)	−7	−12	−11	−10.5	−10	−4.5

Table VI shows the limit to antenna temperature imposed by cosmic noise and the troposphere, disregarding sidelobe effects, and the resulting operating noise figure for a system using a maser. Actual sidelobe effects may add between 20 and 50 degrees to the input temperature.

The data used in Table VI was derived from curves in [7], assuming a constant receiver system temperature of 10°K as would represent a practical maser and its input circuit. The elevation angle of 10° was chosen to indicate the sensitivity which could be attained over a substantial volume of coverage around the radar site. An additional loss factor L of 2 db will be assumed in further calculations to account for plumbing loss, filter matching, and excess noise temperature at the input.

Table VII shows the radar range on a target of 1.0 m² cross section and on a corner reflector of 1.0 m side length, using the radar parameters from the three preceding tables. It can be seen that the considerably higher powers available at the lower frequencies, along with somewhat larger antennas, provide skin-tracking ranges equal to or slightly less than the S-, C- and X-band radars, on noncooperative targets. On cooperative targets, the higher frequencies are considerably better. On beacons (using the earlier assumptions) the higher frequencies have an even greater advantage in that the reduced transmitter power of the radar does not reduce the amplitude of the response.

This tabulation should not be taken as a prediction of "ultimate" performance possible from radar, but as an indication of the relative difficulty of achieving a particular level of performance at different frequencies. Furthermore, as stated earlier, the search and acquisition problem has not been covered, although it obviously favors the lower-frequency bands. For skin tracking, the microwave bands offer significant advantages in equipment size for all targets, and greater range as well on corner reflectors. For beacon tracking, the earlier as

TABLE VII

RADAR RANGE FOR S/N = UNITY

Radar band	UHF	L	S	C	X	K
Wavelength (cm)	75	25	12	6	3	1.5
R_0 on 1.0 m² (mi)	20,000	30,000	30,000	27,000	28,000	22,000
R_0 on corner (mi)	60,000	100,000	120,000	140,000	220,000	200,000
σ of corner (m²)	6	60	250	1000	4000	16,000
R_θ on beacon (10^6 mi)	125	500	750	1400	2600	2000

sumption of 0.01 w-sec pulse energy, independent of frequency, should be revised to indicate the relatively higher efficiency of power generation at lower frequencies. This would approximately double the range at UHF and cut in half the range at *K* band, relative to *S* band as a reference, with intermediate effects on *L*-, *C*- and *X*-band operation. Even so, the relative economy and effectiveness of microwave operation is obvious.

Examples of Tracking Radar Systems for Space Research

As examples of possible tracking radar configurations for use in space research, two systems will be discussed, one at 435 Mc and one at 5800 Mc. Both will be based upon the Haystack Hill class of antenna and pedestal, and upon currently available tubes and components or those under development. The full parameters of the two systems are shown in Table VIII, as compared with AN/FPS-16 and AN/FPQ-6. Obviously the step from AN/FPQ-6 to these proposed trackers is a much greater one than between the original SCR-584 and AN/FPS-16, or between AN/FPS-16 and AN/FPQ-6. For this reason, although antennas are now being built even bigger than the Haystack class, it seems probable that one or more intermediate steps will be implemented between AN/FPQ-6 and the space trackers listed in Table VIII. One such step is an experimental space tracker now under construction at the Moorestown RCA plant. Its characteristics are shown in the last column of Table VIII.

The beacons required for the ranges listed in the table are well within the current state of the art, but are not necessarily desirable items for inclusion in space vehicles with small payload capacity. A 40-kw beacon has been described [5], which would weigh 40 lbs and require 150 watts of input power. This would be suitable for use in a lunar probe with the AN/FPQ-6, but would impose considerable burdens on a space probe of longer life, which would use solar cells as a source of prime power. For such vehicles, further development of tunnel-diode oscillators appears to offer a solution. Present experimental diodes have provided CW output powers of 0.2 mw at 5500 Mc, with efficiency of a few per cent [8]. In the near future, a power level of 10 mw appears attainable, with pulse lengths of 100 μsec. Range on such a beacon would be one-hundredth that shown in Table VIII, or about 6 million miles for the *C*-band tracker. Until higher-power solid-state oscillators become available, it appears that longer response pulses will be necessary, accompanied by narrower-receiver bandwidths in the radar, to achieve sufficient range for probes to Venus and Mars. A tunnel-diode beacon producing 10-mw pulses of 10 msec length would provide a 60-million mile range with the *C*-band tracker of Table VIII. Since the radar-ranging system can interpolate to within about 5 per cent of the received pulse width, a range precision of about forty miles would still be possible for independent measurements taken twenty seconds apart. More accurate data in the vicinity of the planet could be obtained by intermittent operation of a higher-power beacon transmitter, using a pulsed triode at the 100-watt level.

For the UHF space tracker, the 5-watt beacon power listed is only slightly higher than now available from solid-state devices, and ranges of 50 million miles ap-

TABLE VIII
SPACE RANGE RADAR PARAMETERS

Radar type		AN/FPS-16	AN/FPQ-6	Proposed Space Trackers		
						(Exp)
Radar band		*C*	*C*	UHF	*C*	*C*
Radar frequency	Mc	5600	5600	435	5600	5600
Peak power	Mw	1.0	3.0	10	10	3
Pulse width	μsec	1.0	2.4	2000	100	5
Repetition rate	pps	1000	640	30	50	300
Average power	kw	1.0	4.5	600	50	5
Antenna size	ft	12	30	200	120	50
Antenna gain	db	44	52	45	62.5	58
Beamwidth	mils	20	8	16	2	5
Bandwidth	kcps	1600	500	0.5	10	250
Operating noise figure	db	10	8	−7	−10.5	−10
Beacon peak power	w	10,000	4000	5	100	2000
Beacon av. power	w	10	7	0.3	0.5	3.3
R_0: on 1.0 m²	n.m.	150	750	21K	27K	5K
on corner*	n.m.	800	4000	26K	140K	28K
on beacon	n.m.	100K	350K	16M	11M	5.6M
beacon, 3-ft ant.	n.m.	5.6M	20M	90M	600M	300M
Range precision†	ft	5	15	60K	2000	50
Angle precision†	mils	0.2	0.1	1.0	0.1	0.1

* Based on corner reflector with side length $a = 1.0$ m.
† Based on observation for 10 sec at S/N = unity, or for 1 sec at $S/N = 10$.

260

pear feasible for the near future with parallel tunnel diodes or transistors. The over-all accuracy of range measurement would be comparable to the forty-mile figure for the microwave system.

The tracking precision of the radars listed in Table VIII is shown in Figs. 13 and 14. It may be assumed that the angular accuracy is 0.1 mil in all cases, and the range accuracy is limited to the values shown. When using beacons, a serious limitation to be overcome is the uncertainty in the vacuum velocity of light. This can be relieved if the velocity is defined in terms of a molecular standard [9] rather than by the meter bar and the sidereal second.

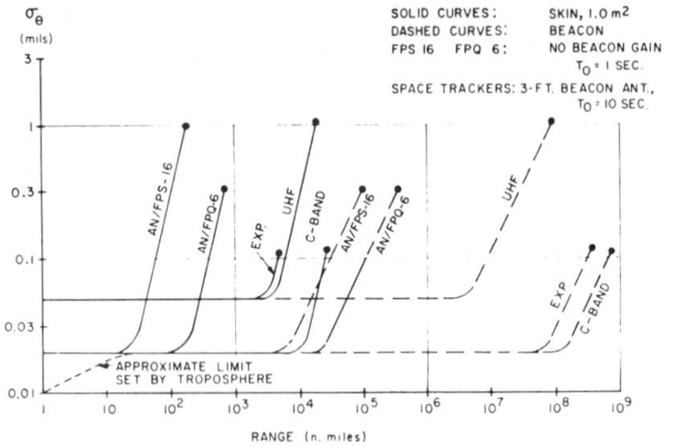

Fig. 13—Angle precision vs range.

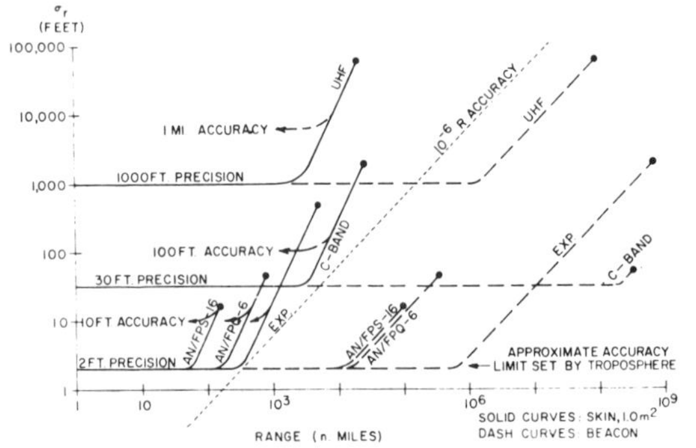

Fig. 14—Range precision and accuracy vs range.

IV. System Applications of Instrumentation Radar to Space Range, Angle and Rate Measurements

The precision and accuracy attained in modern monopulse radars has already been discussed. In space tracking applications, the most important sources of error will usually be receiver noise (at long ranges), atmospheric propagation effects, knowledge of the velocity of light, and systematic errors due to deformation of the large antennas used. Precision figures given in Table VIII were based upon tracking at the maximum range of the radar ($S/N \cong$ unity), with thermal noise as the sole important source of error. In the case of angle tracking, the thermal noise equation from Table III was used with an equivalent servo-noise bandwidth $\beta_n = \frac{1}{2}t_0$ where t_0 is the observation time. An equivalent formula for range precision is

$$\sigma_r = \frac{\tau}{\sqrt{(f_r/\beta_n)(S/N)}} = \frac{\tau}{\sqrt{2t_0 f_r(S/N)}}. \tag{23}$$

Here, τ is the range equivalent of the pulse time (1 μsec = 500 feet) and the other symbols are as used earlier. The equation assumes a matched system $B\tau$ = unity), but may be extended to systems using short-rise time or pulse compression by substituting for τ the reciprocal of the actual system bandwidth used by the transmitted signal and passed by the receiver. (See [10] for a rigorous derivation and detailed discussion.)

Although there may be other sources of error in the system, a good estimate of error in angle or range rate, derived from radar-position data, may be obtained by considering thermal noise alone. As an approximation, if the position data taken over time t_0 has a standard deviation σ_p, the velocity error will be

$$\sigma_v \cong \frac{\sqrt{3}\,\sigma_p}{t_0}. \tag{24}$$

This equation assumes that the velocity is determined by fitting several raw position readings to a straight line (for a constant-velocity target) or to a line of known curvature (for a target with known acceleration). The equations for angle and range rates become

$$\sigma_{v\theta} = \frac{\sqrt{3}\,\theta}{2\sqrt{t_0^3 f_r(S/N)}} \tag{25}$$

$$\sigma_{vr} = \frac{\sqrt{3}\,\tau}{\sqrt{2t_0^3 f_r(S/N)}}. \tag{26}$$

As an example, for the proposed *C*-band space tracker at maximum range and for $t_0 = 10$ seconds, the angle-velocity error would be 0.017 mil/sec (17 μrad/sec), while the radial-velocity error would be 380 ft/sec without pulse compression or Doppler measurement.

If radial velocity is measured by observing Doppler shift, much better results are obtained. A relatively crude method involves use of a discriminator system at IF to measure shift of the envelope of the spectrum (noncoherent or single-pulse Doppler measurement). In this case, the frequency error is

$$\sigma_f = \frac{\beta}{\sqrt{(S/N)(f_r/\beta_n)}} = \frac{\beta}{\sqrt{2t_0 f_r(S/N)}}. \tag{27}$$

Applied to the long-pulse UHF tracker, this would yield a Doppler accuracy of 20 cps (corresponding to 23 ft/sec) at maximum range for $t_0 = 10$ seconds. For the *C*-band tracker, the accuracy would be 316 cps, corresponding to 26 ft/sec. If the system is operated coherently, maintaining consistent phase from pulse to pulse, measurements to within 1 cps are readily achieved, at the expense of ambiguities at the radar-repetition rate. In the *C*-band tracker, these ambiguities would be at intervals of 4 ft/sec, and would require that velocity be measured to within about one ft/sec before the "vernier" Doppler measurement could be interpreted. This would require about 100 seconds tracking, with the parameters listed, but at the end of that time, the Doppler system would provide data to about 0.1 ft/sec accuracy with independent readings every one second.

Long-Arc Measurements

In position or velocity measurements made over long intervals of time t_0, the importance of thermal-noise errors is reduced, and some of the other terms begin to influence results. The atmospheric terms and other fixed or slowly-varying errors put limits on both accuracy and precision of measurement, and (23)–(27) should be used with great caution for values of t_0 beyond a minute. In addition, the assumptions as to constant velocity in radar coordinates or known acceleration in any coordinate system may not hold to the desired accuracy. There are cases, however, where a long series of measurements can yield estimates of target motion far better than those obtained from brief periods of tracking. Generally, in cases where the laws of motion of the target can be specified exactly, the accuracy of position measurement in all three coordinates can be made to approach that of the best measured coordinate. For example, results on a ballistic missile track of 300 seconds duration have been reported [11] where the rms error in azimuth and elevation data is reduced from 0.01° to 0.0015° (corresponding to reduction in space position error from about 650 to 100 ft) by properly matching the raw data to a set of orbital elements. Accuracy at the midpoint of the track is considerably better.

In measurement of orbital elements of satellites, the same reference shows the results of a 50-second track on Sputnik II, using the AN/FPS-16 (see Table IX). The observed elements agree remarkably well with those given by the Smithsonian Astrophysical Observatory, and are, for the most part, well within the expected variations of the short-term or osculating elements from those averaged over extended periods. In this case, the tracking arc was relatively short and the observed range data could be fit within 10 ft rms by the calculated elements. Over longer arcs, the effectiveness of this method improves markedly. The limiting case is represented by space-probe tracking, where tracks over eight to ten hour periods can be matched to orbital elements involving other bodies in the solar system. In such cases, the position accuracy will be determined almost entirely by the range-bias errors in the system and the knowledge of the equations governing motion of the space vehicle. For this reason, the use of pulse-compression systems to refine precision of range measurement may not be warranted in space tracking, unless extremely long pulses (milliseconds or seconds in duration) are used.

TABLE IX
1957 BETA ONE (SPUTNIK II) ON FEB. 4, 1958

Element	Derived from SO Special Rept. No. 13	Derived from Radar Data
Semi-major axis a, in earth radii	1.106790	1.1070834
Eccentricity, e	0.067406	0.068222497
Inclination i, in degrees	65.29	65.216535
Node, in degrees	217.18	217.23710
Argument perigee, in degrees	20.09	21.564365
Time perigee, in hours, min., sec	$11^h33^m57^s.0240$ UT	$11^h34^m19^s.52149$ UT
Nodal period P, sec	5902.84	5904.31
Time of Node T, hrs, min, sec	$11^h29^m8^s.8454$ UT	$11^h29^m9^s.77019$ UT
Range residual, rms, in ft		9.9687
Elevation residual, rms, degrees		0.02288
Azimuth residual, rms, degrees		0.02238

261

In discussing space-tracking systems operating over minutes or hours in measuring orbits, it is important to distinguish between those which provide immediate or "real-time" data and long-arc systems. The latter are good for evaluating orbital constants, but cannot provide data for closed-loop guidance and control systems in the usual sense. An error which builds up during the track can be sensed, but the extent of the correction applied by added thrust will not be measured until long after the thrust is applied. The errors in "real-time" measurement may be higher, except for radial position and velocity, but the control thrust may be sensed within seconds or fractions of a second, and the control loop closed within this time. As objects get farther into space, the delay in propagation introduces a time lag which cannot be overcome, and the definition of real-time needs revision. It can be assumed that the time t_0 allowed for measurement can approach the range delay time $t_r = 2R/c$ without compromising system performance in any important respect.

Range and Doppler Trilateration

The inherent accuracy of radial measurements using pulsed radar can best be exploited in long-range trajectory and orbital problems by combining data from three widely-separated radar stations. Each radar makes an independent measurement of range, Doppler, or both, using the reflected echo or the transponder if one is available. Since each range measurement is completed over the path from one station to the target, no high-performance ground communication links are required and propogation errors are minimized. Having made the measurement and placed it in digital form with a time tag attached to represent the mid-point or end of the measurement period t_0, the range or Doppler-velocity data may be transmitted without distortion to a central point. Here it is combined with data from at least two other stations to provide complete three-coordinate output data. Azimuth and elevation data may be carried along for use at short tracking ranges or in case one station's range data is lost.

The error involved in three-station trilateration measurements can be calculated exactly if the individual station range errors and the over-all station and target geometry are known. However, the range errors are known only statistically, and the expected rms error can only be estimated to within a factor of two for practical systems. It is therefore useful to use approximate methods of evaluating system error and neglect the finer points of analysis involved in the rigorous studies on the subject. A two-dimensional case will be discussed and extended to three-dimensional systems by simple geometrical operations. Fig. 15 shows the geometry of the basic measurement for the conventional radar case and for the trilateration case with target range several times the baseline length. The approximate errors in x and y may be expressed in terms of components due to range, range difference, angle and station survey error as shown in Table X. It is assumed that the survey error σ_s applies to both x and y coordinates of the station with respect to the reference system, so that the baseline length is in error by $\sigma_{\Delta s} \cong \sqrt{2}\sigma_s$, its center by $\sigma_s/\sqrt{2}$, and its orientation (with respect to $\theta=0$) by $\sigma_{\Delta s}/B$. Since there may be systematic errors present in both range and station survey, which are the same at both stations, the difference terms $\sigma_{\Delta s}$ and $\sigma_{\Delta r}$ are introduced to represent the uncorrelated portions of these errors.

It can be seen that the trilateration case is equivalent to a range-angle measurement with a system-range error equal to $1/\sqrt{2}$ times that of the individual station, and a system-angle error in radians equal to $1/B'$ times the error in range difference, where $B'=B \cos \theta$ is the projected length of the baseline normal to the direction of the target. If the range and station survey errors at the two stations are independent, then the error components are approximately as follows (for $\theta \neq 0$):

Component	x error	y error
Range error σ_r	$\sqrt{2}\,\frac{R}{B}\,\sigma_r$	$\sqrt{2}\,\frac{R}{B}\,\sigma_r \tan \theta$
Station error σ_s	$\sqrt{2}\,\frac{R}{B}\,\sigma_s$	$\sqrt{2}\,\frac{R}{B}\,\sigma_s \tan \theta$

Similar relationships apply between radial velocity measurements and x and y components of velocity, if the position triangle is already known.

The tracking-error curves (Figs. 13 and 14) provide most of the data needed for calculating performance of the several systems discussed. As an initial example, assume that two AN/FPQ-6 radars are operated with a separation of 50 miles. The equivalent system-angle accuracy (for targets nearly at right angles to the baseline) will be 14 ft/300,000 ft, or about 50 μrad (0.05

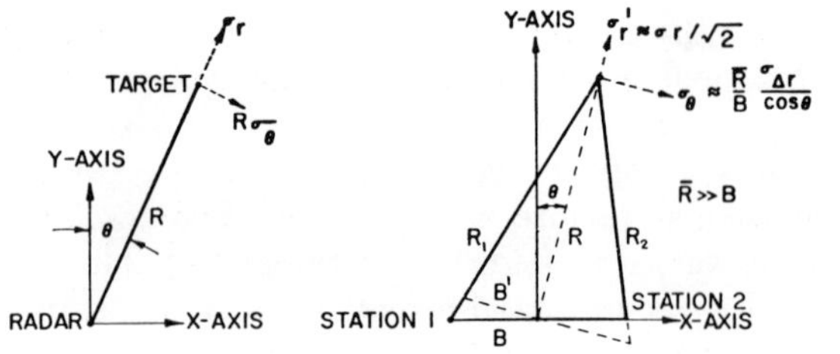

(a) (b)

Fig. 15—Geometry of measurements in two dimensions. (a) Conventional radar solution. (b) Trilateration solution.

TABLE X

COMPARISON OF ERROR COMPONENTS

Component	x error	y error
Radar case		
Range error σ_r	$\sigma_r \sin \theta$	$\sigma_r \cos \theta$
Angle error σ_θ	$R\sigma_\theta \cos \theta$	$R\sigma_\theta \sin \theta$
Station error σ_s	σ_s	σ_s
Trilateration case		
Range error σ_r	$\frac{\sigma_r}{\sqrt{2}} \sin \theta$	$\frac{\sigma_r}{\sqrt{2}} \cos \theta$
Range difference error $\sigma_{\Delta r}$	$\frac{R}{B}\sigma_{\Delta r}$	$\frac{R}{B}\sigma_{\Delta r} \tan \theta$
Station error σ_s	$\sigma_s/\sqrt{2}$	$\sigma_s/\sqrt{2}$
Station difference error	$\frac{R}{B}\sigma_{\Delta s}$	$\frac{R}{B}\sigma_{\Delta s} \tan \theta$

Assumptions:
$R \gg B$
σ_s represents equal but uncorrelated errors in the x and y coordinates, θ in radians,
σ_r equal for two stations.

mil), and the system precision about 10 μrad. This is no better than the performance of the same radars as range-angle trackers. If coherent Doppler measurements are made, the angular velocity component may be measured to $\sigma_r/B = 0.14/300{,}000$, or about 0.5 μrad/sec. This is a good deal better than the performance with differentiated angle data, unless smoothed over many seconds of track.

For longer ranges when beacons or long-range echo trackers are used, the baseline should be extended to approach the diameter of the earth. A practical limit may be set at about 5000 miles. The predominant systematic error, using any but the UHF radar, will now be in station survey, and may amount to about 100 ft using present mapping procedures. The equivalent angle accuracy of the system would therefore be 140/30,000,000 ft, or about 5 μrad. Better position accuracy will be available as mapping techniques are improved. Fortunately, the Doppler measurement accuracy will not be degraded and will approach 0.005 μrad/sec for distant targets. At the range of the moon, this would give an accuracy of about six ft/sec in a direction normal to the radar line of sight with a measurement time of one second or less.

The approximate triangulation errors given for two-dimensional operation are readily extended to three dimensions, where a third station is available and removed from the baseline which joins the first two. A simple, approximate way of estimating the results in the third coordinate is to establish a reference plane through the third station and the target and normal to the plane which includes the first baseline B_1 and the target (Fig. 16). The second baseline B_2 may then be drawn at right angles to the first, connecting the third station with an effective station which represents the combined measurement of the original pair. The z-axis error components may then be estimated as the range-difference and station-difference errors multiplied by $\bar{R}/(B_2 \cos \theta)$ where θ is the angle between the target and the normal at the midpoint of B_2.

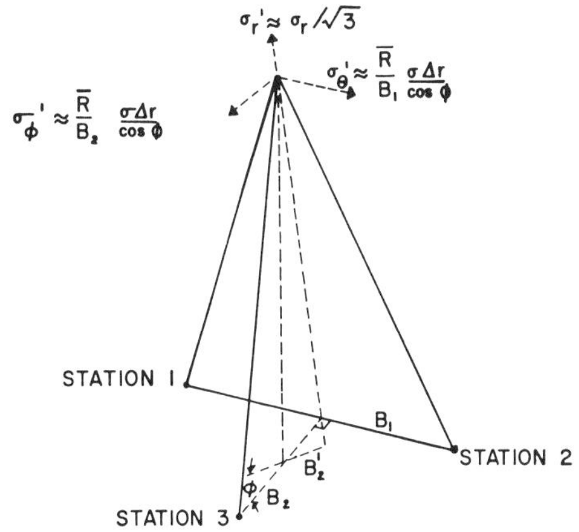

Fig. 16—Extension of trilateration solution to three dimensions.

Target Identification

Before leaving the discussion of radar applications in space tracking, mention should be made of the ability of tracking radar to provide information on target size, shape and motion. Given a record of echo-signal strength taken over a period of a minute or more, it has proven possible to measure, in certain cases, the gross dimensions and the rate and axis of rotation of targets moving through space. Those objects which rotate with periods shorter than the track duration are easiest to evaluate, but stabilized or slowly-tumbling objects may also be measured if long tracks or repeated tracks several hours apart are available.

The general procedure for analyzing a signal-strength record is as follows:

1) Major features in the record are selected and checked for periodicity.
2) Using a crude estimate of target shape, the signal strength and tracking coordinate data are analyzed to relate time to aspect angle, based on an observed constant tumbling rate modified by radar-target geometry.
3) The aspect-angle data is used to establish the widths and separation of lobes in the signal-strength record, and from this information certain dimensions of the object are estimated.
4) Absolute cross-section data is used to estimate other dimensions of the object.
5) All data is reviewed in detail and checked against possible models of the target and its rotational motion to refine the model and establish a consistent explanation for the observed signal-strength variation.

The relationships in Table XI show the major echo characteristics of simple shapes. The application to actual UHF and C-band tracking data on Sputnik II is given in [12] and [13].

Fig. 17 shows the UHF cross-section plot against time, with an aspect angle scale applied. The major features identified by letter correspond to the following:

A) Major lobe, width $2\Delta_a = 1.8°$, peak $\sigma_a = 250$ m^2, indicating a length $L_a = 70$ ft.
B) Repetitive lobing, width $\Delta_b = 1.0°$, peak $\sigma_b = 30$ m^2, indicating two equal-amplitude, broad-lobed reflecting elements with peak reflectivity 30° off mainlobe, separated by $L_b = 63$ ft.
C) Shoulders on mainlobe, width Δ_c between 1.0 and 4.0°, peak $\sigma_c = 50$ m^2, indicating two or more elements separated by L_c between 16 and 63 ft.
D) Lobes early in record, widths Δ_d between 1.0 and 4.0°, peak $\sigma_d = 25$ m^2, indicating two or more elements with peak reflectivity 30° from mainlobe and separated by L_d between 16 and 63 ft.
E) Lobes late in record, width $\Delta_e = 10°$, peak $\sigma_e = 10$ m^2, representing extension of one element of the B pattern beyond the other, and corresponding to a length L_e of about 6 ft.

Fig. 18 shows a model which would explain these five features, with two major areas of reflection plus a small corner reflector at one end of a 70-ft object. Detailed discussions of the basis for this model are given in the references. Although target descriptions from radar data are necessarily uncertain and approximate, they are available at any tracking range and may offer data not otherwise available by any means. Similar procedures for evaluating target motion from transponder or telemetry signal strength records are described in [14].

V. Comparison of Radar with Other Instrumentation Systems

Classification of Systems

Before discussing in detail any of the systems which share with radar the field of trajectory measurement, it is useful to consider in general the various methods of position measurement, and to classify systems with

TABLE XI
ECHO CHARACTERISTICS OF SIMPLE SHAPES

Shape	σ_{max}	Major lobe width 2Δ	Number of lobes	σ_{min}
Sphere	πr^2	2π	1	πr^2
Ellipsoid	πa^2	$\approx \frac{b}{a}$	2	$\pi \frac{b^4}{a^2}$
Dipole	$0.9\lambda^2$	π	2	Null
Cylinder	$\frac{2\pi r L^2}{\lambda}$	$\frac{\lambda}{L}$	$\frac{8L}{\lambda}$	Null
Flat plate	$\frac{4\pi A^2}{\lambda^2}$	$\frac{\lambda}{L}$	$\frac{8L}{\lambda}$	Null
Corner reflector	$\left[\frac{4}{3}\pi\frac{a^4}{\lambda^2}\right.$ +4 flat plate lobes$\left.\right]$	$\frac{\pi}{4}$	4	—

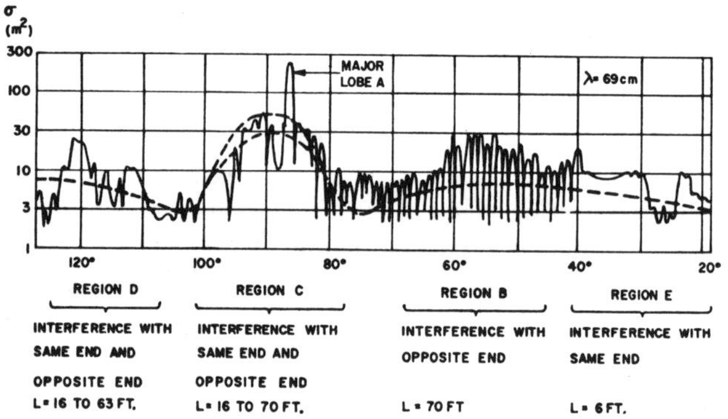

Fig. 17—Comparison of Sputnik II reflection with corner reflector.

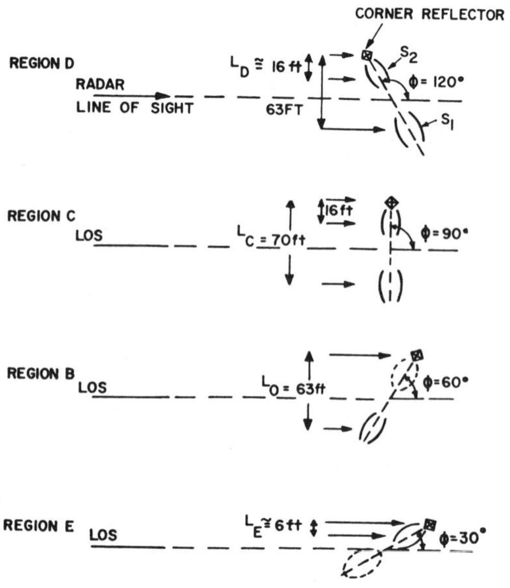

Fig. 18—Location of reflecting surfaces on Sputnik II.

respect to their major characteristics will serve to distinguish between major types of measurement systems:

1) *Frequency:* optical, microwave, or radio. (Infrared and sonic will not be discussed here, and a division between radio and microwave at about 1000 Mc will be used for convenience.)
2) *Coordinate Measured:* angle, range, or a combination of both.
3) *Field of View:* wide (fixed field) or narrow (tracking) angular field of view.
4) *Sensitivity to Motion:* position sensitive, velocity sensitive, or combination of both.
5) *Signal Source:* passive system (receives from target, no transmission) or active (transmits and receives reply).

With the characteristics given above, there are 108 possible combinations that could be considered, but only 15 of these are of immediate interest. The remaining 93 systems can be excluded for various reasons (such as optical systems sensing velocity directly, which are of value to astronomy but too coarse for missile instrumentation). In some cases, the rejected combinations (such as optical ranging systems) appear to merit further investigation for possible future use, but such considerations are beyond the scope of this paper. The 15 systems chosen for comparison are shown in Table XII. Here the five major characteristics are related to practical applications, with specific systems (*e.g.*, AZUSA) given as examples of the technique applicable.

Characteristics of Various Systems

Whenever a choice is made for one of the major characteristics, certain advantages and disadvantages must follow. Tables XIII through XVII outline the chief consequences of choosing among the optional characteristics discussed above. The "right" choice for a particular situation (or choices where more than one system will be used) could be made by assigning a weighting factor to each point given and looking for the system with maximum positive score. No attempt will be made here to assign such factors, but some of the points on which specific information is available will be discussed in more detail and present systems compared in the light of this information.

Propagation Limitations

A tremendous amount of material has been published on the subject of propagation errors and their effects on instrumentation. A survey of much of this material leads to the following conclusions (see [4], [15] and [16] for more detailed discussion and bibliography):

1) Instruments operating over paths longer than 10 miles may make measurements to an accuracy of about 0.05 mil or 2 ft, provided corrections are made for local surface conditions.

TABLE XII
CLASSIFICATION OF FEASIBLE INSTRUMENTATION SYSTEMS

Frequency	Coordinate	Field of view	Sensitivity to motion	Signal source	System example
Optical	Angle	Narrow	Position	Passive	Tracking Theodolite
Optical	Angle	Wide	Position	Passive	Plate Camera
Microwave	Angle	Narrow	Position	Passive	"Radar Theodolite"
Microwave	Angle	Wide	Position	Passive	(Angle portion of AZUSA)
Microwave	Range	Narrow	Position	Active	(Use of 3 radars for ranging)
Microwave	Range	Wide	Position	Active	MIRAN
Microwave	Combination	Narrow	Position	Active	Tracking Radar
Microwave	Combination	Wide	Position	Active	AZUSA
Microwave	Combination	Narrow	Combination	Active	Doppler Tracking Radar
Microwave	Combination	Wide	Combination	Active	(Certain guidance systems)
Radio	Angle	Wide	Position	Passive	COTAR (angles only)
Radio	Range	Wide	Position	Active	DME, LORAN, COTAR (ranging)
Radio	Range	Wide	Velocity	Active	DOVAP
Radio	Range	Wide	Combination	Active	DORAN
Radio	Combination	Wide	Position	Active	COTAR (complete)

TABLE XIII
CHOICE OF FREQUENCY

Choice	Advantages	Disadvantages
Optical	Small, simple equipment High component reliability Proven techniques No frequency allocation problems Attitude and event data available High resolution	Limited range Weather limited Film processing and reading are time consuming and require additional manpower No range data available
Microwave	Medium resolution All-weather use No ionospheric effect Frequencies available Range data available Long-range capability Data in electrical form Small missile antennas	Large complex equipment Relatively new techniques used
Radio	Relatively small, simple and reliable All-weather use Range data available Long-range capability Data in electrical form	Poor resolution Ionospheric effects Frequency allocation difficult New techniques used Missile antenna problems

TABLE XIV
CHOICE OF COORDINATES TO BE MEASURED

Choice	Advantages	Disadvantages
Angle triangulation	Passive system may be used Simple equipment	Geometrical dilution 2 or more stations required Precise timing required Data reduction requires complex processes which are time-consuming and costly in manpower and machine capacity
Range trilateration	Small antennas may be used No mechanical tracking required Minimum propagation error	Geometrical dilution 3 or more stations required Precise timing and wide-band communications may be required Data reduction problem (as with angle triangulation) Transponder required
Combination range plus angle	Only one station No precise timing or unusual communications No geometrical dilution except vs range Simple reduction	Equipment is often larger and more complex than required for angle measurement alone

TABLE XV
CHOICE OF FIELD OF VIEW

Choice	Advantages	Disadvantages
Narrow (tracking)	High resolution High antenna gain or magnification Less interference Less multipath problems Shaft data may be in electrical form	Precise mount required Servo noise and tracking errors present
Wide (fixed)	Mechanically simple No tracking or sevo noise	Antenna gain or magnification is low No resolution in angle (electronic system) Susceptible to interference Multipath problem (electronic system) Film or plate reading (optical system)

TABLE XVI
CHOICE OF SENSITIVITY TO MOTION

Choice	Advantages	Disadvantages
Position sensing	Continuous data not necessary Simple recording	Relatively poor velocity data
Velocity sensing (radial)	Velocity data good	Continuous data necessary to integrate position Complex recording or integration device needed 3 or more stations (to get all components)
Combination	Good velocity (radial) Good position without necessity for continuous data	Complex system 3 stations needed to get good velocity in all components

TABLE XVII

CHOICE OF SIGNAL SOURCE

Choice	Advantages	Disadvantages
Passive (receiving only)	Simple Low-power consumption	Source of radiation in target required Accurate range or Doppler-velocity data not available
Active (transmits and receives response)	Range data available Doppler velocity data available May also operate without transponder	More complex High power required for long range if radar is used without beacon

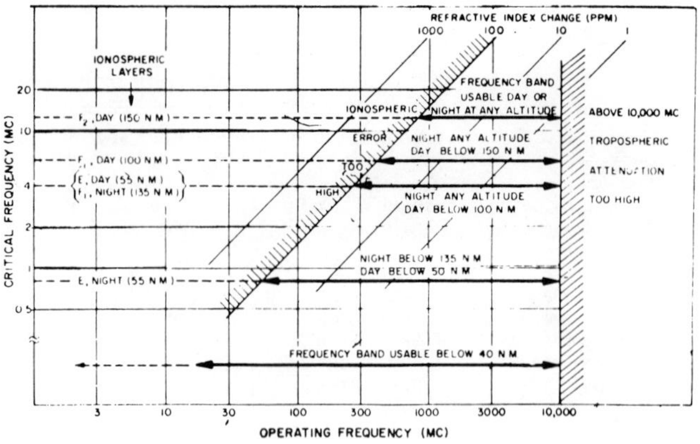

Fig. 19—Frequencies usable for precise instrumentation through ionosphere.

2) Overdetermination by several independent instruments can be used to reduce this propagation error still further.

3) Operating frequencies above 1000 Mc are required to assure 0.1 mil and 100-ft accuracy for targets in or beyond the ionosphere. Errors increase with decreasing frequency, and systems operating at or below 100 Mc are subject to errors of the order of one part per thousand for targets above 50 miles.

4) Only systems operating between 1000 and 6000 Mc can be considered for all-weather operation at ranges extending from 10 miles to beyond 100 miles.

The limitations imposed by tropospheric attenuation and ionospheric refraction are summarized in Fig. 19, which shows the frequency bands usable for precise instrumentation in various altitude regions under daytime and nighttime ionospheric conditions. Critical frequency (the frequency at which total reflection occurs even at normal incidence) is a function of electron density. The refractive-index change is proportional to the square of the ratio of critical frequency to operating frequency. Shown in Fig. 19 are typical values of critical frequency in each ionospheric layer and the operating frequency band satisfactory for operation within or beyond each layer. The criterion used was that the refractive-index change (*i.e.*, the error in incremental position measurement) should not exceed 100 parts per million. This criterion, together with an upper limit of 10,000 Mc due to tropospheric absorption and a maximum critical frequency of 10–14 Mc, defines an operating band of 1000 to 10,000 Mc for long-range precision systems. If the refractive-index change is to be limited to 10 parts per million, the usable frequency band is restricted to 3000 to 10,000 Mc. For operation at low altitudes and/or at night, the lower-frequency limits can be reduced to the extent shown in Fig. 19.

Unpredictable variations in propagation conditions set the ultimate limit to both antenna gain and measurement accuracy. The equation derived by Muchmore and Wheelon [16] and verified approximately by experiments over short tropospheric paths [17], [18], gives the mean square phase variation in propagation through the atmosphere as

$$\overline{\alpha_t^2} \cong \frac{8\pi^2 L_0 L \overline{(\Delta N)^2} \times 10^{-12}}{\lambda^2}, \tag{28}$$

where L_0 is the scale length of the propagation anomalies and $\overline{(\Delta N)^2}$ is the mean square anomaly in N units, or units of $(n-1)\times 10^6$ where n is the refractive index. For the ionosphere, Colin [19] has rewritten (28) in the form

$$\overline{\alpha_i^2} = \frac{2\pi^2 L_0 L \lambda^2}{\lambda_n^4}\left(\frac{\Delta N_e}{N_e}\right)^2, \tag{29}$$

where λ_n is the plasma wavelength, N_e the ionospheric-electron density, and ΔN_e the variation in electron density. Typical values listed by Colin are

	Troposphere	Ionosphere
L_0	200 ft	5 km (occasionally 100 km)
L (vertical ray)	20,000 ft	65 km
$\overline{(\Delta N)^2}$ (clear day)	0.25	—
$\overline{(\Delta N_e/N_e)^2}$	—	3×10^{-4}

leading to the following values for vertical paths:

$$\overline{\alpha_t^2} = \frac{7.34 \times 10^{-2}}{\lambda^2} \qquad \text{(tropospheric)} \tag{30}$$

and

$$\overline{\alpha_i^2} = 2.51 \times 10^{-4}\lambda^2 \qquad \text{(ionospheric)}, \tag{31}$$

where λ is in cm. At elevations near 10°, the mean-square-tropospheric component will be six times as great and the ionospheric perhaps four times as great. For cloudy conditions, $\overline{(\Delta N^2)}$ and L_0 may each reach 1000, increasing $\overline{(\alpha_t)^2}$ by a factor of 20,000. The resulting values of two-way phase variation, converted to equivalent range variation σ_r in feet, are shown in Fig. 20.

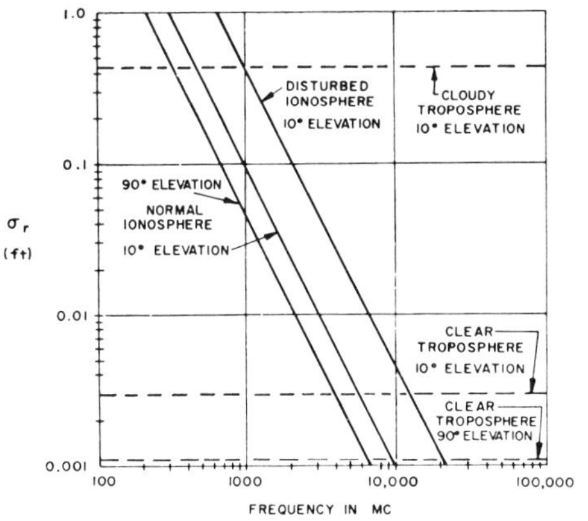

Fig. 20—Range error due to atmospheric fluctuations.

Note that the ionospheric component is appreciable, even at *C* band, and that the worst case (cumulus clouds or thunder storms) causes errors of about one-half foot at all frequencies. These results are particularly important in error analysis of baseline systems, as they set a lower limit an error which is independent of the instrumental precision of the system. Since tropospheric errors will be essentially uncorrelated over paths separated several hundred feet in clear air or several thousand feet under cloudy conditions, the errors shown in Fig. 20 may be considered lower limits to σ_r or $\sigma_{\Delta r}/\sqrt{2}$ in Table X. Ionospheric errors will be uncorrelated for paths spaced tens of miles apart, and will, therefore, affect baseline systems longer than a few miles. In addition to the effects listed, there may be large-amplitude irregularities of very long scale length in both ionosphere and troposphere which will become important where data is smoothed over long tracking periods. The lower limits of error for various baseline lengths are shown in Table XVIII, with the unsmoothed Doppler velocity errors based upon an assumed drift rate of 50 ft/sec for tropospheric anomalies and 1000 ft/sec for ionospheric anomalies (average error periods between 4 sec and 20 sec for both troposphere and ionosphere). It is clear that disturbed conditions of either portion of the atmosphere will drastically affect any of the systems discussed. Only the use of a very long baseline and a frequency above 3000 Mc can provide velocity accuracies approaching one ft/sec in all coordinates at 20,000 miles range.

Factors Favoring Pulsed Radar

The foregoing discussion indicates that pulsed radar should be useful in space-range instrumentation, at least to the same extent that it serves the existing missile ranges. In space work, the use of large, high-gain antennas is mandatory, and at present, the only suitable antennas are of the tracking type. The requirement for low-input noise favors the continued use of such antennas, which require only one maser amplifier (three for monopulse systems) as contrasted with multiple amplifiers for the low-noise electronic-scan types. The required reflector tolerance and pointing accuracy for achieving high gain necessitates the same mechanical design techniques which have been used in the past for precision-tracking radars. Several such radars deployed on two- to four-thousand mile baselines can outperform any of the other schemes described for position and velocity measurements at long range.

267

Applications Unfavorable to Pulsed Radar

There remain certain applications and requirements in missile and space instrumentation for which pulsed radars are not well suited. At launch, there is a need for very accurate tracking on a well-defined portion of the missile, not subject to disturbing effects of ground re-

TABLE XVIII

TYPICAL VELOCITY ERRORS DUE TO ATMOSPHERIC FLUCTUATION

Conditions	Baseline *B* (mi.)	Standard Deviation in lateral velocity					
		$\sigma_{\dot\theta}$(μrad/sec)		Equivalent linear vel error (ft/sec)			
				1000 mi range		20,000 mi range	
		L band	*C* band	*L* band	*C* band	*L* band	*C* band
Clear, normal ionosphere, 90° elevation	1	4.0	0.35	25	2.0	500	40
	10	0.7	0.035	4.2	0.2	84	4
	100	0.07	0.0035	0.42	0.02	8.4	0.4
	1000	0.007	0.00035	0.042	0.002	0.84	0.04
Clear, normal ionosphere, 10° elevation	1	8.0	1.1	50	7.0	1000	140
	10	1.4	0.11	8.4	0.7	170	14
	100	0.14	0.011	0.84	0.07	17	1.4
	1000	0.014	0.0011	0.084	0.007	1.7	0.14
Clear, disturbed ionosphere, 10° elevation	1	3.3	1.1	20	7.0	400	140
	10	3.3	0.11	20	7.0	400	14
	100	0.6	0.02	3.5	0.11	70	2.2
	1000	0.06	0.002	0.35	0.011	7.0	0.22
Cloudy, normal ionosphere, 10° elevation	1	100	100	600	600	12,000	12,000
	10	10	10	60	60	1200	1200
	100	1.1	1	6.5	6.0	130	120
	1000	0.11	0.1	0.65	0.6	13	12

NOTE: Velocity error based on use of any smoothing time less than fluctuation period of atmosphere, 4–20 sec.

flections, clutter or glint. Optical and infra-red devices appear advantageous here. During powered flight, exact measurement of attitude and roll, to the order of one degree in angle, requires optical measurement. Throughout the flight, wide-bandwidth command and telemetry channels are used, which do not lend themselves to use of the low-rate pulsed carrier provided by the radar. Although the radar antenna may provide a high-gain transmission path to the vehicle, separate antennas have the advantage of reducing mutual interference. Furthermore, the relative ease with which low-power solid-state CW oscillators can be designed is favorable to use of command, telemetry and measurement systems operating on the CW carrier. With the present state of the art, and in cases where vehicle weight is of extreme importance, it may well prove more economical to build duplicate ground antennas for separated CW transmitter and receivers.

Ultimately, however, the inherent flexibility and economy of the pulsed radar should make it a prime instrument for space research in passive and active tracking of vehicles and active probing of the surfaces of planets.

Appendix

Comparison of AN/FPS-16 and AN/FPQ-6 Parameters

	AN/FPS-16	*AN/FPQ-6*
Antenna size	12 ft	29-ft Cassegrainian
Antenna feed	4-horn monopulse	5-horn monopulse
Antenna beamwidth	1.2°	0.4°
Antenna gain	44 db	51 db
Antenna polarization	Vertical	Vertical, circular
Pedestal drive	Electric	Hydraulic
Antenna-pedestal rotation-azimuth elevation	Continuous −10° to +190°	Continuous −2° to +182°
Antenna-pedestal weight	16,000 lbs	60,000 lbs
Azimuth bearing	Ball	Hydrostatic
Angle-servo bandwidth	0.25 to 5 cps	0.5 to 5 cps
Angle-tracking rate-azimuth elevation	40°/sec 30°/sec	28°/sec 28°/sec
Angle-tracking precision	0.1 mil rms	0.05 mil rms
Frequency	5400–5900 Mc	5400–5900 Mc
Power output-fixed tunable	1 Mw $\frac{1}{4}$ Mw	3 Mw
Pulse-repetition frequency	Various 341–1707 PPS	160–640 basic, 1707 max
Pulse duration	$\frac{1}{4}$, $\frac{1}{2}$, 1 μsec	$\frac{1}{4}$, $\frac{1}{2}$, 1 and 2.4 μsec
Pulse coding	Up to 5 pulses with duty cycle limits	Up to 5 pulses with duty cycle limits
Receiver noise figure	11 db	8 db
Receiver IF	30 Mc	30 Mc
Receiver bandwidth	Switch with pulse width	1.2 to 1.6/pulse width
Receiver dynamic range	[illegible] db with S.T.C.	110 db with programming
Range-tracking capability	500 nautical miles	32,000 nautical miles Continuous unambiguous
Range-tracking granularity	1 yd	2 yds
Acquisition features	Scan generator variable rep. rates	Scan generator Video integrator "C" scope and joy stick Pulse-coding capability "Aux-track" Multiple-gate array Digital-ranging system
Output—range	20-bit 2-speed binary, dc analog pot	25-bit single-speed binary
Output—angle	17-bit 2-speed binary 16:1 speed synchros Sine-cosine pots	19-bit binary with 8-bit dynamic lag correction 16:1 speed synchros Sine-cosine pots

Bibliography

[1] D. R. Rhodes, "Introduction to Monopulse," McGraw-Hill Book Co., Inc., New York, N. Y.; 1959.

[2] D. K. Barton, "Accuracy of a monopulse radar," *Proc. 3rd Natl. Convention on Military Electronics*, Washington, D. C., June 29-July 1, 1959; pp. 179–186.

[3] D. K. Barton, "Final Report, Instrumentation Radar AN/FPS-16 (XN-2), Evaluation and Analysis of Radar Performance," RCA, Moorestown, N. J., Contract DA36-034-ORD-151, AD212 125; 1957.

[4] D. K. Barton, "Final Report, Instrumentation Radar AN/FPS-16 (XN-2)," RCA, Moorestown, N. J., Contract NOas 55-869c, AD 250 500; 1959.

[5] N. S. Greenberg, "Extending the Range of Radar Beacon Tracking for Lunar Probes," presented at American Astronautical Soc. Lunar Flight Symp., New York, N. Y.; December 27, 1960.

[6] W. W. Ward, "An Account of the Factors that are Important in Tracking of Space Vehicles by Means of Radar," presented at 5th Natl. Symp. on Space Electronics and Telemetry, Washington, D. C.; September 19–21, 1960.

[7] D. C. Hogg and W. W. Mumford, "The effective noise temperature of the sky," *Microwave J.*, vol. 3, pp. 80–84; March, 1960.

[8] F. Sterzer and D. E. Nelson, "Tunnel-diode microwave oscillators," Proc. IRE, vol. 49, pp. 744–753; April, 1961.

[9] M. J. E. Golay, "Velocity of light and measurement of interplanetary distances," *Science*, vol. 131, pp. 31–32; January 1, 1960.

[10] M. I. Skolnik, "Theoretical accuracy of radar measurement," IRE Trans. on Aeronautical and Navigational Electronics, vol. ANE-7, pp. 123–129; December, 1960.

[11] S. Shucker and R. Lieber, "Comparison of short ARC tracking systems for orbital determination," *Proc. 5th Natl. Convention Military Electronics*, Washington, D. C., June 26-28, 1961; pp. 285–301.

[12] D. K. Barton, "Comparison of UHF and *C*-Band Radar Signature Data on Sputnik II," to be published.

[13] D. K. Barton, "Sputnik II as observed by *C*-band radar," 1959 IRE National Convention Record, pt. 5, pp. 67–73.

[14] W. C. Pilkington, "Vehicle motions as inferred from radio-signal strength records," in "Avionics Research, Satellites and Problems of Long Range Detection and Tracking," Pergamon Press, New York, N. Y.; 1960.

[15] G. H. Millman, "Atmospheric effects on VHF and UHF propagation," Proc. IRE, vol. 46, pp. 1492–1501; August, 1958.

[16] R. B. Muchmore and A. D. Wheelon, "Line-of-sight propagation," Proc. IRE, vol. 43, pp. 1437–1458; October, 1955.

[17] J. W. Herbstreit and M. C. Thompson, "Measurements of the phase of radio waves received over transmission paths with electrical lengths varying as a result of atmospheric turbulence," Proc. IRE, vol. 43, pp. 1391–1401; October, 1955.

[18] A. P. Deam and B. M. Fannin, "Phase-difference variations in 9350-megacycle radio signals arriving at spaced antennas," Proc. IRE, vol. 43, pp. 1402–1404; October, 1955.

[19] L. Colin, "The effects of atmospherically induced phase scintillations," pt. IX of final rept., "Upper Atmosphere Clutter Research, Part XIII: Effects of the Atmosphere on Radar Resolution and Accuracy," Stanford Res. Inst., Menlo Park, Calif., RADC-TR-60-44, AD 238 407; April, 1960.

Phase-Amplitude Monopulse System*

W. HAUSZ†, FELLOW, IRE, AND R. A. ZACHARY‡, SENIOR MEMBER, IRE

Summary—**Phase monopulse, amplitude monopulse, and phase-amplitude monopulse are compared. The latter uses only two antenna feeds and by controlling the aperture illumination from these feeds has a separation of the illumination phase centers in one axis and opposed tilts of the phase fronts, or squint, in the other axis. It is described as more fundamental or less redundant than four-horn or other multihorn monopulse designs. In lacking redundancy it is less complex, but has fewer degrees of freedom to optimize performance. The earliest example of phase-amplitude monopulse, the AN/APG-25 (XN-2), is described and its performance given.**

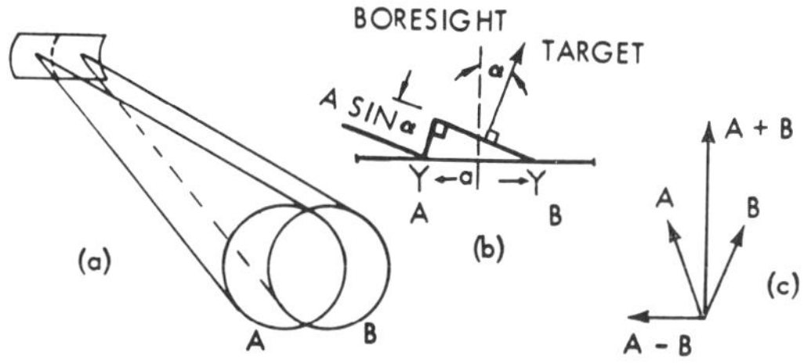

Fig. 1—Phase comparison monopulse.

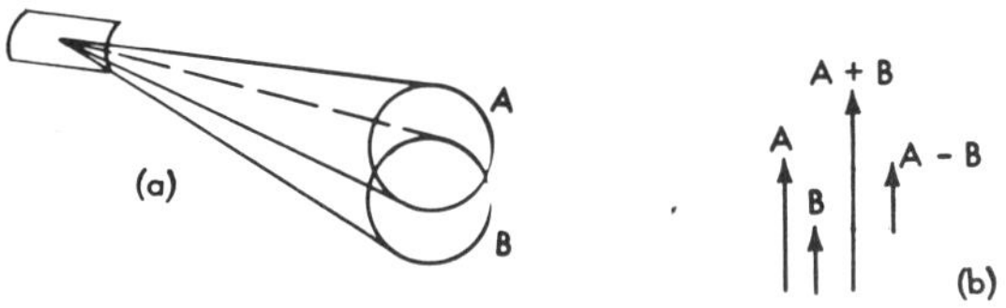

Fig. 2—Amplitude comparison monopulse.

Introduction

THE BASIC CONCEPT of monopulse is that of extracting complete positional information about a target from a single radar pulse. Conventionally, two kinds of monopulse are described:[1–4] phase monopulse and amplitude monopulse, which are not fundamentally distinct in data handling, but differ principally in the hardware concept used to illuminate the antenna aperture with two primary feeds. Phase monopulse, illustrated in Fig. 1(a), subdivides the secondary aperture in half, producing two subapertures which have identical amplitude vs angle pattern responses. For an off-axis signal being received, the amplitude at each feed will be the same, but the time of arrival will be different [Fig. 1(b)], producing a phase difference between the signals received on feeds A and B, as shown in Fig. 1(c). Also shown in Fig. 1(c) are the vector sum $A+B$ and the difference $A-B$, which is in quadrature with $A+B$.

The concept of amplitude comparison, illustrated in Fig. 2(a), is that of two feeds, each illuminating the whole aperture, but displaced symmetrically about the focal point. The patterns produced by A and B are "squinted" to opposite sides of the axis of symmetry, so an off-axis target will receive signals of different amplitudes, as shown in the vector diagram, Fig. 2(b). However, the overlapping illumination of the whole aperture by each feed makes the time of arrival, or phase of the received signal, identical in A and B. Therefore the sum $A+B$ and difference $A-B$ are in phase rather than in quadrature.

From these one-dimensional concepts, the second angular coordinate can be achieved by the simple extrapolation of repeating the aperture-sensing system in the orthogonal direction. For phase monopulse the aperture is divided both horizontally and vertically and four feeds are used. For amplitude monopulse, the squinting is done both vertically and horizontally and four feeds are used.

This descriptive approach is not fundamental, and by so limiting itself to "easily distinguished" means of implemention may miss other approaches inherent in the monopulse concept. Solving for the three coordinates of target position on a single pulse requires enough independent inputs to generate three independent equations. A single antenna feed permits measurement of an amplitude and a time delay of signal received. From the time delay, one unknown, the range, can be found directly. The amplitude cannot give angular information on a single pulse because in addition to the angular unknowns there is the further unknown of system sensitivity which is a composite of the radar parameters, and particularly includes the unknown echoing area of the target.

A second antenna feed which is associated with an antenna aperture illumination differing in phase and/or amplitude from that of the first feed, provides another amplitude and time delay, or two more equations. While the range can also be found from the second time delay, the new information is the small time difference, or the phase difference between the two signals. The system-sensitivity parameters, including the target echoing area, apply identically to both feeds, so taking the ratio of the two amplitudes eliminates them. Two equations are left, in terms of the ratio of the signal amplitudes and the phase difference between them, that can be used to find the two remaining positional unknowns, the elevation angle and the azimuth angle. If for each combination of azimuth and elevation angle in the region of interest there

* Received by the PGMIL, December 1, 1961.
† General Electric Company, Santa Barbara, Calif.
‡ General Electric Company, Utica, N. Y.
[1] J. P. Blewett, S. Hansen, R. Troell and G. Kirkpatrick, "The Multilobe Tracking System," G.E. Res. Lab., Schenectady, N. Y.; January 5, 1944.
[2] R. M. Page, "Monopulse radar," 1955 IRE National Convention Record, pt. 8, pp. 132–134.
[3] H. T. Budenbom, "Monopulse automatic tracking and the thermal bound," *1957 Convention Record*, First Nat'l Conv. on Military Electronics, Washington, D. C., June 17–19, pp. 387–392.
[4] D. R. Rhodes, "Introduction to Monopulse," McGraw-Hill Book Co., Inc., New York, N. Y.; 1959.

is a unique combination of ratio and phase difference between signals, two feeds suffice to completely locate a single target.

Of course the region of interest must be constrained by the practical limitations on the sensitivity with which ratios and phase differences can be measured, the antenna aperture and beamwidth used for system sensitivity, the angular accuracy desired, and the ambiguities which enter for phase differences of more than $\pm 180°$. These, for a tracking radar, generally result in the measurement of angular position from a mechanical reference axis, the boresight, making the ratio of signals unity and the phase difference zero at an angle as close to the boresight axis as possible. The region of interest then is constrained by the beam pattern and phase ambiguities to roughly plus or minus a half beamwidth from the boresight. A further desideratum for maximum angular sensitivity in two orthogonal directions is that orthogonal combinations of phase and amplitude differences correspond to deviations along the two angular axes.

In Fig. 3(a) the vector diagram shows two signals which differ in phase and amplitude. The difference $A-B$ is neither in phase nor in quadrature with the sum $A+B$. The component of $A-B$ that is in phase with $A+B$ (or any other reference axis related to $A+B$ by a definite phase angle) is independent of, or is orthogonal to, the component of $A-B$ in quadrature with $A+B$ (or the reference axis).

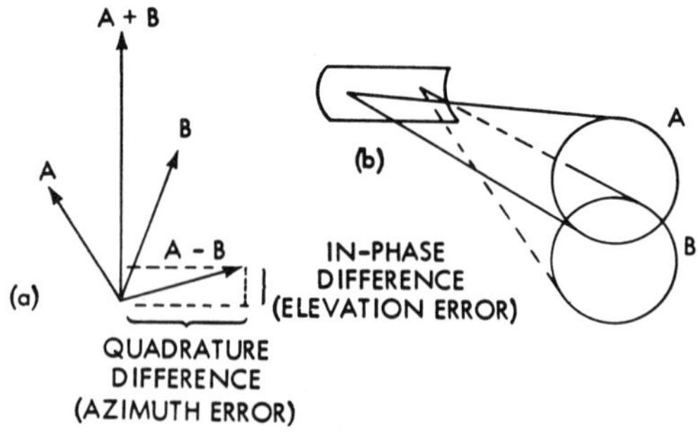

Fig. 3—Phase-amplitude monopulse.

One obvious possible implementation, by extrapolation from Figs. 1 and 2, is to make azimuth error angles from the boresight be represented by the quadrature component and elevation-error angles be represented by the in-phase component. One way of doing this is shown in Fig. 3(b) where two line feeds illuminate the right and left halves of a parabolic cylinder. The aperture illuminations are then separated horizontally. If one feed is above the focal line and the other is below the focal line, the phase fronts on the two halves of the aperture will be tilted down and up, or the two beams will be squinted in elevation as shown.

Since two feeds can suffice to determine the position of a single target, the use of four feeds contains a certain amount of redundancy. Such redundancy, properly used, can be advantageous enough to merit the increase in complexity. It may be used to better control the aperture illumination during transmit-and-receive for better sensitivity, lower sidelobes, or more linear relations between angles and signals. Potentially, if four or more feeds are provided with suitably independent aperture illuminations, enough equations are available to solve simultaneously for the angular positions of two targets at the same range and both within the main beam, that is, to greatly increase angular resolution. Little has been done as yet to implement this usefully in real time.

Phase-Amplitude Monopulse System

To go beyond the conceptual approach to a phase-amplitude monopulse system inplied in the vector diagram and beam configuration of Fig. 3, a system block diagram is shown in Fig. 4. The signals from feeds A and B are combined in a hybrid junction, which has outputs $A+B$, (Σ), and $A-B$, (Δ). Each RF signal is heterodyned with the same local oscillator, preserving the phase and amplitude relationships. Two IF amplifiers raise both sum and difference signals to a useful level. The output of the sum channel is used to determine range, to use in various displays, and as a reference signal for angular-error determination. The requirements on a detector for angular information are that it is sensitive only to the complex ratio between two signals, and not to their magnitude, that it is sensitive to the phase-amplitude function representing one angular coordinate and is insensitive to that representing the other, and that it meet desired sensitivity criteria (such as maximum sensitivity in the neighborhood of the boresight, or alternatively maximum linearity over a specified range).

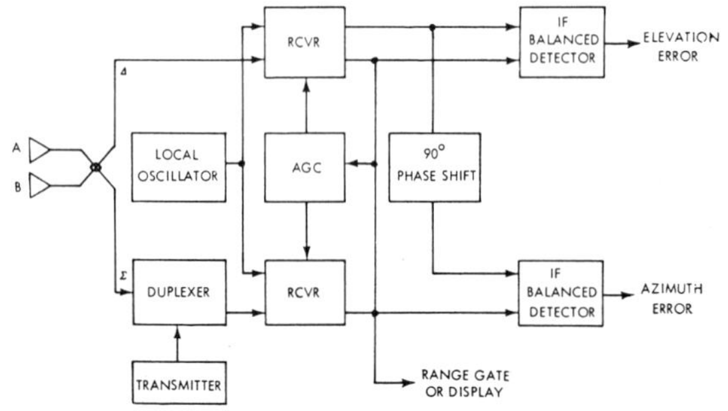

Fig. 4—Block diagram of phase amplitude monopulse system.

A typical detector of this type adds and subtracts the error signal (difference) and the reference signal (sum), limits both resultants and forms an output proportional to the difference between them. In essence it forms two vectors differing in phase and, by limiting before combining them, forms an output dependent on the phase difference alone. However, as was noted, the two error signals differ only in the relative phase of the sum and difference signals. By rotating the error signal by 90° before combining with the reference signal, an identical detector can produce the other error component as output. The two error signals in video form are then used to servo the antenna boresight axis to the target, or to indicate and display the magnitude and direction of the angular deviation between target and boresight.

The transmitter is connected to the sum channel output of the hybrid junction, using a duplexer for receiver protection, so that a symmetrical pattern of maximum gain can be radiated on transmission.

It should be noted that the phase-amplitude monopulse system requires only two antenna feeds, one hybrid junction and two IF channels. In contrast, either phase monopulse or amplitude monopulse normally requires four antenna feeds, three hybrid junctions, and three IF channels. Where light weight and cost are important considerations, the reduced complexity of the phase-amplitude monopulse system proves a marked advantage.

Aperture Control

In all antennas it is necessary to accept compromises in the feed and aperture design to balance maximum gain and minimum beamwidth against tolerable sidelobes (including spillover). It is particularly complex in monopulse, where the feeds are to be designed to give the optimum combination of gain in the sum output channel, angular sensitivity in each of two angular outputs, minimum sidelobes and spillover,[5] and, for electronic tracking, linearity of the error signal over as great an angular range as possible. In this design problem, redundancy of feeds, using four to as many as twelve feeds,[5,6] permits tailoring the aperture illumination for higher performance in all channels, but at the expense of complexity.

For phase-amplitude monopulse, the opposite phase tilts of the aperture illumination on the two halves of the antenna makes it difficult to visualize the antenna pattern in all four quadrants from simple consideration of the sum and difference patterns along the two major axes. Fig. 5(a) and (b) show computer-calculated sidelobe contours for the sum and difference patterns of one extreme of aperture illumination. The illumination for this case comprised uniform amplitude of illumination all over a square aperture which was divided horizontally into two halves with opposing phase tilts in the vertical direction. Only one quadrant is shown. It will be noted that the 3-db beamwidth and the magnitude and extent of sidelobes for angles not on the major axes are considerably poorer than the major axis patterns would indicate.

Even with only two feeds there is considerable flexibility of feed design and aperture control. The antenna configuration may be as simple as two parabolic dishes each fed by a point-source feed, as described later in this paper, or line-source feeds in a parabolic cylinder, or two aperture-controlling feeds located vertically in a single paraboloid. The latter two permit some degree of overlap of the illumination from the two feeds. For the parabolic cylinder, since one line source is above the focal line and the other one is below it, they may both be lengthened to overlap any desired amount, and the illumination along the line feed made nonuniform, both for sidelobe control and to achieve the best compromise between sum channel gain and error sensitivity in the phase-comparison axis. Similarly, two primary feeds in a single paraboloid can be small and displaced in the vertical direction, but sufficiently large in the horizontal direction that they can be oriented to illuminate

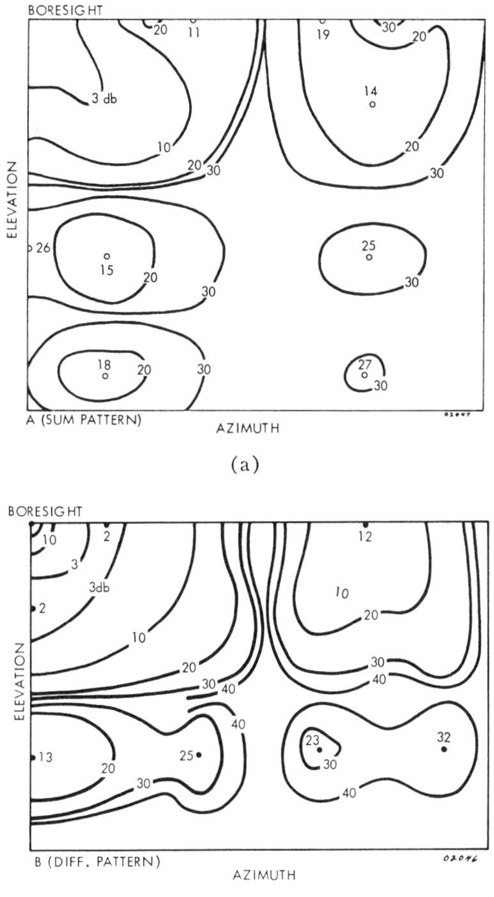

(a)

(b)

Fig. 5—Computed sidelobe contours for a square aperture, uniform illumination, horizontal phase comparison, vertical amplitude comparison. (a) Sum pattern (b) Difference pattern.

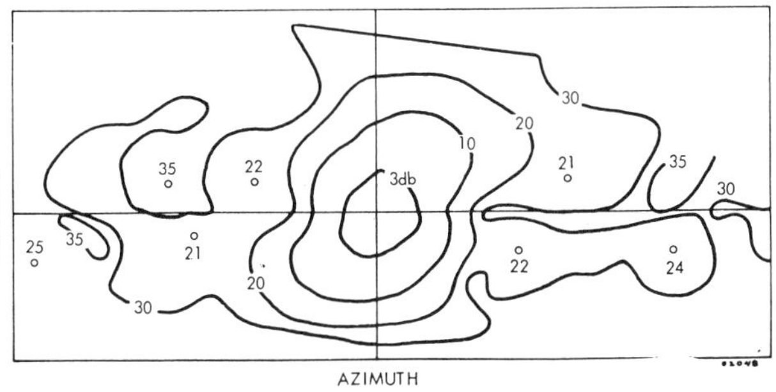

Fig. 6—Experimental sum pattern of sidelobe contours for parabolic cylinder with phase-amplitude monopulse line feeds 34 per cent overlapped.

[5] P. W. Hannan, "Optimum feeds for all three modes of a monopulse antenna," Pts. I and II, IRE Trans. on Antennas and Propagation, vol. AP-9, pp. 444–460; September, 1961.

[6] L. J. Ricardi and L. Niro, "Design of a twelve-horn monopulse feed," 1961 IRE International Convention Record, pt. 1, pp. 49–56.

only one half of the aperture, but with appropriate overlap. To show that a considerably better sum pattern can be achieved, Fig. 6 gives an experimentally measured field of sidelobe contours for a 34 per cent overlapped illumination of two line feeds in a parabolic cylinder.

Because of the fewer degrees of freedom available in a two-feed, phase-amplitude monopulse, its performance can never equal the maximum available with an unlimited number of feeds, but the sum gain and error sensitivities can each be brought to within about 3 db of the optimum performance. For beacon-type tracking this loss of sensitivity is seldom a disadvantage. Even for a skin tracking radar, a complete systems analysis must consider that the extra plumbing complexity involved in four or more feed systems is not lossless and reduces the performance differences; also the reduction in weight and cost of microwave plumbing and IF strips can be traded off to some extent for other increases in performance such as a larger dish, or more transmitter power so that, particularly for small airborne radars, phase-amplitude monopulse may often be the cheapest and best in performance.

The AN/APG-25 Test Radar

A number of different radars based on phase-amplitude monopluse have been built, tested and used. The first known example of this form of monopulse was the AN/APG-25 and the ASTIA document describing it has recently been declassified.[7] It will be discussed as an example.

The AN/APG-25 radar was conceived as part of a tail-turret, fire-control system for a Navy patrol aircraft. Work on this radar, designed to exceed significantly the tracking performance of similar equipment used in World War II, was begun in earnest in late 1946.

Because system design called for separate search and track radars, the APG-25's function was limited only to tracking. Target tracking was based on the stabilized line-of-sight approach in which the radar-tracking loop furnishes present position and angular rate data, and the computer predicts future target position. In this system a gyro mounted on the tracking antenna structure stabilizes the antenna line-of-sight against ownship's motion. Target tracking is then performed by feeding the radar angular-error signal to precession coils in the gyro, which in turn supplies the driving signal to the antenna-tracking motors. Because of the stabilizing action of the gyro in the over-all servo loop, the radar signal need not furnish stabilization against ownship's motion. The radar signal was, therefore, suitable for use by the computer as a measure of target angular rates.

It was clear from the outset of the program that reduction of radar-tracking noise would be a significant problem. Some form of monopulse radar was in order to provide a low-noise, servo-error signal.

Initially, an amplitude-comparison, four-lobe monopulse scheme was investigated although it did impose severe requirements on antenna size and microwave system complexity. To reduce the complexity of such a system somewhat, the azimuth and elevation-difference signals (which are developed separately in four-lobe systems) were added in quadrature. This eliminated one of the two difference-channel IF amplifiers. This idea led to the dual-beam, phase-amplitude concept of monopulse which was promptly adopted because it promised a generous payoff in antenna and plumbing simplification.

Electrical parameters for the system were fairly conventional for that period. RF power was 40-kw peak at a PRF of 2000 and a pulse width of 0.5 μsec. The transmitter operated at a single frequency in the X band.

Tracking Loop Configuration

Deviations of the target from the antenna boresight axis appear at the output of the phase discriminators (see Fig. 4). Amplified, these error signals act on the gyros, which in turn drive the antenna-positioning motors to return the boresight axis to the target. Stabilization against ownship's motion is provided by the gyros alone. To avoid the problems and complexity of four rotating joints for the two microwave signals, it was decided that the first experimental model of the AN/APG-25 would have the microwave structure, the two IF amplifiers, and the detector or phase discriminator unit, which generated the tracking-error signals, on the antenna.

This entire unit, called the tracking head, is shown in Fig. 7. This head supplied dc error signals to the servo via shielded leads. It was unnecessary to transfer any RF signals from the tracking head although a video line, carrying

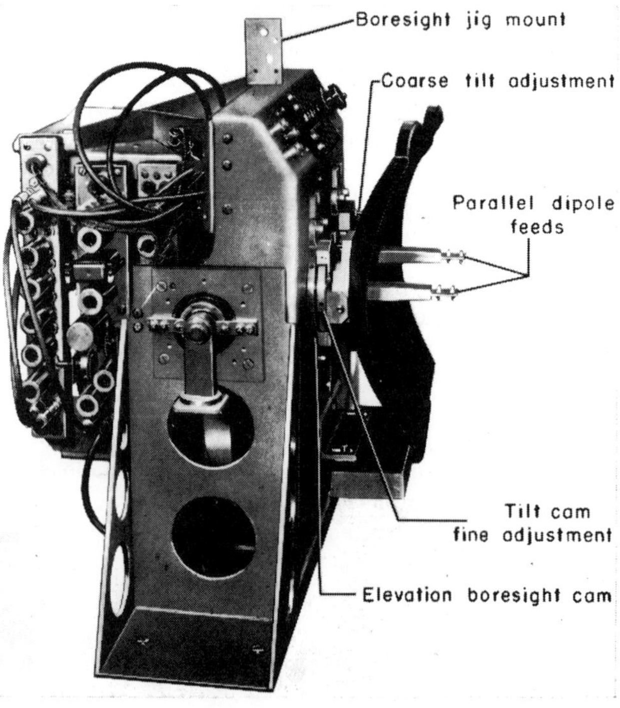

Fig. 7—Tracking head AN/APG-25 (XN-2)

[7] "Automatic Gun Laying Turret, Tail, Aero X54: Design Study Report," vol. 2, G.E. Co., Schenectady, N. Y., ASTIA doc. no. 85310 (U); 1949.

the detected-sum signal for use in the range-tracking circuits, was transferred from the tracking head by a flexible coaxial cable.

Antenna Design

It was decided, principally for ease in construction and in design, to use a dual parabolic antenna with separated feeds. Each paraboloid was cut from a 12-inch dish and mounted separately on the tracking-head structure with a mechanical arrangement for adjusting tilt of each cut dish separately to vary the squint angle. This angle was normally set for 5°, permitting a 70 per cent power crossover. It was felt that adjustment of this was desirable for the experimental model, but that future designs would provide for casting of the entire paraboloid structure as a single piece with a fixed squint angle. The antenna was vertically polarized. Patterns taken in the H plane (azimuth) and the E plane (elevation) are shown in Fig. 8(a) and (b).

Microwave System

The microwave system was much simpler than that needed in four-lobe monopulse systems and is shown as an isometric drawing in Fig. 9. The antenna feeds couple into the two arms of a hybrid T which provides microwave sum and difference signals. Each of these feeds to balanced mixers for conversion to the 30-Mc intermediate frequency.

Besides the balanced mixers, the plumbing system includes a local oscillator, AFC mixer, TR tubes, and phase shifter. This last item adjusts the phase shift between the sum and difference channels, thereby compensating for any relative difference in phase of the sum and difference signals in either the microwave or IF portions of the system. Because the experimental model of the APG-25 was designed as a single-frequency, X-band radar, no particular attempt was made to obtain broad-banding.

IF Amplifiers

The two IF amplifiers are identical except that the sum-channel amplifier provides a detector and video amplifier to furnish a sum-video signal to the range-tracking circuits.

The IF portions of these two amplifiers developed more than 100-db gain in five stages of transformer-coupled pentodes. Over-all bandwidth was four Mc. Gain control is accomplished by means of a common AGC line coupled to the grids of stages two, three, and four of each of the amplifiers.

The input stage was an early version of the low-noise, dual-triode circuit. The only dual triode available was the 6J6, which had a common cathode. Of necessity, this circuit emerged as a grounded-grid stage driven by a cathode follower. Although it provided a better noise figure than could have been obtained with a pentode input, performance was not as good as that of the now-conventional cathode circuit.

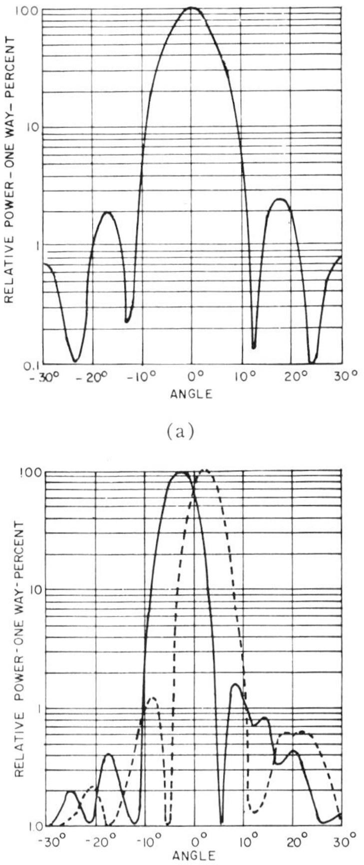

(a)

(b)

Fig. 8—Antenna patterns AN/APG-25 (XN-2); (a) H plane. (b) E plane.

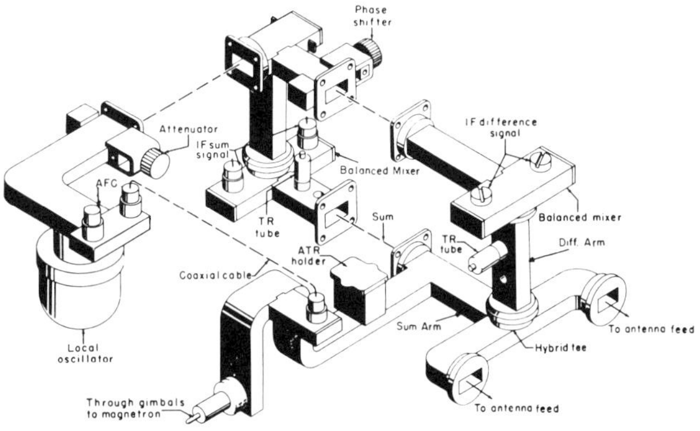

Fig. 9—Microwave structure for AN/APG-25 (XN-2).

Phase Discriminator

Early in the design of the APG-25 it was found that available phase discriminators for use at 30 Mc were extremely frequency-sensitive. Since it was felt that the automatic frequency control could not hold the IF frequency closer than one Mc, it became necessary to develop a new discriminator circuit with reduced sensitivity to IF frequency changes.

A special circuit developed for this purpose is shown

schematically in Fig. 10. This circuit consists of a quarter-wave transmission line driven at each end by a matched IF amplifier. In the absence of a difference signal, the sum signal appears with equal voltages at all points on the matched line. The detector outputs, which are equal, are subtracted across the resistor network to yield zero discriminator output.

Presence of a difference signal in quadrature phase results in a standing wave along the line producing unequal IF voltages at each end of the transmission line. These voltages detected and subtracted from each other across R_3 and R_4 yield an output curve as shown in Fig. 11. IF difference signals in phase with the sum signal can be shown to yield equal voltages at each of the detectors and, hence, a zero output.

Since this circuit is sensitive to difference signals in quadrature with the sum signal, it detects azimuth or phase-comparison channel errors; it is insensitive to elevation or amplitude-comparison channel errors. The same circuit is made sensitive to elevation-signal components by means of a 90° phase shift provided by a second quarter-wave transmission line. This line is connected between one of the pentode line drivers and one of the IF amplifiers. To separate from all other targets the one target to be tracked, the range gate generated in the range-tracking circuits was used to gate the IF signal ahead of the phase discriminator.

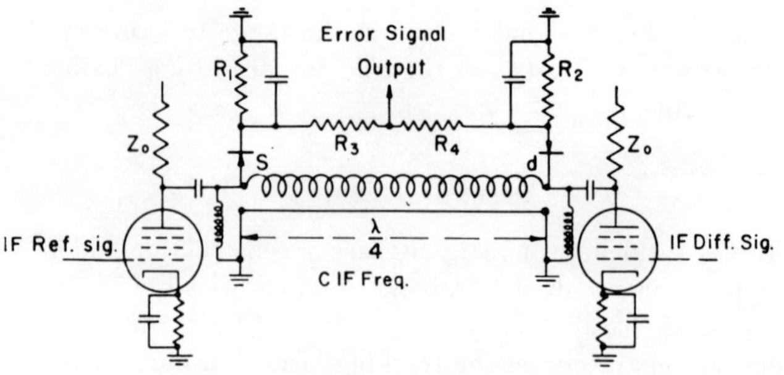

Fig. 10—Transmission line phase discriminator.

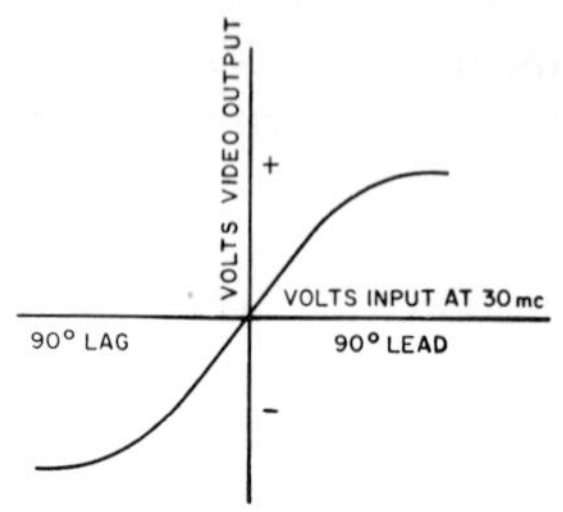

Fig. 11—Output characteristic of phase discriminator.

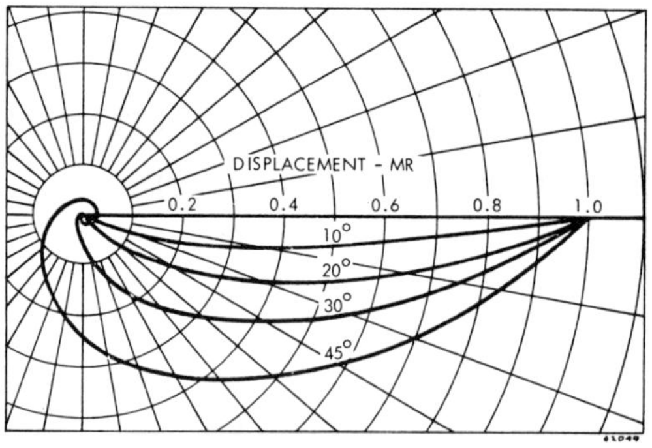

Fig. 12—Return to boresight paths of phase amplitude monopulse for indicated phase errors.

Predicted Trouble Spots

Whenever a new system is designed around a novel technique, there are bound to be hazards, some of which can be predicted with reasonable assurance. One of the first of these was the problem of relative phase shift between the sum and difference channels. This could be caused by unequal phase changes in aging IF amplifiers or the effect of the automatic gain control which was capable of changing IF gain over a range of some 60 db. Another source of such interchannel phase shift was the phase recovery of the TR tubes. Since the sum-channel TR tube is subjected to considerably more RF power than the difference-channel TR tube, its phase recovery could also be expected to differ. Although in the phase-amplitude monopulse system relative phase change between channels does not cause the boresight axis to shift, it can cause cross-coupling between the azimuth and elevation-error signals. Calculations based on a static situation in which the antenna is pulled off the target deliberately showed that phase shifts of the order of 10° or 20° would cause the boresight axis to describe a spiral path in returning to the target as shown in Fig. 12.

Test results from the research model of the AN/APG-25 laid this problem to rest. The IF channels appeared to be quite stable both with variations of AGC voltage and with time. Some phase shift with target range was noticed with certain TR tubes, but it was possible to match these tubes in such fashion that targets could be tracked from maximum range down to ranges as small as 100 yd. Deliberately induced phase shift showed little effect on tracking performance until 45° of phase shift was reached. At this point the antenna would continue to track but would circle the target.

Beam skew, an interesting geometrical peculiarity of the phase-amplitude system, caused some concern at the beginning of the project. The amplitude-comparison plane, which is the plane generated from the lines joining the beam centers, is horizontal at the antenna feeds but becomes nearly vertical at ranges in excess of a few yards. For the antenna used in the experimental model, the "horizontal" and "vertical" axes were shown to be orthogonal to within 1.1° at a 100-yd range. Any effect of beam skew is probably the same as IF channel relative-phase shift. During the testing of the experimental model, no peculiarities in tracking performance could be attributed to geometrical beam skew.

Protection of the difference- and sum-channel balanced mixers required the use of two TR tubes. In the experimental model, type 1B63A TR's were used. These were designed for normal radar use in the protection of crystals against peak transmitter powers of the order of kilowatts. Because the difference-channel TR tube is subjected to much less power than the sum-channel TR tube, there was some concern that the difference-channel tube would not fire and that such RF power as did leak through during the main "bang" would burn out the difference-channel mixer crystals. Test runs on several types of TR tubes showed that sufficient protection was provided, even at low-power level, to prevent crystal failure.

Test Results

The experimental model was never intended for flight test, but it did see extensive ground-to-air testing. Since the monopulse feature was expected to give a considerable reduction in tracking noise, most of the testing centered around noise measurement in the error-signal channels.

In these tests a motion-picture camera, operated at 32 frames/sec, was mounted on the tracking head and adjusted to point along the boresight axis to measure true-target error. A Brush recorder was used to record the actual phase-disciminator output in both the azimuth and elevation-error signal channels. The target deviation from boresight was measured on each frame of the motion-picture film. The true error in terms of phase-discriminator volts was obtained by comparison with the phase-discriminator output-voltage characteristic. This was then subtracted from the actual Brush recording of the phase-discriminator servo signal. The difference yielded error-signal noise.

Similar tests were run on a comparable X-band, conical-scan radar with a 12-inch dish. A comparison of the error signals is shown in Fig. 13.

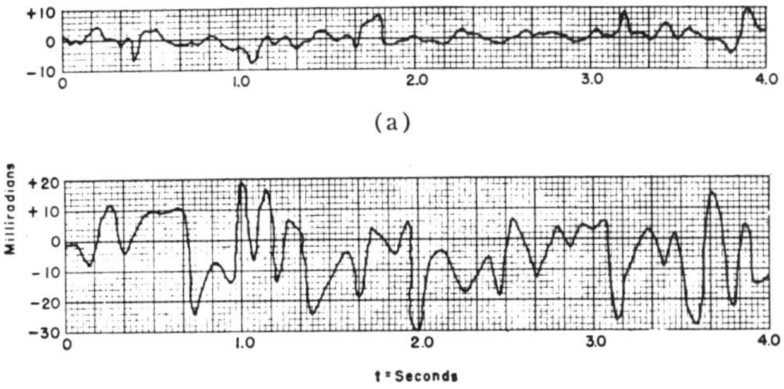

(b)

Fig. 13—Comparative error signal noise. (a) AN/APG-25 (XN-2). (b) Conical scan radar of comparable size.

Autocorrelation curves from this data were used to obtain the rms noise in the tracking channels. A typical figure for the APG-25 was 2.75 angular mils; a comparable figure for the conical scan radar was 9.85 angular mils.

In all respects the servo-noise improvement demonstrated in the AN/APG-25 Research Model tests met expectations. The system performance resulting from improved radar-tracking easily justified the use of phase-amplitude monopulse in this fire-control system.

Acknowledgment

The authors gratefully acknowledge the help and counsel of the many members of the General Electric Electronics Laboratory and General Electric Light Military Electronics Department, particularly O. H. Winn, who contributed to the theoretical development of this form of monopulse and to the construction of the described example.

Monopulse Difference Slope and Gain Standards*

It has been shown that the maximum theoretical angular sensitivity (difference mode gain slope at boresight) results from a linear odd antenna aperture illumination having no spillover.[1] This maximum sensitivity provides a suitable reference standard for rating the angular sensitivity of a monopulse antenna.

Alternatively, Hannan has shown that the maximum theoretical difference mode gain for a rectangular aperture is 2.15 db below the gain obtained with uniform equiphased illumination.[2] As his note suggests, a monopulse antenna might also be rated on this basis. It is the purpose of this communication to examine these two possible monopulse standards in somewhat greater detail.

First of all, it should be noted that both the angular sensitivity and difference gain standards are based on explicit aperture illuminations without regard for the means by which these illuminations are achieved. Both standards apply, therefore, to all of the so-called types of monopulse (amplitude-amplitude, phase-phase, etc.). Secondly, both the angular sensitivity and difference mode gain are affected by the aperture outline. Perhaps this can best be illustrated by Fig. 1, which graphs the relation between the aperture power from the odd incident field (normalized by the total aperture field power) and the angle of incidence of a plane wave in both circular and rectangular apertures.

At least two observations seem pertinent at this point. By making the aperture "transmit" illumination identical to the complex conjugate of the odd incident field distribution for the angle which maximizes odd field aperture power, the difference mode received power will achieve its maximum possible value. In the case of a rectangular aperture, this can be seen to be 2.15 db below the maximum possible sum mode received power (fields matched at $u=0$), the result previously derived by Hannan.[2] Similarly, for a circular aperture, maximum theoretical difference mode gain will be down 2.47 db from the maximum theoretical sum mode gain. The corresponding diffraction pattern can be obtained by noting that a unit amplitude plane wave incident in the ϕ_0 plane on a circular aperture has an odd field distribution expressed by

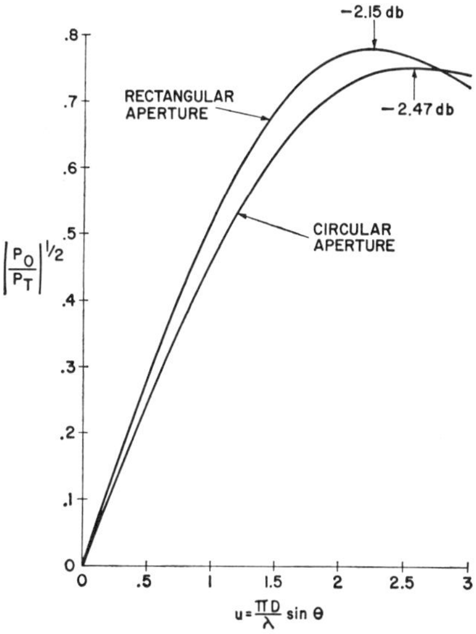

Fig. 1—Normalized odd incident field power vs angle of incidence.

* Received by the PGAP, December 13, 1961.

[1] G. M. Kirkpatrick, "Aperture illuminations for radar angle-of-arrival measurements," IRE Trans. on Aeronautical and Navigational Electronics, vol. AE-9, pp. 20–27; September, 1953.

[2] P. W. Hannan, "Maximum gain in monopulse difference mode," IRE Trans. on Antennas and Propagation, vol. AP-9, pp. 314–315; May, 1961.

$$f(r, \phi') = j \sin \left[u_1 r \cos (\phi_0 - \phi')\right],$$

where

$$u_1 = \frac{\pi D}{\lambda} \sin \theta_1.$$

Matching the odd field illumination to the odd field distribution of the incident wave ($A(r, \phi') = f(r, \phi')^*$) and solving the far-field diffraction integral yields the expression

$$E(u) = 2\pi a^2 \frac{J_1(u - u_1)}{(u - u_1)} - 2\pi a^2 \frac{J_1(u + u_1)}{(u + u_1)}. \quad (1)$$

The pattern having the maximum peak gain will be obtained if $u_1 = 2.568$, the angle of maximum odd field aperture power as indicated in Fig. 1. This pattern, normalized by its peak value, is shown in Fig. 2. The first sidelobe is -18.4 db compared with -12.6 db for the maximum difference gain pattern of a rectangular aperture.[2]

A second observation in regard to Fig. 1 is that the normalized odd field power increases more rapidly in a rectangular aperture than in a circular aperture, which is to say that the maximum angular sensitivity is greater (same aperture areas). Regardless of aperture outline, however, a linear odd

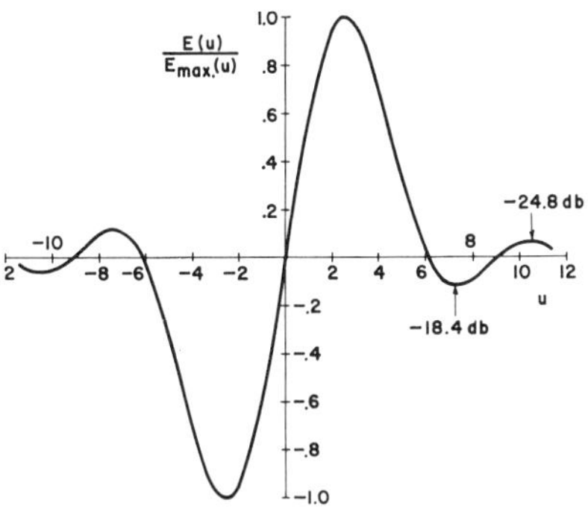

Fig. 2—Circular aperture difference pattern having maximum peak gain.

illumination results in the maximum theoretical sensitivity. This can be seen in general through the "matching fields" approach or derived for any particular case through a Calculus of Variations solution.

If angular sensitivity is defined by

$$K = \left| \frac{\partial G_\Delta^{1/2}}{\partial u} \right|_{u=0} \text{ relative volts/radian unit} \quad (2)$$

and K_0 represents the maximum theoretical sensitivity, while G_0 represents the maximum theoretical sum mode gain, the following relations are easily derived:

[3] S. Silver, "Microwave Antenna Theory and Design," MIT Rad. Lab. Ser., McGraw-Hill Book Co., Inc., New York, N. Y., vol. 12, ch. 6; 1949.

Rectangular Aperture

$$K_0 = \frac{G_0^{1/2}}{\sqrt{3}} \text{ relative volts/radian unit.} \quad (3)$$

Circular Aperture

$$K_0 = \frac{G_0^{1/2}}{2} \text{ relative volts/radian unit.} \quad (4)$$

Since the half-power point of the diffraction pattern from a uniformly illuminated rectangular aperture occurs at $u = 1.392$ radians,[3] (3) can be written

Rectangular Aperture

$$\frac{K_0}{G_0^{1/2}} = 1.607 \text{ volts/volt/U.B.W.}, \quad (5)$$

where U.B.W. emphasizes the fact that the B.W. is that for uniform illumination.

The half-power point of the circular aperture diffraction pattern occurs at $u = 1.617$ radians,[3] so that (4) becomes

Circular Aperture

$$\frac{K_0}{G_0^{1/2}} = 1.617 \text{ volts/volt/U.B.W.} \quad (6)$$

TABLE I

Aperture Illumination	1st S.L.	$\frac{G_\Delta}{G_0}$	$\frac{K}{K_0}$
Linear odd	− 8.3 db	−2.45 db	0 db
Half-sine	−10.0 db	−2.26 db	−0.08 db
Max. Δ gain (2.15-db taper)	−12.6 db	−2.15 db	−0.35 db
Full-sine	−18.3 db	−2.64 db	−2.17 db

Comparison of (5) and (6) reveals the interesting fact that, in terms of the U.B.W., the normalized maximum angular sensitivities differ by less than 1 per cent.

From normally measured antenna characteristics (sum gain, normalized sensitivity and aperture dimensions) it is a simple matter to determine the db loss in angular sensitivity from the theoretical maximum.

$$20 \log \frac{K}{K_0} = 20 \log \frac{K/G^{1/2}}{K_0/G_0^{1/2}} + 10 \log G/G_0. \quad (7)$$

$\frac{K}{G^{1/2}}$ = normalized angular sensitivity, volts Δ/volt Σ/angular unit, obtained from sum and difference patterns.

$\frac{K_0}{G_0^{1/2}}$ = value determined from (5) or (6).

G/G_0 = antenna gain factor (measured).

If beacon tracking were of interest, the sensitivity loss given by (7) would be the required figure-of-merit. For skin tracking, the round-trip sum gain enters by requiring an additional 10 log G/G_0 loss to be included in the figure-of-merit. The total loss figures so obtained represent the reduction in tracking range as compared to an antenna with both maximum gain and maximum angular sensitivity, assuming the same angular errors due to noise and a free-space environment. Thus, rating a monopulse antenna on the basis of angular sensitivity loss also provides a quantitative measure of boresight tracking performance. As in the

the sensitivity standard over the difference mode gain standard might also be mentioned. The former rating provides a more sensitive indication of monopulse antenna performance for the tapered aperture illuminations usually encountered in practice. To illustrate this effect, the normalized peak difference gain and angular sensitivity loss for a rectangular aperture have been computed with aperture illuminations which match the odd field distribution for several different angles of incidence. These are tabulated in Table I along with the first difference pattern sidelobe.

In the case of horn-fed reflectors, practical considerations such as spillover loss and diffraction pattern sidelobes usually require an illumination taper within the bounds of the last two tabulations (which exclude spillover loss). In such cases, the angular sensitivity can be seen to provide a more exact measure of antenna performance.

Richard R. Kinsey
Tracking Radar Development Engrg.
Heavy Military Electronics Dept.
General Electric Co.
Syracuse, N. Y.

20.

A Relationship Between Slope Functions for Array and Aperture Monopulse Antennas*

In a recent paper[1] on monopulse radar, Brennan has derived a function, designated as $F(\Psi)$ and shown in Fig. 3 of his paper. If Ψ is the monopulse squint angle, then $F(\Psi)$ is proportional to the mean-square error in measurement of incremental phase shift and is subject to limitations described in his paper. It is the purpose of this note to give the derivation of an alternate slope function $K(\alpha_s)$ which is related to $F(\Psi)$ as follows:

$$K(\alpha_s) \sim \frac{1}{[F(\psi)]^{1/2}}. \tag{1}$$

This alternate function $K(\alpha_s)$ is for an aperture antenna, whereas Brennan has derived the $F(\Psi)$ function for a phased-array radar. Therefore, it may be that the $F(\Psi)$ or $K(\alpha_s)$ functions are of broader significance than the examination of either separately would indicate.

The function $K(\alpha_s)$ is defined[2] as the slope of the normalized voltage pattern of the antenna (at boresight)

$$K(\alpha_s) = \left.\frac{\partial G(\alpha;\,\alpha_s)^{1/2}}{\partial \alpha}\right|_{\alpha=0} \tag{2}$$

where

$G(\alpha;\,\alpha_s)$ = power gain of the antenna as a function of α and α_s
$K(\alpha_s)$ = slope, in units of volts/volt/radian
α = normalized angle, $2\pi d \sin\theta/\lambda$
α_s = normalized squint angle $2\pi d \sin\theta_s/\lambda$
h, d = antenna dimensions for assumed rectangular aperture
λ = wavelength
θ = far field angle

The model antenna used for analysis consists of two dipoles illuminating a rectangular collimator. However, to eliminate spillover it is assumed that the dipoles are located at the focal points of an ellipsoid and that the radiating aperture is the surface of the rectangular collimator in the side of the ellipsoid. This configuration is sketched in Fig. 1.

It is further assumed that the collimated aperture field from each dipole is of the form

$$A_a = Ae^{j(\alpha_s x/d)}$$

and

$$A_b = Ae^{-j(\alpha_s x/d)}. \tag{3}$$

The fields from the dipoles a and b superimpose in the aperture, and two conditions are of interest. The sum and difference of the fields are

$$A_e = A_{a1} + A_{b1} = 2A_1 \cos\left(\frac{\alpha_s x}{d}\right),$$

and

$$A_0 = A_{a2} - A_{b2} = j2A_2 \sin\left(\frac{\alpha_s x}{d}\right).$$

For α_s (and s) approaching zero the even aperture field approaches a uniform $2A_1$, and the odd a linear odd function $j2A_2\alpha_s x/d$. It appears that the odd function field will vanish as $\alpha_s \to 0$; however, it is assumed that the individual dipole fields A_{a2} and A_{b2} are increased as $\alpha_s \to 0$ so as to maintain a constant radiated power. The confining ellipsoid, dipoles, and collimator are assumed to be lossless. The field over the surface of the rectangular aperture can be replaced by an equivalent current sheet in order to determine an expression for the far field gain.[3] Using the gain expression in (2), the slope is

$K(\alpha_s)$

$$= G_0^{1/2}\frac{\sqrt{2}}{\alpha_s}\,\frac{\left[\cos(\alpha_s/2) - \dfrac{\sin(\alpha_s/2)}{(\alpha_s/2)}\right]}{[1 - \sin\alpha_s/\alpha_s]}. \tag{5}$$

The use of appropriate trigonometric identities to alter the right side of (5) will demonstrate the relationship (1) of $F(\Psi)$ and $K(\alpha_s)$. The variation of $K(\alpha_s)$ with α_s is given by Fig. 2. There is a second function $\Delta'(\alpha_s)$ also given in Fig. 2. This second function $\Delta'(\alpha_s)$ results from assuming A_2 independent of squint angle instead of adjusting A_2 for constant radiated power, and for this case

$\Delta'(\alpha_s)$

$$= G_0^{1/2}\frac{\sqrt{2}}{\alpha_s}\left[\cos(\alpha_s/2) - \frac{\sin(\alpha_s/2)}{\alpha_s/2}\right]. \tag{6}$$

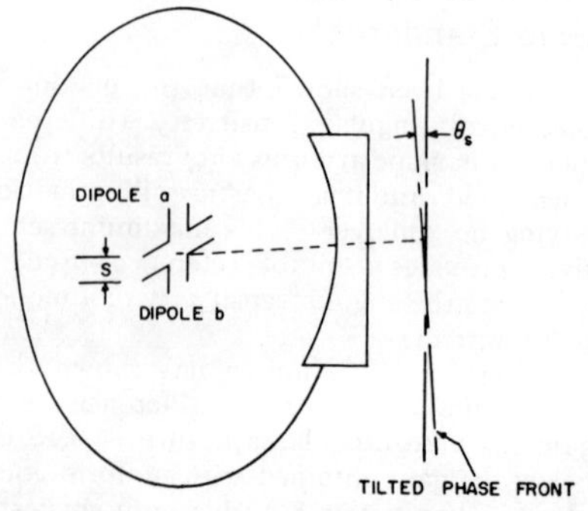

Fig. 1—Aperture illuminated by two dipoles.

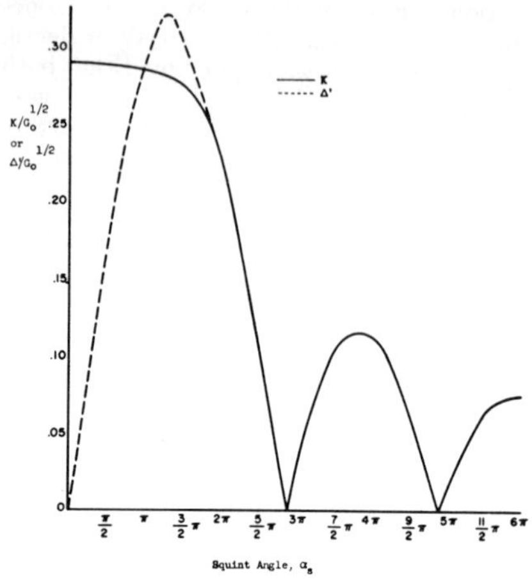

Fig. 2—Comparison of K and Δ' as measures of angular sensitivity.

As shown in Fig. 2, the two functions K and Δ' merge for $\alpha_s > 2\pi$, whereas there are significant differences for $\alpha_s < 3\pi/2$. Brennan has pointed out the relatively minor change in $F(\Psi)$ for α_s less than π. The small change in $F(\Psi)$ and $K(\alpha_s)$ for α_s less than π coupled with the difficulty of constructing adequate laboratory models of either the phased array or aperture antenna for $\alpha_s < \pi$ has undoubtedly resulted in a lack of substantiating data for this interesting region. The requirement for a large A_{a2} and A_{b2} for $\alpha_s \to 0$ was pointed out above, and it should also be noted that for the phased array considered by Brennan there is a corresponding requirement that the individual amplifier gains (or a single output amplifier gain) be very large as $\Psi_s \to 0$ if a useful output is to result.

George M. Kirkpatrick
Electronics Lab.
General Electric Co.
Syracuse, N. Y.

* Received by the PGAP, January 19, 1962.

[1] L. E. Brennan, "Angular accuracy of a phased array radar," IRE Trans. on Antennas and Propagation, vol. AP-9, pp. 268–275; May, 1961.

[2] G. M. Kirkpatrick, "Aperture illuminations for radar angle-of-arrival measurements," IRE Trans. on Aeronautical and Navigational Electronics, vol. ANE-3, pp. 20–24; September, 1953.

[3] S. Silver, *et al.*, "Microwave Antenna Theory and Design," M.I.T. Rad. Lab. Ser., McGraw-Hill Book Co., Inc., New York, N. Y., vol. 12; 1949. (See ch. 6, "Aperture Illumination and Antenna Patterns.")

New performance records for instrumentation radar

by J. T. Nessmith,
Manager, Advanced Systems Projects,
Missile & Surface Radar Div., Radio Corp. of America*

THE TRADEX radar is designed to provide a maximum amount of target information with a minimum of signal degradation. Through a combination of narrow range gating and Doppler sorting, small resolution elements are obtained so that each of the signal characteristics can be attributed to the correct target in a multi-target environment. The system will also provide simultaneous multi-parameter information on all the targets in the radar beam.

For each target in a multi-target complex the Tradex can simultaneously obtain approximately 17 bits of data in each of the azimuth and elevation angular axes, 21 bits in range, 15 bits in Doppler, and 10 bits in relative target size—all in two space-orthogonal polarizations whose time phases can be maintained and adjusted, and all read out at least 100 times a second.

For any large number of targets, the direct processing of such data on a real-time basis normally requires sizeable computational facilities. For the Tradex, therefore, a special IF recorder was developed for complete information storage on all targets. In real-time opera-

*Missile & Surface Radar Div., Radio Corp. of America, Moorestown, N. J.

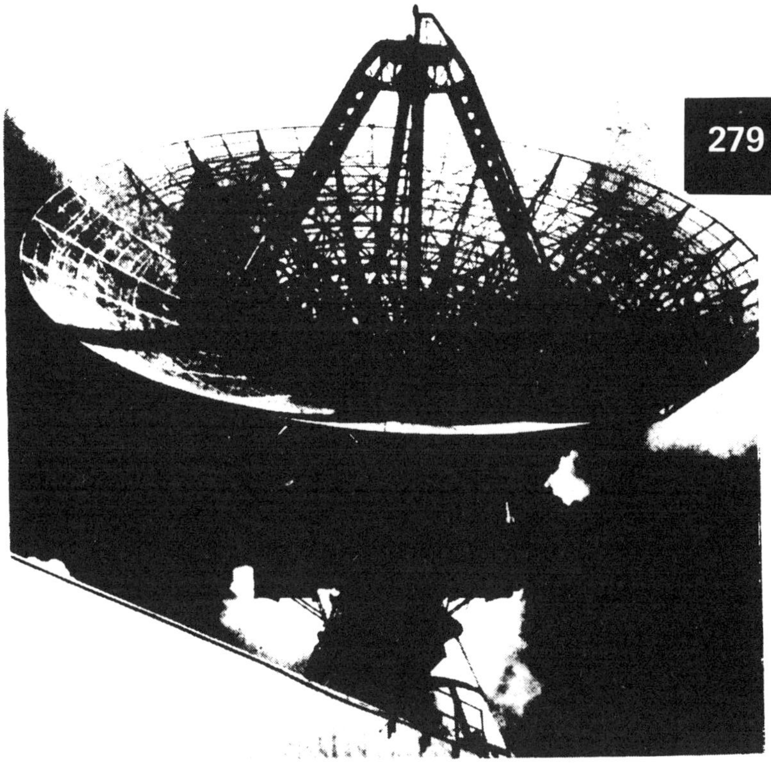

TRADEX with its 84-ft dish, installed on Kwajalein in the Pacific for anti-ICBM tests.

IN BRIEF

A member of the same basic family as the FPS-16 and the BMEWS tracking radars, RCA's Tradex is setting new records for instrumentation radars. Combining high S/N, high repetition rates, small range-Doppler resolution cells, and an IF tape recorder, it is claimed to collect "far more detailed information than any other known radar."

This article reviews the basic Tradex system concept, describes a number of advanced techniques combined for the first time in a single radar system, and analyzes the signal flow in the RCA design. In addition, test results are outlined that have verified the performance expected of the Tradex.

tion, the target complex is tracked by selecting a single, centrally located element. All target elements in the radar beam during real-time tracking are recorded simultaneously. In the playback mode, the transmitter and antenna-pedestal are turned off. Each element of a multi-body target is then tracked independently from the recording (and within the resolution capability of the radar) to obtain fine-grain data. For five minutes of recording time and a 10-element target, the serial readout to a computer during playback takes 50 minutes times the number of polarizations.

Tradex continuously gathers data

Unlike a search or track-while-scan radar, the Tradex continuously gathers data on a target complex and translates its observation into digital and analog forms readily usable by computers and human observers. With a large aperture-power product and a low-noise front end, it produces a high S/N ratio for very small targets.

High repetition rates, (about 1500 PPS) make it possible to use very wide information bandwidths. Sequential use of the frequency and time domains—on the basis of range-gating and narrow-band IF filtering—together with the two dimensions of azimuth and elevation given by the antenna aperture lead to resolutions otherwise obtainable only with a larger aperture-bandwidth combination (which limits the volumetric coverage) or with much higher transmitter powers.

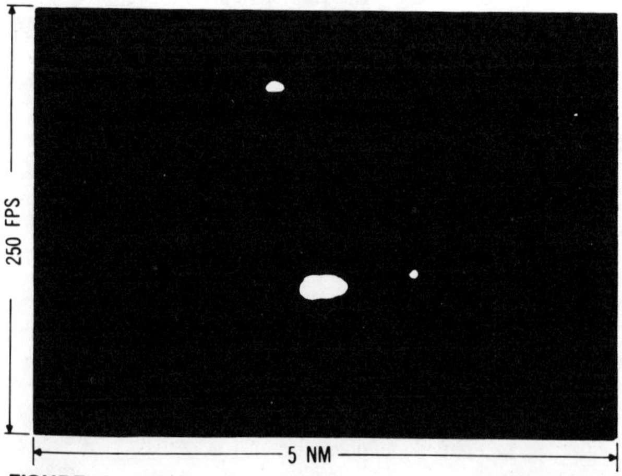

FIGURE 1: Display of range vs range rate shows the effectiveness of the Tradex's coherent integration. The single-pulse S/N ratios are about 16 db for the largest target, about 5 db for the second-largest, and about −5 db for the smallest. Two of the targets are at almost the same range but separated in Doppler by 100 fps, and two of the targets have almost zero Doppler difference but are separated in range by about one mile.

Tradex Design Parameters

	UHF	L-Band
Frequency	425 mc ± 5%	1320 mc ± 5%
Peak power		
Required	4 Mw	1.0 Mw, 60 kw
Goal		5.0 Mw, 300 kw
Pulse repetition frequency	1112-1482 PPS	
Transmitted pulsewidth	50 usec	40 usec
Unweighted pulse compression ratio	50	32
Antenna beamwidth per gain	2 deg per 38 db	¾ deg per 47 db
Polarization (for transmission & reception)	RH circular, LH circular, 45-deg linear, 135-deg linear, vertical, horizontal	
Noise figure (including losses)	3.0 db	3.5 db
Information bandwidth (tracking filters)	10/35 CPS	10/35 CPS
Doppler tracking range	±40 kc	±125 kc
Pulse-Doppler tracking resolution	1.5 usec × 10 CPS	

	Range	Azimuth	Elevation
Velocity	60,000 FPS	12 deg/sec	12 deg/sec
Maximum acceleration	3200 ft/sec²	30 deg/sec²	24 deg/sec²
Nominal servo bandwidth	1 CPS	1 CPS	1 CPS
Granularity	2 yd	0.24 mil	0.024 mil

The operation of a Tradex starts with the radar's assignment to a particular target complex. Once a single, centrally located target is selected in this complex, the radar tracks automatically. An IF recording of all returns is made at a radar intermediate frequency in a form that preserves all the information.

As "live" track operation starts, the antenna is set in angle to the expected location of the target at some suitable range, and predetermined range, range-rate, and angle rates are inserted manually. Automatic tracking begins after manual assignment of the tracking loops to the signal return of a particular target. At a precalculated range, the IF tape recorder begins recording all target returns within the radar beam.

IF recording replayed for analysis

Post-flight tracking of each target in the target complex is done much as is the initial tracking. The IF recording is replayed through the same receiving system for a comparative analysis on a target-by-target basis. In this mode antenna and transmitter naturally are inoperative, since all necessary reference data are already recorded.

The Tradex represents an advance in the state of the art of radar system design not because it uses any particular technique for the first time but because it does for the first time combine several important advanced radar techniques:

- *Dual-Frequency Operation*—The Tradex transmits simultaneously with two high-power transmitters, one operating at UHF and the other at L-band. Either linearly or circularly polarized transmission and reception with simultaneous pulse operation are possible at both frequencies. UHF monopulse techniques are used for tracking, with illumination at L-band to obtain data on the frequency sensitivity of the targets through correlation of the UHF and L-band returns.
- *Monopulse Operation*—The UHF receiver system

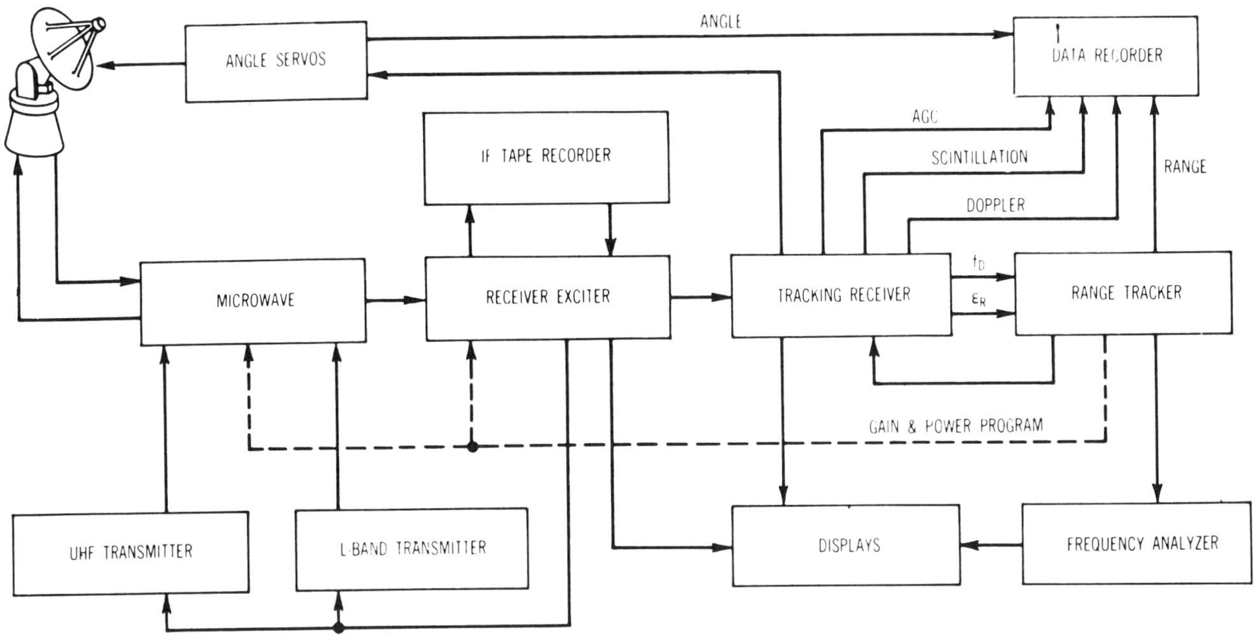

FIGURE 2: Signal flow in the Tradex system.

operates in a five-horn amplitude monopulse mode. Its central horn is used for transmission and can be adjusted for 45- or 135-deg linear polarization or left- or right-hand circular polarization. On reception, it is also used for the monopulse reference signal. The vertically and horizontally polarized signals out of the horn into the receiving system are separate. They are combined at IF after reception to introduce the desired polarization components of the signal.

The outputs from a pair of elevation error horns are combined at RF to provide a difference signal for angle estimates in the vertical plane (phase-correlated to the reference signal). Similarly, cross-elevation (azimuth) angle estimates are derived from a pair of horizontal horns. On reception, error and reference signals are processed from the horns in the same polarization sense.

L-band horn transmits search beam

The search, or illumination, beam at L-band is axially aligned with the UHF beam, and transmitted and received, by a concentric L-band horn. Independently, the same variety of transmitting and receiving polarizations is attainable as at UHF.

• *Pulse Expansion-Compression*—By means of long-pulse operation, this technique provides high average power despite limited peak power and with a resolution equivalent to that of a short pulse. At an expansion ratio of about 50, the pulse energy in transmission is about 50 times that of a short pulse of the same transmitter operating at peak power. After reception and following compression to minimize time side-lobes, the pulse is further compressed by filtering.

The present system design uses linear FM techniques. However, other forms of modulation are readily achievable within the bandwidths available in the transmission and receiving subsystems.

• *Phase-Coherent Operation* (Pre-Detection Integration)—The transmitted signals at both UHF and L-band have pulse-to-pulse phase coherence. In reception, the system performs coherent integration of the signal return of a large number of pulses by means of a narrow-band pre-detection integration filter. The nominal filter bandwidth of the signal tracking channels for range, angle, amplitude, and Doppler frequency is 10 CPS. For targets with high scintillation characteristics or high accelerations, 36 CPS is available.

• *Magnetic-Disk Spectrum Analysis*—During acquisition, a system of 30 consecutive range gates is generated. The signals in a single gate interval are stored pulse-to-pulse in consecutive positions on a magnetic disk. The information is then made available in the form of continuous signals corresponding to the Doppler frequencies of the targets in that gate. From the signal analyzer, signals are generated so that range versus range rate (Doppler) can be displayed on a CRT (*Fig. 1*).

• *Range and Velocity Ambiguity Resolution*—Targets at long range require a low repetition rate for unambiguous range measurements, while rotating targets traveling at high velocity require a high repetition rate for unambiguous Doppler-frequency and scintillation-spectral measurements around a single spectral line. Tradex uses an intermediate repetition rate that eliminates neither range nor Doppler-frequency ambiguities but is high enough to insure complete target scintillation information and low enough so that the range ambiguity can readily be resolved. A major advantage of the lowered range interval is that the range search time during acquisition is shortened. Velocity and range ambiguities are resolved automatically once automatic tracking has started, and the output is presented unambiguously in both range and velocity.

• *IF Tape Recording*—For real-time tracking, eight channels are now used for a single target, but this can easily be increased. For the primary purpose of instrumentation, though, all received radar information is recorded at an intermediate frequency on a wide-band, multi-channel IF recorder. On separate tracks this unit concurrently records the signal returns from each UHF reference and error channel (in both polarizations) and from the two L-band channels (again in both polarizations) as well as pedestal reference signals, timing signals, and radar operating signals. During readout, each target is played back through the receiving Doppler- and range-tracking systems to provide separate tracks in all four coordinates and any desired polarization.

The signal flow through the Tradex system starts with the RF energy supplied to the antenna by the UHF and L-band transmitters (*Fig. 2*). Expanded, coherent pulses are accepted by the transmitters from the receiver exciter and processed to provide the proper frequencies and power levels at the output. Power dividers at the transmitter outputs control the amount of energy to be fed to the antenna.

In the microwave system this energy is separated into two channels for the generation of space and phase quadrature components at the antenna. With the two channels properly phased, either of two linearly or of two circularly polarized signals may be transmitted by the antenna feed. The feed horn is made up of a five-horn, dual-polarization monopulse array for UHF and a single, dual-polarization horn for L-band, and serves as the illuminating and receiving source for the antenna reflector. The radiated energy from the radar is directed by hydraulically moving the antenna in azimuth and elevation.

The received signals are modulated in frequency by the radial velocity of the target, in time by range, and in amplitude by the effective electromagnetic cross-section related to target aspect and range. Received in space and phase quadrature, they are handled independently and not recombined until accepted by the tracking receiver. There are eight receiving channels—vertically and horizontally polarized reference and error (elevation and azimuth) channels for monopulse operation at UHF and a reference channel for each polarization at L-band.

Exciter uses matched-filter methods

The receiver exciter compresses the received pulses, using matched-filter techniques, and removes the coarse amplitude variation due to the range of the target (by making the gain of the amplifiers a function of the target range as determined by the range tracker). The outputs of the receiver exciter include information on all the targets seen by the radar. It is this information that is stored by the IF tape recorder.

The tracking receiver accepts vertically and horizontally polarized data on all targets from either the receiver exciter (during live track) or the tape recorder (during playback) and recombines the signals into the desired polarization. A single target is selected from the aggregate and separated from the others in range and Doppler frequency by electronic gating and filtering techniques. All signals not within the range gate are excluded.

At this point, the conventional tracking radar by time discrimination develops the necessary signals to close a range-tracking loop. The Tradex in addition uses a velocity-tracking loop for differentiation in case several lie within the range gate. This loop is an interconnected automatic frequency control that tracks targets in Doppler frequency. Until this loop is closed, signals are not available to the range- and angle-tracking systems.

Separation in frequency is accomplished by controlling the frequency of an oscillator, which is heterodyned with the target signals, so that only information on the specific derived target is passed through a high-Q filter. A wide-band AGC system within the servo loop eliminates the amplitude variations of the signal and so provides information on the amplitude characteristics of the target.

All tracking signals are modified by the AGC system and the narrow-band Doppler filters. Prior to this modification, scintillation data may be extracted through separate receiver channels.

From the receivers, the angle servos receive error signals (initially generated by the five-horn monopulse system). These signals are processed and ultimately used to control the displacement of the hydraulic pumps which drive the pedestal in angle with a velocity proportional to the tracking error.

Operator establishes track criteria

A target is tracked only if it meets the range and Doppler-frequency (range-rate) criteria established by an operator on the basis of the displays. All targets are displayed by means of a multi-gate acquisition system in the range tracker and the magnetic-disk spectrum analyzer. The display shows range rate (generated from the outputs of the spectrum analyzer) as the ordinate, range as the abscissa, and the strength of a signal as the intensity of the blip (*Fig. 1*). By placing the frequency and range cursors on an indicated target, the operator can adjust the tracking range gates and the tracking oscillator to exclude all targets but the desired one.

The outputs of the radar are in both analog and digital form. Analog signals are recorded for quick-look information and as a means of screening the digital data during post-mission analysis. The digital outputs are in a format suitable for direct introduction into a computer in real time and/or recorded for post-mission analysis.

Tests of the Tradex in tracking satellites and the moon have confirmed the performance expected of the system. The residual data after reduction of the results for azimuth, elevation, range, and Doppler vs time of track for tests with the Echo 1 showed that the orbit

Recorded Digital Target-Data Signals

	Bits
Pedestal Position	
Azimuth angle	18
Elevation angle	18
Angle offset	
Azimuth	10
Elevation	10
Real time of day	24
Range	22
Programed gain control	10
Tracking (AGC)	10
Doppler frequency	17
Transmitted power	
UHF	10
L-Band	10

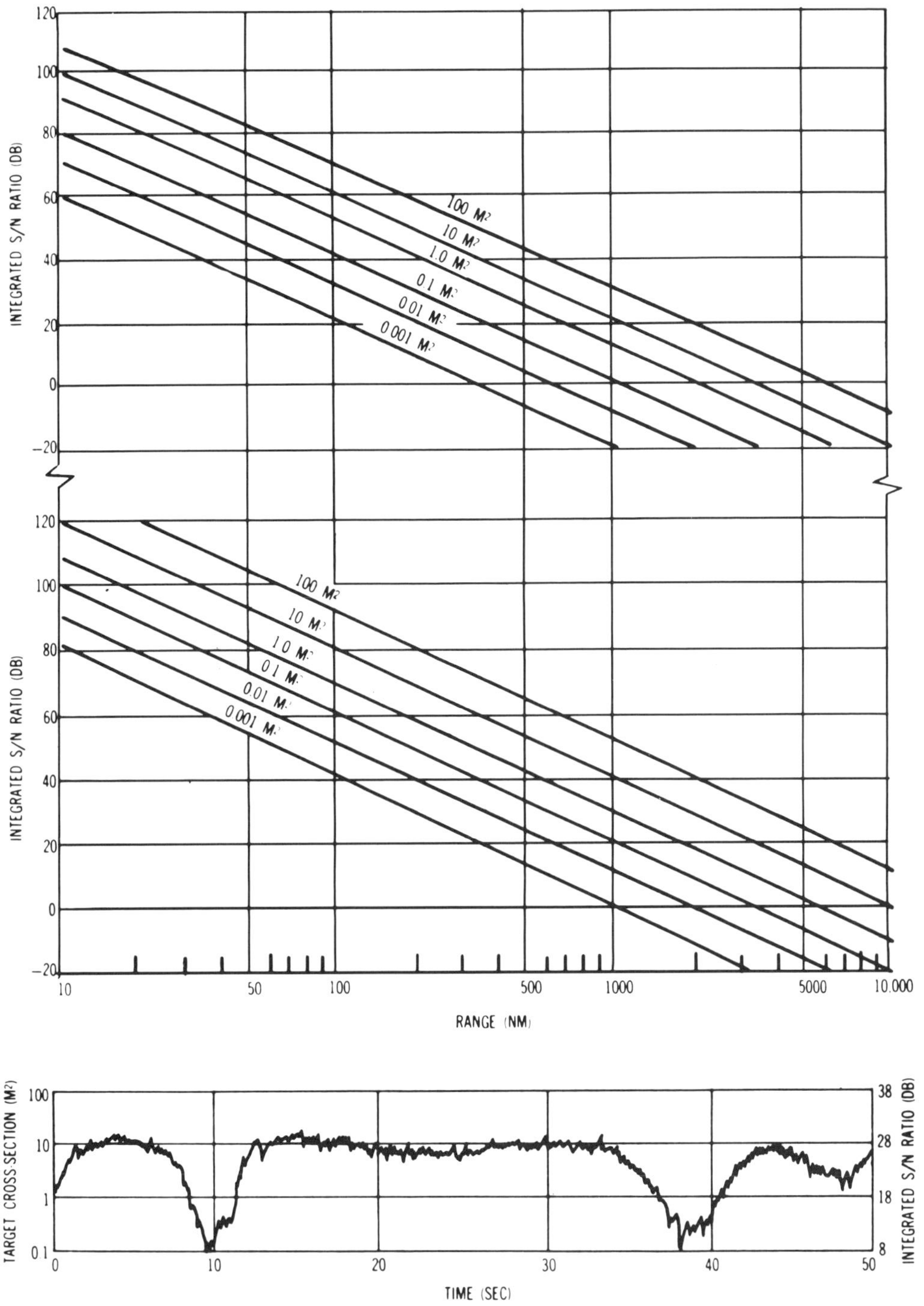

FIGURE 3: Calculated curves of UHF S/N vs range for single-pulse operation (top) and integrated 150-pulse operation have been exceeded in tests by 1-2 db (center). Bottom: Tests with the Midas 3 proved the effectiveness of the Tradex's coherent integration. The integrated S/N as shown here is about 17 db higher than the single-pulse S/N, which therefore was less than zero decibels for all S/N values on this plot of target cross-section vs time that lie above 17 db. This test result was confirmed by calculations of the range and optical cross-section of the satellite as well as by observations made with other systems.

fit was virtually unbiased and that the radar noise (1 σ) performance was: range, 14.70 ft; azimuth, 0.33 mils; elevation, 0.34 mils; range rate, 1.40 FPS.

The sensitivity of the system has been confirmed in a number of tests involving balloon-borne metal spheres and Echo 1 and other satellites of known cross-section while both good sensitivity and the effectiveness of coherent integration were proved by tests with the Midas 3 (*Fig. 3*).

The Midas 3's track of about 2700 nm represents the longest range over which any man-made satellite has been tracked to date. At a high single-pulse S/N ratio, the Tradex has also been used to track the moon in a so far inconclusive attempt to use the radar for boresighting. However, this track to some extent again confirmed the sensitivity of the radar.

microwave journal
Tele-Communications Systems

SCAMP-- Single-Channel Monopulse Processor

A NEW RATIO COMPUTING TECHNIQUE WITH APPLICATION TO MONOPULSE

W. L. RUBIN and S. K. KAMEN

WILLIAM L. RUBIN†

Dr. Rubin received a BS degree in Electrical Engineering from Columbia University in 1948 and an MS and PhD in Electrical Engineering in 1953 and 1960 respectively from the Polytechnic Institute of Brooklyn. In 1957 he joined the Advanced Studies Department of the Sperry Gyroscope Company as a research engineer. In his present position as research department head responsible for the Advanced Radar Studies Department, he has been principally engaged in applied research and systems analysis in the fields of signal processing and detection theory. Dr. Rubin is a member of the IEEE, Eta Kappa Nu and Sigma Xi.

SUSAN K. KAMEN†

Mrs. Kamen received a BA degree in Physics from Smith College, Northampton, Mass., in 1957 and an MS in Physics from Yale University in 1958. Since 1958 she has been employed in the Advanced Studies Department of the Sperry Gyroscope Company, engaged primarily in the fields of noise and detection theory, signal processing and radar systems analysis. She is a member of the IEEE and Sigma Xi.

† ADVANCED STUDIES DEPARTMENT, SPERRY GYROSCOPE COMPANY, GREAT NECK, NEW YORK

INTRODUCTION

A new signal processing technique which instantaneously computes the ratio of two input signals will be described. Because of its dynamic range and signal-to-noise performance, the technique can be useful in many applications. Emphasis in this paper will be placed on its application to monopulse radar.

The basic monopulse concept is the determination of angular information for a distant target by comparing radar signals which are simultaneously received on two or more antenna patterns. Such a comparison relies on the fact that, although the absolute values of the amplitude and phase of the received signals may vary with changing characteristics of the source and the propagation medium, their relative values are functions only of the angle of arrival.

The monopulse concept can be applied to both "track" and "search" radars. In track applications, the monopulse receiver output has been utilized to reposition the antenna in order to reduce the deviation from boresight of an incoming wave to an average value of zero. In search, the angular position of each target falling within the antenna beamwidth can be determined relative to boresight, provided such targets do not occur in overlapping range resolution cells.

In order to extract pure angle information from a search or track monopulse system, the monopulse signal processor must be capable of removing large variations in received signal amplitude which result from such factors as differences in target cross section, target scintillation, range, etc. Ideally, this should be accomplished instantaneously and accurately for any number of (nonoverlapping) target echoes. Previous circuit realizations have achieved limited success in meeting these requirements.

A new significantly improved monopulse signal processing technique, which has been named "SCAMP" (Single-Channel Monopulse Processor), is described in this paper. By appropriately combining signals in a single channel, monopulse data from all received target returns over an extremely wide dynamic range of input signals can be instantaneously processed. The method can be instrumented with existing components and is applicable to search, track and track-while-scan monopulse radar systems.

The basic signal processor is a novel broadband real-time analog divider. In the next section, the physical basis for its operation is described and justified analytically, followed by a discussion of its performance in the presence of noise. Its application to three basic monopulse receiver configurations is discussed in the following section. In the last part of the paper, an error analysis is performed for a SCAMP monopulse receiver which relates target angle accuracy to input SNR.

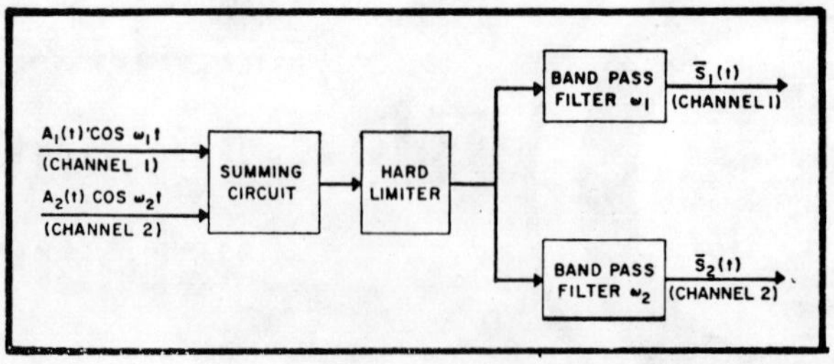

Figure 1 — Basic signal processor.

BASIC SIGNAL PROCESSOR

A block diagram of the signal processor is shown in Figure 1. The input consists of two narrow band amplitude and/or phase-modulated signals whose carrier frequencies are sufficiently displaced so that their spectra do not overlap. It will be shown that if the sum input signal is hard-limited, the limiter output in the spectral region of the smaller input signal is proportional under certain conditions to the ratio of the amplitudes of the small signal to the large signal. If necessary, a fixed gain amplifier can be incorporated into the block diagram to insure that the desired normalizing signal is always the larger of the two inputs. The output in the spectral region of the larger signal consists essentially of a hard-limited (constant amplitude) version of that signal.

The mathematical analysis which leads to the above conclusions is based on the well-known result[1] that when the input to an ideal bandpass limiter is a narrowband signal, the output is a constant amplitude waveform with phase modulation identical to that of the input. In other words, if the input signal $S(t)$ is expressed as

$$S(t) = A(t)\cos[w_c t + \phi(t)] \quad (1)$$

then the output $L(t)$ from an ideal full-wave (odd) bandpass limiter is found to be

$$L(t) = \frac{4a}{\pi}\cos[w_c t + \phi(t)] \quad (2)$$

where a is the limiter threshold.

Consider the input to an ideal limiter to be the sum $S(t)$ of two narrowband signals $S_1(t)$ and $S_2(t)$, where

$$S_1(t) = A_1(t)\cos[w_1 t + \phi_1(t)] \quad (3)$$

$$S_2(t) = A_2(t)\cos[w_2 t + \phi_2(t)] \quad (4)$$

By simple trigonometric operations, $S(t)$ can be expressed as a single narrowband function with carrier frequency w_1, as follows

$$S(t) = [A_1^2(t) + A_2^2(t) + 2A_1(t)A_2(t) \cdot \cos\{(w_2 - w_1)t + \phi_2(t) - \phi_1(t)\}]^{1/2} \times \cos\left\{ w_1 t + \tan^{-1} \cdot \frac{A_1(t)\sin\phi_1(t) + A_2(t)\sin[(w_2 - w_1)t + \phi_2(t)]}{A_1(t)\cos\phi_1(t) + A_2(t)\cos[(w_2 - w_1)t + \phi_2(t)]} \right\} \quad (5)$$

From (1), (2) and (5), the limiter output becomes

$$L(t) = \frac{4a}{\pi}\cos\left\{ w_1 t + \tan^{-1} \frac{A_1(t)\sin\phi_1(t) + A_2(t)\sin[(w_2 - w_1)t + \phi_2(t)]}{A_1(t)\cos\phi_1(t) + A_2(t)\cos[(w_2 - w_1)t + \phi_2(t)]} \right\} \quad (6)$$

Using trigonometric identities, (6) can also be written as

$$L(t) = \frac{4a}{\pi} \left\{ \frac{A_1\cos[w_1 t + \phi_1] + A_2\cos[w_2 t + \phi_2]}{[A_1^2 + A_2^2 + 2A_1A_2\cos\{(w_2 - w_1)t + \phi_2 - \phi_1\}]^{1/2}} \right\} \quad (7)$$

Dividing numerator and denominator by A_2 yields

$$L(t) = \frac{4a}{\pi}\left[1 + \frac{A_1^2}{A_2^2} + \frac{2A_1}{A_2} \cdot \cos\{(w_2 - w_1)t + \phi_2 - \phi_1\}\right]^{-1/2} \times \left\{\frac{A_1}{A_2}\cos[w_1 t + \phi_1] + \cos[w_2 t + \phi_2]\right\} \quad (8)$$

For $A_1 < A_2$, the quantity under the square-root sign in (8) can be expressed by a power series (using the binomial expansion) as follows

$$
\begin{aligned}
&[1+A_1^2/A_2^2]^{-1/2} \\
&\cdot\left[1+\frac{2A_1A_2}{A_1^2+A_2^2}\cos\{(w_2-w_1)t+\phi_2-\phi_1\}\right]^{-1/2} \\
&=\left(1-\frac{1}{2}\frac{A_1^2}{A_2^2}+\frac{3}{8}\frac{A_1^4}{A_2^4}-\frac{5}{16}\frac{A_1^6}{A_2^6}+\dots\right) \\
&-\left(\frac{A_1}{A_2}-\frac{3}{2}\frac{A_1^3}{A_2^3}+\frac{15}{8}\frac{A_1^5}{A_2^5}-\dots\right) \\
&\qquad\cdot\cos[(w_2-w_1)t+(\phi_2-\phi_1)] \\
&+\left(\frac{3}{2}\frac{A_1^2}{A_2^2}-\frac{15}{4}\frac{A_1^4}{A_2^4}+\frac{105}{16}\frac{A_1^6}{A_2^6}-\dots\right) \\
&\qquad\cdot\cos^2[(w_2-w_1)t+\phi_2-\phi_1] \\
&\qquad -\text{etc.} \qquad\qquad (9)
\end{aligned}
$$

Substituting (9) into (8), expanding by means of trigonometric identities and collecting terms, the limiter output becomes

$$
\begin{aligned}
L(t)=\frac{4a}{\pi}\Bigg[&\left\{\frac{A_1}{2A_2}+\frac{1}{16}\frac{A_1^3}{A_2^3}+\frac{3}{128}\frac{A_1^5}{A_2^5}+\frac{25}{2048}\frac{A_1^7}{A_2^7}+\dots\right\}\cos(w_1t+\phi_1) \\
&+\left\{1-\frac{1}{4}\frac{A_1^2}{A_2^2}-\frac{3}{64}\frac{A_1^4}{A_2^4}-\frac{5}{256}\frac{A_1^6}{A_2^6}-\dots\right\} \\
&+\{\text{other terms at carrier frequencies } w_1\pm n(w_1-w_2)\}\Bigg] \\
&n=2,3,4\dots \qquad (10)
\end{aligned}
$$

By appropriately spacing the input carrier frequencies, it is possible to individually filter the limiter output narrowband signals $\overline{S}_1(t)$ and $\overline{S}_2(t)$ whose carrier frequencies are w_1 and w_2. The magnitudes of each of these signals, computed from (10), are plotted in Figure 2 for $0 \leq [A_1(t)/A_2(t)] \leq 1$ (which is the region of interest for monopulse systems).

When $A_1(t)/A_2(t) \ll 1$, $\overline{S}_1(t)$ and $\overline{S}_2(t)$ are given with negligible error by

$$\overline{S}_1(t) \cong \frac{2a}{\pi}\left[\frac{A_1(t)}{A_2(t)}\right]\cos[w_1t+\phi_1(t)] \qquad (11)$$

$$\overline{S}_2(t) \cong \frac{4a}{\pi}\cos[w_2t+\phi_2(t)] \qquad (12)$$

Equation (11) indicates that for small A_1/A_2 the envelope of $\overline{S}_1(t)$ is very nearly linearly proportional to the instantaneous ratio of the amplitudes of the two input signals. As the ratio A_1/A_2 increases, it can be shown that the magnitude of $\overline{S}_1(t)$ remains a monotonic (although not linear) function of A_1/A_2 for all values of this variable.

Figure 3 shows the dynamic range performance of an experimental signal processor, which consisted of several cascaded wideband limiters followed by the appropriate bandpass filters. Note that the percentage deviation of the experimental points from an ideal horizontal straight line is approximately the same for each value of A_1/A_2. The experimental data deviate significantly from the horizontal line when signal $S_2(t)$ ceases to be of sufficient magnitude to control the limiter action. Dynamic range can be further increased by cascading additional limiter stages.

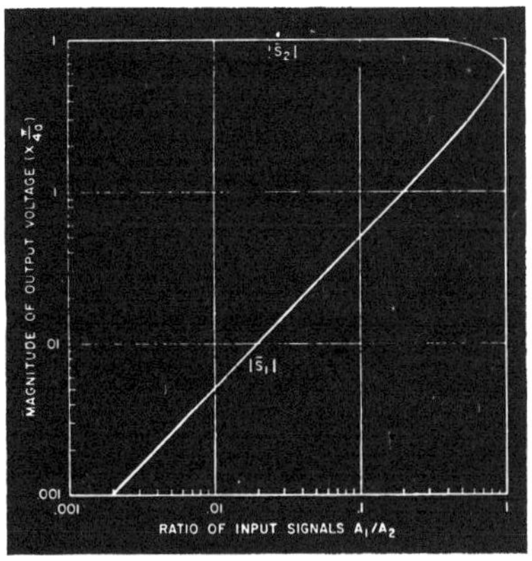

Figure 2 — Theoretical performance of signal processor.

287

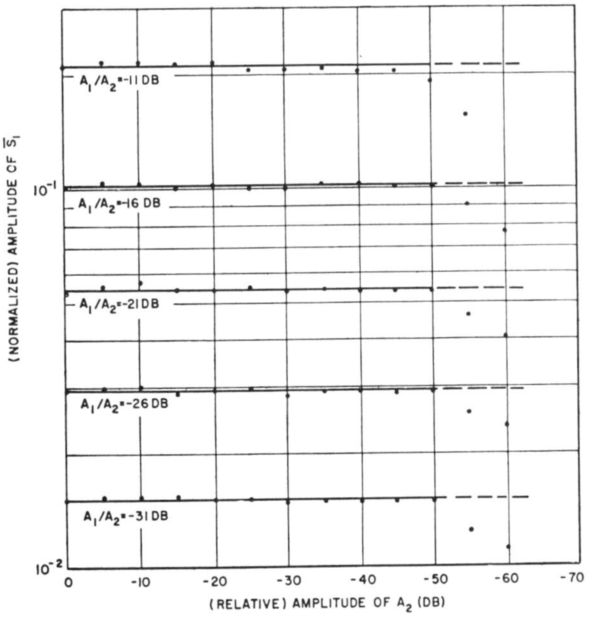

Figure 3 — (Experimental) dynamic range performance of basic signal processor.

Figure 4 compares the experimental and theoretical performance of the signal processor for various ratios of A_1/A_2 with A_2 as a parameter. Deviation between the two curves which occurs for very small values of A_1 is due to leakage of large signal energy into the small signal channel. The deviation can be reduced by improving output filter skirt selectivity and by increasing the frequency separation between the two input carrier frequencies.

An additional discrepancy between the experimental and theoretical curves occurs for large values of A_1. This can be explained by noting that A_1/A_2 is approaching unity in this region and that the output of the first limiter stage consists of a wideband FM signal whose furthest sidebands are not passed by subsequent limiter stages.

A detailed study has been made of the performance of the signal processor when Gaussian noise is present in each input channel. Because the analysis is very long, only the principal results are described here. Output signal power and output noise power in each channel have been calculated as a function of the input SNR in the large signal (A_2) channel and the ratio A_1/A_2. From these results, each channel output SNR has also been computed. The analysis was carried out by first deriving the auto-correlation function of the output of a hard limiter for two sinusoidal input signals plus narrowband Gaussian noise.

The power spectrum of the limiter output was then found by taking the Fourier transform of this function. Assuming a rectangular bandpass filter following the limiter, expressions for the signal and noise spectral components in the channel of interest were derived. Due to the complexity of the final expressions, they were evaluated with the aid of a high speed digital computer.

Although the analysis assumes that the input signal consists of two sine waves, the results are also applicable to pulsed signals by the following reasoning. A hard limiter, being a zero memory device, cannot discriminate on an instantaneous basis between a pulsed or continuous sinusoidal train of waves. If the transfer function of the bandpass filter following the limiter is matched to the input pulse, the peak signal out of the filter is equal in magnitude to the filter output for an infinite sine wave input. Hence the results obtained for CW signals are applicable to pulsed signals when the limiter is followed by a matched filter.

The results of the analysis are shown in Figures 5 and 6. Figure 5 contains plots of the rms voltage out of each channel as a function of A_1/A_2 for several values of large channel input SNR from 0.02 to infinity. It will be noted from the curves that the output of the small signal channel is very close to that of the noise-free case for all values of input SNR's greater than 0.5.

Figure 6 contains plots of the gain or loss in output SNR's in each channel as a function of the large channel input SNR. When $A_1/A_2 \leq 0.1$, the plot of the output SNR in the larg signal channel (A_2) conforms closely to the results obtained by Davenport,[2] who analyzed the noise performance of a bandpass limiter for a single sinusoidal input signal. When the input SNR exceeds 10, the loss in SNR through the small signal channel is negligible. This is explained by the fact that for this condition both signal and noise in this channel are equally suppressed by the large channel signal in the limiter.

When the ratio A_1/A_2 approaches unity, there is a significant loss in SNR through both channels. This loss occurs because the waveform corresponding to the sum of two almost

288

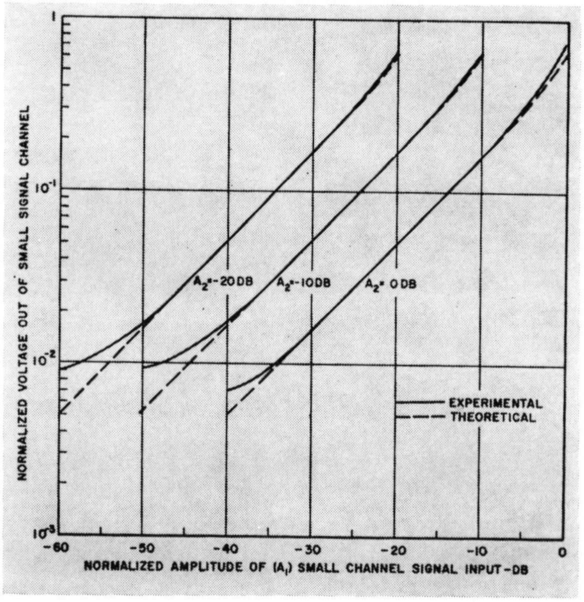

Figure 4—Comparison of experimental and theoretical performance of basic signal processor.

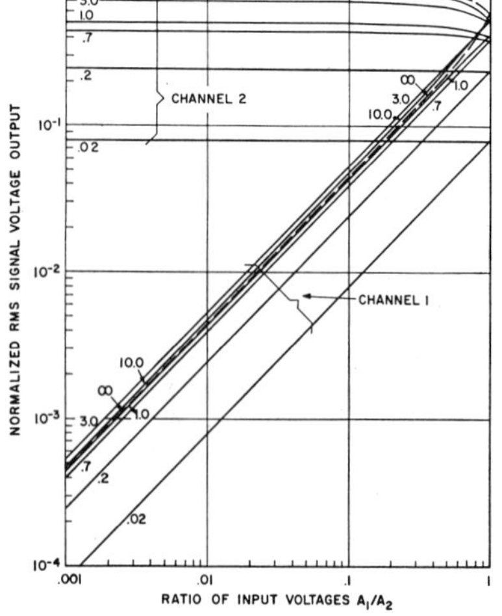

Figure 5 — Signal processor performance in presence of noise.

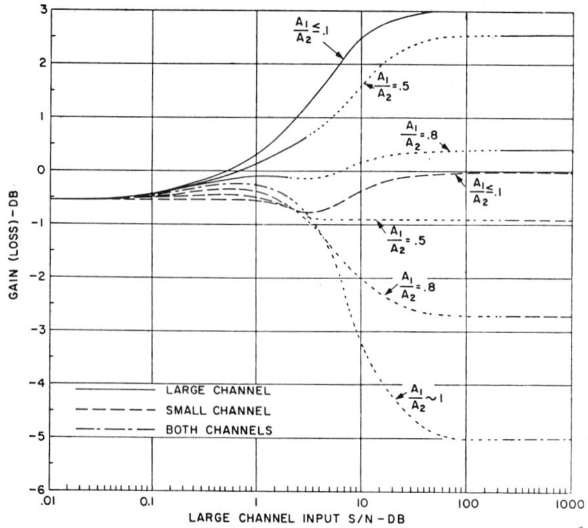

Figure 6 — Gain (loss) in SNR through basic signal processor.

equal sinusoids which are separated in frequency is similar to that of an overmodulated signal. Such signal waveforms are characterized by appreciable time intervals during which signal amplitude is small and hence subject to suppression by noise peaks in the limiter. This condition can be prevented by preamplifying the larger signal by a known amount prior to limiting.

SCAMP MONOPULSE REALIZATIONS

The SCAMP technique can be applied to two basic monopulse receiver configurations. One system which is widely used compares the sum (Σ) and difference (△) of the signals received by a phase-sensing or amplitude-sensing monopulse antenna. This is accomplished by combining the two antenna outputs in a microwave comparator to produce a difference signal which is an odd function and a sum signal which is an even function about the boresight axis. The ratio of the difference to the sum signal, △/Σ, is always independent of received signal strength for either monopulse antenna pattern, and therefore bears a fixed relationship to the magnitude and sense of the target angle relative to boresight.

In the monopulse realization shown in Figure 7(a), the ratio △/Σ is evaluated by deriving an automatic gain control signal (AGC) from the sum channel amplifier which is then used to control the gain of both Σ and △ channels. The net effect is to normalize both the sum and difference signal amplitudes with respect to the sum amplitude. The normalized difference signal can be detected by a conventional phase-sensitive detector, whose output contains both magnitude and sensing information, or by an amplitude detector which provides accurate magnitude information and a phase-sensitive detector which provides only sensing information. The latter realization is less sensitive to phase distortion.

The block diagram in Figure 7(a) is easily modified for three-dimensional monopulse by adding a second △ channel whose gain is also controlled by the Σ-channel AGC. Proper system operation requires that the gain of the two (or three) amplifying channels track closely as a function of AGC voltage over a wide dynamic range. For search or multitarget track applications, this technique necessitates separate processing channels for each received target return due to the relatively long AGC time constant.

The corresponding SCAMP realization, which eliminates these problems, is shown in Figure 8. In this case instantaneous normalization of the △ signal by the sum signal is accomplished by means of the basic signal processor previously described. The RF sum and difference input signals are mixed to two closely spaced intermediate carrier frequencies, summed and hard-limited in a multistage wideband amplifier-limiter. The output of the limiter is given by (10), where △ replaces A_1 and Σ replaces A_2.

In the vicinity of boresight, the difference and sum frequency signals are given by (11) and (12), respectively. After separating the two output signals by filtering, the sum signal is mixed to the difference signal frequency to provide the normal reference signal in the phase-sensitive detector. Angle information can be obtained by either of the two techniques described above. Calibration of the output signal in terms of angle deviation from boresight includes both the dependence on antenna pattern as well as the SCAMP signal processing characteristic. SCAMP can also be extended to include a second angle error channel. This is accomplished by mixing the three inputs to three closely spaced intermediate frequencies such that their spectra do not overlap and permitting the sum signal to simultaneously normalize both difference signals exactly as before. Proper filtering permits the recovery of the three signals at the limiter output.

A second basic monopulse receiver configuration is shown in Figure 7(b). In this system the ratio of the input signals derived from an amplitude monopulse antenna pattern is determined by amplifying each signal in an IF amplifier with a logarithmic gain characteristic and finding the difference between their amplitudes after detection. Ideally, the difference signal is independent of echo amplitude and is a monotonic function of the angle deviation from boresight. Although there are no limitations on the number of targets which can be processed in real time by this method, there still remains the difficult problem of matching the logarithmic characteristics of two (or more) amplifiers over a wide dynamic range. A SCAMP realization similar to that in Figure 8 can also be applied to this monopulse configuration. In this case, however, only an amplitude detector is required to obtain both the magnitude and sense of target angle deviation from boresight.

Another ratio-taking scheme is shown in Figure 7(c). This has in common with SCAMP the use of limiters, which makes the ratio computation instantaneous. However, it has critical requirements on phase and amplitude tracking between the two limiters. This difficulty is avoided in SCAMP by employing a single limiter.

SCAMP ANGLE ACCURACY

Since boresight is more accurately defined in a Σ, △ monopulse configuration than in a straight amplitude monopulse system, it has been extensively utilized in the past in track radar applications and is presently under study for application in electronic scanning array monopulse systems. For

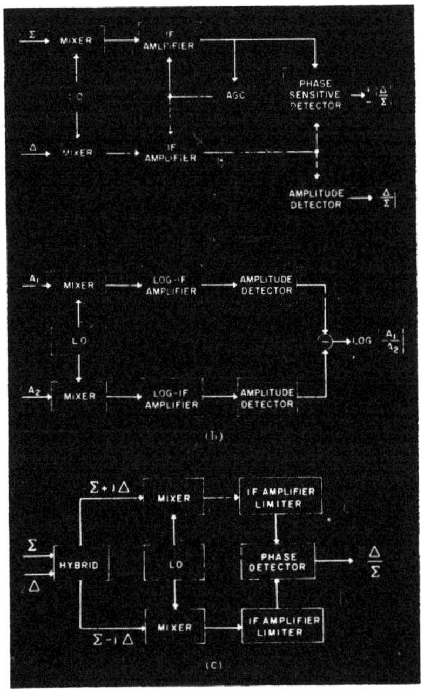

Figure 7 — Conventional monopulse realizations.

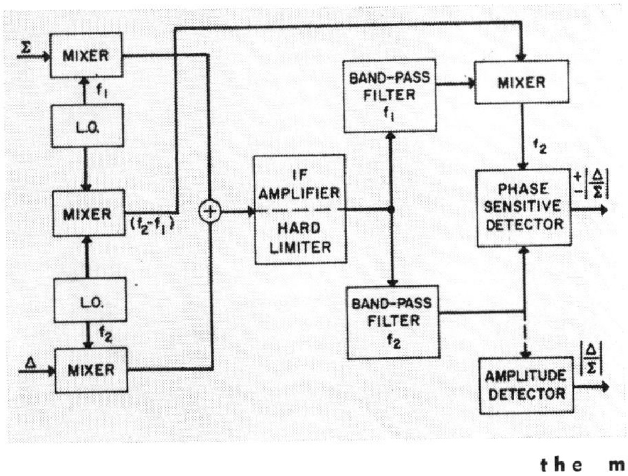

Figure 8 — (Σ, △) - SCAMP monopulse realization.

this reason it is of considerable interest to determine angle accuracy as a function of input SNR for a SCAMP Σ, Δ monopulse system. An analysis has been made for the block diagram shown in Figure 8, assuming a one-axis monopulse antenna whose main-lobe pattern is approximated by two Gaussian shaped antenna beams crossing at the 3 db points.

For white Gaussian noise at the system input, the signal plus noise in the sum and difference channels are respectively given by

$$E_\Delta = (\Delta + x_\Delta)\cos w_1 t + y_\Delta \sin w_1 t \quad (13)$$

$$E_\Sigma = (\Sigma + x_\Sigma)\cos w_2 t + y_\Sigma \sin w_2 t \quad (14)$$

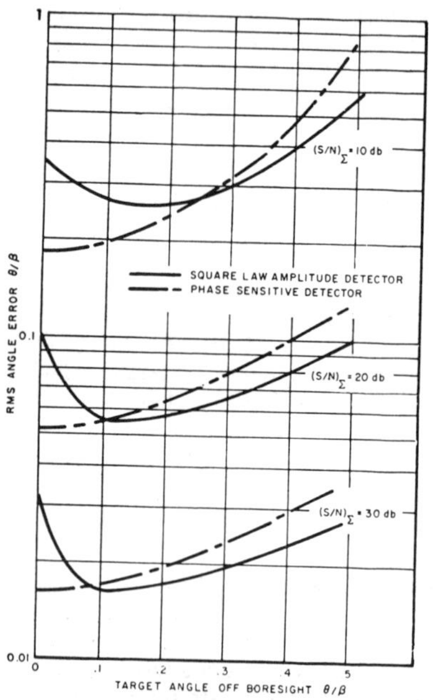

Figure 9 — Single hit rms angle error for (Σ, Δ)-SCAMP monopulse.

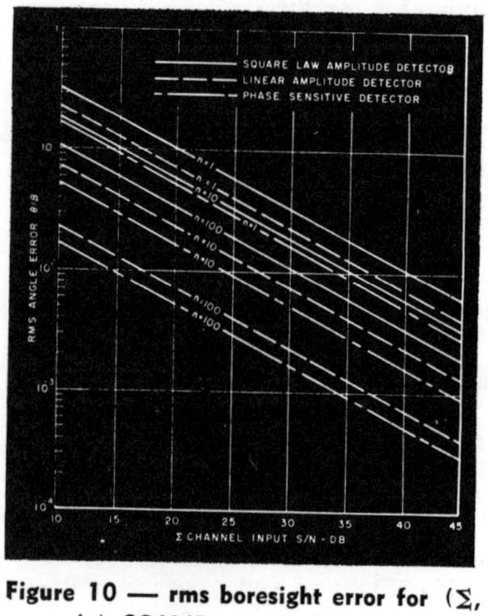

Figure 10 — rms boresight error for (Σ, Δ)-SCAMP monopulse for 1, 10 and 100 hits.

where x_Δ, x_Σ, y_Δ and y_Σ are independent Gaussian random variables and

$$\overline{x_\Sigma} = \overline{y_\Sigma} = \overline{x_\Delta} = \overline{y_\Delta} = 0 \quad (15)$$

$$\overline{x_\Sigma^2} = \overline{y^2} = \overline{x_\Delta^2} = \overline{y_\Delta^2} = \sigma^2 \quad (16)$$

In order to obtain reasonably accurate monopulse data on a single hit, input SNR's greater than 10 db are normally required. It is further assumed that the signal processor outputs are approximately given by (11) and (12), which become in the present notation,

$$E_{\Delta/\Sigma} \cong \left(\frac{(\Delta/\Sigma) + (x_\Delta/\Sigma)}{1 + (x_\Sigma/\Sigma)} \right) \cos w_1 t + \frac{y_\Delta/\Sigma}{1 + (x_\Sigma/\Sigma)} \sin w_1 t \quad (17)$$

$$E_\Sigma \cong (1) \cos w_2 t + \frac{y_\Sigma/\Sigma}{1 + (x_\Sigma/\Sigma)} \sin w_2 t \quad (18)$$

Equations (17) and (18) indicate that the output distribution of signal plus noise in each channel is approximately Gaussian. The output noise results in angle pointing and jitter errors, where the pointing error is defined as the average or mean deviation from true target angle and the jitter is the rms angle fluctuation about the mean.

Angle accuracy as a function of target angle deviation from boresight and input SNR has been calculated for both the amplitude and phase-sensitive detector configurations in Figure 8. In each case, only square-law detection characteristics were considered. The pointing and jitter errors were found by relating the mean and variance of the post-detection distribution of signal plus noise to system angle sensitivity.

Figure 9 contains plots of total rms angle error for a single hit as a function of target angle off boresight for sum channel input SNR's of 10, 20 and 30 db. The error curves for the phase-sensitive detector represent only rms jitter since the pointing error in this case is zero. For comparative purposes, the jitter and pointing errors associated with amplitude detection have been combined into a single rms error curve. The results indicate that the average (single hit) errors over a beamwidth are of the same order of magnitude for both configurations.

Figure 10 shows the improvement in angle accuracy at boresight for linear and square-law characteristics as a function of sum channel input SNR. For large SNR, the performance of a phase-sensitive detector is the same for both detection laws. These curves demonstrate that a phase-sensitive detector performs better around boresight. The plot also shows that the rms angle error decreases almost linearly for increasing input SNR.

Post-detection integration of n hits reduces the mean and rms values of the output distributions of signal plus noise by the square root of n. Figure 10 shows the expected improvements at boresight for n=10 and 100 hits as a function of input SNR. It can be observed that the relative improvemen in angle accuracy for a linear envelope detector and a phase-sensitive detector is approximately the same for large n. The improvement is not as great for a square-law amplitude detector. For n=100 and an input SNR of 10 db the rms angle error will be in the order of 0.01 beamwidth for the phase-sensitive and amplitude detectors.

CONCLUSIONS

SCAMP promises a number of important advantages over existing monopulse signal processors. The key feature of the SCAMP technique is the ability to process monopulse angle information instantaneously in a single channel over a wide dynamic range of received echo amplitudes. This is accomplished by employing a single wideband amplifier-limiter channel followed by appropriate bandpass filters. A vacuum-tube realization has been tested with varying length pulses down to one microsecond. An RF realization using a TWT as a combined amplifier-limiter is presently being investigated. The performance of SCAMP with respect to angle accuracy is comparable to that of conventional monopulse realizations.

ACKNOWLEDGMENTS

The authors wish to acknowledge the contribution made by J. DeLorenzo in the design, construction and testing of the experimental equipment. The equations relating to the noise performance of the basic signal processor were programmed for computer solution by the Computer Services Engineering Department of the Sperry Gyroscope Co.

REFERENCES

1. Davenport, W. B., Jr. and W. L. Root, *An Introduction to the Theory of Random Signals and Noise*, McGraw-Hill Book Co., Inc., New York, N.Y., 1958 p. 288.

2. Davenport, W. B., Jr., "Signal-to-Noise Ratios in Band-Pass Limiters," *J. Appl. Phys.*, Vol. 24, June 1953, pp. 720-727.

Reprinted from:
Proceedings of the National Electronics Conference
Volume XXII, 1966

THE USE OF "COMPLEX INDICATED ANGLES" IN MONOPULSE RADAR TO LOCATE UNRESOLVED TARGETS

Samuel M. Sherman

Missile and Surface Radar Division
Radio Corporation of America
Moorestown, N. J.

I. INTRODUCTION

Monopulse radar is a highly developed, widely used form of tracking radar which gives excellent accuracy when tracking a single target. But when two or more unresolved targets are present, the direction indicated by a monopulse radar (or any other existing tracking radar) wanders, sometimes well beyond the angular interval subtended by the targets, and track may be lost. No satisfactory solution to this problem has been found in the past.

A new method of attack shows that there is additional useful information in the arriving wave which previous radar designs and analyses have ignored. This method offers the possibility of a practical solution under a useful range of conditions. By a solution is meant determination of the angular locations of two unresolved targets (as well as their amplitude ratio and relative phase) or tracking of some stable "centroid".

A particular new feature is the introduction of the "complex indicated angle" in each angle coordinate. Previous analyses and designs have dealt only with the conventional normalized error signal, which is the real part of the complex indicated angle (when the proper scale factor is applied). In the case of a single point target, the imaginary part is zero. Multiple unresolved targets, however, produce an imaginary part as well as a distorted real part. The imaginary part can be measured by a simple addition to the radar. The real and imaginary parts are both essential to the solution.

For two unresolved targets, it is shown that measurements taken on two pulses separated by a short time interval, during which the relative phase of the two targets changes, will theoretically yield the angular locations, amplitude ratio, and relative phases.

A very important special case of two unresolved targets is that of a low-angle target and its image below the surface of the earth. It is shown that in this case, because of certain known relationships, a single-pulse measurement of the complex indicated angle in elevation theoretically suffices to determine the target elevation angle.

A more detailed derivation of some of the results and conclusions stated in this paper will be found in Ref. 1.

II. REVIEW OF SINGLE-TARGET OPERATION

Refs. 2-5, among others, contain much excellent information on the theory, design, and performance of monopulse radars. In this section a basic equation for single-target tracking will be reviewed, since it is an essential foundation for the two-target analysis that will follow. The discussion will be in terms of amplitude-comparison monopulse, but it will be pointed out later how the results can be applied to phase-comparison monopulse as well.

A typical amplitude-comparison monopulse radar has four antenna feeds symmetrically offset from the axis in traverse and elevation. From these feeds, by the use of microwave combining circuitry, the following signal voltages are formed:

S = sum signal,
D_{tr} = traverse difference signal,
and D_{el} = elevation difference signal.

Suppose there is a point target T_A at traverse angle A_{tr} and elevation angle A_{el} measured from the beam axis. The angles are identified in Fig. 1, in which the origin is the beam axis. The second target T_B shown in Fig. 1 is to be ignored for the present.

The radar is so designed that to a close approximation out to about one-half beamwidth from the axis

$$D/S = pA, \qquad (1)$$

where p is a constant angle sensitivity coefficient. The subscript "tr" or "el" is to be inserted after D and A and also after p if necessary, since p_{tr} and p_{el} need not be identical. Wherever these subscripts are omitted, it is to be understood that the equation or discussion holds for each angle component individually.

The LHS of (1) is called the normalized difference signal or error signal. It is used either as the error input to the angle servo for closed-loop tracking, or as an open-loop indication of target direction.

Dividing (1) through by p yields

$$\frac{1}{p}\frac{D}{S} = A, \qquad (2)$$

which is a more convenient form for the purpose of this paper. We will refer henceforth to the quantity (1/p)(D/S) on the LHS of (2) as the indicated angle, while A is the true angle. In the single-target case the two are equal, but in the multiple-target case the relationship is not so simple.

Strict linearity between D/S and A is assumed in this paper for mathematical convenience, but if there is appreciable nonlinearity, it can be compensated by insertion of a linearizing function.

III. TWO UNRESOLVED TARGETS

PHASE RELATIONSHIPS: AN EXAMPLE

In the single-target case, division of D by S in (1) yields a real ratio, since D and S are in phase (or 180° out of phase). In the general case, however, we must note that D and S are phasors, expressed by complex numbers, and that they may have arbitrary phase relative to each other.

Consider an example: Introduce the second target T_B shown in Fig. 1 and confine attention to one angle component, say traverse. Though the two targets are assumed to be in the same resolution cell, in general there will be some differential phase between them. Let S_A and S_B be their individual sum signal phasors, as shown in Fig. 2. Then the total sum signal is the resultant S. Since the two targets lie on opposite sides of the axis in traverse, the difference signal D_A is in phase with S_A while D_B is in phase opposition to S_B. The total difference signal D is their resultant (the subscript "tr" is to be understood after D_A, D_B, and D). It is clear that in general D has a component in phase quadrature to S as well as an in-phase component. In other words, the ratio D/S is complex.

What the conventional monopulse radar actually does in normalizing the difference signal is this: By deriving an AGC voltage from the sum channel and applying it to both the sum and difference channels, it produces voltages $S/|S|$ and $D/|S|$. These are applied as inputs to a product detector or phase detector, the output of which is the component of D that is in phase with S, divided by $|S|$. This output, expressed in another way, is the real part of the complex ratio D/S. The imaginary part of D/S, which is ignored in present radars, is essential to the solution presented in this paper. To measure it would require the addition of a second product detector, with either the sum or difference signal input phase-shifted 90°.

292

It is often erroneously assumed that the radar cannot tell the difference between two unresolved targets and a single target located at the "centroid" angle. The radar can tell the difference if it measures the imaginary part as well as the real part of D/S.

THE COMPLEX INDICATED ANGLE

Since the indicated angle (1/p)(D/S) is complex in general, we will hereafter call it the complex indicated angle.

As before, the analysis will consider only one angle component, since the analysis for the other is identical. To simplify notation, therefore, we will omit the subscripts "tr" and "el".

Let S_A and S_B be the individual sum signals of two unresolved targets at angles A and B with respect to the beam axis, as in Fig. 1. Then

$$S = S_A + S_B. \tag{3}$$

From (2), the individual difference signals are

$$D_A = pAS_A \quad \text{and} \quad D_B = pBS_B, \tag{4}$$

so that $$D = D_A + D_B = p(AS_A + BS_B). \tag{5}$$

Let $$S_B/S_A = gr\, e^{i\phi}, \tag{6}$$

where g = ratio of antenna voltage gains in the directions of the two targets,
r = ratio of backscatter voltage coefficients of the two targets (i.e., the square root of the ratio of their backscatter cross-sections), and
ϕ = relative phase of returns from the two targets.

From (3) and (5) we obtain

$$\frac{1}{p}\frac{D}{S} = \frac{AS_A + BS_B}{S_A + S_B} \tag{7}$$

and dividing through by S_A and using (6), we obtain

$$\frac{1}{p}\frac{D}{S} = \frac{A + gr\, e^{i\phi}B}{1 + gr\, e^{i\phi}} \tag{8}$$

Equation (8) is the basic formula for the complex indicated angle of two unresolved targets. It is instructive to examine its nature.

If we consider ϕ as a variable while the other quantities on the RHS of (8) remain constant, the quantity $\exp(i\phi)$ traces out a circle in its complex plane as ϕ varies through 360°, and since the RHS of (8) is a bilinear transformation of $\exp(i\phi)$, the indicated angle (1/p)(D/S) also traces out a circle in its complex plane. Different values of constant gr give different circles, which form a family. The circles for several values of gr are drawn in solid lines in Fig. 3. This figure has been normalized by putting the origin at the midpoint of A and B and taking (B-A)/2 as the unit of angle. The circle for gr = 1 degenerates into a straight line, the imaginary axis. The circles for gr = 0 and infinity are the points A and B respectively.

If we assign constant values to the relative phase ϕ while varying gr, we obtain another family of circles, orthogonal to the constant-gr family. These are drawn as broken circles in Fig. 3.

It must be kept in mind that the equations and curves derived so far pertain to a single angle coordinate, say traverse. The vertical coordinate in Fig. 3, then, represents the imaginary part of the indicated traverse angle, not the indicated elevation angle. The same set of normalized curves given in Fig. 3 would apply to elevation also.

Equation (8) is equivalent to a pair of real equations. The LHS is a measured complex quantity (i.e., a pair of measured real numbers) derived from a single-pulse return in one angle channel. On the RHS there are four unknowns: A and B (the angles of the two targets in the one coordinate), gr (their amplitude ratio), and ϕ (their relative phase). (At this point we will regard gr as a single factor, but it should be remembered that g is a known function of A_{tr}, A_{el}, B_{tr}, and B_{el}, so that when these angles have been solved, g is determined from a knowledge of the beam pattern.) To solve for the unknowns, we must make additional measurements that will yield the required number of independent equations.

If we make simultaneous single-pulse measurements of the complex indicated angle in both angle components, we then have two complex equations like (8), distinguished by subscripts "tr" and "el", equivalent to four real equations. The unknowns are two components each of A and B, a common value of ϕ, and a common value of gr. Thus we have 6 unknowns but only 4 equations, still not enough for a solution.

We conclude that a single-pulse solution is not possible.

SOLUTION USING TWO PULSES

A solution can be obtained with two pulses in sequence provided either the radar or target parameters change in some way between pulses in order to provide independent equations. The method presented in this paper is based on the change in relative phase ϕ which almost always occurs between pulses.

The time interval between the pulses is assumed to be so short that the values of all the parameters except the relative phase angle ϕ can be considered constant, but it is assumed that ϕ changes appreciably from the first pulse to the second. As an example, suppose two targets with a differential radial velocity of 1 m per second are being observed by a radar of 10 cm wavelength. In one-hundredth of a second the change in relative phase is 72°, while the other parameters (A, B, and gr)can be considered constant over such a short interval. This is true even if the directions of the two targets are changing rapidly (but not relative to each other) provided the antenna is moving with them, since A and B are measured with respect to the axis. We assume that the average rate of change of relative phase is too small to permit separation of the targets by doppler filtering, or that the radar is not equipped to do such filtering.

For convenience, we introduce the following notation for the real and imaginary parts of the complex indicated angle:

$$\frac{1}{p}\frac{D}{S} = x + iy. \qquad (9)$$

As before, we will deal with only one angle component at a time, and will therefore omit the subscript "tr" or "el". We will, however, use subscripts "1" and "2" to denote the first pulse and the second pulse.

The measurements of $x_1 + iy_1$ and $x_2 + iy_2$ represent two points in the complex plane of Fig. 3. Since they are produced by the same values of A, B, and gr, but different values of ϕ, then they lie on one of the solid circles shown in Fig. 3, and the two points are sufficient to determine which circle, provided $x_1 \neq x_2$. However, fixing the circle is not sufficient to give the values of A, B, and gr, since the same circle could have been produced by an infinite number of combinations of A, B, and gr. An additional measurement is needed to make the solution unique. This measurement is $|S_2/S_1|$, the ratio of sum-signal amplitudes of the two pulses. With this measurement, in addition to x_1, y_1, x_2, and y_2, we have five measured quantities from which we can set up five equations in the five unknowns A, B, gr, ϕ_1, and ϕ_2, the last two being the relative phases on the first and second pulses. Four of the five equations are obtained by rewriting (8) as

$$x + iy = \frac{A + gr\, e^{i\phi}B}{1 + gr\, e^{i\phi}}, \qquad (10)$$

equating reals and imaginaries separately, and inserting subscripts 1 and 2 after x, y, and ϕ for the first and second pulses. The fifth equation is

$$\left|\frac{S_2}{S_1}\right|^2 = \frac{1 + g^2r^2 + 2gr\cos\phi_2}{1 + g^2r^2 + 2gr\cos\phi_1}. \qquad (11)$$

An algebraic-graphical method has been devised for manual solution and tested on a numerical example. This can be converted into a computer program for real-time solution.

Since A and B can be determined, their midpoint or their "centroid" (that is, their weighted mean position, with the weighting proportional to amplitude or power) could be computed, and by closing the loop, the antenna could be made to track that direction, thus making good the assumption that A and B do not change during the interval between the two pulses.

Although the method just described would theoretically give a solution based on only two pulses, a practical implementation would preferably obtain an overdetermined solution by using measurements made on many pulses, thus improving accuracy.

A display of the complex indicated angle can be presented on a CRT by applying the real and imaginary parts to the horizontal and vertical deflection inputs. If the changes in relative phase are rapid compared with changes in amplitude ratio and angular separation, the display of the complex indicated angle will be observed to trace out a circle, in accordance with Fig. 3, and this will be evidence of the presence of two targets.

A conceptual form of implementation of the two-target solution is indicated in Fig. 4.

ERRORS DUE TO NOISE

The analysis of thermal noise errors is too involved to be presented here but a numerical example will provide some feeling for the order of magnitude. Considering only one angle component, let targets A and B be at angles 0.2 and -0.1 beamwidth from the axis and let gr = 0.5 (i.e., let the signal from target B have half the amplitude of the signal from target A). Let p = 5/3 per beamwidth (a representative value). Suppose the S/N ratio is 30 db when the two targets are in phase; then it will be about 20.5 db when they are in phase opposition. Target A alone would have a S/N ratio of 26.5 db and target B alone 20.5 db. If the measurements of x_1, y_1, x_2, y_2, and $|S_2/S_1|$ are made on one pulse at the time when the two targets are in phase and the other pulse when they are in phase opposition, the rms errors in the solutions are: 0.015 beamwidth for angle A (compared with 0.014 beamwidth if A were present alone) and 0.034 beamwidth for angle B (compared with 0.028 beamwidth if B were present alone). However, the errors are heavily dependent on the change in relative phase between the two pulses, the most favorable case being the one just cited (i.e., maximum difference between x_1 and x_2). At the other extreme, the solution becomes indeterminate if x_1 and x_2 approach equality. Therefore it is important that the two pulses be properly selected or that a best estimate be formed from several pulses.

293

ERRORS DUE TO GLINT OF EACH TARGET

The analysis above assumes that each target behaves like a point target. Actually each target has non-zero angular dimensions and the resulting angle glint of each individual target produces errors of the order of 1/4 of the angular span of the target. However, for small targets and/or long ranges the glint error is negligible in comparison with thermal noise. If not, glint of the individual targets must be included in the error analysis.

IV. LOW-ANGLE TARGET: THE MULTIPATH PROBLEM

We now take up the case of a target at a low elevation angle, say within a beamwidth of the horizon. This problem, also known as the multipath problem, is of great practical importance because elevation tracking of such a target by present means is erratic and sometimes impossible. Traverse (azimuth) tracking is usually not seriously affected.

The reradiation from the target reaches the antenna both by a direct path and by reflection from the earth's surface, the latter being equivalent to radiation from the image of the target. The target and its image constitute a special case of two unresolved scatterers. In this case certain known relationships between the amplitudes, phases, and locations of the two scatterers aid in determining the elevation angle of the target. We shall show that a solution is theoretically possible with a single pulse. This takes a different form from the two-target solution presented in the preceding section.

We assume a flat earth in order to obtain a relatively simple, approximate picture of the behavior. The geometry is shown in Fig. 5. In place of the symbols A and B used above for the angles of two targets measured from the antenna axis, we will find it more convenient in the present problem to use the new symbols E, the elevation angle of the target, and E_o, the elevation angle of the beam axis, both measured from the horizontal. The relationships between the two sets of angles are

$$A = E - E_o \tag{12}$$

and

$$B = -E - E_o. \tag{13}$$

The presence of a plane earth can be represented by a complex reflection coefficient Γ :

$$\Gamma = re^{i\phi_a}, \quad 0 \leq r \leq 1. \tag{14}$$

The magnitude r is equivalent to the image-to-target amplitude ratio and ϕ_a is one component of their relative phase ϕ.

We now calculate (1/p)(D/S), the complex indicated angle in elevation, for the target-image combination. When substitutions given in (12) and (13) are made in (8), this becomes

$$\frac{1}{p}\frac{D}{S} = -E_o + E\,\frac{1 - gr\,e^{i\phi}}{1 + gr\,e^{i\phi}}, \tag{15}$$

from which we obtain the real and imaginary parts

$$x = \mathrm{Re}\left(\frac{1}{p}\frac{D}{S}\right) = -E_o + E\,\frac{1 - g^2r^2}{1 + g^2r^2 + 2gr\cos\phi} \tag{16}$$

and

$$y = \mathrm{Im}\left(\frac{1}{p}\frac{D}{S}\right) = -E\,\frac{2gr\sin\phi}{1 + g^2r^2 + 2gr\cos\phi}. \tag{17}$$

To compute x and y as functions of E we must express g, r, and ϕ as functions of E, as follows.

The phase angle ϕ of the image relative to the target has two components:

$$\phi = \phi_a + \phi_b, \tag{18}$$

where ϕ_a is due to the fact that Γ is complex and ϕ_b is due to the geometric path length difference.

From Fig. 5 it is seen that

$$\phi_b = -4\pi h \sin E, \tag{19}$$

where h equals z divided by the wavelength.

The values of r and ϕ_a as functions of frequency, polarization, and elevation angle can be read from Figs. 5.3 to 5.6 in Ref. 6, assuming a smooth reflecting surface. To calculate g as a function of E it is necessary to know the antenna beam pattern for the sum channel.

We will present an example based on the following conditions:

(a) The target is being tracked normally in traverse; in other words, the beam axis, the target, and the image lie in the same vertical plane.

(b) The beam is fixed in elevation at some low angle, and open-loop measurements of the complex indicated angle are made as the target passes through the beam.

(c) The reflecting surface is smooth.

The following set of parameters has been selected for the example:

Beamwidth: $\theta_o = 1^o$.

Antenna height: h = 100 wavelengths.

Beam axis elevation angle: $E_o = 0.5^o$.

Sum voltage pattern: $|S| = \cos^2(1.14\,\theta/\theta_o)$.

In the last equation θ is the angle off axis and θ_o is the half-power beamwidth. The quantity in parentheses is in radians. This particular pattern is selected because it is a typical monopulse sum voltage pattern, as reported in Ref. 2. From the given pattern and beam axis elevation angle E_o, g can be computed as a function of target elevation angle E.

With this set of parameters, the complex indicated angle in elevation has been computed as a function of target elevation angle E. The results are plotted in Figure 6. The x and y coordinates are the real and imaginary parts respectively of the complex indicated angle in elevation.

This plot is the locus of the complex indicated angle in the complex plane as the target elevation angle E rises from the horizon to the beam axis. Values of E are spotted along the locus.

For each radar installation and for a selected fixed elevation angle of the beam axis, a curve similar to Fig. 6 could be plotted in advance and used as a scope overlay (or computer look-up table). An open-loop measurement of the complex indicated angle by a radar modified for this purpose would then fall somewhere on the curve and the value of E at that point could be read off. This could be accomplished in theory on a single pulse. In practice, integration of several pulses to improve signal-to-noise ratio, as in conventional radar detection or tracking, would certainly be desirable. Also, observation of the time history of the indicated angle (for example, in the form of a CRT display with an overlay similar to Fig. 6) would be helpful.

The corrections for earth's curvature can be computed with the aid of numerical and graphical procedures presented in Ref. 6. These corrections will not be carried through here; they would alter the detailed shape of the curve in Fig. 6 but not its general form.

For the low-angle application the added product detector is needed only in the elevation channel.

Fig. 7 is a plot of only the real part of the complex indicated angle as a function of E, for the same set of parameters used in Fig. 6. In the absence of the reflecting surface, this would be simply a 45° line, which is shown dotted. The actual curve, drawn solid, is radically different. Even if a curve such as Fig. 7 were available for "calibration", the conversion is ambiguous. A horizontal line representing a measurement of Re[(1/p)(D/S)] will usually intersect the curve of Fig. 7 in at least two points, sometimes three.

In contrast, the complex plot shown in Fig. 6 has no ambiguities (points at which the curve crosses or touches itself). Such ambiguities can occur for some combinations of parameters, but generally they will be limited to a few cross-over points and easily resolved.

In addition to the multipath effect, there are two other factors that sometimes cause difficulty in low-elevation tracking. These are ground (or sea) clutter and masking by terrain obstructions. The method described here will not solve these two problems. However, there are many situations where masking is not a problem and where clutter is not present at the range of the target or can be removed by doppler methods.

V. PHASE-COMPARISON MONOPULSE

The results derived for amplitude-comparison apply as well to phase-comparison monopulse, provided:

(a) Amplitude as well as phase information is preserved instead of destroying the amplitude information by limiting, as is usually done,

(b) Sum and difference signals are formed just as in amplitude-comparison monopulse, and

(c) The real and imaginary parts of the complex indicated angle are interchanged.

VI. COMPARISON WITH PREVIOUS WORK

Other authors have analyzed the problem of two unresolved targets, using only the real part of the complex indicated angle, and offering no method of solving for the locations of the individual targets. Plots of the real part vs. ϕ are given in Fig. 3.10 of Ref. 2, Fig. 5.15 of Ref. 5, and similar figures in Refs. 7 and 8. The complex plots in Fig. 3 of this paper, which this author believes to be new, contain the same information plus the imaginary part in addition.

Ref. 9 does recognize the existence of the imaginary part but does not offer any approach or suggestion for using it. Ref. 10 suggests the possible use of the imaginary part to derive information about target extent, rotation, and shape, but gives no details or specific method.

Refs. 11-13 offer methods of separating two or more plane waves received from different directions, but these methods require special antenna installations and are not suited to tracking of moving targets.

VII. PRACTICAL CONSIDERATIONS

Perfect normalization of the difference signals with respect to the sum signal is assumed in the analysis. However, when normalization is obtained by AGC action, it becomes inaccurate at the extremes of the dynamic range and it usually has a time lag. This is a matter of implementation, not a fundamental limitation. For any particular radar, it is necessary to determine whether the existing normalization is acceptable or whether improvements are needed.

The selection and programming of the computer are important considerations in implementing the two-pulse tracking solution for two targets.

A related problem is how best to use the measurements available from many pulses for an optimum overdetermined solution.

Factors to be examined in more detail for the single-pulse method of locating low-altitude targets include the earth's reflectivity (not usually as simple as assumed here), surface irregularities, and imperfect knowledge of the antenna pattern and open-loop angle sensitivity coefficients. The important thing is not whether these factors conform precisely to a theoretical model but whether they are stable and repeatable. If they are, the appropriate curve of the form of Fig. 6 can be plotted from actual measurements on a target whose position is varying in a known manner, and then used as a calibration curve to determine unknown elevation angles.

Other potential sources of error to be considered include amplitude and phase distortions in the radar, calibration drifts, interaction of elevation and traverse channels, antenna sidelobe returns, and departures from the assumption used in the two-target solution that A, B, and gr do not change during the interpulse period.

VIII. CONCLUSIONS

The extension of the monopulse normalized error signal into the complex domain, in the form of the complex indicated angle, makes it theoretically possible for a monopulse radar to determine the individual angular locations of two targets that are unresolved in the conventional sense, or to determine the elevation angle of a low-angle target when present methods fail because of multipath errors. The required radar modifications are not very extensive.

An analysis of errors due to thermal noise and individual target glint indicates that these errors are within acceptable bounds under a reasonable range of practical conditions.

Further work, including an experimental program, is needed to verify some of the assumptions and approximations made, analyze or measure additional sources of error, and work out the practical implementation.

REFERENCES

1. S. M. Sherman, Complex Indicated Angles in Monopulse Radar, Ph.D. Dissertation, University of Pennsylvania, Dec. 1965.
2. D. K. Barton, Radar System Analysis. Englewood Cliffs, N.J.: Prentice-Hall, Inc., 1964.
3. R. S. Berkowitz (ed.), Modern Radar: Analysis, Evaluation, and System Design. New York: John Wiley & Sons, 1965.
4. D. R. Rhodes, Introduction to Monopulse. New York: McGraw-Hill Book Co., Inc., 1959.
5. M. I. Skolnik, Introduction to Radar Systems. New York: McGraw-Hill Book Co., Inc. 1962.
6. D. E. Kerr, Propagation of Short Radio Waves. New York: McGraw-Hill Book Co., 1947.
7. J. H. Dunn, D. D. Howard, and A. M. King, Phenomena of scintillation noise in radar-tracking systems, Proc IRE 47, No. 5, May 1959, pp 855-863.
8. J. E. Meade, Target Considerations, Chapter 11 of book, Guidance by A. S. Locke et al. Princeton, N. J.: D. Van Nostrand Co., Inc., 1955.
9. L. Peters, Jr., Accuracy of tracking radar systems, Ohio State U. Research Foundation Report 601-29, Dec. 31, 1957, ASTIA Document AD 200,027.
10. D. B. Anderson and D. R. Wells, A note on the spatial information available from monopulse radar, Proc 5th Mil-E-Con, 1961, p. 268.
11. E. W. Hamlin, P. A. Seay, and W. E. Gordon, A new solution to the problem of vertical angle-of-arrival of radio waves, J. Appl. Phys. 20, March 1949, pp 248-250.
12. F. E. Brooks, Jr., A receiver for measuring angle-of-arrival in a complex wave, Proc IRE, Vol. 39, Apr 1951.
13. C. C. Watterson, Field strength measurements in a multipath field, U.S. National Bureau of Standards Report 7600, June 28, 1962.

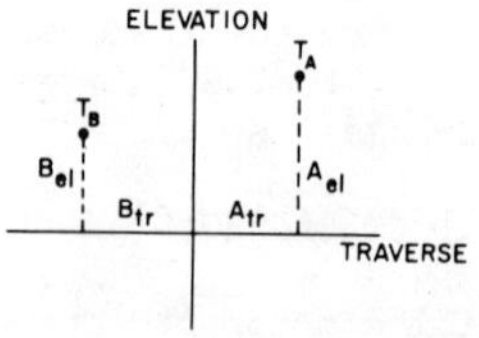

Figure 1. Two-Target Geometry

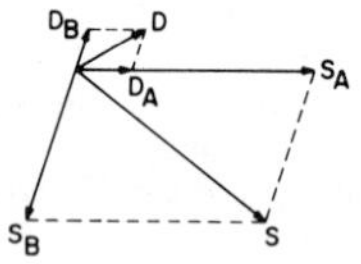

Figure 2. Phasor Diagram For Two Targets

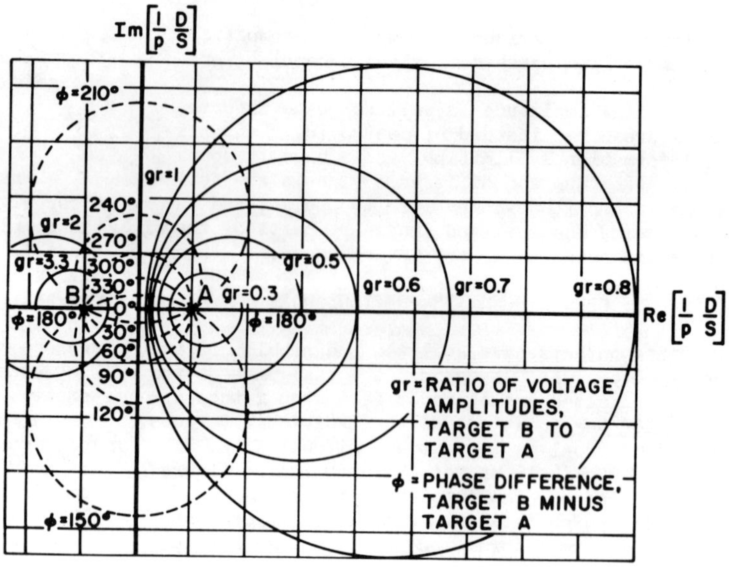

Figure 3. Complex Indicated Angle as Function of gr and ϕ

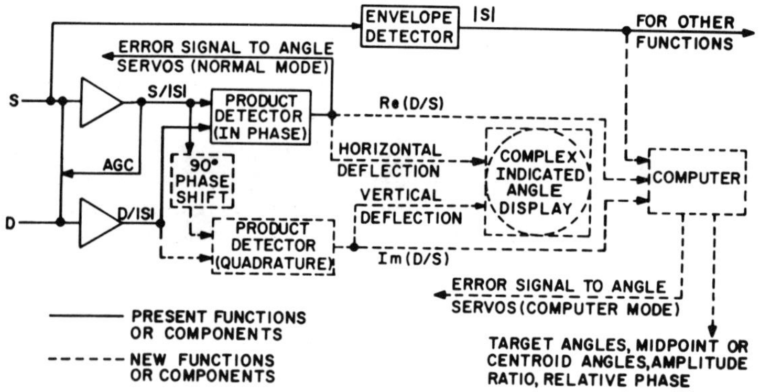

Figure 4. Functional Diagram for Two-Target Solution (one angle component)

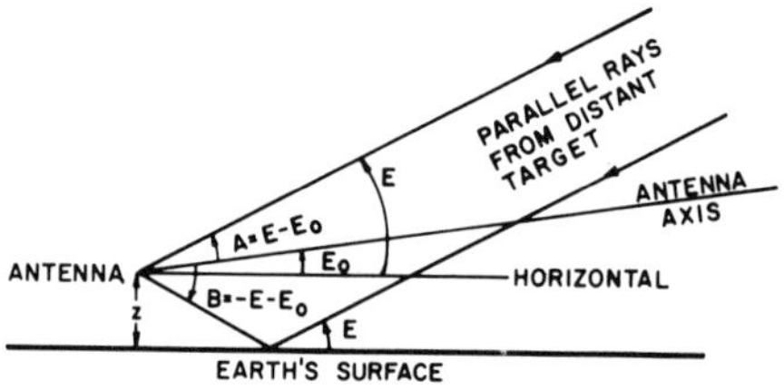

Figure 5. Geometry of Target and Image, Flat-Earth Approximation

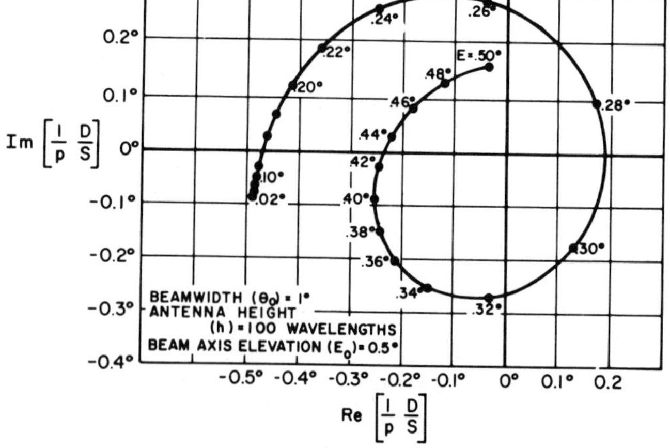

Figure 6. Complex Indicated Angle As Function of Target Elevation Angle E

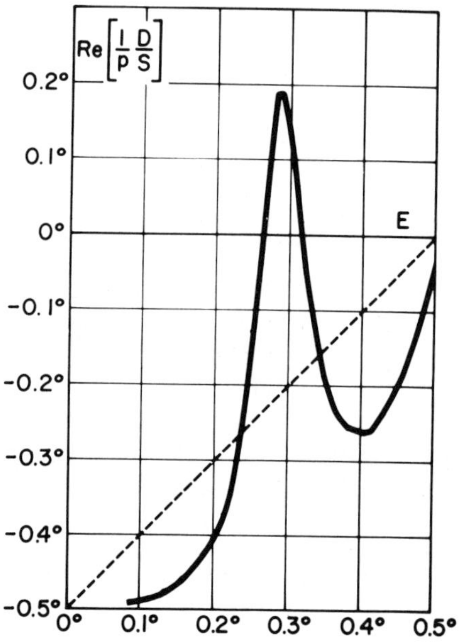

Figure 7. Real Part of Indicated Angle as Function of Target Elevation Angle

Monopulse Operation with Continuously Variable Beamwidth by Antenna Defocusing

HENRY W. REDLIEN, JR., SENIOR MEMBER, IEEE

Abstract—Monopulse tracking operation with continuously variable beamwidth may be obtained by antenna defocusing. To demonstrate this property, antenna patterns (including phase) of a square aperture for large amounts of defocusing have been computed by diffraction theory for the case of highly tapered illumination in both the sum and difference antenna modes.

From an examination of the computed patterns over a 10 to 1 beamwidth range, it is shown that with the increasing defocusing, the patterns in the sum and difference modes exhibit smoothly increasing width. Some change of shape is noted, but the pattern amplitudes and phase relationships are still suitable for on-axis tracking or off-axis location by monopulse techniques.

To achieve monopulse operation with the defocused patterns two modifications to ordinary monopulse receivers are necessary. A phase shifter, inserted in the difference or sum channel and ganged to the amount of defocusing, is needed to cancel the phase variation with defocusing. In addition, amplitude correction is required to keep the error-signal slope constant.

I. Introduction

IN MODERN radar systems, monopulse tracking with variable beamwidth is a recent requirement. One application that may be anticipated requires the coverage of a fixed area in space at a continuously changing distance from the radar. Another application for variable beamwidth is in the acquisition of targets with a wide beam for subsequent high-precision monopulse tracking with a narrow beam.

To design an antenna with variable beamwidth, there are methods which might be employed which do not require defocusing. For example, an iris which covers part of the radiating aperture would widen the beam, but this method would lower the efficiency of the antenna. Also, an antenna with variable diameter, variable focus (curvature), and variable feed position might be imagined. Such an arrangement could introduce many mechanical problems. An antenna array with a variable illumination size could also provide variable beamwidth, but the complexities associated with a phased array may not be justified for the particular application. Each of the systems described operates with a focused beam from a reduced aperture size to provide the beam widening, and each example cited is either mechanically complex or has low efficiency.

An alternate way to provide beam widening is to defocus an antenna while retaining the same physical aperture size. Several arrangements appear to be mechanically simple and also retain antenna efficiency. Some examples are a lens (or reflector) antenna with a movable feed (Fig. 1), a Cassegrain double-reflector antenna with a movable sub dish, and an array with defocusing phase shifts. Although these arrangements provided beam widening, the question arises as to whether the amplitude and phase relationship in the antenna patterns required for monopulse tracking are maintained as the antenna is defocused.

The present paper is an investigation of the monopulse tracking properties of variable-beamwidth antennas which operate in many defocused conditions. Although many specific examples of symmetric defocused antenna patterns have been published (selected examples are references [9] and [10]), there has not been sufficient information available to evaluate greatly defocused antennas for monopulse tracking operation, which require antisymmetric patterns. To make this evaluation, the amplitude and the phase of both the sum and the difference patterns are needed. It was necessary to have these patterns calculated from physical optics diffraction theory as is described in Section IV.

Since these computations have been performed, H. A. Wheeler has published a paper describing the defocusing of idealized aperture excitation functions which provide widened beam antenna patterns.[15] These functions are the symmetrical Gaussian distribution function of probability theory and the antisymmetrical Rayleigh distribution function. The Fourier transforms of these functions when defocused provide antenna pattern shapes which show that monopulse operation may be achieved under the condition of a large amount of defocusing. The limitation on this approach is in the fact that infinite apertures are required for these aperture excitation functions. The present paper presents defocused performance with a finite aperture for excitation functions which may be obtained, or at least approximated, in practice.

II. Symbols

The following voltage ratios represent the received signals induced in the various channels at the output of the comparator relative to those induced in an isotropic antenna. They represent the standard one-way radiation patterns of the antenna as a function of target angle.

E_s = the complex signal in the sum channel;

Manuscript received January 10, 1966; revised December 4, 1967, and March 4, 1968. The work reported here has been performed under a subcontract with Bell Telephone Laboratories, Inc., Whippany, N. J., Purchase Order D-352813, under Prime Contract DA-30-069-ORD-1955 between the Army Ordnance Corps and the Western Electric Company

The author is with Wheeler Laboratories, Inc., Smithtown, N.Y.

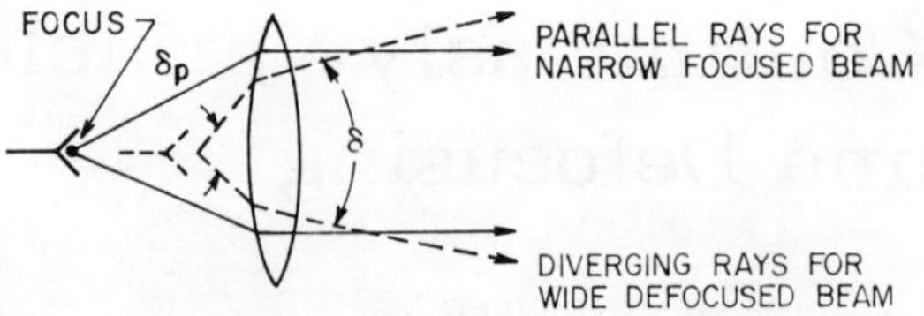

Fig. 1. Lens antenna with defocusing by feed displacement.

E_d = the complex signal in the difference channel;

S_d = the slope of the difference-signal pattern at zero target angle;
 = $d|E_d|/d\delta$, voltage ratio per unit angle, at $\delta = 0$;

ψ = phase angle of the difference-mode signal relative to the sum-mode signal;

ψ_3 = phase angle, ψ, measured at target angle where sum pattern is 3 dB below its peak value.

The following quantities represent characteristics derived from the basic parameters described above.

$2E_d/E_s$ = relative difference signal;

E_0 = error signal;
 = direct voltage output of product detector;

$2S_d/E_s$ = the slope of the relative difference signal at zero target angle;
 = $d|E_d/E_s|/d\delta|\ \delta = 0$, voltage ratio per unit angle;

ψ_0 = hybrid monopulse angle;
 = angle of the locus of the relative difference-signal vector at zero target angle;

θ = phase angle of difference signal relative to sum signal at input to product detector;
 = $(\psi + \phi)$.

The following exponential relative gain parameters represent asymmetry between the sum and difference channels. Positive phase shifts indicate a leading time phase angle.

ϵ = magnitude of relative complex gain of the difference channel with respect to the sum channel;

ϕ = relative phase shift of the difference channel with respect to the sum channel.

The following quantities are needed to represent antenna geometry and aperture illumination.

D = aperture width;

F = focal length;

ΔF = change of focal length;

$d_{x,y}$ = aperture coordinate;

x = normalized aperture coordinate;
 = $2d_x/D$;

u = normalized pattern angle;
 = $\pi D/\lambda \sin \delta$;

δ = target angle or pattern angle;
 = the angle between the direction of the target and the direction of the antenna axis;

δ_p = angle of primary pattern;

$S(u)$ = antenna pattern;

ζ = phase deviation in aperture plane;

ζ_0 = maximum phase deviation (MPD).

III. Antenna Requirements for Monopulse Operation

A monopulse tracking radar develops tracking signals by the proper combination of a "sum" pattern and a "difference" pattern. The difference pattern is nominally in phase with the sum pattern for tracking angles on one side of the antenna axis and out of phase on the other. The error signal derived from a monopulse product detector[11] for driving the tracking servos or for off-axis location may be formulated in terms of the sum and difference patterns and circuit parameters as follows. This equation assumes automatic gain control derived from the sum signal and operating on both the sum and difference channels:

$$E_0 = 2\left|\frac{E_d}{E_s}\right| \cos(\psi + \phi) \exp \epsilon \qquad (1)$$

where

$$\psi = \underline{/E_d} - \underline{/E_s}. \qquad (2)$$

The factors ϵ and ϕ represent asymmetries of channel gain and phase shift, respectively. The phase angle ψ represents the contribution of the antenna patterns to the phase angle between the sum and difference signals at the product detector. The value near zero target angle is called the hybrid monopulse angle and is given the symbol ψ_0.

From an examination of the error signal formula the antenna requirements may be stated. The sum antenna pattern is required to be symmetric to provide a reference for the difference signal in both amplitude and phase. The difference signal is required to be antisymmetric (have a phase reversal) to provide a reversal of the sign of the error signal when it is combined with the sum signal in the product detector. The fact that ψ is not zero is not significant, since monopulse operation may be achieved by adjusting ϕ, with a phase shifter in the channels, to have $\psi + \phi = 0$. This is in fact the condition for optimum phase adjustment.[8] Another requirement is that ψ remain relatively constant with target angle. If ψ varies, the error signal is reduced by the cosine of the angle of variation, $\Delta\psi$. In the case where $\Delta\psi$ is greater than 90°, a false reversal of sense would be indicated. System requirements for tracking typically require that the change of phase caused by both antenna and channels not exceed ±30°.

It is also required that the amplitude ratio of the difference signal divided by the sum signal (called "relative difference signal") be approximately linear or at least unambiguous (single valued) with target angle. The absolute value of the slope of the relative difference signal is not necessarily important, since it may be adjusted by the amplitude factor ϵ. With a suitable gain adjustment, the error-signal slope may be held constant, as required for stable tracking.

IV. Basis for a Computation of Defocused Antenna Patterns

To evaluate the proposed defocusing method against monopulse requirements, antenna patterns which include amplitude and phase information have been computed for both the monopulse sum and difference modes. The computations were set up on the basis of diffraction theory or physical optics for a case simply to analyze, but chosen to be representative of practical antennas. The following assumptions were made and are illustrated in Fig. 2.

1) Aperture is square;
2) Illumination amplitude is as follows: a) sum mode is half cosine vertically and horizontally; and b) difference mode is a full sine vertically by a half cosine horizontally;
3) Illumination phase varies with the square of radius.

Some comment about the above assumptions is needed. The rectangular aperture was chosen since it allows separation of variables in the radiation pattern formula. The sine and cosine illuminations are close approximations to those which may be obtained in practice and their patterns are easily evaluated. The square-law phase deviation with radius is obtained in practice for long focal-length antennas, and it also is separable mathematically in rectangular coordinates. The illumination patterns are completely tapered in both the sum and difference modes. This type of illumination requires the newly developed feed systems which provide independent control of the sum and difference mode patterns.[2],[5] It is to be noted that with highly tapered illuminations the illumination in the corners of the rectangular aperture is small and would not be expected to contribute significantly to the far-field pattern. Therefore, it is expected that these patterns would also be good approximation to the patterns of a circular aperture.

The radiation pattern formula based on the above assumptions is

$$S(u) = \int_{-1}^{+1} \cos\frac{\pi}{2}x \exp j\zeta_0 x^2 dx \cdot \int_{-1}^{+1} \begin{Bmatrix} \cos\frac{\pi}{2}y \\ \sin \pi y \end{Bmatrix} \cdot \exp j[uy + \zeta_0 y^2] dy, \qquad (3)$$

where the cosine function is for the monopulse sum mode and the sine function for the monopulse difference mode. The first integral is a constant term, independent of u, which is needed to compute the reduction in gain with defocusing for the two-dimensional rectangular aperture. For long-focus systems an approximate relation may be derived for the illumination phase. The value of the angle at the edge of the aperture is called the maximum phase deviation (MPD) and is given the symbol ζ_0; it is to appear as a parameter later in this paper.

$$\zeta_0 = \frac{\pi}{4}\frac{\Delta F}{\lambda}\left(\frac{D}{F}\right)^2. \qquad (4)$$

The widening of the antenna patterns by defocusing may also be understood from geometric optics by associating the increased angle of the diverging rays with increased antenna pattern width (Fig. 1). For the focused antenna the rays are parallel, indicating a very narrow beam which is widened only by diffraction to a half-power beamwidth approximately equal to λ/D radians. Defocusing causes the rays to diverge

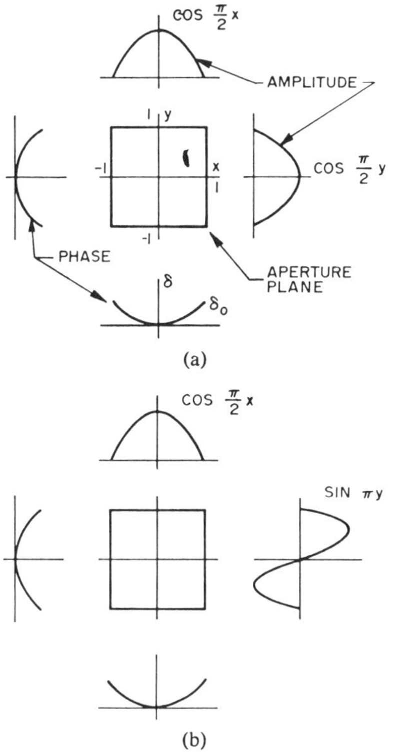

Fig. 2. Aperture illuminations assumed for computations. (a) Sum-mode aperture illumination. (b) Difference-mode aperture illumination.

to form beams wider than the diffraction width. For long focal-length parabolic antennas, the ratio of the secondary pattern angle δ to the primary feed angle δ_p may be simply derived by paraxial theory to be

$$\frac{\delta}{\delta_p} = \frac{\Delta F}{F}, \qquad (5)$$

where ΔF is the change in feed position and F is the focal length of the parabola. From this viewpoint the secondary pattern has the same shape as the primary pattern of the feed with only a change of angle scale.

V. Computed Antenna Patterns

Examples of defocused antenna patterns based on the assumptions listed in the previous section are presented for a wide range of phase deviations. The patterns include the sum-mode amplitude, the difference-mode amplitude, and the amplitude and phase of the relative difference signal. The patterns are presented for four values of the maximum phase deviation from 0 to 6π radians.

Fig. 3 shows the normalized sum-mode patterns; the decrease in gain accompanied with defocusing is shown later in Section VII. The focused pattern, where the maximum phase deviation (MPD) is zero, exhibits a narrow beam and a typical side-lobe structure. The defocused patterns exhibit progressively wider beams, no side-lobe structure down to −34 dB, and fairly smooth patterns over the 10 to 1 beamwidth range. (The 10 to 1 range does not represent a theoretical limit but is discussed here for convenience.) It is

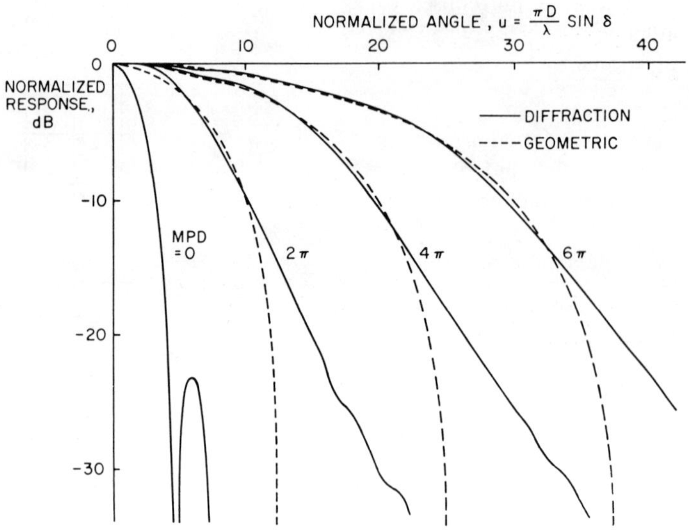

Fig. 3. Normalized sum-mode patterns.

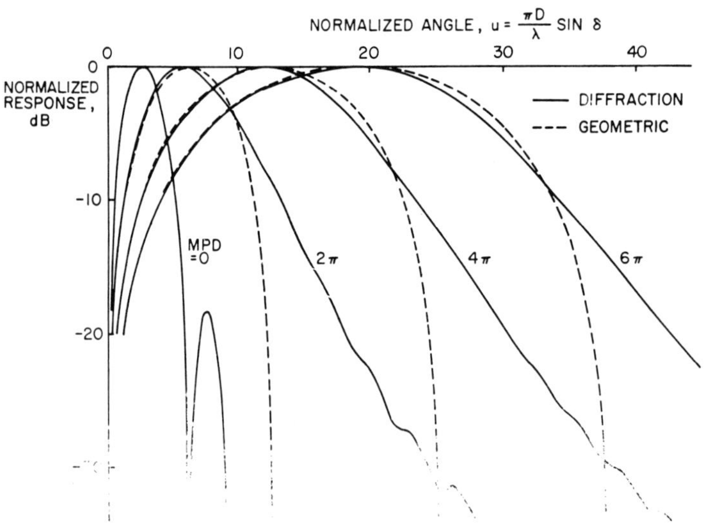

Fig. 4. Normalized difference-mode patterns.

believed that the smooth patterns were obtained because of the highly tapered illumination; for less tapered illuminations, ripples are observed.[9]

Fig. 4 shows the normalized defocused difference patterns; these exhibit a smooth behavior similar to that of the sum patterns. It has been observed experimentally that patterns with less edge taper exhibit ripples, and that the smooth patterns obtained here were a consequence of the extreme tapering of the illumination.

It was found possible to graph the sum and difference patterns computed from the theory of geometric optics in terms of the coordinates of Figs. 3 and 4 by an interpretation of formulas (4) and (5). These "geometric" patterns are shown as dashed lines. For the focused case (MPD$=0$), the geometric pattern is a vertical straight line at $u=0$, which obviously does not fit the pattern shown. In this case the pattern is formed completely by diffraction effects. However, for MPD $=2\pi$ and greater, the geometric patterns closely follow the computed patterns for levels from the peak to about the -10-dB level. At these pattern levels and MPD the diffraction effect appears as a small perturbation about the geometric pattern. At lower levels the diffraction effect is seen to widen the patterns.

Fig. 5 shows the amplitude of the relative difference signal. This is the primary amplitude factor in formula (1) for the error signal. The shape of the patterns change with increased defocusing, but within the angle which is 3 dB down on the sum patterns they are smooth and approximately

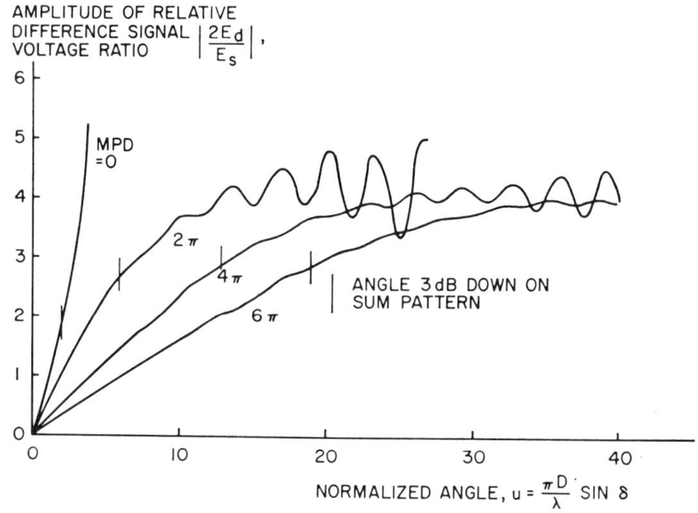

Fig. 5. Amplitude of relative difference signal.

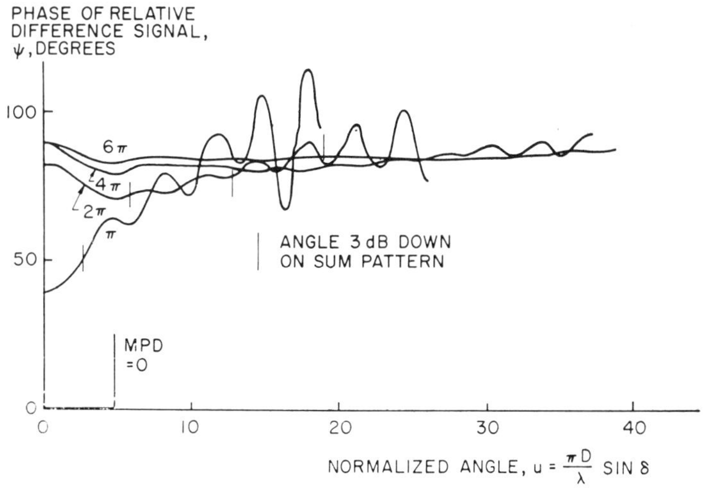

Fig. 6. Phase of the relative difference signal.

linear. The slope of the patterns at zero pattern angle (pattern slope factor) changes with defocusing; this change will be discussed in Section VI.

Fig. 6 shows the phase of the relative difference signal. This phase angle also appears in formula (1) for the error signal. The average phase angle changes from 0° at focus to 90° for the defocused case where the MPD is 6π. For any fixed amount of defocusing, the change in ψ with pattern angle out to the 3 dB-sum pattern angle is small. The extent of the change is discussed later in Section VII.

VI. Compensation of Defocusing Effects for Monopulse Operation

In Section V patterns were presented which have suitable amplitude shapes for monopulse operation over a wide range of beamwidths. However, variations were indicated for two parameters which, if left uncorrected, would degrade the performance. These parameters are the change of the phase angle and the change of the amplitude of the relative difference signal with defocusing.

As explained in Section III, the phase angle between the difference and sum signal at the input to the product detector must be either 0 or 180° to provide maximum error signal. Therefore, the phase angle of the defocused patterns, which varies from 0 to 90° with defocusing, must be compensated by a variable phase shifter which is ganged to the defocusing mechanism. The phase shifter must be inserted in the monopulse channels between the comparator and product detector, but may be placed in either the sum or difference channels. Since the azimuth and elevation difference channels may have different defocusing characteristics, a phase shifter in each difference channel may be required. A more detailed discussion of the variation of phase shift with defocusing is given in Section VII.

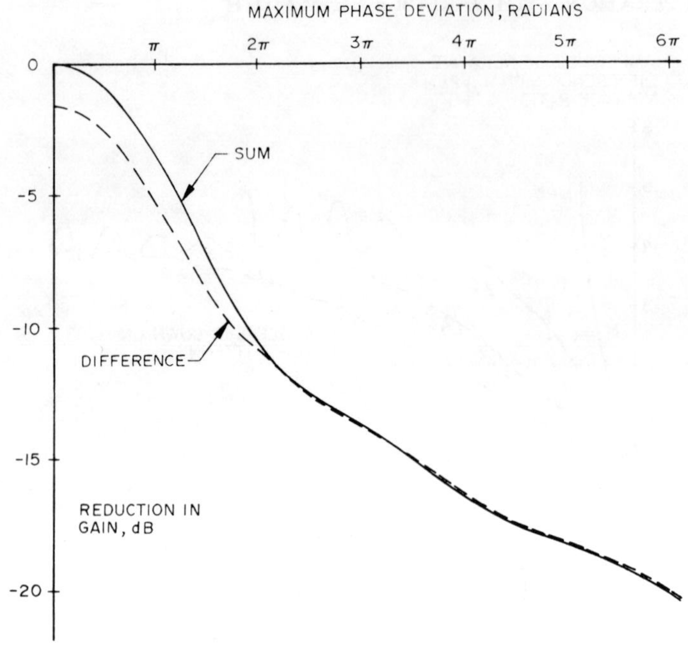

302

Fig. 7. Sum- and difference-mode gain. Focused rectangular aperture is reference for gain (0 dB). Focused gain $=4\pi A$, λ^2 A = aperture area, η = efficiency = 0.66 for cosine by cosine tapered illumination with no spillover or losses.

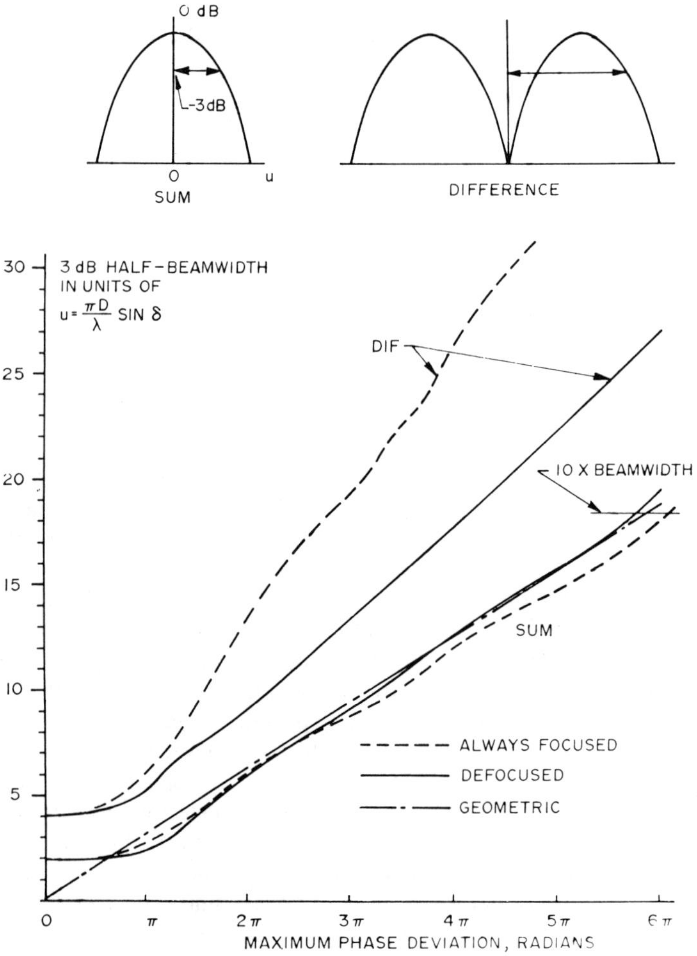

Fig. 8. Sum and difference 3-dB half beamwidth.

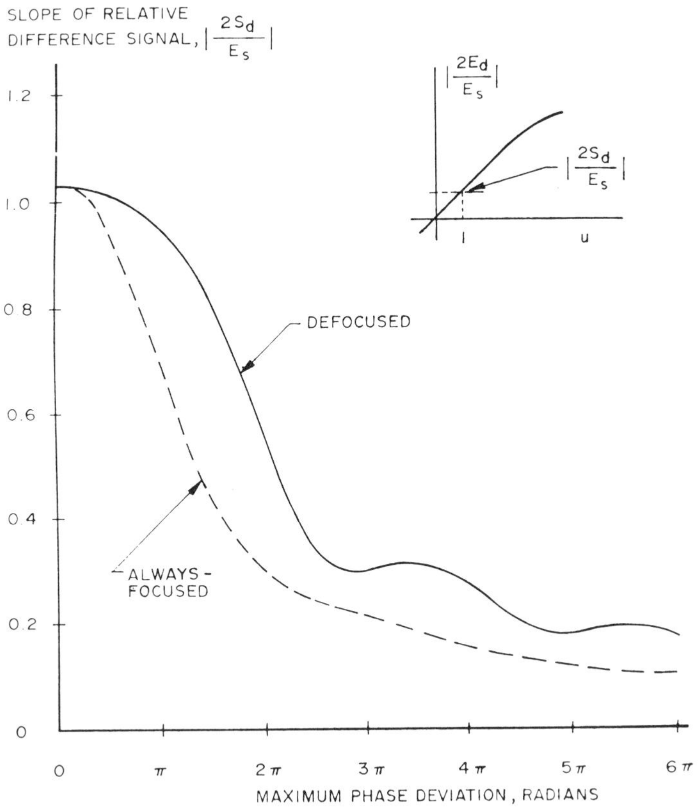

Fig. 9. Slope of relative difference signal.

In addition to the phase-angle variations with defocusing, the amplitude of the relative difference signal slope in the linear region is reduced. This reduction is inherent in variable beam systems and would also be observed in variable beam systems which are always focused. Without correction, this reduction results in a reduction of the servo loop gain when applied to automatic tracking systems. Variable gain ganged to the defocusing mechanism is one solution, with the voltage amplification being approximately proportional to beamwidth. This variable gain may be inserted between the comparator and product detector, but may also be inserted in the error channel following the product detector. More detailed information on the amplitude variation is also presented in Section VII.

VII. Summary of Significant Monopulse Parameters

This section summarizes monopulse parameters derived from the computed patterns to indicate specific values for evaluating monopulse operation. The summary includes numerical data in addition to the few examples presented in Section V. All the parameters are graphed versus the maximum phase deviation (MPD).

Fig. 7 shows the reduction in sum-mode and difference-mode gain as a function of MPD. The 0-dB reference is the gain of the focused antenna which may be computed as indicated in Fig. 7. There is a 19.8-dB reduction in gain from 0 to 6π MPD. The difference gain is 1.8 dB lower than the sum gain for the focused case and is equal to the sum gain at the 6π defocused case.

Fig. 8 shows the sum and difference mode beamwidth versus MPD, and an "always-focused" beamwidth for reference. This reference is an equivalent "always-focused" antenna whose aperture has been reduced to have the same sum-mode gain as the defocused antenna. The efficiency of the reference antenna has been assumed constant. The sum-mode beamwidth computed by geometric optics is also graphed. A similar graph for sum-mode patterns from a circular aperture is presented in DeSize and Woodard.[14] The sum-mode beamwidth for the defocused antenna increases by a factor of about 10 over the range shown in a smooth fashion. The reference beamwidth varies in almost the same manner. At 10 times the focused beamwidth ($u=18.5$), the "equivalent MPD" for the reference antenna is 6.1π which, as expected, corresponds to -19.4-dB reduction in gain, indicating approximately the same gain reduction for corresponding changes in beamwidth for the two cases. It is also interesting to note that there is only a small change in beamwidth over the first π radians of MPD, but that the gain is reduced by 3.5 dB. Presumably, in this range of defocusing, power is taken from the peak of the pattern to fill in the nulls of the defocused pattern, while maintaining constant beamwidth.

The sum-mode beamwidth computed from geometric optics is a straight line, which indicates zero beamwidth at zero MPD. This computation starts to approximate that com-

304

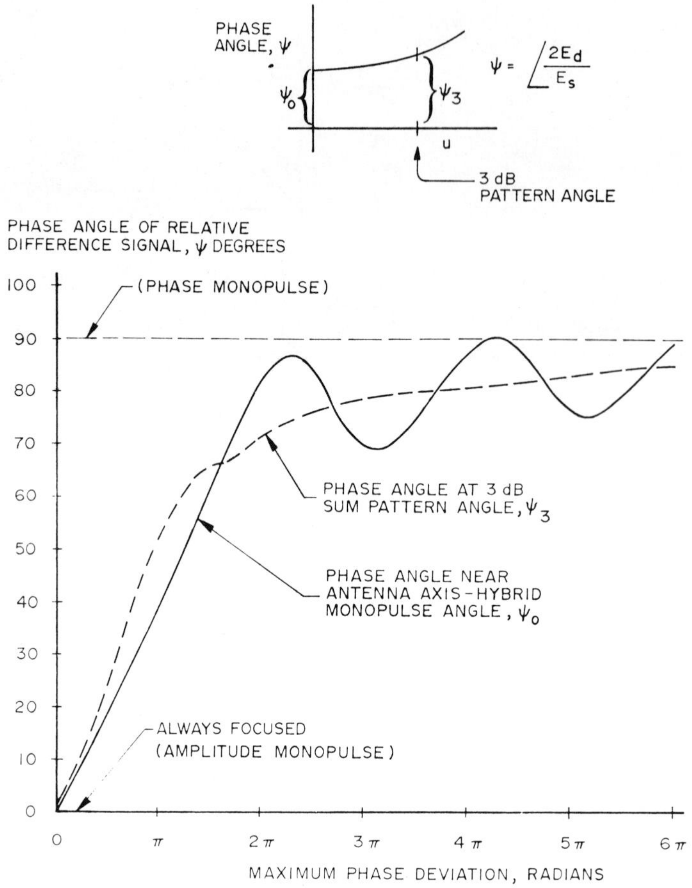

Fig. 10. Phase of the relative difference signal.

puted by diffraction roughly at an MPD of about π radians. Thus, for MPD less than π, the width of the beam is caused by diffraction, but for greater MPD, the width of the beam may be determined by geometric optics.

The difference-mode beamwidth changes by a factor of 6.5 for the 6π range of MPD. Thus, for the same defocusing, the difference-mode beamwidth widens less than the sum-mode. This nonequivalent performance for the two modes is presumably dependent on the choice of the relative sum and difference illumination shapes. However, discussion of the optimum illumination shapes is beyond the scope of this paper.

Fig. 9 shows a detailed graph of the slope of the relative difference-signal amplitude, or pattern slope factor, versus MPD, which supplements the information previously presented in Fig. 5. The pattern slope factor is reduced with increasing defocusing and exhibits periodic ripples for MPD greater than 2π. As described in Section VI, pattern slope factor reduction requires an increase in error-channel gain if the slope of the error signal is to be held constant. The slope factor of the defocused antenna is always greater than that of the equivalent focused antenna, which means that less of an increase in error-channel gain is required; this results in higher signal-to-noise ratios for the defocused antenna. (Signal-to-noise ratio depends on the product of the difference and sum signal and not on the pattern slope factor, which is a ratio. However, in Fig. 9, since the sum gain of the defocused and reference antennas are the same, the greater pattern slope factor indicates a greater defocused difference-signal gain and therefore a greater signal-to-noise ratio.)

Fig. 10 shows a detailed graph of the phase of the relative difference signal versus MPD for pattern angles near zero (hybrid monopulse angle), and for the pattern angle which is 3 dB down on the sum pattern. The hybrid monopulse angle varies from zero at focus to near 90° when defocused with periodic ripples for MPD greater than 2π; the 3-dB phase angle is seen to vary smoothly. The hybrid monopulse angle of zero indicates that the antenna operates as amplitude monopulse; the 90° angle indicates phase monopulse.[1] As discussed previously, the phase of the relative difference signal must be corrected to be near zero by the addition of a phase shifter in the difference channel. The change in phase from zero pattern angle to the 3-dB angle for any fixed amount of defocusing is limited to about 13°. If the correction is made to the average of the phase at zero and the 3-dB angle, the resulting change of phase with target angle is within $\pm 6.5°$.

To conclude, the patterns and curves presented indicate that with suitable amplitude and phase correction, defocused monopulse operation may be achieved with performance which approaches focused monopulse operation. It is believed that this quality of operation is obtained only with highly tapered aperture illuminations in both the sum and difference modes.

ACKNOWLEDGMENT

This work has been performed under the direction of M. R. Dungan at Bell Telephone Laboratories, Inc. The mathematical reduction of the antenna-pattern formulas and the programming for the computer were performed by A. M. Kay with the advice of S. P. Morgan, both of Bell Telephone Laboratories, Inc.

The work at Wheeler Laboratories was carried out by H. W. Redlien with the advice of H. A. Wheeler and P. W. Hannan.

REFERENCES

[1] H. W. Redlien, "A unified viewpoint for amplitude and phase comparison monopulse tracking radar," Wheeler Labs., Smithtown, N. Y., Rept. 845 to Bell Telephone Labs., Inc., March 1959.

[2] J. P. Shelton, "Improved feed design for amplitude monopulse radar antenna," *IRE Nat'l Conv. Rec.*, pt. I, pp. 93–102, March 1959.

[3] P. W. Hannan, "Optimum feeds for all three modes of a monopulse antenna—I: Theory, II: Practice," *IRE Trans. Antennas and Propagation*, vol. AP-9, pp. 444–461, September 1961.

[4] ——, "Microwave antennas derived from the Cassegrain telescope," *IRE Trans. Antennas and Propagation*, vol. AP-9, pp. 140–153, March 1961.

[5] P. W. Hannan and P. A. Loth, "A monopulse antenna having independent optimization of the sum and difference modes," *IRE Internat'l Conv. Rec.*, vol. 9, pt. 1, pp. 57–60, 1961.

[6] J. H. Dunn and D. D. Howard, "Precision tracking with monopulse radar," *Electronics*, vol. 33, no. 17, April 1960.

[7] H. W. Redlien, "Theory of focusing action of a microwave lens," Wheeler Labs., Smithtown, N. Y., Rept. 460A to Bell Telephone Labs., Inc., March 1952; revised August 1960 (see also ASTIA Doc. AD-59289).

[8] ——, "Theory of monopulse operation—effects of a defocused antenna," Wheeler Labs., Smithtown, N. Y., Rept. 495A to Bell Telephone Labs., Inc., August 1953; revised August 1960 (see ASTIA Doc. AD-59288).

[9] L. W. Lechtreck, "Fresnel antenna patterns," *IRE Trans. Antennas and Propagation (Communications)*, vol. AP-3, pp. 138–139, July 1955.

[10] Report by Airborne Instruments Laboratory, "Computed diffraction patterns of a circular aperture with square-law deviation and uniform, cosine and cosine-squared amplitude distribution," vols. I–III, December 1957; ASTIA Doc. AD-148 756–148 758.

[11] R. M. Page, "Monopulse radar," *IRE Conv. Rec.*, pt. 8, p. 132, 1955.

[12] D. R. Rhodes, *Introduction to Monopulse*. New York: McGraw-Hill, 1959.

[13] H. W. Redlien, "Theory of monopulse operation—introduction," Wheeler Labs., Smithtown, N. Y., Rept. 434B to Bell Telephone Labs., Inc., January 1953; revised August 1960 (see also ASTIA Doc. AD-305 996).

[14] L. K. DeSize and B. A. Woodard, "An investigation of the feasibility of obtaining a constant beamwidth Luneberg lens," *Proc. Nat'l Electronics Conf.*, vol. 15, p. 958, 1959.

[15] H. A. Wheeler, "Antenna beam patterns which retain shape with defocusing," *IRE Trans. Antennas and Propagation*, vol. AP-10, pp. 573–580, September 1962.

Monopulse Networks for Series Feeding an Array Antenna

ALFRED R. LOPEZ, MEMBER, IEEE

Abstract—**Series networks for feeding an array antenna are attractive because they can be designed compactly. A series "ladder" network is available which is particularly applicable for feeding a monopulse array antenna since it has complete independent control of the sum and difference excitations, minimum dissipative loss, and physical symmetry. This network consists of primary and secondary transmission lines feeding two sets of directional couplers that are interconnected to resemble a ladder. To obtain physical symmetry, the ladder is fed at the center by two hybrid junctions and an additional directional coupler.**

An experimental array fed by a ladder network was fabricated and tested. The design objectives of 25 dB sidelobe suppression for the sum and difference mode patterns were achieved.

I. Introduction

Since the inception of monopulse antennas, it has been a continuing goal to optimize performance in terms of antenna gain, sidelobe suppression, and tracking accuracy. At the outset, in the design of feeds for reflector antennas, it became clear that physical symmetry, minimum dissipative loss, and independent control of the required sum and difference mode aperture excitations were desirable characteristics.[1] During the development of electronically scanned array antennas, some of the previously developed techniques were applied directly to the design of networks for feeding a monopulse array antenna. Series-feed networks appeared attractive because they could be designed compactly;[2],[3] unfortunately, complete independent control of the excitations did not appear to be simply achievable. Recently,[4] a series network has been described which has complete independent control of the aperture excitations, but lacks physical symmetry and has an inherent dissipative loss. The following paper describes a series-feed network which possesses all three of the desired monopulse characteristics.

II. Series-Feed Networks

A series feed for an array antenna is one in which power to each radiating element is tapped off sequentially from a main feed line. A major feature with respect to a compact design is that, once the basic series-feed configuration has been established, continued growth in the number of array elements will not increase the thickness of the feed network. In this paper, feeds for a linear array will be discussed; the principles may be easily extended to a planar array.

Fig. 1(a) is a sketch of the simplest form of a series-feed network for a monopulse antenna. It is referred to as a "two-module" feed since it is divided into two sections; each consists of a transmission line feeding a set of power dividers. A hybrid junction at the center of the network generates the sum and difference excitations. Directional couplers are utilized as power dividers, with the unused ports terminated; this minimizes the generation of spurious radiation lobes caused by reflections from the radiating elements. This network is ideally lossless (this assumes ideal directional couplers, an ideal hybrid junction, and lossless transmission lines). It also has physical symmetry. However, it lacks any degree of independent control of the sum and difference excitations. That is, if the couplers are set for a good sum excitation, the difference excitation is necessarily identical to the sum in amplitude, as shown in the figure.

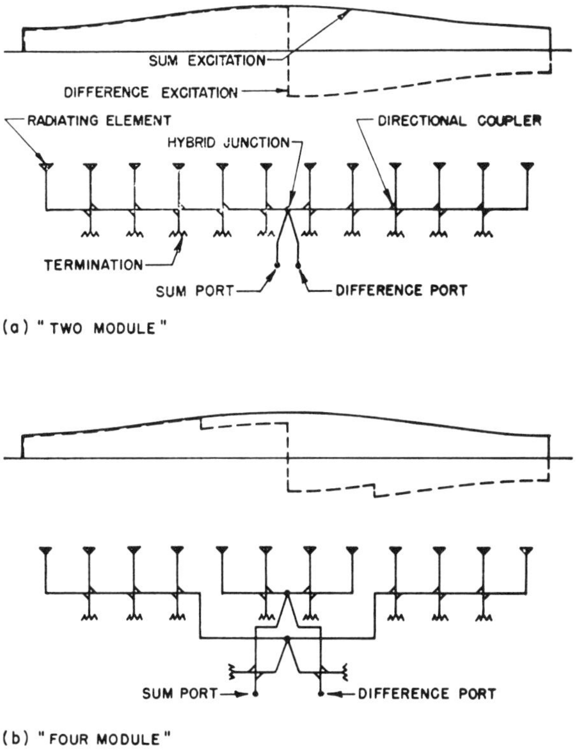

Fig. 1. Monopulse series-feed networks.

Some independent control is achieved by the network shown in Fig. 1(b). This network consists of four modules fed by two hybrid junctions at the center. Controlling the ratio of the inputs to the sum and difference ports of the hybrid junctions permits a limited degree of independence.

Manuscript received December 28, 1967; revised March 18, 1968. This work was done under subcontract to the Raytheon Company.

The author is with Wheeler Laboratories, Inc., Smithtown, N. Y.

Extending this process so that the number of modules equals the number of elements yields complete mode independence,[1] but the series feed then becomes actually a complex parallel feed.

Another class of series-feed networks are available which provide complete independent control. These are referred to here as "ladder" networks, and are described and analyzed below.

III. Analysis of Ladder Networks

The basic ladder network consists of primary and secondary transmission lines feeding two sets of directional couplers, as shown in Fig. 2(b). The primary-line and secondary-line directional couplers are interconnected by short sections of transmission lines to resemble a ladder. The primary-line coupler outputs are fed to the radiating elements while the unused ports of the secondary-line couplers are terminated. Two types of ladder networks are described and analyzed below; but first, some general principles are discussed.

A. Proof of Excitation Independence

The following theorem for ladder networks is now proved by induction.

Theorem: If it is assumed that lossless transmission lines and ideal directional couplers are utilized, then ladder networks can be designed that are lossless and provide two independently specified excitations of the elements of an array antenna.

Consider first the simple network shown in Fig. 2(a), consisting of two radiating elements and a single ideal directional coupler. An ideal directional coupler is defined so that the coupler outputs, B_n, are related to the coupler inputs, A_n, by the following matrix relationships:

$$\begin{bmatrix} B_1 \\ B_2 \\ B_3 \\ B_4 \end{bmatrix} = \begin{bmatrix} c & \sqrt{1-c^2} & 0 & 0 \\ \sqrt{1-c^2} & -c & 0 & 0 \\ 0 & 0 & -c & \sqrt{1-c^2} \\ 0 & 0 & \sqrt{1-c^2} & c \end{bmatrix} \cdot \begin{bmatrix} A_1 \\ A_2 \\ A_3 \\ A_4 \end{bmatrix} \tag{1}$$

$$\begin{bmatrix} A_1 \\ A_2 \\ A_3 \\ A_4 \end{bmatrix} = \begin{bmatrix} c & \sqrt{1-c^2} & 0 & 0 \\ \sqrt{1-c^2} & -c & 0 & 0 \\ 0 & 0 & -c & \sqrt{1-c^2} \\ 0 & 0 & \sqrt{1-c^2} & c \end{bmatrix} \cdot \begin{bmatrix} B_1 \\ B_2 \\ B_3 \\ B_4 \end{bmatrix}. \tag{2}$$

Let B_1 and B_2 be one desired excitation ($B_3 = B_4 = 0$) of the two-element array for a signal applied to the first input port.

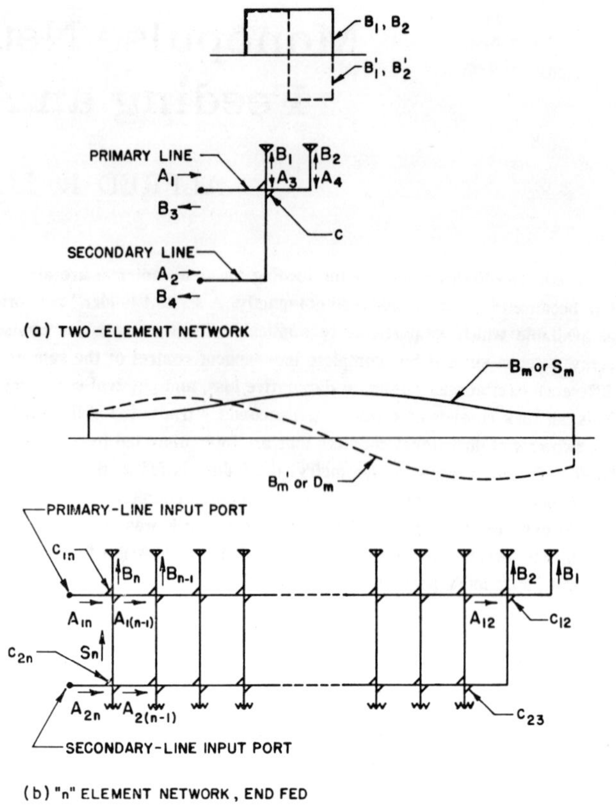

Fig. 2. Basic "ladder" networks.

Then in (1), A_2, A_3, and A_4 are equal to zero and

$$B_1 = cA_1 \tag{3}$$

$$B_2 = \sqrt{1-c^2}\, A_1. \tag{4}$$

With these two equations it is possible to solve for the two unknowns, A_1 and c, to provide the first excitation. Let B_1' and B_2' be a second desired excitation ($B_3' = B_4' = 0$) for *both* A_1' and A_2' applied to the input ports ($A_3' = A_4' = 0$). Equation (2) and c as just determined gives

$$A_1' = cB_1' + \sqrt{1-c^2}\, B_2' \tag{5}$$

$$A_2' = \sqrt{1-c^2}\, B_1' - cB_2'. \tag{6}$$

Thus, A_1' and A_2' are determined and provide the second desired excitation. *Therefore, setting the coupling value c and feeding one input port gives one excitation; providing the proper relative inputs to the coupler gives the other desired excitation.*

It is now possible to prove that if the theorem is true for a two-element array network, it is also true for a three-element array network and, in general, if it is true for an $(n-1)$-element network, it is also true for an n-element network. Fig. 2(b) shows a general n-element array network consisting of primary- and secondary-line directional couplers. Let B_m ($m = 1, 2, \cdots, n$) be one desired excitation of the elements produced by A_{1n}, the input to the primary line.

As now assumed, B_{m-1} is excited by $A_{1(n-1)}$, the primary-line input to the $(n-1)$ primary-line coupler. Equations (3) and (4) are used, as before, to determine c_{1n} and A_{1n} to produce B_n, the desired excitation of the nth element, and $A_{1(n-1)}$. Consequently, B_m, the complete desired excitation is obtained.

Let B_m' be a second desired excitation produced by A_{1n}' and A_{2n}', the inputs to the primary and secondary lines, respectively. As now assumed, B_{m-1}' is excited by $A_{1(n-1)}'$ and $A_{(2n-1)}'$. Equations (5) and (6) are used to determine A_{1n}' and S_n, since c_{1n}, B_n' and $A_{1(n-1)}'$ are known, and (3) and (4) are used to determine A_{2n}' and c_{2n}, since S_n and $A_{2(n-1)}'$ are known. Thus A_{1i}' and A_{2i}' will excite the complete desired excitation B_m'. Since ideal couplers are used, no power goes into the terminations when the array is perfectly matched; therefore, there is no loss in the network. This completes the proof of the theorem.

B. *Orthogonality Condition*

If a multi-port network is passive, reciprocal, and lossless, the output and input voltages are related by an orthogonality condition.[2],[5],[6] This condition is utilized in the following analysis of ladder networks, and is derived and expressed in a form which is more directly applicable to this type of network.

Let A_1 and A_2 be the inputs to the primary and secondary lines, and B_{1m} and B_{2m} be their respective excitations of the elements. Since the input ports and terminated ports are decoupled (transfer coefficient$=0$) and the excitations are real (phase at each element port$=0°$ or $180°$), then conservation of energy gives

$$A_1^2 = \sum_{m=1}^{n} B_{1m}^2 \tag{7}$$

$$A_2^2 = \sum_{m=1}^{n} B_{2m}^2. \tag{8}$$

If A_1 and A_2 are simultaneous inputs, then

$$A_1^2 + A_2^2 = \sum_{m=1}^{n} (B_{1m} + B_{2m})^2. \tag{9}$$

Combining these three equations results in

$$\sum_{m=1}^{n} B_{1m}B_{2m} = 0 \tag{10}$$

which is the orthogonality condition. Thus, *two excitation input ports are decoupled if and only if the corresponding excitations they produce are orthogonal.*

C. *"End-Fed" Ladder Network*

Fig. 2(b) shows the general "end-fed" ladder network for series feeding a monopulse array antenna. Also shown in the figure are the envelopes for typical sum and difference mode excitations. These excitations, S_m and D_m, are specified independently of each other. The network can be designed so that an input S to the primary line excites S_m. As indicated in the preceding discussion, the network can be designed to provide a second independent excitation; in this case, D_m. Since S_m and D_m are inherently orthogonal, the ports that generate the excitations must be decoupled. Thus, the D_m excitation is provided by feeding an input D to the secondary line only.

This network, even though it is ideally lossless and has independent control of the sum and difference modes, lacks the physical symmetry which is desirable for monopulse operation. The next network to be described does not have this shortcoming.

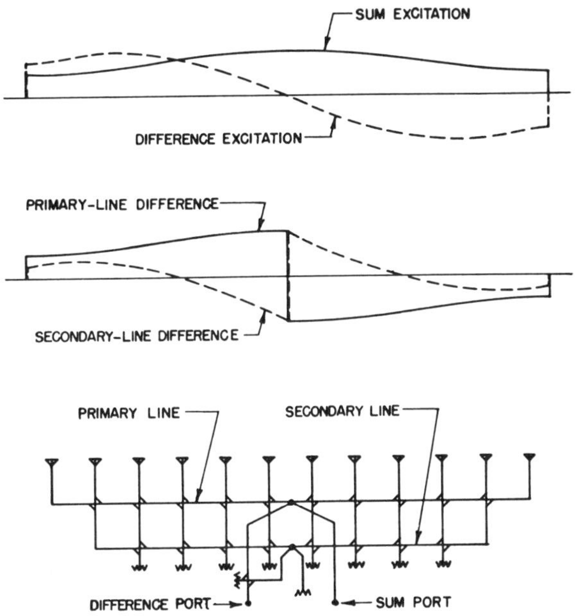

Fig. 3. "Center-fed" ladder network.

D. *"Center-Fed" Ladder Network*

Fig. 3 shows a "center-fed" ladder network for series feeding a monopulse array antenna. This network consists of two end-fed ladder networks fed at the center by two hybrid junctions and an additional directional coupler. Also shown in the figure are typical sum and difference excitations. The sum mode excitation is again provided by setting the primary-line couplers and feeding an input to the primary line. It is noted that the sum and difference excitations are *not* orthogonal in the region between the centerline and the edge of the network. Thus, to excite the difference mode, it is necessary to simultaneously feed the primary and secondary input ports. Since the input ports are inherently decoupled, feeding only the secondary line would produce an orthogonal excitation. The required combination of signals to the input ports is achieved by means of the additional directional coupler feeding the required inputs to the difference ports of the primary and secondary hybrid junctions. Fig. 3 shows the difference excitations produced by inputs to the difference ports of the primary- and secondary-line hybrid junctions. It is noted that these component excitations are orthogonal and that their combination results in the desired difference pattern.

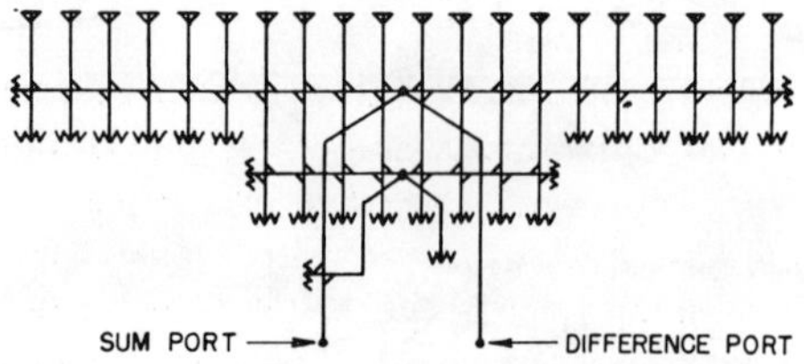

Fig. 4. Experimental array and modified "center-fed" ladder network.

Fig. 5. Photograph of experimental array and modified "center-fed" ladder network.

The "center-fed" ladder network thus has all three desirable monopulse characteristics: 1) it provides complete independent control of the excitations, 2) it is ideally lossless, and 3) it has physical symmetry.

IV. Experimental Array

To verify some of the characteristics of a "center-fed" ladder network, and to gain some insight into its operation, an experimental 20-element array was built and tested. The design objective was 25 dB sidelobe suppression for both the sum and difference modes.

Fig. 4 is a sketch of the network. It is noted that less than half of the available secondary-line directional couplers were actually utilized. For 25 dB sidelobe suppression, the outer portions of the sum and difference excitations are nearly identical; consequently, the objective can be achieved without the outer section of the secondary line. Note also that the difference mode, instead of the sum mode, is excited by the primary line. This is an alternative arrangement that favors the difference pattern over the sum pattern. It is also noted that the primary and secondary lines are terminated by a lossy termination instead of being terminated by the last element. In the short time available for the development of the array, it was impractical to develop the near 3 dB directional couplers required for the ideally lossless design.

Fig. 5 is a photograph of the experimental array. The network utilizes waveguide transmission lines and cross-guide directional couplers spaced one guide-wavelength apart. The radiating elements are waveguide horns. A phase shifter is included to control the relative phase between the primary and secondary lines. Figs. 6 and 7 are the measured sum and

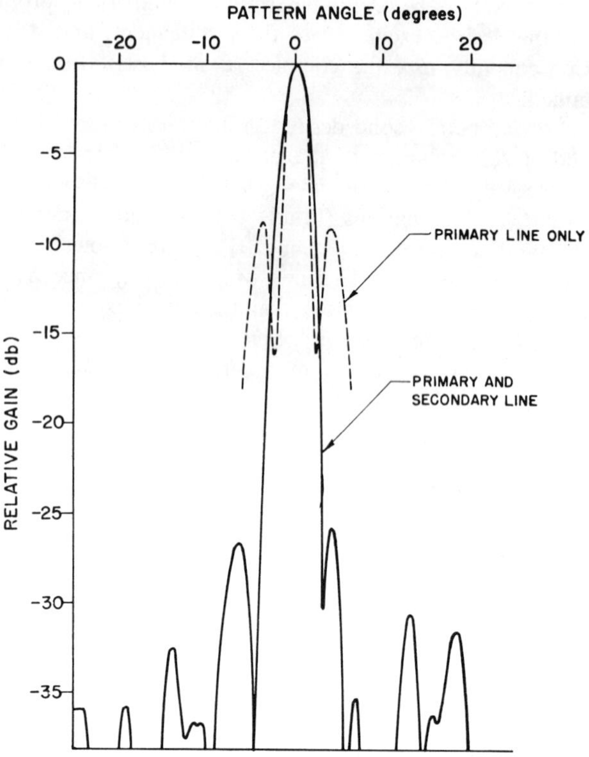

Fig. 6. Sum pattern of experimental array and modified "center-fed" ladder network.

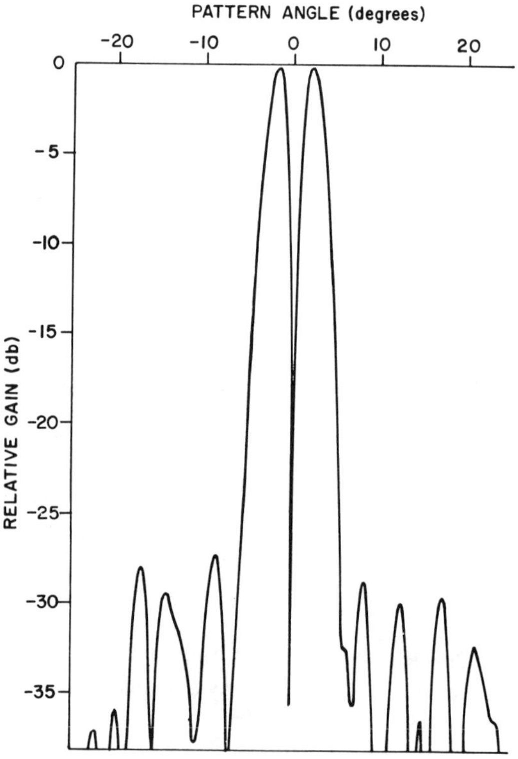

Fig. 7. Difference pattern of experimental array and modified "center-fed" ladder network.

difference patterns. Fig. 6 also shows the central portion of the sum pattern that is obtained when only the primary line is fed in the sum mode. As expected, a significant reduction in the sidelobe level is observed when both the primary and secondary lines are fed together with the proper relative amplitude and phase. The measured difference pattern is seen to have a deep null at the center; this was achieved without any adjustment of the network. In both the sum and difference modes, the 25 dB sidelobe-suppression objective was achieved.

V. Discussion

It has been shown that the "center-fed" ladder network is a direct way of designing a series feed for a monopulse array antenna when it is desirable to have complete independence of the sum and difference mode excitations, ideally no dissipative loss, and physical symmetry. As is characteristic of series networks, it can be extended indefinitely to feed more elements without increasing its thickness.

With respect to operating bandwidth, it is noted that, at any radiating element port, the path length and, consequently, the phase of signal components arriving via various paths are identical for all signal components. Thus, for any frequency in the operating bandwidth, the amplitude excitation of the elements is dependent only on the bandwidth characteristics of the directional couplers and hybrid junctions. In practice, the network components introduce some path inequalities; these, however, can easily be adjusted to be within tolerable limits.

With respect to instantaneous bandwidth, it is noted that, at any radiating element port, the path length and, consequently, the time delay of signal components arriving via various paths are identical for all signal components. Thus, the instantaneous bandwidth characteristics are the same as would be expected for a simple single-line series feed. It is also noted that, for broadside scan, the differential time delay for a "center-fed" ladder network is half that of an "end-fed" ladder network.

The principles of a ladder network can be extended to more than two main transmission lines. If a third line is utilized, three independently specified excitations can be obtained; the first by setting the primary-line couplers and providing an input to the primary line; the second, by setting the secondary-line couplers and feeding the proper relative inputs to the primary and secondary lines; and the third by setting the trinary-line couplers and feeding the proper relative inputs to the primary, secondary, and trinary lines. For an n element array, the procedure can be extended to provide n independently specified excitations.

Acknowledgment

The author wishes to express his gratitude to G. D. M. Peeler of the Raytheon Company for his contributions and encouragement toward the completion of this work. At Wheeler Laboratories, P. W. Hannan, R. J. Hanratty, R. A. Lodwig, R. J. Giannini and P. J. Sroka also contributed significantly to the work. 311

References

[1] P. W. Hannan, "Optimum feeds for all three modes of a monopulse antenna: I—Theory; II—Practice, "*IRE Trans. Antennas and Propagation*, vol. AP-9, pp. 444–461, September 1961.

[2] R. C. Hansen, *Microwave Scanning Antennas*, vol. 3. New York: Academic Press, 1966, p. 6 and p. 258.

[3] J. Blass, "Multidirectional antenna—A new approach to stacked beams," *1960 IRE Nat'l Conv. Rec.*, pt. 1, pp. 48–50, 1960.

[4] J. D'Agostino and R. J. Timms, "Three dimensional lightweight transportable surveillance antenna," *IEEE G-AP Internat'l Symp. Digest*, pp. 454–462, December 1966. (Utilizes an "end-fed" ladder network with terminations on the ends of the main feed lines.)

[5] N. Marcuvitz, *Waveguide Handbook*, vol. 10, M.I.T. Rad Lab. Series. New York: McGraw-Hill, 1951, p. 108.

[6] J. L. Allen, "A theoretical limitation on the formation of lossless multiple beams in linear arrays," *IRE Trans. Antennas and Propagation*, vol. AP-9, pp. 350–352, July 1961.

[7] A. R. Lopez, "Monopulse networks for series feeding an array antenna," *IEEE G-AP Internat'l Symp. Digest*, pp. 93–101, October 1967.

CONTOUR PATTERN ANALYSIS OF A MONOPULSE RADAR CASSEGRAINIAN ANTENNA

DEAN D. HOWARD, Tracking Branch Radar Division
Naval Research Laboratory
Washington, D.C.

INTRODUCTION

Contour mapping of antenna patterns is a well known technique but it is seldom used because of the time and labor involved in their preparation. However, the contour presentations provide complete information on the characteristics of an antenna pattern. This includes the off-axis regions that are not shown by the conventional azimuth and elevation patterns and which are rarely even measured. The information in the off-axis regions is necessary for fully evaluating the antenna performance particularly in the case of monopulse radar with its complex set of reference and error sensing patterns. They also provide an overall picture of the phenomena of highly directive antennas and monopulse angle error sensing.

As a part of the evaluation of the AN/FPQ-6 monopulse instrumentation tracking radar[1] shown in Fig. 1, contour patterns of the antenna were prepared for the main lobe and their near-in sidelobes. The antenna is a 29 foot diameter Cassegrainian system with an f/d ratio of 0.3 and a 30 inch diameter sub-reflector. The evaluation was performed by RCA, Moorestown, N.J. under Contract Now-61-0428d supported by the Navy and NASA (Goddard Space Flight Center and Wallops Island Station) to establish the basic limitations and precision of the tracking capability of this radar. The contours and associated 3-D (three-dimensional) drawings were prepared by the Naval Research Laboratory as a part of the evaluation to provide both qualitative and quantitative data relating to radar performance as described below.

Fig. 1 — Monopulse instrumentation radar set AN/FPQ-6 installation at NASA Station, Wallops Island, Virginia.

AN/FPQ-6 RADAR ANTENNA AND ITS CONTOUR PATTERNS AND 3-D DRAWINGS

The antenna of the C-band (5.4 to 5.9 gc) AN/FPQ-6 radar, Fig. 1, consists of a five-horn monopulse, Cassegrainian feed and a 29-foot-diameter solid-surface paraboloidal reflector giving a nominal gain of 51-dB. The sub-reflector of the Cassegrainian feed is a convex hyperboloid 30 inches in diameter on an adjustable mount. It is supported by three spars which extend forward from the substructure to form an apex on the paraboloid axis. The Cassegrainian approach exploits the principle of folded optics, attaining an effective focal-length-to-diameter ratio of 1.59 although the paraboloid f/d ratio is only 0.3. The five-horn feed is located in the end of the "ice cream cone" projection from the center of the 29-foot paraboloid (Fig. 1) and is covered by a small radome. This system transmits and receives either linearly or circularly polarized signals (the choice is made at the console) and was the first monopulse feed designed for this purpose. The central horn of the feed is used for transmission and develops the RF reference signal upon reception. The two pairs of passive error horns are arranged symmetrically about the reference/transmission aperture and create the RF azimuth and elevation angle-tracking signals.

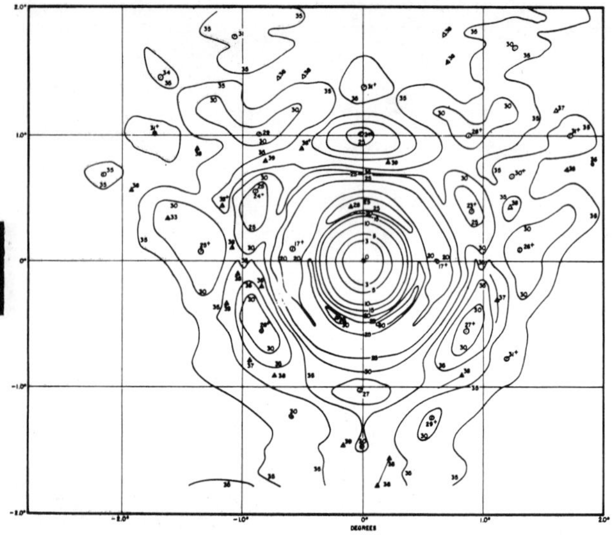

Fig. 2 — Monopulse antenna reference signal contour pattern where each line represents a constant signal level contour of the level indicated, in dB below the main lobe peak.

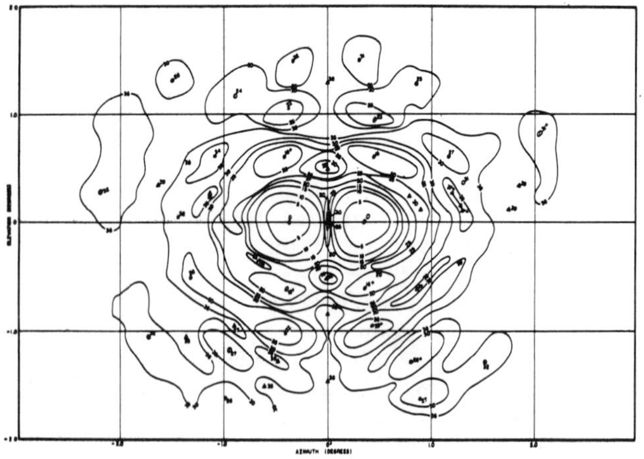

Fig. 3 — Monopulse antenna difference signal contour plot where each line represents a constant signal level contour of the level indicated, in dB below the peak of the major lobes.

The patterns and drawings shown in Figs. 2-6 were taken with vertical linear polarization. Figures 2 and 3 are contour plots which are essentially a topographical map where the elevation lines represent constant signal level contours designated in dB measured with respect to the pattern peak. Figure 2 is the contour plot of the monopulse reference signal pattern which, for this antenna, is the output of the central horn of the five-horn monopulse feed cluster. Figure 3 is the contour plot of the monopulse azimuth difference signal which is a signal obtained by subtracting the outputs of the horns on either side of the center horn in the five horn cluster. The elevation difference signal, not shown, is similarly obtained by subtracting the outputs of the horn above and the horn below the central horn.

The difference signals are used for sensing angle error, being zero amplitude on-axis and increasing with increasing displacement of the signal source from the antenna axis. The two main peaks of the difference signal have a 180° relative phase which is used in the angle error detector to determine the direction of the error. At I-F, when the difference signal is compared to the reference signal in a phase detector, a video output is obtained which is proportional to error in the region near the antenna axis and has a polarity corresponding to the direction of the error from the antenna axis.[2]

Figures 4 and 5 are 3-D drawings of the contour plots of Figs. 2 and 3 respectively. The three dimensions are azimuth angle and elevation angle in the horizontal plane and the vertical axis represents amplitude with a dB scale.

TECHNIQUE FOR PREPARING CONTOUR PLOTS AND 3-D DRAWINGS

The conventional technique was used for preparing the contour plots. Azimuth patterns were taken at incremental steps of two milliradians through the region of interest. Each pattern corresponds to a horizontal line on the contour plots. On each of these lines a point was placed corresponding to the azimuth angle where the pattern crossed a given level such as the 3-dB level, for example. The 3-dB contour is then constructed by drawing connecting lines between adjacent 3-dB points.

To prepare the 3-D drawings of the patterns the general outline of the figure was obtained from the contour plots. Each contour level, such as the 3-dB, 5-dB and 10-dB levels, were plotted on separate transparent sheets of drawing paper. Each sheet is then displaced along a vertical axis by an amount proportional to a linear scale in dB so that 0-dB falls at the top, the 3-dB contour displaced 3 units down, the 5-dB contour displaced 5 units down, etc. This procedure for arrangement of the transparent drawings of each contour level produces an outline of the 3-D shape of the antenna pattern. The completion of the drawing is mainly artistic effects of appropriate shading to emphasize the three dimensional shape of the pattern.

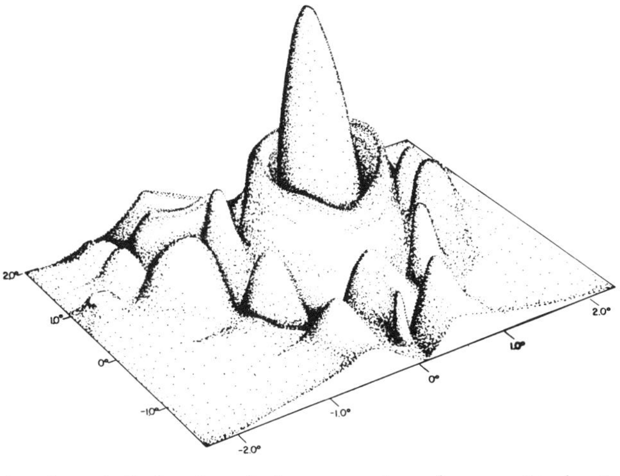

Fig. 4 — 3-D drawing of the monopulse reference signal pattern described in Fig. 2 where the vertical scale is proportional to dB.

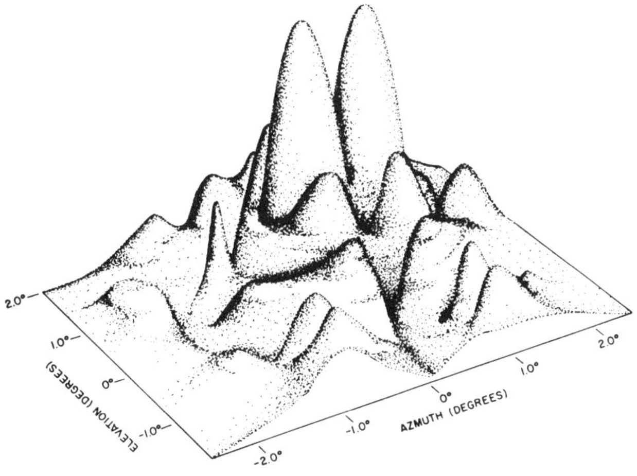

Fig. 5 — 3-D drawing of the monopulse difference signal pattern described in Fig. 3 where the vertical scale is proportional to dB.

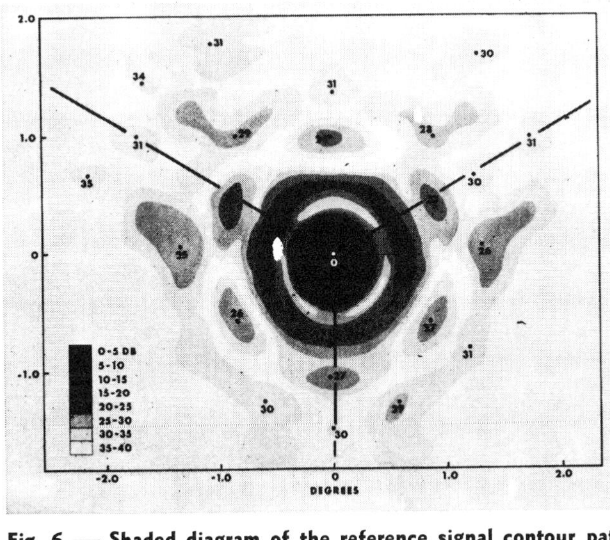

Fig. 6 — Shaded diagram of the reference signal contour patterns of Fig. 2 which emphasizes the degree of symmetry of the high and low gain regions.

USE OF THE ANTENNA PATTERN CONTOUR AND 3-D DRAWINGS

These presentations of an antenna beam provide both qualitative and quantitative information that is difficult to obtain from conventional antenna paterns. Examples of the uses of the contour and 3-D antenna pattern are: (1) As a tutorial aid — The 3-D drawings provide a three dimension picture of the antenna beam in space which may be difficult to visualize from conventional major axes pattern. It also demonstrates the off-major-axis tracking limitations of tracking radars showing the loss of angle tracking sensitivity (related to the slope near the null between peaks, Fig. 5) with increasing off-axis angle. (2) Measure of degree of circularity of antenna patterns — Antenna pattern integration techniques frequently assume circular contour patterns. The pattern of Fig. 2 generally verifies good circularity of the main lobe and major sidelobe of this antenna structure. The gain of this antenna was calculated by pattern integration techniques assuming circular symmetry further verifying the accuracy of the assumption by a calculated value of 51.1-dB compared with a measured gain of 51.2-dB[1] averaged over the full 5.4 to 5.9 GHz band. (3) Effects of the hyperbola tripod support — The contours show negligible effect of the tripod structure in the significant areas of the antenna pattern; however, the effects of the tripod can be observed in the low signal level regions. Figure 6 shows a grey level presentation of the contour pattern which emphasizes the effects of the tripod which is located along the 3 radial lines drawn on the figure. (4) Multi-path errors — Severe angle tracking errors occur for low elevation angle targets because of the reflection of RF energy from the earth or ocean surface. This error may be calculated from a known geometry and a knowledge of the antenna gain at the angle of the image of the target reflected from the surface.[3] Over land paths, the reflected image of the target can be off the major antenna axes and contour patterns are necessary to determine the values for antenna gain in the direction of the target image that are necessary for calculating the multi-path error.

SUMMARY

Antenna pattern contour and 3-D drawings are given for a modern Cassegrainian precision monopulse tracking radar. These presentations of antenna performance give needed qualitative and quantitative data for evaluation of the radar performance as well as establishing general characteristics to be expected from the modern Cassegrainian antenna.

REFERENCES

1. Mitchell, R., et al, "Final Report, Measurements and Analysis of Performance of MIPIR (Missile Precision Instrumentation Radar Set AN/FPQ-6)". Under Navy Contract NOw 61-0428d to RCA, Missile and Surface Radar Division, Moorestown, N.J., December 1964.
2. Dunn, J. H. and D. D. Howard, "Precision Tracking with Monopulse Radar", *Electronics*, April 22, 1960.
3. Barton, D. K., *Radar Systems Analysis*, Prentice-Hall, Inc., Englewood Cliffs, N.J., 1964, pp. 327-331.

Complex Indicated Angles Applied to Unresolved Radar Targets and Multipath

SAMUEL M. SHERMAN, Senior Member, IEEE
RCA, Government and Commercial Systems
Moorestown, N.J. 08057

Abstract

The off-axis angle indicated by a conventional monopulse radar is only the real part of a "complex indicated angle." The presence of unresolved targets or multipath distorts the real part (causing an erroneous angle indication) and also produces an imaginary part, which can easily be measured by processing the normally unused quadrature-phase component of the difference signal. Under certain conditions the angles, amplitude ratio, and relative phase of two unresolved targets can theoretically be determined by measurements of the complex indicated angle on two pulses separated by a short interval. In the special case of multipath, known relationships between the unresolved target and image theoretically permit determination of target elevation with a single pulse.

Manuscript received October 5, 1970.

I. Introduction

Monopulse is a widely used technique for measuring or tracking the direction of arrival of a radio wave from a radiating source or a radar target. This paper is concerned with radar applications, but some of the results to be presented are pertinent to other applications as well. The monopulse technique is highly developed, well understood, and capable of accurate angular measurements of a single target.

It is well known, however, that the presence of two or more unresolved targets or a single target with multipath causes the indicated direction to wander, sometimes far beyond the angular interval subtended by the targets. Over the years many approaches have been proposed to combat this problem, and some have been tried, without great success. Most of them have been "fixes" rather than true solutions; that is, they have attempted to reduce the wander and oscillation by techniques such as filtering, inhibiting the servo response to "wild" error signals, etc.

More recently there has been increased interest in the possibility of a true solution, particularly for the case of two unresolved targets. By a solution is meant determination of the angular locations of the two targets, with their amplitude ratio and relative phase as by-products. Methods have been proposed which theoretically could derive a solution from a single pulse, but they require special antenna configurations, more complicated than those normally used in monopulse radars. The practicality and accuracy of these methods have yet to be demonstrated.

In the investigation reported in this paper, the aim has been to see what can be done to achieve a solution by moderate additions to monopulse radars of present design, without modifying or redesigning their beam patterns, and without affecting the normal single-target mode of operation. Another ground rule has been that literal monopulse, that is, solution with only one pulse, is desirable but not necessary, since time is usually available for more than one pulse.

The key to the approach is the use of the component of the normalized difference signal that is in phase quadrature with the sum signal. In conventional monopulse radars only the in-phase component is used. The quadrature-phase component can readily be measured in addition. The measurements of the two components can be neatly combined in a quantity which has been named the complex indicated angle, of which conventional radars measure only the real part. Expression of the indicated angle as a complex quantity simplifies the analysis in some respects, gives better insight into the problem, and offers the possibility of solutions.

It will be shown that for two independent unresolved targets, measurements taken on two pulses separated by a short time interval can yield a solution, provided certain assumed conditions are met.

In the important special case of multipath, in which a low-angle target and its image below the surface of the earth are equivalent to two unresolved targets, known relationships between target and image permit a simpler, single-pulse solution for the target elevation angle.

Errors and practical considerations of both types of solution will be discussed.

The content of this paper has been selected and adapted from a much lengthier source [1], not as readily available to the average reader. A brief condensation [2] is available, but the present paper includes steps and additional material omitted from [2] because of space limitations, as well as minor revisions and updating. A companion paper by Peebles and Goldman [3] in this issue presents results of simulation of the multipath solution described here.

II. Review of Single-Target Monopulse Operation

It is assumed that the reader has some familiarity with the principles, design, and performance of monopulse radars. References [4] to [7], among others, contain much information on the subject. Only enough will be reviewed here to serve as a foundation for analysis of the two-target problem.

The analysis will be presented in terms of a steerable-reflector type of antenna. The Appendix shows how the results are easily adapted to phased arrays, the main difference in the latter case being the use of direction cosines instead of angles.

A typical sum-and-difference monopulse radar produces the following signal voltages:

S = sum signal
D_{tr} = traverse difference signal
D_{el} = elevation difference signal.

Definitions of traverse and elevation are included in the Appendix.

In this paper, S, D_{tr}, and D_{el} are complex numbers representing phasors. Depending on the type of monopulse configuration (e.g., amplitude or phase comparison), on the RF combiners, and on other details of the radar design, D_{tr} and D_{el} in their "raw" form may have some fixed phase other than 0° or 180° relative to S. In that case it is customary to insert a phase correction so that for a single point target D_{tr} and D_{el} do have 0° or 180° phase relative to S at the product detector (to be discussed later). The convention followed in this paper is that D_{tr} and D_{el} are defined after this phase correction.

Suppose there is a point target at traverse angle A_{tr} and elevation angle A_{el} measured from the beam axis. The radar is usually so designed that to a close approximation out to one-half beamwidth or more from the axis,

$$D/S = pA, \tag{1}$$

where p is an angle sensitivity coefficient. The subscript "tr" or "el" is to be inserted after D and A, and also after p, if necessary, since p_{tr} and p_{el} need not be the same. Wherever these subscripts are omitted, it is to be understood that the equation or discussion holds for each angle component individually.

The left-hand side of (1) is called the normalized difference signal or normalized error signal. It is used either as the error input to an angle servo for closed-loop tracking, or as an open-loop indication of target direction. Normalization is necessary to make the angle indication independent of target strength.

Dividing (1) through by p yields

$$\frac{1}{p}\frac{D}{S} = A, \tag{2}$$

which is a more convenient form for the purpose of this paper. The quantity $(1/p)(D/S)$ on the left-hand side of (2) is the measured quantity, which will be referred to as the indicated angle, while A is the geometric off-axis angle.

To allow for cases where the angles are outside the linear range expressed by (2), the relationship between the indicated angle and the true off-axis angle can be generalized by the equation

$$\frac{1}{p}\frac{D}{S} = A' = f(A), \tag{3}$$

in which A' can be thought of as a fictitious geometric angle equal by definition to the indicated angle, and $f(A)$ is a known real function of the actual off-axis angle. The constant p in this case is the slope of D/S versus A when $A = 0$. Then A' is practically equal to A for small angles, while for larger angles a known calibration correction can be applied to convert one to the other.

In this paper the analysis will be expressed in terms of the assumed linear relationship (2) because the simplified model aids physical interpretation. It is emphasized, however, that the results that will be derived are easily adapted to the nonlinear case by the linearizing conversion expressed by (3). The specific manner of applying the conversion, when needed, will be described in Sections III and IV in connection with the solutions developed there for two unresolved targets and for multipath, respectively.

III. Two Independent Unresolved Targets

A. Phase Relationships

In the single-target case, the ratio D/S in (1) is real, since D and S, by definition, are either in phase or 180° out of phase. In the case of unresolved targets, however, D and S may have any relative phase, and their ratio is therefore complex, in general.

Consider an example of two independent[1] unresolved targets. Let A, with appropriate subscripts, denote the angle components of the first target, as in the preceding

[1] As distinguished from the multipath problem, to be discussed in Section IV.

section. Now let B, with subscripts, denote the angle components of the second target. Though the two targets are in the same range resolution cell by definition, in general there will be some phase difference between them. Let S_A and S_B be their individual sum-signal phasors, as shown in Fig. 1. Then the total sum signal is the resultant S.[2] Confine attention to one angle component, say traverse, and suppose the two targets lie on opposite sides of the beam axis, the first target being on the side that makes its individual normalized traverse difference signal positive. Then the difference signal D_A is in phase with S_A, while D_B is in phase opposition to S_B. The total difference signal D is their resultant. The subscript "tr" is to be understood after D_A, D_B, and D. It is clear from the figure that, in general, D has a component in phase quadrature with S, as well as an in-phase component. In other words, the ratio D/S is complex. It is easy to see also that if S_A and S_B are 180° out of phase and nearly equal in magnitude, the ratio D/S can become very large.

The foregoing result is not confined to cases where the two targets lie on opposite sides of the beam axis. It is quite general.

Conventional monopulse radars measure only the real part of the complex ratio D/S. For example, one method that is used is to derive an AGC voltage from the sum channel and apply it to both the sum and difference channels, producing voltages $S/|S|$ and $D/|S|$. These voltages are applied as inputs to a product detector or phase detector, the output of which is a bipolar signal equal (except for a scale factor) to the product of the amplitudes of the two inputs and the cosine of their relative phase angle. Expressed another way, the output is the component of D that is in phase with S, divided by $|S|$, and this is equivalent to the real part of D/S. To measure the imaginary part would require only the addition of a second product detector and associated circuitry for each angle component, with either the sum or difference signal input phase-shifted by 90°.

Other radars achieve the normalization not by AGC, but by measuring pulse amplitudes and performing division in an associated computer or digital processor. This method is superior to AGC in that it gives pulse-by-pulse normalization and avoids errors due to time lag and limited dynamic range of AGC. In cases where the radar system computes the real part of the complex ratio D/S, the imaginary part can easily be computed in a corresponding manner. Some radars, however, measure only the *magnitude* of the ratio D/S, and the *sign* of the real part. Such radars would require more extensive modification to produce the real and imaginary parts of D/S.

It is often erroneously believed that from a single-pulse measurement a radar cannot tell the difference between

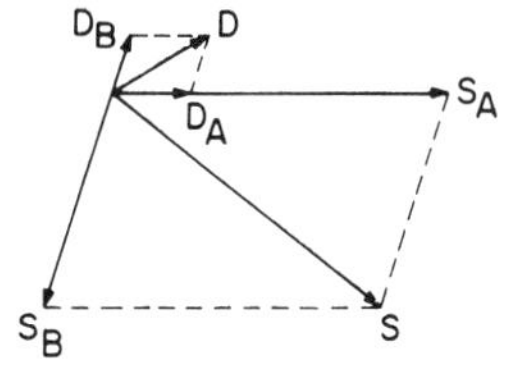

Fig. 1. Sum and difference phasors for two targets.

unresolved targets and a single target located at the "centroid" angle. The radar *can* usually tell the difference if it measures the imaginary part as well as the real part of D/S.

B. The Complex Indicated Angle

Since the indicated angle $(1/p)(D/S)$ is complex, in general, we will hereafter call it the "complex indicated angle."

As before, the analysis will consider only one angle component, since the analysis for the other is identical. To simplify notation, therefore, we will omit the subscripts tr and el .

Let S_A and S_B be the individual sum signals of two unresolved targets, at angles A and B with respect to the beam axis. Then

$$S = S_A + S_B \;. \tag{4}$$

From (2), the individual difference signals are

$$D_A = pAS_A \text{ and } D_B = pBS_B ,$$

so that

$$D = D_A + D_B = p(AS_A + BS_B) \;. \tag{5}$$

Let

$$S_B/S_A = gre^{j\phi}$$

where

g = ratio of antenna voltage gains in the directions of the two targets

r = ratio of backscatter voltage coefficients of the two targets (i.e., the square root of the ratio of their backscatter cross sections)

ϕ = relative phase of returns from the two targets.

The ratios g and r and the relative phase ϕ are arbitrarily defined to refer to the second target (at angle B) relative to the first (at angle A), not vice versa. The quantity g means the two-way voltage gain ratio in the usual case of passive radar targets. In the case of active sources or beacons it refers to one-way gain. Both g and r are defined as non-negative and real. In most of the analysis that follows, gr will be treated as a single factor.

[2]Strictly speaking, superposition is not valid, since the presence of each target modifies the field scattered by the other. However, this coupling effect is negligible unless the distance between the targets is so small as to be comparable to their dimensions.

From (4) and (5) we obtain

$$\frac{1}{p}\frac{D}{S} = \frac{AS_A + BS_B}{S_A + S_B} \tag{7}$$

and dividing through by S_A and using (6), we obtain

$$\frac{1}{p}\frac{D}{S} = \frac{A + Bgr\, e^{j\phi}}{1 + gr\, e^{j\phi}}\,. \tag{8}$$

Equation (8) is the basic formula for the complex indicated angle of two unresolved targets. Before attempting solutions it is instructive to examine its nature.

If the relative phase ϕ is considered as a variable while the other quantities on the right-hand side of (8) remain constant, the quantity exp $(i\phi)$ traces out a circle in its complex plane as ϕ varies through 360°, and since the right-hand side of (8) is a bilinear transformation of exp $(i\phi)$, the indicated angle $(1/p)(D/S)$ also traces out a circle in its complex plane. Different constant values of *gr* give different circles, which form a family.

The circles for several values of *gr* are drawn in solid lines in Fig. 2. This figure has been normalized by putting the origin at the midpoint of A and B and taking $(B - A)/2$ as the unit of angle.

The circles for *gr* less than 1 lie on the side closer to A. The circles for *gr* greater than 1 lie on the side closer to B. Reciprocal values of *gr* produce pairs of circles symmetrically located with respect to the imaginary axis. The circle for $gr = 1$ degenerates into a straight line, the imaginary axis. The circles for $gr = 0$ and infinity are the points A and B, respectively.

If *gr* is varied while ϕ and the other quantities on the right-hand side of (8) are kept constant, another family of circles, orthogonal to the constant-*gr* family, is obtained. These are drawn as broken circles in Fig. 2.

It must be remembered that Fig. 2 pertains to a single angle coordinate, say traverse. The vertical coordinate in the figure represents the imaginary part of the indicated traverse angle, not the indicated elevation angle. The same set of normalized curves would apply to elevation also.

If the amplitude ratio, the relative phase, and the angles of the two targets relative to the beam axis are known, Fig. 2 can be used to read off the complex indicated angle in the following manner. Let unit distance on the real and imaginary axes represent one-half the algebraic difference between the (real) indicated angles of the two targets individually. Shift the entire figure to the left or right so that the distances of points A and B from the origin equal their respective individual indicated angles. The intersection of the appropriate *gr* circle and ϕ circle is the value of the complex indicated angle measured by the radar.

If D/S versus angle is linear, a shift in beam pointing angle relative to the two targets means a simple horizontal shift of the figure (or of the origin) by a corresponding amount, with no change of scale, keeping in mind, however, that the value of g will change and that the correct value of *gr* for the actual beam position must be used. If D/S versus angle is not linear, a shift in beam pointing angle generally means a change of scale in addition to a horizontal shift of the figure.

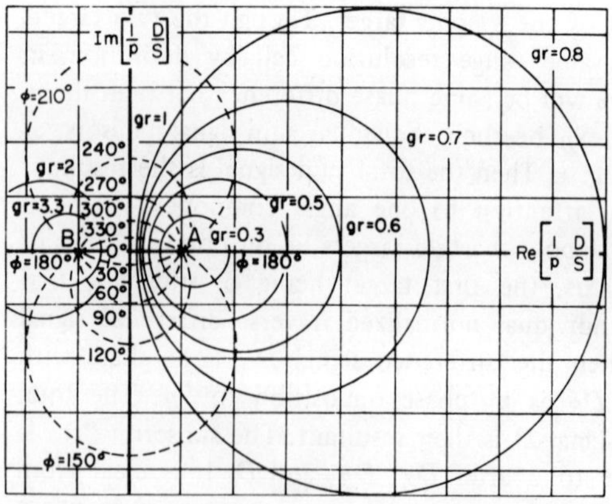

Fig. 2. Complex indicated angle in one coordinate. The first target is at angle A, second target at angle B. Amplitude ratio and phase of the second target relative to the first are *gr* and ϕ, respectively. The figure is normalized so that origin is at midpoint of A and B and unit distance is $(A - B)/2$.

Plots of the real part of the complex indicated angle versus relative phase are found in several places in the literature (though not called by that name). See, for example, Fig. 3.10 of [4], Fig. 5.15 of [7], and similar figures in [8], [9], and [10]. It is interesting to note that those rather complicated-looking families of curves take on the more elegant form of families of circles in the complex plane, as in Fig. 2 of this paper.

Now, consider the possible use of (8) to solve for unknown parameters of two targets by measuring their complex indicated angle. Equation (8) is equivalent to a pair of real equations. The left-hand side is a measured complex quantity (i.e., a pair of measured real numbers) derived from a single-pulse return in one angle channel. On the right-hand side there are four unknowns: A and B (the angles of the two targets in the one coordinate), *gr* (their amplitude ratio), and ϕ (their relative phase). Though *gr* is treated as a single factor here, g is actually a known function of A_{tr}, A_{el}, B_{tr}, and B_{el}, so that if solutions can be obtained for the angles and for *gr*, the values of g and r are determined. The absolute amplitudes of the two targets are not of interest in this problem because only their ratio affects the angle measurements.

By equating reals and imaginaries, we obtain two equations; but since there are four unknowns, a solution based on a single pulse is impossible.

If we make simultaneous single-pulse measurements of the complex indicated angle in both angle components, we then have two complex equations like (8), distinguished by subscripts "tr" and "el", equivalent to four real equations. The unknowns are two components each of A and B, a common value of ϕ, and a common value of gr. Thus we have six unknowns, but only four equations, still not enough for a solution.

The fact that the primary interest is in the target angles rather than in their amplitude ratio and relative phase does not help, since the solutions are not separable.

In some monopulse radars there is an unused by-product of the RF combiners, sometimes called the quadrupolar signal, which is terminated in a dummy load. For example, in a 4-feed monopulse system, where D_{tr} is formed by the difference of the right-hand pair and the left-hand pair and D_{el} is formed by the difference of the upper pair and the lower pair, the quadrupolar signal is formed by the difference of the two diagonal pairs. The question arises whether this signal contains additional information that might make a single-pulse solution possible. It is shown in [1] that this signal is not useful, because it has zero or low sensitivity to target position in the vicinity of the beam axis and along the traverse and elevation principal planes.

Therefore, in the general case of two arbitrary targets, and with a conventional monopulse antenna configuration, a single-pulse solution is impossible. In Section IV it will be shown that in the special case of multipath, where there are known relationships between a low-angle target and its image, a single-pulse solution is theoretically possible; but, for the present, we continue with the general case.

C. Solution Using Two Pulses

A solution can be obtained with two pulses in sequence provided either the radar or target parameters change in some way between pulses so as to provide independent equations. Several ways in which this change might be produced are mentioned in [1]. The method presented in this paper is based on a change in relative phase ϕ between pulses, due to relative motion of the two targets, while other parameters remain essentially constant. For example, suppose the radar has a wavelength of 10 cm and that during the interval between pulses, say 0.01 second, the range difference of the two targets changes by 1 cm. Then the change in ϕ is 72°, while the other parameters (A, B, and gr) can usually be considered constant over such a short interval. This is true even if the target directions are changing rapidly (but not relative to each other), provided the antenna beam is moving with them, since angles are measured with respect to the beam axis.

If the relative phase is changing at a steady rate, there is a Doppler difference, which amounts to 1 m/s in this particular case. It might be argued that there is already a known solution for this type of problem, namely resolution of the two targets by Doppler filtering. However, the average rate of change of relative phase may be too small to permit Doppler resolution of the targets by a particular radar, or the radar may not be equipped to do Doppler resolution. Furthermore, the pulse-to-pulse change may be random or transient rather than steady, so that Doppler resolution is impossible. For example, two aircraft attempting to maintain a constant separation may have zero relative Doppler on the average, but random perturbations in their positions will produce pulse-to-pulse changes in relative phase. The method to be described will work with random as well as steady phase changes.

The method does, however, have some problems and limitations, which will be discussed later.

We now introduce the following simplified notation for the real and imaginary parts of the complex indicated angle:

$$\frac{1}{p}\frac{D}{S} = x + jy, \tag{9}$$

in which the subscripts "tr" and "el" can be introduced as needed. For the time being, we will deal with only one angle component and will, therefore, omit these subscripts. We will, however, use subscripts "1" and "2" to denote the first pulse and the second pulse.

In the analysis that follows, the references to Fig. 2 are intended to aid in visualizing the problem and its solution, though Fig. 2 is not actually used in the solution. The measurements of $x_1 + jy_1$ and $x_2 + jy_2$ are represented by two points in the complex plane of Fig. 2. Since they are produced by the same values of A, B, and gr, but different values of ϕ, they lie on one of the solid circles of Fig. 2, and since the center is known to be on the real axis, the two points are sufficient to determine which circle, provided $x_1 \neq x_2$. However, the same circle could have been produced by an infinite number of combinations of A, B, and gr. An additional measurement is needed to make the solution unique. This measurement is $|S_2/S_1|$, the ratio of sum-signal amplitudes of the two pulses. With this measurement, in addition to x_1, y_1, x_2, and y_2, we have five measured quantities from which we can set up five equations in the five unknowns, A, B, gr, ϕ_1, and ϕ_2, the last two being the relative phases on the first and second pulses. Four of the five equations are obtained by rewriting (8) as

$$x + jy = \frac{A + Bgr\, e^{j\phi}}{1 + gr\, e^{j\phi}}, \tag{10}$$

equating reals and imaginaries separately, and inserting subscripts 1 and 2 after x, y, and ϕ for the first and second pulses. The fifth equation, based on the fact that the sum signal is the phasor resultant of the contributions from the individual targets, is

$$\left|\frac{S_2}{S_1}\right|^2 = \frac{1 + g^2r^2 + 2gr\cos\phi_2}{1 + g^2r^2 + 2gr\cos\phi_1}. \tag{11}$$

Solution of this set of equations by hand is rather laborious. One algebraic-graphical method, which could be converted into a computer program for repetitive or real-time calculations, is derived in [1].

Since A and B can be determined, their midpoint or their "centroid" (that is, their weighted mean position, with the weighting proportional to amplitude or power) could be computed, and by closing the loop through the computer, the antenna could be made to track the selected point, thus making good the assumption that A and B do not change during the interval between the two pulses.

The solution just described is for a single angle coordinate. When measurements are made in both angle channels simultaneously, the number of measurables is increased by four (x_1, y_1, x_2, and y_2 in the second angle coordinate), but the number of unknowns is increased by only two (A and B in the second angle coordinate). No doubt an estimation procedure can be devised to take advantage of this overdetermination to improve the accuracy.

Although the method described would theoretically give a solution based on only two pulses, some pairs of pulses will give good accuracy and others poor accuracy. In fact, if it happens that $\cos\phi_1 = \cos\phi_2$, the solution is indeterminate. Therefore, in practice, as many pulses as possible should be used.

Recall that in deriving the solution, linearity between D/S and the off-axis angle has been assumed, as expressed by (2). If the values of the angles obtained from the solution are found to lie outside the linear range, they are to be interpreted not as the true geometric angles, but as the "raw" indicated angles A' and B' of the two targets if they were measured individually. Conversion to the true geometric angles is then a simple matter of inverting the known calibration function expressed by (3): $A = f^{1}(A')$, $B = f^{-1}(B')$.

A display of the complex indicated angle can be presented on a CRT by applying the real and imaginary parts to the horizontal and vertical deflection inputs. If the changes in relative phase are rapid compared with changes in amplitude ratio and angular separation, as it is believed they usually will be, the display will be observed to trace out a circle, like the solid circles of Fig. 2, and this will be evidence of the presence of two unresolved targets. This information in itself is useful in some situations. The circle will be traced out regardless of whether the phase changes randomly or at a steady rate. Even on a single pulse the presence of an imaginary part above some threshold, set so as to be rarely exceeded by noise, would serve as a warning that the angle measurement obtained from the real part is "contaminated" by unresolved targets or extraneous signals and should be given reduced (or zero) credence.

D. Sources of Error in the Two-Pulse Solution

A partial analysis of noise errors has been carried out in [1]. It is too involved to be presented here, but a numerical example will provide some feeling for the order of magnitude of error. Considering only one angle component, let the two targets be at angles 0.2 and -0.1 beamwidth from the axis and let $gr = 0.5$ (i.e., let the signal from target B have half the amplitude of the signal from target A). Let $p = 5/3$ per beamwidth (a representative value). Suppose the S/N ratio is 30 dB when the two targets are in phase; then it will be about 20.5 dB when they are in phase opposition. The first target alone would have a 26.5-dB S/N ratio, and the second target alone 20.5 dB. If the measurements of x_1, y_1, x_2, y_2, and $|S_2/S_1|$ are made on one pulse at the time when the two targets are in phase, and on the other pulse when they are in phase opposition, the rms errors in the angle solutions are: 0.015 beamwidth for angle A (compared with 0.014 beamwidth if the first target were present alone) and 0.034 beamwidth for angle B (compared with 0.028 beamwidth if the second target were present alone). However, the errors are heavily dependent on the change in relative phase between the two pulses, the most favorable case being the one just cited (i.e., maximum difference between x_1 and x_2). At the other extreme, the solution becomes indeterminate if x_1 and x_2 approach equality. Therefore, it is important that the two pulses be properly selected or that a best estimate be formed from several pulses. The overdetermination provided by simultaneous measurements in both angle coordinates should also be helpful, as pointed out in Section III-C.

The analysis has assumed that each target behaves like a point target. Actually, each target has nonzero angular dimensions and the resulting angle glint of each individual target produces errors of the order of some fraction of the angular span of the target. If the angular extent of each target is small compared with the angular separation of the two targets, the glint error is generally negligible. If not, glint of the individual targets must be included in the error analysis.

A difficulty could arise if the relative phase were changing slowly while one or both of the targets had rapid scintillation (amplitude variation), in which case it might not be possible to select an interval that would satisfy the assumed conditions.

Perfect normalization of the difference signal with respect to the sum signal has been assumed. If normalization is obtained by conventional AGC action, it becomes inaccurate at the extremes of the dynamic range and has a time lag. Techniques such as instantaneous AGC or pulse-by-pulse digital processing and division are preferable.

Additional errors can arise from imperfect knowledge of the antenna patterns, circuit imbalance or calibration drifts, interaction of traverse and elevation channels, antenna sidelobe returns, and departures from the assumption that A, B, and gr do not change during the interpulse interval. A known departure from linearity of D/S versus off-axis angle is not a source of error, since it can be accounted for in the solution.

Since single-target D/S will generally have the same range of values in the sidelobes as in the main lobe, ambiguities in the solution may occur. A strong target in a sidelobe could be mistaken for a target in the main lobe, or vice versa. Continued tracking or a shift in beam pointing direction could resolve the ambiguity.

IV. Low-Angle Target: The Multipath Problem

A. Theory and Approach

We now take up the multipath problem of a target at a low elevation angle, say within a beamwidth of the horizon. This problem is of great practical importance because elevation tracking of such a target by present means is erratic and sometimes impossible. Traverse (azimuth) tracking is usually not seriously affected.

The reradiation from the target reaches the antenna both by a direct path and by reflection from the surface of the earth (or sea). The reflected path is equivalent to radiation from the image of the target below the surface of the earth.[3] The target and its image constitute a special case of two unresolved scatterers. In this case certain known relationships between the amplitudes, phases, and locations of the two scatterers aid in determining the elevation angle of the target. We shall show that a solution is theoretically possible with a single pulse, in contrast to the problem of two independent targets, which has been shown to require at least two pulses for solution.

To illustrate the approach in a relatively simple manner, we assume for the present a flat, smooth, level earth. Corrections for earth's curvature and surface roughness will be discussed in Section IV-C. The geometry is shown in Fig. 3. The distance to the target is assumed to be so great that the direct and reflected rays are parallel. The reflected ray reaching the radar can be considered to originate from the image, which is below the surface at a point symmetrical to that of the target.

In place of the symbols A_{el} and B_{el} used in Section III for the elevation angles of two targets relative to the antenna axis, it is more convenient in the present problem to use new symbols, E and E_0, the elevation angles of the target and of the beam axis, respectively, measured from the horizontal. The relationships between the two sets of angles are

$$A_{el} = E - E_0 \tag{12}$$

and

$$B_{el} = -E - E_0 \tag{13}$$

To calculate $(1/p)(D/S)$, the complex indicated angle in elevation for the target and image, substitute (12) and (13) in (8), obtaining

$$\frac{1}{p}\,\frac{D}{S} = \frac{(E - E_0) - gr\,e^{j\phi}(E + E_0)}{1 + gr\,e^{j\phi}}$$

or (14)

$$\frac{1}{p}\,\frac{D}{S} = -E_0 + E\,\frac{1 - gr\,e^{j\phi}}{1 + gr\,e^{j\phi}},$$

from which are obtained the real and imaginary parts

[3] Of course, the target is also illuminated by two paths, but these are equivalent to a single resultant illuminating field.

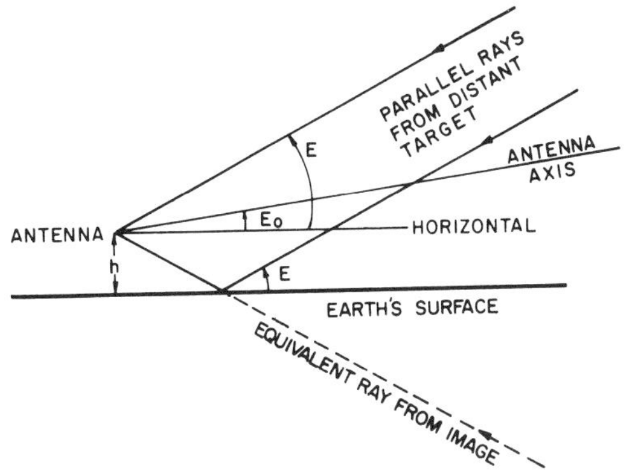

Fig. 3. Geometry of target and image, flat-earth approximation.

$$x = \mathrm{Re}\left(\frac{1}{p}\,\frac{D}{S}\right) = -E_0 + E\,\frac{1 - g^2r^2}{1 + g^2r^2 + 2gr\cos\phi} \tag{15}$$

and

$$y = \mathrm{Im}\left(\frac{1}{p}\,\frac{D}{S}\right) = -E\,\frac{2gr\sin\phi}{1 + g^2r^2 + 2gr\cos\phi}. \tag{16}$$

The quantities g, r, and ϕ are all functions of E. The amplitude ratio r, which has a value between 0 and 1, is the magnitude of the surface reflection coefficient Γ. The relative phase ϕ has two components,

$$\phi = \phi_s + \Delta\phi\,, \tag{17}$$

where ϕ_s is the phase of the surface reflection coefficient Γ and $\Delta\phi$ is due to the geometric path length difference. From Fig. 3 it is seen that

$$\Delta\phi = -4\pi(h/\lambda)\sin E, \tag{18}$$

where h is the height of the center of the antenna above the earth and λ is the wavelength.

The values of r and ϕ_s as functions of frequency, polarization, and elevation angle (grazing angle) can be read from Figs. 5.3 to 5.6 in [12], assuming a smooth reflecting surface (the reduction of the reflection coefficient due to roughness can be taken into account in the manner shown in [3]). When the elevation angle is small (say, not over 1°), r is very close to 1 and ϕ_s is very close to 180°.

To calculate g as a function of E for a specified E_0, when the target is on-axis in traverse, the sum-channel elevation beam pattern of the antenna must be known.

A method of solution will now be described on the basis of the following mode of operation:

1) The target is tracked normally in traverse, so that the beam axis, the target, and the image lie in the same vertical plane.

2) The beam is fixed in elevation at some low angle, and open-loop measurements of the complex indicated elevation angle are made as the target passes through the beam.

Under these conditions all quantities on the right-hand side of (14), (15), and (16) are either constants or functions of E only. Therefore, if x and y are measured on a single pulse, it should be possible to solve for the unknown E. It will be shown that this is the case and that the solution is usually unique.

The method of solution will be illustrated by an example, in which the following parameters are assumed:

beamwidth: $\theta_0 = 1°$.
antenna height: $h/\lambda = 100$.
beam axis elevation angle: $E_0 = 0.5°$.
sum voltage pattern: $|S| = \cos^2 (1.14\ \theta/\theta_0)$.

In the last equation, θ is the angle off axis and θ_0 is the half-power beamwidth. The quantity in parentheses is in radians. This particular pattern is selected as a typical monopulse sum voltage pattern, as reported in [4]. From the given pattern and beam-axis elevation angle E_0, g can be computed as a function of target elevation angle E.

With this illustrative set of parameters, the complex indicated elevation angle has been computed as a function of target elevation angle E. The results are plotted in Fig. 4. The x and y coordinates are the real and imaginary parts, respectively.

This plot is the locus of the complex indicated angle in the complex plane as the target elevation angle E varies. Values of E are spotted along the locus. This curve would be traced out by the moving spot on a CRT if x and y were connected to the horizontal and vertical deflection plates, respectively.

The modification of the radar to measure the imaginary part y, for the multipath solution, is needed only in elevation. For each radar installation and for a selected fixed elevation angle of the beam axis, a curve similar to Fig. 4 could be computed in advance, or obtained by flight-test calibration, and used as a scope overlay or computer look-up table. An open-loop measurement of the complex indicated elevation angle by a radar modified for this purpose would then fall at some point on the curve (if there were no noise or other sources of error), and the value of E at that point could be read off. The effects of errors will be discussed in Section IV-C.

The illustrative spiral-shaped plot of Fig. 4 has been arbitrarily terminated at an elevation angle of 0.5°, equal to half the beamwidth, in order to simplify the calculations by assuming a linear relationship between single-target indicated angle and true angle, as in (2). In practice, the spiral should be extended further, say up to one beamwidth above the horizon. This means that the actual functional relationship, as in (3), must be used rather than the linear approximation, but this causes no problem other than additional computation, provided the patterns are known. In some cases, depending on the parameters, the spiral will cross itself at one or more points, creating ambiguities in the solution. It should be possible to resolve these occasional temporary ambiguities by knowing the time history of the target elevation angle. In fact, even one additional pulse preceding or following the ambiguous one would provide another point and thus determine which of the two intersecting arcs of the curve is the correct one, provided the target elevation changes significantly in the interval. An alternative would be an additional pulse at a slightly different beam elevation angle (in the case of a phased array) or at a different frequency, in order to shift the point of ambiguity.

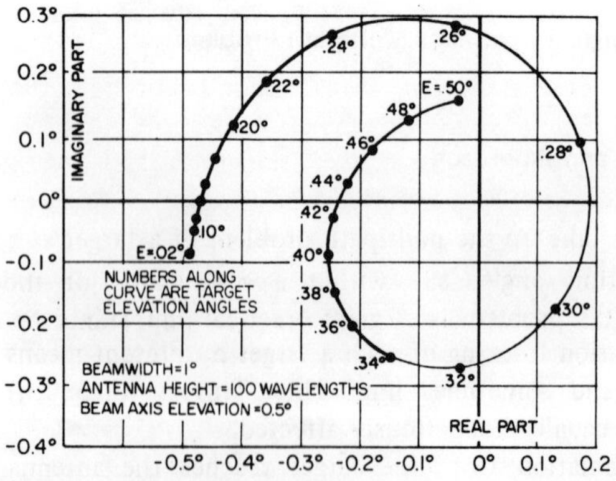

Fig. 4. The complex indicated angle in elevation when multipath is present. Numbers along curve are values of target elevation angle E.

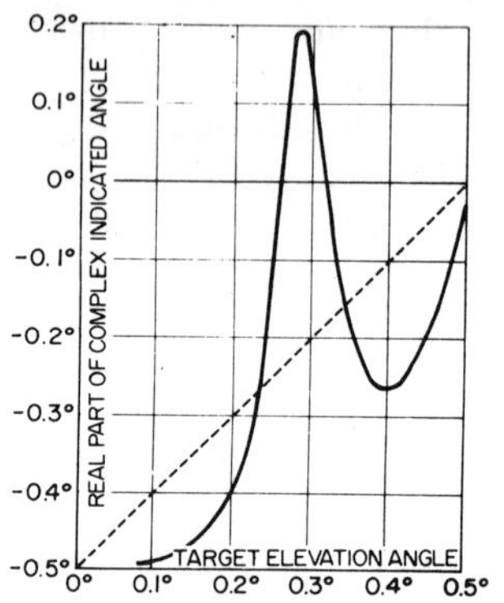

Fig. 5. Real part of the complex indicated angle as a function of target elevation angle E. Same parameters as in Fig. 4.

As the antenna height in wavelengths increases, the number of turns of the spiral increases, producing more ambiguities. The single-pulse method, therefore, appears questionable for very high antennas. However, the two-pulse method described in Section III-C, though more complicated, should be applicable in this case.

B. Comparison With Use of Real Part Alone

To see the advantage of using both the real and imaginary parts, compare Fig. 4 with Fig. 5, which is a plot of the real part alone as a function of true elevation angle. Both figures are plotted for the same set of illustrative parameters. In the absence of the reflecting surface, the plot in Fig. 5 would be simply a 45° straight line, which is shown dotted. The actual curve due to multipath, drawn solid, is radically different, and this is what a conventional monopulse radar would measure. Even if a curve such as Fig. 5 were available for calibration, the conversion is ambiguous. Suppose, for example, that the real part was measured as -0.2°. A horizontal line at this height drawn on Fig. 5 intersects the curve at three points, yielding three possible values of true elevation angle. In Fig. 4, if a *vertical* line is drawn at an abscissa of -0.2°, it also intersects the spiral curve at three points. However, if the imaginary part has been measured in addition, it uniquely identifies the correct point on the spiral. Ambiguities are the exception, whereas with the real part alone they are the rule.

An additional advantage of using both the real and imaginary parts is improvement in sensitivity, which can be defined as the increment in the measured quantities resulting from a specified increment in true elevation angle. The higher the sensitivity, the smaller the error due to noise. For example, when the true elevation is in the vicinity of 0.29°, the curve of Fig. 5 has poor sensitivity because it is near a maximum. The curve of Fig. 4 has good sensitivity at this point because, although the real part is near a maximum, the imaginary part is changing rapidly.

C. Practical Considerations and Errors in the Multipath Solution

In general, because of noise and other errors, the measured point in the complex plane will not fall exactly on the curve, but somewhere near it, and the estimate of target elevation is then obtained from the point on the curve closest to the measured point. One method of calculating the estimate is described by Peebles and Goldman [3].

The sources of error can be grouped in three categories: random noise, deviations of the reflecting surface from the assumed model, and uncertainties in the radar parameters.

An evaluation of noise errors is reported in [3]. Noise errors can be reduced by integration of several pulses or averaging of the single-pulse estimates. Observation of the time history of the indicated angle would also be helpful.

The solution as described so far applies only to specular reflection (i.e., a point image), which is the only kind of reflection that would be produced if the surface were perfectly smooth. Surface roughness reduces the specular reflection coefficient by some factor which can be estimated and taken into account in the solution if the nature of the reflecting surface is known. Error sensitivities due to use of incorrect values of surface roughness, conductivity, and dielectric constant, all of which are factors in the reflection coefficient, are evaluated in [3]. The sensitivity of the solution to these factors is important because, even if they were known perfectly under particular conditions, they might vary with azimuth direction, season, weather, and possibly other conditions. To the extent that these variations are predictable, they could be taken into account in the solution.

Surface roughness also causes a diffuse reflection component (i.e., a broadened image), which is a source of error pointed out by Barton and Ward [11]. The magnitude of the diffuse reflection depends not only on the surface roughness, measured in wavelengths, but also on the grazing angle. The smaller the angle, the higher the ratio of specular to diffuse reflection. It appears that there are many situations where the reflection is predominantly specular, and where the solution presented here would therefore be applicable; but experiments are needed to establish quantitatively the errors due to roughness and other causes.

Additional possible sources of error are earth curvature and variation of effective antenna height due to tide (if the radar is on land overlooking a coast). The errors due to the assumption of a flat earth will usually be small. If needed, corrections for earth curvature can be computed by methods presented in [12] and [13].

The errors and uncertainties in radar characteristics mentioned in Section III-D also apply in the multipath application, but only in elevation. Interaction between traverse and elevation channels is no problem if closed-loop tracking is used to keep the beam axis on or near the target in traverse, but could be a source of error in a search radar that makes open-loop measurements in both coordinates.

The most important question is not whether the radar parameters and surface reflection characteristics conform precisely to a predicted theoretical model, but whether they are stable and repeatable. If they are, the appropriate curve of the form of Fig. 4 can be plotted from actual measurements rather than calculations, using a calibration target whose elevation angle varies in a known manner.

In addition to the multipath effect, there are two other factors that sometimes cause difficulty in low-elevation tracking. These are ground (or sea) clutter and masking by terrain obstructions. The method described here will not solve these two problems. However, there are many situations where masking is not a problem, and where clutter is not present at the range of the target or can be removed by Doppler methods.

V. Comparison with Other Investigations

The fact that most previous analyses of unresolved targets have ignored the imaginary part of the complex indicated angle has already been mentioned in Section III-B. The use of complex plots, as in Figs. 2 and 4, is believed to be new.

Peters [14] recognized the existence of the imaginary part, but offered no approach or suggestion for using it. Anderson and Wells [15] briefly suggested the possible use of the imaginary part to derive information about extent, rotation, and shape of a single target, but gave no details or specific method.

References [16] to [18] offer methods of separating two or more plane waves received from different directions, but these methods require special antenna installations and do not appear to be suited to tracking of moving targets.

More recently, Peebles and Berkowitz [19] and Peebles [20] have presented true monopulse (i.e., single-pulse) solutions for unresolved targets, with emphasis on the two-target problem, including multipath (without assuming *a priori* knowledge concerning the reflecting surface). Pollon and Lank [21] have proposed a method of individually tracking two targets that are unresolved in the usual sense. However, these solutions require special monopulse beam configurations, whereas the aim in the present paper has been to utilize conventional monopulse radars without changing the beam patterns. Furthermore, the other methods obtain all the information needed to update the solution from each pulse (that is, they sample the wave front across the aperture at one instant) and are, therefore, limited in accuracy by the Cramer-Rao lower bound on variance, as derived in [21] and [22]. Although the lower bound for the two-pulse method described in this paper has not been investigated, it may turn out to be lower than for the single-pulse methods, because during the interpulse interval the wave front "slides" across the aperture, producing the same effect as simultaneous sampling of the wave front over an aperture larger than the physical aperture [1]. This assumes, however, that nothing changes between the two pulses except the relative phase of the two targets.

VI. Conclusions

The off-axis angle indicated by a conventional monopulse radar can be considered as the real part of a complex quantity, which has been named the complex indicated angle. Just as the real part is derived from the component of the normalized difference signal that is in phase with the sum signal, the imaginary part can be obtained from the quadrature-phase component, which is usually ignored. The required additional circuitry is relatively simple, involving mainly an added product detector or equivalent in each angle component, with one input shifted 90° in phase.

A single point target produces only a real part, the measurement of which gives the angular location of the target. The presence of unresolved targets or unresolved propagation paths (multipath) not only distorts the real part (causing the well-known problem of erroneous or meaningless angle indications in such cases), but also produces an imaginary part, which can serve as a warning of "contamination." Extension of the indicated angle into the complex domain gives added insight into the problem, simplifies the analysis in some respects, and offers possible solutions.

Using measurements of the complex indicated angle and of sum-signal amplitudes on two pulses separated by a short time interval, it is theoretically possible to determine the individual angular locations, the amplitude ratio, and the relative phases of two independent targets that are unresolved in the usual sense, provided a change in relative phase occurs between the pulses while other parameters remain essentially constant (a not unusual situation). Since two pulses are required, the technique is not monopulse in the literal sense, whereas other techniques described in the literature do, in theory, obtain single-pulse solutions. However, time is usually available for more than one pulse. A set of five simultaneous equations must be solved, requiring a computer if the solution is to be obtained in real time. If desired, a tracking loop in each angle coordinate could be closed through the computer so as to track either of the targets, their midpoint, or their amplitude or power centroid, while the additional information on the two targets is read out or recorded. A display of the real and imaginary parts applied to the horizontal and vertical deflection plates of a CRT would be useful for monitoring. A preliminary analysis of thermal noise errors in the two-target solution offers hope for usable accuracy in some applications. Additional sources of error and limitations are recognized but remain to be investigated.

In the special case of multipath, a low-angle target and its image are equivalent to a pair of unresolved targets, causing erratic measurements of elevation angle. The solution in this case takes a different form, requiring only a single pulse. The radar beam tracks normally in traverse, but is kept at a fixed low elevation angle. Because of approximately known relationships between target and image, the complex indicated elevation angle as a function of actual target elevation angle can be computed in advance. This can be plotted on a CRT overlay, forming a characteristic spiral curve, or put into computer memory. A single-pulse measurement of the complex indicated angle of an unknown target will fall on or near the curve, and the closest point on the curve gives an estimate of elevation angle. For multipath, the modification to measure the imaginary part is needed only in the elevation angle channel. The computer solution for the multipath problem is simpler than for the two-target problem. An evaluation of some of the anticipated sources of error has been conducted and the results [3] appear promising.

In contrast to other solutions that have been proposed for unresolved targets and multipath, the complex indicated angle approach requires no modification of the antenna beam patterns or of the basic signal processing. It is a selectable mode. When it is not in use, the radar operates in its normal mode, unaffected by the modifications.

Further work, including an experimental program, is needed to verify some of the assumptions and approximations, analyze or measure additional sources of error, and work out practical implementations.

Appendix

Definition of Angles and Application to Phased Arrays

Traverse and elevation angles are defined with respect to a vertical plane perpendicular to the antenna aperture plane and a slant plane perpendicular to the vertical plane and the aperture plane. The traverse and elevation angles of a target are the angles that the target line of sight makes with the vertical plane and slant plane, respectively.

A more "natural" system of coordinates, which appears in beam pattern equations, consists of direction cosines of the angles that the line of sight makes with a pair of reference axes in the aperture plane, perpendicular, respectively, to the two planes defined above. These angles, often designated α and β, are the complements of traverse and elevation, respectively. For small deviations from the aperture normal, traverse and elevation are very nearly equal to cos α and cos β, respectively. For a beam wider than a few degrees, formed by a movable-reflector antenna, direction cosines are preferable to traverse and elevation, but either set can be used provided the nonlinearity of D/S, if any, is taken into account, as indicated in (3).

In phased arrays the direction cosines must be used. The discussion and equations given in this paper are modified in the following manner when applied to phased arrays. Traverse and elevation angles are replaced, respectively, by $(\cos\alpha - \cos\alpha_0)$ and $(\cos\beta - \cos\beta_0)$, where α_0 and β_0 identify the direction to which the beam is steered. When solutions have been obtained for the quantities in parentheses, α and β are thereby determined, since α_0 and β_0 are known.

References

[1] S.M. Sherman, "Complex indicated angles in monopulse radar," Ph.D. dissertation, University of Pennsylvania, Philadelphia, December 1965.

[2] , "The use of 'complex indicated angles' in monopulse radar to locate unresolved targets," *Proc. 1966 Natl. Electron. Conf.*, vol. 22, pp. 243-248.

[3] P.Z. Peebles, Jr., and L. Goldman, Jr., "Radar performance with multipath using the complex angle," this issue, pp. 000.

[4] D.K. Barton, *Radar System Analysis.* Englewood Cliffs, N.J.: Prentice-Hall, 1964.

[5] R.S. Berkowitz, Ed., *Modern Radar: Analysis, Evaluation, and System Design.* New York: Wiley, 1965.

[6] D.R. Rhodes, *Introduction to Monopulse.* New York: McGraw-Hill, 1959.

[7] M.I. Skolnik, *Introduction to Radar Systems.* New York: McGraw-Hill, 1962.

[8] J.H. Dunn, D.D. Howard, and A.M. King, "Phenomena of scintillation noise in radar-tracking systems," *Proc. IRE*, vol. 47, pp. 855-863, May 1959.

[9] D.D. Howard, "Radar target angular scintillation in tracking and guidance systems based on echo signal phase front distortion," *Proc. 1959 Natl. Electron. Conf.*, vol. 15, pp. 840-849.

[10] J.E. Meade, "Target considerations," chapter 11 of *Guidance.* Princeton, N.J.: Van Nostrand, 1955.

[11] D.K. Barton and H.R. Ward, *Handbook of Radar Measurement.* Englewood Cliffs, N.J.: Prentice-Hall, 1969, pp. 147-151.

[12] D.E. Kerr, *Propagation of Short Radio Waves.* New York: McGraw-Hill, 1947, pp. 112-122.

[13] P. Beckmann and A. Spizzichino, *The Scattering of Electromagnetic Waves from Rough Surfaces.* New York: MacMillan, 1963, pp. 222-238.

[14] L. Peters, Jr., "Accuracy of tracking radar systems," Ohio State University Research Foundation Rept. 601-29, December 31, 1957, ASTIA Doc. AD 200,027.

[15] D.B. Anderson and D.R. Wells, "A note on the spatial information available from monopulse radar," *Proc. 5th Natl. Conv. on Military Electron.*, 1961, p. 268.

[16] E.W. Hamlin, P.A. Seay, and W.E. Gordon, "A new solution to the problem of vertical angle-of-arrival of radio waves," *J. Appl. Phys.*, vol. 20, pp. 248-250, March 1949.

[17] F.E. Brooks, Jr., "A receiver for measuring angle-of-arrival in a complex wave," *Proc. IRE*, vol. 39, April 1951.

[18] C.C. Watterson, "Field strength measurements in a multipath field," U.S. National Bureau of Standards, Rept. 7600, June 28, 1962.

[19] P.Z. Peebles, Jr., and R.S. Berkowitz, "Multiple-target monopulse radar processing techniques," *IEEE Trans. Aerospace and Electronic Systems*, vol. AES-4, pp. 845-854, November 1968.

[20] P.Z. Peebles, Jr., "Methods of reducing radar multipath tracking errors," *Proc. 8th Ann. Allerton Conf. on Circuit and System Theory* (October 1970).

[21] G.E. Pollon and G.W. Lank, "Angular tracking of two closely spaced radar targets," *IEEE Trans. Aerospace and Electronic Systems*, vol. AES-4, pp. 541-550, July 1968.

[22] J.R. Sklar and F.C. Schweppe, "The angular resolution of multiple targets," M.I.T. Lincoln Lab., Lexington, Mass., Group Rept. 1964-2, January 14, 1964.

Samuel M. Sherman (M'49-SM'53) was born in Camden, N.J., on September 12, 1914. He received the B.A. and M.A. degrees in physics and the Ph.D. degree in electrical engineering from the University of Pennsylvania, Philadelphia, in 1934, 1939, and 1965, respectively.

During World War II he was a Radar Design and Development Officer in the U.S. Air Force. After the war he held positions as a Research Associate at the University of Pennsylvania and as Superintendent of the Control Equipment Division at the U.S. Naval Air Development Center, Johnsville, Pa. Since joining RCA, Moorestown, N.J., in 1955 he has been a member of the Missile and Surface Radar Division in supervisory and staff positions. He is now a Senior Staff Scientist, reporting to the Manager of Systems Engineering. He has directed or participated in numerous projects relating to advanced radar systems and techniques and has served as a member of government study panels evaluating radar system concepts.

Dr. Sherman is a member of Phi Beta Kappa, Sigma Xi, and Pi Mu Epsilon.